Physical analysis for tribolog

Physical analysis
for tribology

Terence F. J. Quinn

Professor of Engineering
School of Engineering and Applied Science
United States International University
(European Campus)
Bushey, Hertfordshire

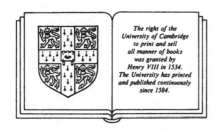

The right of the
University of Cambridge
to print and sell
all manner of books
was granted by
Henry VIII in 1534.
The University has printed
and published continuously
since 1584.

CAMBRIDGE UNIVERSITY PRESS

Cambridge

New York   Port Chester

Melbourne   Sydney

CAMBRIDGE UNIVERSITY PRESS
Cambridge, New York, Melbourne, Madrid, Cape Town, Singapore, São Paulo

Cambridge University Press
The Edinburgh Building, Cambridge CB2 2RU, UK

Published in the United States of America by Cambridge University Press, New York

www.cambridge.org
Information on this title: www.cambridge.org/9780521326025

First published 1991
This digitally printed first paperback version 2005

A catalogue record for this publication is available from the British Library

ISBN-13 978-0-521-32602-5 hardback
ISBN-10 0-521-32602-8 hardback

ISBN-13 978-0-521-01963-7 paperback
ISBN-10 0-521-01963-X paperback

Contents

Preface

This book has been written with the aim of demonstrating the power of using modern techniques of physical analysis for studying the complex interactions that often occur between the contacting surfaces of tribo-systems, that is, systems involving relative motion between the various elements.

It is an *interdisciplinary* book, which should be of interest to tribologists whose major discipline is physics, chemistry, metallurgy or any branch of engineering involved with moving parts. Obviously, mechanical and production engineers will have more interest in tribology than (say) civil engineers or electrical engineers, but even with these disciplines, tribological problems can occur for example traction between asphalt and rubber and the wear of carbon brushes.

The book aims to be understandable by readers at all levels of technical competence with an interest in tribology. It should be of special interest to those final year undergraduate students intending to make a career in the research and development laboratories associated with the oil companies, the electricity generating industry, the aerospace companies, the steel-making and steel-forming enterprises, the automotive and diesel engine manufacturers, the railways, or any other industrial concern heavily dependent upon good tribological knowledge and practice.

Even those undergraduates destined for the production side of these industries, should find the introductory chapter on tribology useful background reading, especially if they are involved with design. Certain aspects of the chapters on the analysis of lubricant films, surface temperatures pitting failures and ceramic tribo-systems should also be of interest to these prospective industrial engineers.

The introductory chapters on the various physical methods of analysis have been written so that the chapters on applications can be understood by non-specialists *without reference* to any other sources of information. The master's degree student, or first year doctoral degree student, should find these chapters of use in their studies, even if their interest in tribology is only peripheral.

Obviously, the applications chapters are aimed at committed tribologists and those engineers and other technologists involved with the successful functioning of tribo-systems. Even if the particular tribo-systems chosen for this book do not match with the systems actually of interest to these people, the methods of application and the interpretation of the analytical results will be relevant.

The chapters dealing with the basic aspects of the physical analytical techniques have been deliberately placed separately from the applications, rather than discuss each technique together with its applications. It was considered a more suitable way to deal with all of the thirteen physical analytical techniques covered by this book and their application to the study of tribo-systems, especially as some techniques are of universal applicability. In this way, some of the more important topics of tribology are discussed in a more comprehensive and unified manner. The choice of topics ensures that most aspects of tribological endeavour are covered in this book. There is some bias towards the research interests of the author. For instance, the chapter on the analysis of oxidational wear in tribo-systems contains several references to my work over the past 20 years or so. Such bias is inevitable in a book on a subject which has interested me since I first began research in 1953 at the research laboratories of Associated Electrical Industries at Aldermaston. There has been, however, an attempt to include as much reference as possible to the contributions of other investigators using physical analaysis in their approach to tribology.

Some of the more specialized surface analytical techniques such as time-of-flight secondary ion mass spectrometry (TOF-SIMS), static secondary ion mass spectrometry (SIMS), liquid metal ion source, secondary ion mass spectrometry (LMIS-SIMS), scanning tunnelling microscopy (STM) and Fourier-transform infra-red (FTIR) spectroscopy, have not been included, chiefly on the grounds that they have not been applied to tribological problems.

It is possible that some of the *nuclear techniques*, such as Rutherford back-scattering (RBS), nuclear reaction analysis (NRA) and proton-induced X-ray analysis (PIXE) will *eventually* be used more frequently than heretofore in the study of tribosystems. Certainly, radio-active tracers were applied with some success in the early days of nuclear energy. The lack of availability of particle accelerators and/or the problems of providing protection from ionizing radiations, has tended to limit the use of these techniques in tribological investigations. To include *all possible* physical techniques for analysing the products of surface interactions between tribo-elements would be *prohibitive* (and not very relevant to the practising tribologist). Hence the compromise solution embodied by this book.

It is hoped that this compromise will not prevent all readers of this book from gaining *some* new knowledge, regardless of their main interest. If, as a result of reading this book, those readers with strong tribologist interests can appreciate the power of using physical analysis in their investigations, and can talk *knowingly* with the experts carrying out that analysis, then I will feel I have accomplished my original aims in writing the book. If, on the other hand, the analytical experts also begin to appreciate the fundamental tribology of the tribo-systems they are called upon to analyse as a result of reading this book, then this will be a definite bonus as far as the furtherance of tribological knowledge is concerned.

Nomenclature

$A_1 =$ Area of densitometer aperture (in Equation 2.38)

$A_2 =$ Area covered by opaque particles (in Equation 2.38)

$(A_b) =$ Absorption coefficient (light).

$A =$ General Arrhenius constant.

$A_{real} =$ Real area of contact $(= W/p_m)$.

$(A_e)_s =$ Elastic area of single constant.

$(A_p)_m =$ Real area of multiple contact.

$(A'_p)_m =$ Interpenetration area (normal to sliding direction).

$(A_e)_m =$ Multiple area of elastic contact.

$(A_n) =$ Nominal (apparent) area of contact.

$(A_p) =$ Arrhenius constant for parabolic oxidation of tribo-elements during sliding.

$AA =$ Arithmetical average deviation from the centre line of a surface profile.

$(A)_\parallel =$ Real area of contact between *parallel* members of graphite/graphite sliding interfaces.

$(A)_{PE} =$ Area of sample from which photoelectrons are detected.

$(A)_T =$ Real area of contact between *tilted* members of graphite/graphite interfaces.

$(A)_R =$ Real area of contact between *randomly* oriented members of graphite/graphite interfaces.

$\delta A_r =$ Contribution to the total contact area from an asperity in the rth level (Archard's simple model)

$(A)_w =$ Atomic weight.

$AES =$ Auger Electron Spectroscopy.

$AGR =$ Advanced Gas Cooled Reactor.

$a =$ Radius of contact (generally).

$(a_e), (a_p) =$ Radii of elastic, plastic contact.

$(a_i) =$ Inverse sensitivity factor for element (i) (AES).

$(a_L) =$ Amplitude of light wave.

$(a_W) =$ Scale factor of Weibull Equation (2.35).

$(a_{uc}) =$ Side of unit cell in the X-direction [100] of a crystal

$(a^*_{uc}) =$ Side of unit cell in the X*-direction, (d_{100}) in the reciprocal lattice.

$(a_i^H) =$ Handbook value for (I_s^H/I_i^H) (Auger analysis).

$(a_i^0) =$ Relative inverse sensitivity factor between standard element(s) and pure element (i) under the same conditions [equals (I_s^H/I_i^H)].

$(\alpha_{uc})=$ Angle between $\mathbf{b}_{uc}$ and $\mathbf{c}_{uc}$ in the unit cell.

$(\alpha_{uc}^*)=$ Angle between $\mathbf{b}_{uc}^*$ and $\mathbf{c}_{uc}^*$ in the reciprocal unit cell.

$\alpha'=$ Modified Auger parameter [Equation (4.29)].

$(\alpha_P)=$ Pressure/viscosity coefficient.

$(\alpha_{Pe})=$ Collection of terms related to the Peclet Number [see Equation (6.4)].

$(\alpha_1)=$ Part of function growth Equation (6.28).

$(\alpha_{SA})=$ Semi-aperture angle.

$B=$ Constant of *inverse* logarithmic and *direct* logarithmic 'oxide growth with time' laws.

$(B_{DW})=$ Debye–Waller temperature factor $(=B_0+B_T)$.

$BE=$ Binding Energy.

$B_{10}=$ Life which 90% of bearings will reach *without* failure.

$b=$ Diameter of point in reciprocal lattice.

$(\mathbf{b}_{uc})=$ Side of unit cell in the Y-direction [010] of a crystal.

$(\mathbf{b}_{uc}^*)=$ Side of unit cell in the Y^*-direction, $\mathbf{d}_{010}$, in the reciprocal lattice of a crystal.

$(b_W)=$ Slope of Weibull plot.

$\beta=$ One of the constants of Equation (1.6).

$(\beta_{uc})=$ Angle between a_{uc} and c_{uc} in the unit cell.

$(\beta_{uc}^*)=$ Angle between a_{uc}^* and $\mathbf{c}_{uc}^*$ in the reciprocal lattice unit cell.

$C=$ Contrast of an electron micrograph.

$(C_A)=$ True concentration of constituent A.

$(C_B)=$ Contrast of background adjacent to a feature in an electron micrograph.

$(C_f)=$ Contrast of a feature in an electron micrograph.

$(C_C)=$ Chromatic aberration constant of an electron microscope lens systems.

$(C_{dyn})=$ Basic dynamic capacity of a bearing (ie load corresponding to the B_{10} life).

$CLA=$ Centre Line Average.

$CPT=$ Crack Propagation Time.

$(C_S)=$ Spherical aberration constant of an electron microscope lens system.

$(C_x)=$ Atomic fraction of the x-component.

$(C_w)=$ Constant of Equation (6.33) related to ξ, k_p and t_o.

$c=$ Velocity of light.

$(c_o)=$ Specific heat of the oxide formed on the surface of tribo-elements.

$(c_{uc})=$ Side of unit cell in the Z-direction [001] of a crystal.

$(c_{uc}^*)=$ Side of unit cell in the Z^*-direction, $\mathbf{d}_{001}$, in the reciprocal lattice of a crystal.

$(c_x)=$ Constant of Equation (4.12).

$(c_Z)=$ Distance between the slits of Young's Double-Slit Interferometer.

$D =$ Ratio of combined surface roughness to the minimum thickness of a lubricant film.

$(D_{min}) =$ Least distance of distinct vision (optical systems).

$(D_{ss}) =$ 'Slits-to-Screen' Distance in Young's double slits.

$(D_o) =$ Optical density of a 'Ferrogram'.

$(D_p) =$ Density of photographic blackening.

$(D_1), (D_2), (D_{F3}), (D_{F4}) =$ Different definitions of the D-ratio [Equations 7.5(a)–(d)].

$(D_s) =$ 'Specimen-to-film distance' in the flat plate X-ray diffraction method.

$d =$ Distance of sliding between two model asperities ($= 2a$).

$(d_a) =$ Thickness of air film (in Newton's Rings and Multiple Beam Interferometers).

$(d_g) =$ Thickness of plane-parallel film (glass).

$(d_t) =$ Thickness (generally).

$(d_i) =$ Interplanar spacing of each atomic layer in a wear particle (according to Holm, 1946).

$(d_{min}) =$ Minimum resolvable distance in the object.

$(d_p) =$ Thickness of crystalline specimen.

$(d_p)_c =$ Critical thickness of crystalline specimen when *both* kinematic and dynamical theories of electron diffraction are relevant.

$(d_s) =$ Diameter of spherical crystal.

$(\mathbf{d}_{hkl}) =$ Interplanar spacing vector of the (hkl) planes in a crystal lattice.

$(\mathbf{d}_{hkl}^*) =$ Reciprocal lattice vector related to $\mathbf{d}_{hkl}$ by Equation (3.7).

$\Delta =$ Phase difference.

$\Delta r_i =$ Radius of disc of confusion (optics).

$\Delta\sigma =$ 'Relaxation' angle for Bragy reflection.

$(\delta_{expt}) =$ Experimentally determined division of heat at the interface ($= H_1/UF$).

$(\delta_{theory}) =$ Theoretical division of heat at the sliding interface.

$(E_1), (E_2) =$ Young's Modulus of Elasticity for a *hard hemi-spherical asperity* and a *soft flat*.

$E(\theta) =$ *Differential* cross-section for elastic scattering of electrons.

$E'' =$ Reduced elastic modulus.

$E_c =$ Critical energy for K X-ray line excitation.

$E_i =$ Energy of inner shell of an atom.

$E_p =$ Energy of primary X-ray beam.

$(E_n), (E_m) =$ Energy of nth and mth, orbit of electron in an atom.

EPMA $=$ Electron Probe Microanalysis.

$e =$ Electronic charge.

$\varepsilon =$ Crystallite size.

$(\varepsilon_p) =$ Crystal(lite) size parallel to the incident beam of radiation.

$(\varepsilon_0) =$ The permittivity of vacuo.

$F =$ Force of friction.

$(F_{kin}) =$ Kinetic force of friction.

$(F_{stat}) =$ Static force of friction.

$(F_e)_m =$ Frictional force between multiple elastic contacts.

$(F_p)_m =$ Frictional force between multiple plastic contacts.

$(F_\parallel) =$ Force required to shear *parallel* members of twinned crystallites.

$(F_T) =$ Force required to shear *tilted* members of twinned crystallites.

$(F_P) =$ Force required for *ploughing* of asperities.

$(F_e)_{hkl} =$ Structure factor for electrons.

$F(t) =$ Probability of failure of a percentage of the population in a life-time (t_B) of a bearing.

$F(\lambda_R) =$ Fraction of total elastic area of contact across which asperity contacts occur.

$(F_x)_{hkl} =$ Amplitude of resultant of all the X-ray waves scattered by the n-atoms in the unit cell into the angle $(2\theta_{hkl})$ [i.e. Structure factor for X-rays].

$(F_R) =$ $4(R_R)/(1 - R_R)^2$, [see Equation (2.21)].

$f =$ X-ray flux (XPS).

$f(\chi) =$ Absorption function [see Equations (4.21) and (4.22)].

$f_X(f_e) =$ Atomic scattering factor for X-rays (electrons).

$(f_X)_T =$ Atomic scattering factor for X-rays at temperature (T).

$(f_o) =$ Mass fraction of oxide that is composed of oxygen.

$(f_{eye}) =$ Focal length of eyepiece lens.

$(f_{obj}) =$ Focal length of objective lens.

$\Phi =$ One of the constants of Equation (1.6).

$\Phi_i =$ Phase difference between X-ray wave scattered by an atom at fractional coordinates, $x_i\,y_i\,z_i$, in the unit cell, compared with the wave scattered by an atom at the origin.

$\phi =$ Phase of light wave.

$\phi_e =$ *Exponential* distribution of asperity heights.

$\phi_R =$ Angle of specimen rotation (figure 5.17).

$\phi_G =$ *Gaussian* distribution of asperity heights.

$\phi_{hkl} =$ Scattering angle (equals to $2\theta_{hkl}$).

$\phi_p =$ Angle between incident electron beam and specimen surface (AES).

$\phi_s =$ Sliding direction (Table 9.1).

$\phi_{sp} =$ Spectrometer work function (XPS and AES).

$\gamma_{uc} =$ Angle between $\mathbf{a}_{uc}$ and $\mathbf{b}_{uc}$ in the unit cell.

$\gamma_{uc}^* =$ Angle between a_{uc}^* and $\mathbf{b}_{uc}^*$ in the reciprocal lattice unit cell.

$\gamma_F =$ Fluorescent correction [Equation (4.23)].

$H =$ Constant of Equation [4.22(a)] equals [$1.2(A)_w/Z^2$].

$H_1 =$ Rate of heat flow along the pin (at the interspace between the pin and the disc).

$H_2 =$ Heat flow rate entering section of pin where thermocouple measuring T_A was conducting heat away (see Figure 6.1).

$\text{HEED} =$ High Energy Electron Diffraction (see also THEED and RHEED).

$h =$ Planck's constant ($= 6.626 \times 10^{-34}$ J-s).

$(h^*) = (h_{\text{lub}})/(h_{\text{min}})_0$.

$(h_c) =$ Convective heat transfer coefficient.

$(h_{\text{lub}}) =$ Lubricant film thickness.

$(h_{\text{min}}) = $ *Minimum* lubricant film thickness.

$(h_{\text{min}})_0 = $ *Central* lubricant film thickness.

$(h_Z) =$ Increment of asperity heights in Z-direction for Archard's (1953) simple model.

$h, k, l =$ Miller indices describing the (hkl) planes in a crystal (never separated from each other).

$\eta_0 =$ Dynamic viscosity coefficient (at normal atmospheric pressure, P_0).

η or $\eta(P) =$ Dynamic viscosity coefficient as a function of pressure (i.e. $\eta(P) = \eta_0 \exp(\alpha_p)(P - P_0)$).

$I =$ Incident intensity (generally).

$I_R =$ Reflected intensity (generally).

$I_T =$ Transmitted intensity (generally).

$I_a =$ Intensity of X-rays scattered by an atom.

$I_e =$ Intensity of X-rays scattered by an electron.

$I_{hkl} =$ Intensity of X-rays scattered into a cone of semi-angle $(2\theta_{hkl})$ from original direction.

$(I_{hkl})_{\text{ED}} =$ Intensity of electrons diffracted from original direction.

$I_A =$ Auger electron current from element A.

$I_i = $ *Measured* Auger current from element i in a multi-element specimen.

$I_0 =$ Electron intensity *incident* upon a transmission electron microscope specimen (or HEED specimen).

$I_c =$ Electron intensity *emerging* from a transmission electron microscope specimen.

$(I_c)_0 =$ Electron intensity transmitted through an electron microscope specimen when *no* scattered electrons are assumed to return to the original beam direction.

$(I_c)_f =$ Intensity of electron beam in the region of a feature of the electron micrograph.

$(I_c)_B =$ Intensity of electron beam in the region of the background adjacent to the feature.

$I(x) =$ Intensity as a function of distance (x) into specimen.

$I_F =$ Intensity of X-rays produced by characteristic line fluorescence.

$I_1, I_2 =$ Intermediate and final images in a simple optical microscope.

$I_K =$ Total number of electrons scattered through *small* angles on passing through an electron microscope specimen.

$I_n =$ Relative Integrated Intensity from the nth component of a multi-component specimen (X-ray diffraction).

$I_{pe} =$ Peak intensity (electron diffraction patterns).

$I_{we} =$ Integrated intensity (electron diffraction patterns.)

$I_{px} =$ Intensity of X-rays due to primary excitation by electrons (electron probe microanalysis).

$(I_{hkl})_w =$ Relative integrated intensity of X-rays scattered into semi-angle $(2\theta_{hkl})$.

$(I'_{hkl})_{ED} =$ Intensity of electrons diffracted into a complete diffraction ring.

$(I_x) =$ Combined intensity of interfering light waves.

$J =$ Mean ionization potential.

$K =$ K-factor that is the probability of producing a wear particle per asperity encounter.

$K_e =$ Equivalent thermal conductivity, which allows for the effect of anoxide film upon the transfer of heat.

$K_S =$ Thermal conductivity of steel.

$K_i =$ Constant of Equation (4.35) relating to I_A.

$KE =$ Kinetic Energy.

$K_s =$ Fraction of electrons *singly* scattered into *all* angles/unit thickness (TEM).

$K_R =$ Constant of Rayleigh's Criterion [Equation (2.4)].

$K_r =$ Constant of Equation (7.15).

$(K_S)_p =$ Thermal conductivity of steel (pin).

$(K_S)_d =$ Thermal conductivity of steel (disc).

$K_o =$ Thermal conductivity of oxide.

$K_{xe} =$ Constant of the definition of the reciprocal lattice vector (equals λ or 10λ for XRD and unity for electron diffraction).

$k =$ General oxidation rate constant or the constant introduced in Equation (4.33) to preserve its dimensional stability.

$k_D =$ Constant of the Gaussian distribution of asperity heights.

$k_c =$ Constant of *cubic* growth (of oxide) law.

$k_d =$ Constant of *direct logarithmic* growth (of oxide) law.

$k_i =$ Constant of *inverse logarithmic* growth (of oxide) law.

$k_l =$ Constant of *linear* growth (of oxide) law.

$k_p =$ Constant of *parabolic* growth (of oxide) law.

$L =$ Specimen to photographic film/plate distance in the electron diffraction camera.

$L_a =$ Crystallite size in the X- or Y-direction of graphite.

$L_c =$ Crystallite size in the Z-direction of graphite.

$L_{co} =$ Roughness width cut-off.

$LEED =$ Low-Energy Electron Diffraction.

$L_F=$ Latent Heat of fusion/unit volume.

$LTP=$ Life to pitting failure.

$L_1=$ Length of pin between thermocouples reading T_A and T_B (see Figure 6.1).

$\lambda=$ Wavelength (generally).

$\lambda_b=$ Wavelength of *blue* light.

$\lambda_e=$ Mean free path of *elastically* scattered electrons.

$\lambda_i=$ Mean free path of *inelastically* scattered electrons.

$\lambda_p=$ Wavelength of X-rays modified by interaction with matter.

$\lambda_R=$ Lambda ratio (equals $1/D$).

$\lambda_r=$ Wavelength of *red* light.

$\lambda_T=$ Mean free path for scattering of electrons by both inelastic and elastic interactions.

$M=$ Magnification.

$m^*=$ $\ln(r^*)$.

$m=$ Multiplicity factor (or quantum number).

$m_e=$ Mass of the electron.

$\mu=$ Refractive index (optical wavelengths).

$\mu_{kin}=$ Kinetic coefficient of friction.

$\mu_l=$ Linear absorption coefficient for X-rays.

$\mu_m=$ Mass absorption coefficient for X-rays.

$\mu_\parallel=$ Contribution to the kinetic coefficient of friction from crystallites oriented *parallel* to sliding surface.

$\mu_R=$ Contribution to the Kinetic coefficient of friction from the crystallites oriented *randomly* with respect to sliding surface.

$N=$ Number of circular contacts (of radius a) between two surfaces in relative motion.

$N(E)=$ Number of electrons with energy (E).

$N_A=$ Number of atoms per unit volume.

$N_i=$ Number of atoms per unit area on site i of the surface (AES).

$N_R=$ Number of revolutions (cycles) to first pit (in a contact fatigue experiment).

$N_T=$ Total number of asperities in contact between two surfaces (contact mechanics insofaras it relates to stationary surfaces).

$NA=$ Numerical Aperture [equals $\mu \sin(\alpha_{sA})$].

$n_{CL}=$ Total number of elemental areas within the interval L_{CO} (surface roughness analysis).

$n=$ Order of interference (optical interferogram), *or* quantum number (Bohr theory).

$n_C=$ Number of contacts (N) per unit area of surface.

$n_A=$ Fraction of emitted Auger electrons actually collected by analyser (AES).

$n(\sigma)\,d\sigma =$ Number of [001] poles lying between σ and $(\sigma + d\sigma)$ with respect to the electron beam.

$v =$ Kinematic viscosity or frequency (generally).

$v_{nm} =$ Frequency of (optical) radiation given off (or absorbed) when electron goes from energy level (E_n) to energy level (E_m) (or vice versa).

$\bar{v}, \bar{v}_{nm} =$ Wave number (reciprocal of frequency).

$v_{\max} =$ Maximum frequency.

$v_1, v_2 =$ Poisson's ratios for surfaces 1 and 2.

OPD $=$ Optical path difference between two parts of the wavefront impingeing upon the dispersing element.

$P_q =$ A P-fold rotation axis associated with a translation of (a/P) times the repeat distance along that axis.

$P_R =$ *Percentage ratio* of randomly-oriented to preferentially-oriented graphitic crystallites.

$P_A =$ Peak-to-Peak Height (AES) [see Equation (4.35)].

$P =$ Pressure (generally).

$P_P =$ Force required to plough (or deform) asperities.

PSO $=$ Point surface origin (Pitting through contact fatigue).

$P_T =$ Transparency of AES analyser.

$P(Z, Z + dZ) =$ Amplitude probability.

$p =$ Probability of mis-match between adjacent (001) planes in graphitic crystallites.

$(p_s) =$ Fraction of electrons singly scattered *through small angles* per unit specimen thickness.

$(p_{\max}) =$ Maximum Hertzian pressure.

$(p_m) =$ Hardness (or mean pressure, if less than yield strength).

$(p_m)_{\text{local}} =$ Local hardness.

$(p_m)_\perp =$ Hardness measured in a direction *perpendicular* to the basal planes of graphite.

$(p_m)_\parallel =$ Hardness measured in a direction *parallel* to the basal planes of graphite.

$(p_m)_0 =$ Room temperature hardness.

$p_l =$ Linear momentum.

$p_\phi =$ Angular momentum.

$Q =$ General activation energy for oxidation.

$Q_p =$ Activation energy for *parabolic oxidation* of tribo-elements during sliding.

$q =$ Reduced pressure $= [1/(\alpha_P)]\{1 - \exp[-(\alpha_P)P]\}$.

$q^* = q(h_{\min})_0^2/[12(\eta_0)(\bar{\mu})(a)]$.

$q_i =$ Screening factor for element A on site i (AES).

$R =$ Radius of hemispherical asperity (also used for electrical *resistance*).

R_S = Radius of reference sphere (crystallography), or, parameter related to the radius of curvature of asperities on a *real* surface.

$\bar{R}$ = Universal gas constant ($=8.314\,kJ/(kmol–K)$).

R_B = Rydberg constant.

R_{BS} = Back-scattering coefficient of electrons in the electron probe microanalyser.

R_{ES} = Radius of the Ewald sphere.

R_{eff} = Effective radius (of the cylindrical camera used in the glancing-angle, edge-irradiated X-ray diffraction film technique).

R_f = Fraction of crystallites *randomly-oriented* with respect to surface of graphitic contact film.

R_L = Radius of curvature (of an optical lens).

RMS = Root-mean-square surface roughness.

R_P = Levelling depth.

R_a = Average roughness height (see also *AA*, CLA).

R_R = Fraction of light intensity reflected at an interface between optical media.

R_t = Radius of pin (in cylindrical pin on disc wear machine).

R_{RC} = Radius of cylindrical single crystal camera.

R_T = 'Peak-to-valley' surface roughness.

R_r = Reduced radius[given by Equation (5.44)].

R'_1, R'_2 = Radii of the two contacting bodies involved in (R_r).

RHEED = Reflection High Energy Electron Diffraction.

R_{ED} = Distance of an electron diffraction maximum from centre spot of the electron diffraction pattern.

r' = Radius of disc of confusion (see also Δr_i).

r_i = Back-scattering factor for element A on site i of the surface (AES).

r_n = Radius of the nth orbit of an atom.

r = Radius of *sector* of spherical ball drawn parallel to flat against which ball is elastically loaded.

$r^* = r/a$.

ρ = Density (generally).

$\rho_A(Z)$ = Amplitude density function.

(ρ_{hkl}) = Angle between normals to (hkl) and (001) planes.

ρ_0 = Density of oxide.

ρ_z = Radius of the [100] zone of the electron diffraction pattern from a single crystal of cubic material.

$(\mu\lambda_e)$ = Transparency thickness (electron scattering).

S = Adhesive component of F_{kin} [see Equation 1.16].

$S_A(E_p, E_i)$ = Electron impact ionization cross-section of element A (AES).

S_S = Atomic sensitivity factor.

$S_{||}$ = Shear strength of junctions between opposing surfaces when crystallites have their basal planes oriented *parallel* to the interface.

$S_R =$ Shear strength of junctions between opposing surfaces when crystallites have their basal planes oriented *randomly* to the interface.

$SEM =$ Scanning Electron Microscopy.

$SP =$ Stopping Power of a specimen (in the Electron Probe Microanalyser).

$s =$ Mean tangential shear stress (in a solid).

$s_f =$ Shear rate (for a liquid).

$\Sigma =$ Exposure of a photographic film or plate.

$\sigma =$ Orientation of [001] poles with respect to electron beam (or standard deviation of asperity heights).

$\sigma_C =$ Modified Lenard coefficient [see Equation (4.22(b))].

$\sigma_e =$ Elastic electron scattering cross-section.

$\sigma_i =$ Inelastic electron scattering cross-section.

$\sigma_{PE} =$ Photoelectric cross-section [see Equation (4.25)].

$T =$ Temperature (generally).

$T_2 =$ Transition load beyond which severe metallic wear changes to mild-oxidational wear (Welsh, 1965).

$T_3 =$ Transition load beyond which mild-oxidational wear changes to severe-oxidational wear (Welsh, 1965).

$T^* = [(\chi s)(p_m)_0/K_S]$ [see Equation (6.20)].

$T(x) =$ Temperature at distance (x) from heat source (H_1).

$(T_A) =$ Thermocouple reading where pin emerges from the insulated portion of the calorimeter (see Figure 6.1).

$(T_B) =$ Thermocouple reading at a distance (L_1) from the thermocouple reading (T_A) (see Figure 6.1).

$(T_B)_m =$ Measured temperature as given by a thermocouple embedded just below the wearing surface of the pin.

$(T_{AE}) =$ Transparency of the analyzer (AES).

$(T_c) =$ Temperature of the real areas of contact during sliding (sometimes called the 'flash temperature').

$(T_D) =$ Detector efficiency for electrons emitted from the sample of the X-ray photoelectron spectrometer.

$(T_E) =$ Temperature of air flowing past the exposed portion of the pin (in Figure 6.1).

$(T_c^*) = [(\chi_s)(p_m)_0/K_e]$ [see Equation (6.26)].

$(T_m) =$ Melting point.

$TEM =$ Transmission Electron Microscopy.

$THEED =$ Transmission High Energy Electron Diffraction.

$T_0 =$ Temperature of oxidation at the real areas of contact during sliding.

$T_{OS} =$ Thermocouple reading of the *outside surface* of the insulator surrounding the pin (see Figure 6.1).

$T_S =$ General surface temperature.

$T_{SA}, T_{SB} =$ General surface temperatures of bodies A and B in sliding contact.

$(T_S)_d =$ General Surface Temperature of disc.

$(T_S)_p =$ General Surface Temperature of pin.

$T_T =$ Fraction of the incident light intensity that is *transmitted* at an interface between optical media.

$t =$ time (generally).

$(t/t_0) =$ Average time fraction during which *no* contact occurs (Section 7.5.2).

$t_B =$ Individual bearing life.

$t_{bl} =$ Bearing length ratio.

$t_c =$ Characteristic life.

$t_d =$ Fictitious *excess* temperature of disc, assuming all heat generated at the interface goes into the disc (of a pin-on-disc wear machine).

$t_0 =$ Time to build up a critical oxide film thickness during oxidational wear.

$t_p =$ Fictitious *excess* temperature of pin, assuming all heat goes into the pin.

$\tau_e =$ *Elastic* shear stress between each surface asperity junction.

$\tau_f =$ Shear stress in liquid.

$U =$ Speed of a moving surface.

$\tilde{U} =$ Normalized speed [see Equation (1.34(a))].

$\bar{U} = (U_1 + U_2)/2$, the mean speed of the speeds (U_1 and U_2) of the bearing surfaces.

$(U_0) = (E_P)/(E_C)$ (Electron probe microanalysis).

$u_0 =$ Distance of *object* from objective lens (working distance).

$V =$ Volume (generally) or valence levels (in AES).

$(V_A) =$ Accelerating voltage.

$[(\Delta V_A)/(V_A)] =$ Stability of accelerating voltage.

$(V_C) =$ Critical electron accelerating voltage required for the excitation of a given series of X-ray spectra.

$(V_f) =$ Visibility of a feature in an electron.

$(V_n) =$ Volume percentage of (nth) component in a multi-component specimen.

$v =$ Speed of an elementary particle moving in a field-free environment.

$v_0 =$ Distance of *image* from *objective* lens.

$v_{ac} =$ Volume of the unit cell.

$W =$ Load (generally).

$\tilde{W} =$ Normalized pressure [see Equation (1.34(b))].

$W_A =$ Half-width of an Auger peak (i.e. the separation between the positive and negative peaks in the $(dN(E)/dE)$ spectrum.

$(W_e)_s =$ Normal load causing *elastic* deformation of a *single* contact.

$(W_p)_s =$ Normal load causing *plastic (as well* as *elastic)* deformation (at a single contact).

$(W_p)_m =$ Normal load-carrying *multiple, plastic,* contact.

$w =$ Wear rate (volume removed per unit sliding distance).

$\tilde{w} =$ normalized wear rate [see Equation (1.34(c))].

$w_\alpha, w_\beta, w_\gamma =$ First, second and third terms in the general expression for oxidational wear [see Equations (1.40), (1.41) and (1.42)].

$w_m =$ Mass removed per unit rolling distance.

$w_X =$ X-ray fluorescent yield (in AES).

$w_c =$ *True* width of diffraction line (due to limited crystallite size).

$(w_c)_s =$ Width of (sharp) lines of standard material ($E > 1000 \text{Å}$).

$(w_c)_m =$ *Measured* width of diffraction line.

$X_d =$ Thermal diffusivity of disc material.

$X_i =$ Atomic percentage concentration of element i (AES).

$X_o =$ Thermal diffusivity of oxide ($= K_o/(\rho_o c_o)$).

XPS $=$ X-ray Photoelectron Spectroscopy.

XRD $=$ X-ray Diffraction.

$x_i, y_i, z_i =$ Fractional coordinates of ith atom in the unit cell.

$\xi =$ Critical oxide film thickness.

$\xi_d, \xi_p =$ Thickness of critical oxide film formed on disc and pin surfaces respectively.

$\xi_N =$ Number of atomic layers in a wear particle (according to Holm, 1946).

$\theta_A =$ Arc angle (see Figure 5.17).

$\theta =$ Angle between c-axis and the sliding interface (or sometimes used as an abbreviated form for θ_{hkl}).

$\theta_E =$ Angular efficiency function (XPS).

$(\theta_{hkl}) =$ Bragg angle (sometimes written without subscripts).

$\theta_m =$ 'Hot-spot' temperature (excess over the general surface temperature, T_S).

$(\theta_p) =$ Incident angle of primary beam (AES).

$\psi =$ Take-off angle (EPMA).

$(\psi_p) =$ Reduced flow pressure.

$Y =$ Elastic limit.

$(y_E) =$ Efficiency of producing photoelectrons.

$Z =$ Atomic number.

$(Z_n) =$ Number of atoms removed per atomic encounter.

$(Z_h) =$ Collection of terms relating to heat flow in a pin-and-disc tribo-system [see Equation (6.9)].

$\omega_0 =$ Angular velocity of specimen at 0.

1 Tribology

1.1 Definitions of common tribological terms

1.1.1 Tribology

Tribology is a new word based on the Greek word 'tribo', which means 'rubbing'. Hence tribology is the 'study of rubbing'. The word was first used by a British Government committee (chaired by Dr Peter Jost and hence known as the 'Jost Committee') that produced a report, in 1966, calling for increased education and research into a subject that was estimated (at 1966 prices) to be costing the United Kingdom about £300 million per year.

The Jost Committee defined tribology as 'the study of the science and technology of interacting surfaces in relative motion'. It was hoped that the new word might provide the basis of a more unified approach to subjects previously studied separately under titles such as 'friction', 'adhesion', 'lubrication' and 'wear'. It is indeed unfortunate that, to date, the Jost Committee's awareness of the need for a more unified (that is, interdisciplinary) approach has not been shared by many tribologists. This resistance to the calls for a change in our partisan approaches to the subject is illustrated quite neatly by the fact that it has taken nearly 20 years for the *Journal of Lubrication Technology* (*JOLT*) to change its name to the *Journal of Tribology* (*JOT*), namely from 1966 to 1985. It is hoped that this change in name will open the journal to a new generation of papers written by physicists, chemists, chemical engineers and materials engineers (each with their own particular approach to tribology), as well as mechanical engineers (with their special interests in rheology and lubricant pressures).

1.1.2 Friction

The earliest studies of relative motion between two contacting bodies were carried out by Leonardo da Vinci, in the fifteenth century. He showed that the tangential force needed to instigate sliding between two loaded surfaces was proportional to the normal load being applied at the contact, but was independent of the apparent area of that contact. These are the laws of friction, where friction is the resistance to shearing force between the two solid bodies. The ratio of the frictional force to the normal force is called the 'coefficient of friction'. For surfaces sliding without lubrication, this ratio tends to take on values between about 0.3 and 0.9. Leonardo was, in fact, measuring the static friction, whereas most modern tribologists are more interested in the tangential force required to maintain motion, that is the dynamic friction. For surfaces that roll with very little or no sliding, the frictional force is much less. This type of friction is known as 'traction' and is normally measured under lubricated conditions.

1.1.3 Lubrication

The study of lubrication had reached a fairly advanced stage at the time the new word 'tribology' was introduced in the 1960s. Consequently, much of the definitive work in tribology has occurred in lubrication; in particular, the lubrication of gears, cams and tappets and other non-conforming contacts. The prime task of a lubricant is to prevent damage of the interacting surfaces, while they are performing their designed function. For gears, this function is the transmission of power. For plain bearings, the function is to support the load or, sometimes, to provide location, by restricting the number of degrees of freedom. A secondary, but equally important, task of the lubricant is to provide a means of cooling the interacting surfaces. A lubricant delivery system is often required as an aid to this cooling, which thereby provides a means of removing extraneous particles in the lubricant by means of suitable filtration.

Lubrication theory has tended to concentrate on the complexities involved in applying corrections to the classical equation of hydrodynamics, namely Reynold's equation. These corrections arise from the very strong dependence of the coefficient of viscosity upon the *pressures* generated in the non-conforming contact, such as between gear teeth and between cams and tappets. There has been considerable progress in formulating the elastic and hydrodynamic equations governing these pressures as a function of position within the contact zone. This is the topic known as 'elasto-hydrodynamics'. There is, however, an equally strong influence of temperatures within the contact zone upon the coefficient of viscosity of the lubricant. Very little *new* research has been carried out on this aspect of the lubrication of non-conforming geometries.

1.1.4 Wear

Neale (1973) has defined wear as the progressive loss of substance resulting from *mechanical* interaction between two contacting surfaces. In general, these surfaces will be in relative motion, either sliding or rolling, and under load. This definition is too restricted. It should include the possibility that wear can occur through the combined effects of *chemical reaction* with the ambient fluid (whether that fluid be air or oil) *and* mechanical interaction. It should also distinguish between the deliberate removal of one surface (e.g. the workpiece) and the unwanted wear of the other surface (i.e. the cutting tool). Even for two specimens (*A* and *B*) of *identical bulk* composition and hardness, where one might expect equal wear on either face, it is found that differences in the rate of surface removal may occur, especially where there exist differences in *nominal* surface configurations. Different surface configurations will lead to different surface temperatures (T_{SA} and T_{SB}) which, in turn, lead to (a) differences in oxidation states of both surfaces (if metallic); (b) differences in surface and substrate hardnesses (if metallic or polymeric); and (c) differences in the type and amount of interaction between the lubricant and the oxide film on the two surfaces (if metallic or ceramic).

From the previous paragraph, the reader may have realized that wear is really only definable in terms of specimen geometry, specimen hardness and a precise knowledge of the ambient conditions under which the tribo-system is wearing.

There does not exist a single number which *universally* holds valid as the number of cubic metres removed per metre sliding distance for a given pair of surfaces. When a wear rate is quoted, it should also be made clear to which surface (A or B) this rate is relevant. The surface temperature (T_{SA} or T_{SB}) should also be quoted, together with the hardness value concomitant with this temperature. The constitution of the atmosphere or fluid around each surface should also be quoted in terms of its oxygen and additive content.

1.1.5 Pitting

For many years the materials engineer has been aware that repeatedly compressing and stretching a metal in a typical rotating-bending machine will cause rupture and fatigue, even though the stresses are well below the yield strength. A similar alternation of elastic stretching and compressing occurs in the contact of gear teeth and cams and tappets. Essentially, there is always a position in the relative movement between the opposing members where *pure* rolling occurs. On either side of that extremely narrow region of pure rolling, relative sliding will occur. It is the combination of sliding contact and rolling contact that can sometimes cause a failure of one of the surfaces by pitting, that is a characteristic cratering of the surface, due to the removal of relatively large pieces of that surface by a contact fatigue process. As the surfaces deform each other elastically (in the rolling contact region) over many millions of cycles (for the tribo-system using oil as the lubricant) or many thousands of cycles (for tribo-systems lubricated with water emulsions) fatigue cracks are initiated, which eventually lead to fracture and the formation of pits. Much work has been carried out on contact mechanics (the analysis of the contact stresses occurring in tribo-systems), and even more on the fatigue of materials, but very little has been done towards bringing these two relevant areas together to describe 'contact fatigue' analytically. Pitting, or contact fatigue, is still awaiting its complete elucidation.

1.1.6 Fretting wear

This is sometimes known as 'fretting corrosion', mainly because this type of wear manifests itself through the formation of oxides and/or hydroxides in the region of the fretting contact. It was originally thought that this implied that corrosion had occurred rather than wear. Essentially, fretting wear arises due to the *microscopic* oscillations that occur between two surfaces in contact (whether that contact merely be due to the weight of one of the members of the system or a nut and bolt) when the whole system is subject to vibration. For amplitudes of vibration greater than about $10\,\mu m$, it is usual to refer to the system as one in which 'reciprocated sliding' is taking place. Possibly the most important characteristic of fretting is the fact that the wear debris is trapped between the vibrating surfaces and hence plays an important part in the subsequent wear processes. Fretting wear is normally only a nuisance as, for instance, in the fretting damage caused during the transportation of cars on transporters from the factory to the dealer. Even with the engine and wheel bearings fully lubricated, it is not normally possible to exclude all metal-to-metal contact, especially after being subjected to several hours of vibration on the back of the transporting vehicle.

One of the most irritating things about fretting is its unexpected occurrence in situations not normally considered to be tribosystems. In the Advanced Gas-Cooled Nuclear Reactor (the AGR), for instance, the clamps that hold the cooling rods can suffer fretting damage arising directly from the vibrations caused by the passage of the cooling gas (a mixture of CO and CO_2). This damage reduced the replacement time from the projected 30 years down to 2 or 3 years.

1.1.7 Abrasion

'Abrasion', as normally perceived by the engineer on the macroscopic scale, truly is 'the removal of substance resulting from mechanical interaction between two contacting surfaces' (Neale, 1973). This definition of wear describes 'abrasion' most adequately, especially where it involves 'two body abrasive' contact. This is where one of the bodies is so hard (e.g. a diamond wheel) that all the material removed originates from the workpiece. Sometimes, a third body becomes involved either deliberately, for an abrasive grit embedded into a lead lap, or accidentally, in the case of a very hard piece of one of the surfaces (perhaps an oxide) acting analogously to the abrasive grit particle. 'Three body abrasion' is considered to be rather complex and is, therefore, not the subject of very much reliable hypothesising. Abrasion of drills and mechanical diggers by rock and soil and abrasion of the guideways and belts of mineral conveyor systems by the minerals they are conveying are just a few examples of the relevance of abrasion to industry, particularly the construction industry. Because of their size, the economic losses incurred by abrasive wear of systems like the mechanical digger form a significant part of the losses due to all types of wear. Typically, however, these systems do not readily lend themselves to simulation in the laboratory, nor to the application of physical analytical techniques. In fact, it will be shown (in Section 1.4) that abrasion is a wear *mechanism* rather than a wear *classification*, a mechanism that seems to be involved in some form or another in several types of wear. We will see that physical techniques can be effectively used in elucidating these abrasive wear mechanisms when they occur on the microscopic scale.

1.1.8 Adhesion

Strictly speaking, adhesion is a term relating to the force required to separate two bodies in contact with each other. Much of the early work in tribology was involved with the force required to separate two metallic surfaces *in static contact*. Bowden and Tabor (1954) at the Physics and Chemistry of Surfaces Laboratory at Cambridge, England, were responsible for the hypothesis (which is still accepted by many tribologists) that the force of *static friction* between two sliding surfaces is, in fact, the force required to break the 'cold welds' which momentarily form between the contacting high spots of the opposing surfaces. Bowden and Tabor's (1954) hypothesis, which will be described in some detail in Section 1.2 of this book, does not necessarily remain valid for materials which form surface films (e.g. oxide films on steels) and there are some formidable difficulties in extending the hypothesis to *dynamic friction*. Essentially, it will be shown that adhesion probably plays an important role in wear processes which involve only *mechanical* interactions between two contacting surfaces, so that it is these wear processes that should truly be described as 'adhesive wear'. When chemical reactions are also involved, then it seems reasonable to describe those forms of wear according to the type of film formed *after* 'running-in', for example oxidative wear or, more generally, reactive film wear. Undoubtedly, adhesion is involved during 'run-in'. Quite often, investigators have only run their experiments over relatively short times; that is, over part of the initial 'running-in' period, thereby concluding (sometimes incorrectly) that 'adhesive wear' is characteristic of the materials they had been running against each other.

1.1.9 Grooving, scoring, scuffing and spalling

These are just four of the many words often found in the vocabulary of the materials engineer when he/she wishes to describe what has happened to the surfaces that have failed. 'Galling' is another such word. There is no satisfactory standard usage. Typically, 'scuffing' is the term applied to gears that, quite early in their operational career, fail in a fairly catastrophic manner. 'Spalling' relates to a form of failure in which large flat areas of one or both of the surfaces in contact become detached. 'Scoring' and 'grooving' are merely words describing the appearance of one or both of the wearing surfaces in contact. In this book we will only use these words when we speak of the work of other investigators. In all such cases, however, we will also try to establish the basic wear mechanism. The science of tribology has been hindered by these imprecise terms. It is suggested (in Section 1.4) that the results of sliding and/or rolling wear experiments should be discussed in terms of just two mechanisms, namely severe (i.e. mechanical) wear and mild (i.e. chemically reactive) wear. With this simpler classification, it will be much easier for the reader to understand how and why one uses one or more given physical analytical techniques for a given system.

1.1.10 Dry bearings/solid lubricants

Some of the earliest work with solid lubricants involves graphite. Graphite is a typical solid lubricant since:
(a) it depends on crystallite re-orientation for its effectiveness in lowering the friction between sliding surfaces;
(b) it involves a layer-like structure which, when oriented in a certain way relative to the sliding interface, provides surfaces of easy-glide;
(c) it bonds effectively to most metal/metal oxide surfaces.
Both graphite and molybdenite (MoS_2) are often used as solid lubricants, graphite being used in high temperature/normal atmospheric pressure situations, whilst molybdenite is used in normal temperature/high vacuum situations. Dry bearings typically use composite materials, generally a fibre-reinforced plastic, where the fibre is often composed of graphite 'whiskers' (or 'carbon fibres', as they are usually called). For light loads, the dry bearing is often made of a low-friction polymer (such as polytetra-fluoroethylene, PTFE). The wear rate of PTFE is unacceptably high, which is why it cannot be used for highly-loaded tribo-systems without some fibre reinforcement. We shall see, in Chapter 5, that a high

wear rate is often associated with a low friction coefficient between two surfaces treated with solid lubricants. Hence there is always the problem of lubricant replenishment when using solid lubrication.

1.1.11 Bearings

Practically all tribo-systems involve bearings of some sort or another, whether it be a plain bearing, a rolling contact (ball or roller) bearing or an intermittent sliding contact (such as a hinge, lock, circuit breaker, etc.).

The typical plain bearing consists of a shaft rotating inside a journal of slightly larger diameter than the shaft. When stationary, the load of the shaft acts vertically upon the journal, causing wedge-shaped gaps between the two surfaces on either side of the vertical.

Under these conditions, intermetallic contact is prevented by the boundary lubrication properties of the lubricant, that is by the chain-like molecules adsorbed upon the metal surfaces. When rotation occurs, these wedge-shapes become filled with lubricant and actually support the load. This is known as 'hydrodynamic lubrication' and is not now the subject of much research, the basic principles having been laid down by Reynolds at the end of the last century.

The rolling contact bearing, on the other hand, is a much more complicated situation, since it always involves *slip*, as well as pure rolling. Even the enmeshment of gear teeth involves sliding into the region of pure rolling (i.e. where the pitch lines of both gears are tangential to each other), followed by sliding (in the opposite direction) as the gears become disengaged. These contacts are known as 'non-conforming' contacts (in contrast to the conforming contact of the plain journal bearing), and, because of their geometry, the lubricating film within such contacts possesses special properties implied by the term 'elasto-hydrodynamics'. These concepts will be discussed in Section 1.3 rather than in this sub-section, the main purpose of which is to acquaint the newcomer to tribology with some of the terms constantly used in this book.

1.1.12 Tribo-system

We have seen that both the form of lubrication and the type of wear depends very strongly on the geometry of the *system* being subjected to sliding or rolling. Only recently has the systems approach of Czichos and Salomon (1974) begun to be applied to tribology. These authors strongly recommend that we consider all the characteristics and parameters relevant to tribo-testing in terms of (a) operating variables (load, velocity, temperature, duration of test and type of motion); (b) the tribometer test system (the tribo-elements, lubricant and atmosphere); and (c) the tribometric characteristics (the friction force, friction coefficient, noise and vibrations, the temperature, wear rate and contact conditions).

The authors also include surface characteristics as part of characterizing the tribo-elements. The word 'surface characteristics' can, of course, mean everything necessary to characterize a surface, for example the surface topography, its chemical composition, the hardness of the surface and the layers immediately below the surface films normally formed during sliding. Surface characterization of the tribo-elements before, during and after the interaction between moving surfaces is, of course, the main reason for writing this book. It is the *changes* in surface characteristics caused by the tribometric characteristics that enable us to understand the friction and wear mechanisms that operate in the tribosystem. Czichos and Salomon (1974) do not *explicitly* mention the division of heat at the contacting interface as an important tribometric characteristic. Presumably one could include this under the general heading of 'contact conditions'. We shall see, in Chapter 6, how heat flow analysis can provide independent information about the validity of our hypothetical surface models and hence the validity of our estimates of surface temperatures.

1.1.13 Rheology

This is the study of the flow of materials under stress. In tribology, the flow properties of liquid lubricants are of primary interest. The tribologist wishes to know how a lubricating fluid will behave in the regions of the contacting surface. He is especially interested in the effects of pressure in the contact region. Unfortunately, the effects of temperature upon the properties of the lubricating fluid within the interfacial region has not received as much attention as the effects of pressure. Blok's 'flash temperature' criterion (1937) might be considered to be the pioneering research into the effects of temperature upon the lubrication of moving machinery, in particular gears. In fact, Blok postulates that a plain mineral oil will fail at a certain critical 'flash' temperature, the value of which can be calculated from the experimental conditions. However, his work can tell us nothing about how the lubricating properties of a mineral oil change up to the point at which it fails to function as an effective lubricant.

Essentially there are two ways of defining the property governing the rheology of lubricants, namely the dynamic viscosity (preferred by engineers who have to design bearings) and the kinematic viscosity (preferred by lubricant manufacturers and users). The dynamic viscosity coefficient (η_0) is the constant of proportionality between the shear stress (τ_f) in a liquid and the velocity gradient ($\partial U/\partial y$) between the stationary surface and the moving surface (as shown in Figure 1.1). This can be written as follows:

$$\tau_f = \eta_0(\partial U/\partial y) \tag{1.1}$$

For a velocity profile similar to the one shown in Figure 1.1, namely a linear increase in velocity parallel to the surface from zero at the stationary surface to U at the moving surface, we can write $(\partial U/\partial y) = U_m/h_{lub}$, where h_{lub} is the distance between the parallel plates. Hence

$$\tau_f = \eta_0 \frac{U_m}{h_{lub}} \tag{1.2}$$

If τ_f is in pascals (i.e. N/m^2), U_m is in m/s and h_{lub} is measured in metres, then the units of dynamic viscosity will be (Ns/m^2). The unit of dynamic viscosity is still taken to be the poise (P), which was originally designated as '1 gram/centimetre-second'. Hence we may write:

$$1(\text{Ns/m}^2) = 10 \text{ poise} \tag{1.3}$$

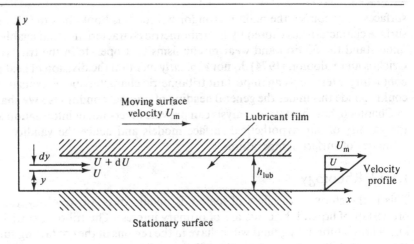

Figure 1.1 *Lubricant film between parallel plates*

In practice, we often talk about 'centipoise' (cP) which is (1/100) of 1 P, that is 10^{-3} Ns/m^2. Sometimes, we call the SI unit a 'pascal-second', since one (N/m^2) is one pascal. Kinematic viscosity (v) is defined as:

$$v = \eta_0/\rho \qquad (1.4)$$

where ρ is the density of the liquid. We still use the 'stokes' as the unit of kinematic viscosity, which was originally designated as '1 cm^2/s'. Hence we may write the SI unit (1 m^2/s) as:

$$1\,\text{m}^2/\text{s} = 10^4 \text{ stokes} \qquad (1.5)$$

Again, we tend to use a much smaller unit in practice, namely the 'centistokes' (cSt). 1 cSt = 1 mm^2/s is approximately the value that one obtains for the viscosity of water at 20 °C.

Any fluid whose viscosity is independent of the rate of shear (or velocity gradient) is said to be 'Newtonian'. Equation (1.1) was, in fact, suggested by Newton. Mineral oils and synthetic oils of low molecular weight are Newtonian under almost all practical working conditions. Polymeric liquids of high molecular weight (e.g. silicones, molten plastics) may exhibit non-Newtonian behaviour at relatively low rates of shear. An approximate relation has been proposed for non-Newtonian liquids, namely

$$\tau_f = [\Phi(s_f)]^\beta \qquad (1.6)$$

where Φ and β are constants, and s_f is the shear rate. For a Newtonian fluid, $\Phi = 1$ and $\beta = \eta_0$ and $s_f = (\partial U/\partial y)$, so that Equation (1.6) reduces to Equation (1.1) in the limit. For a silicone fluid, the typical value of β is approximately 0.95 (Naylor, 1973).

1.1.14 Summary

In this sub-section, thirteen or so terms have been defined that are in common use amongst practicing tribologists. One of the factors that has hindered progress in tribology must be the imprecise nature of some of the terminology. I have tried to be consistent with my use of these terms within this book. Unless there is a consistent set of definitions, accepted by all concerned, it will become impossible for any technology transfer to occur in this very complex subject. Let us now discuss the basics of the frictional behaviour of materials, since this was, historically, where the first advances were made in tribological knowledge.

1.2   Frictional behaviour of materials

1.2.1   Introduction

The laws of friction have already been mentioned (if only briefly) in this chapter (Section 1.1.2). Let us now discuss these laws more fully. When we try to slide a rectangular block along a plane surface, there is a force of friction between the base of the block and the surface of the plane that opposes motion. If the only other force acting between the block and the plane is the weight of the block (which, in this case, acts at right angles to the general surface between them), then it is found that the external force required to maintain a constant velocity in the plane of motion is always proportional to the normal component of the reaction of the plane to the weight of the block. This external force must, of course, be equal to the kinetic force of friction (F_{kin}). We distinguish between this force and the force opposing that force necessary to start the block moving from the static position, by calling it the static force of friction (F_{stat}). The mechanical engineer, with his/her bias towards rotating machinery and bearings is mainly interested in F_{kin}, whereas it is feasible that the civil engineer would be more interested in F_{stat}.

It is interesting to note that most of the early research has been concerned with F_{stat}. For instance, the pioneer work of the Cambridge School, led by the late Professor Bowden, was almost entirely related to adhesion and the force required to break the cold welds that were said to occur between metal asperities on opposing surfaces in contact. In this section, we will review some of the classical work carried out by the Cambridge School involving plastic deformation and its relation to the laws of friction. This school was mainly responsible for the concept of the real area of contact (A_{real}) being related to the load (W) and the hardness (p_m) of the softer of the two materials in contact, namely:

$$A_{real} = W/p_m \qquad (1.7)$$

We will also review some of the equally classical research of the Aldermaston School, namely the Surface Physics Department of the Research Laboratories of the Associated Electrical Industries group of companies in the United Kingdom. The Aldermaston School was responsible for much of the pioneering research work in what was later called 'tribology'. In particular, their alternative explanation of the laws of friction in terms of purely elastic deformation is well worth being given another 'airing'! This sub-section also describes some of the later work, carried out in other laboratories, which goes some way towards producing an expression containing both plastic and elastic parameters. Finally, some recent work carried out at the Georgia Institute of Technology in Atlanta is briefly reviewed insofar as it provides experimental evidence that possibly we should *not* be using Equation (1.7) as our basis for the real area of contact while the surfaces are actually in sliding motion.

1.2.2 Plastic deformation of surfaces and its relation to the basic laws of friction

Any hypothesis of friction must take into account the fact that even the smoothest surface obtainable by mechanical polishing comprises many valleys and hills, as revealed by profilometry (see Section 2.3), optical microscopy (Section 2.1) and scanning electron microscopy (see Section 2.3). Hence when two surfaces are in stationary contact over an *apparent* (or nominal) area of contact, they are, in fact, in contact only where the tips of the hills (or 'asperities', as these irregularities are called) on both surfaces actually touch. Clearly, this area must be an area less than the apparent area of contact. We call it the *real* area of contact (A_{real}), already mentioned in Section 1.2.1. Without defining this area more closely for the present, let us consider what the area of contact should be for the idealised case of a single point contact of a hemispherical asperity on a smooth flat surface. Let E_1 and E_2 be the elastic (Young's) moduli for the asperity and flat surface respectively and assume the flat surface is made of softer material than that of the spherical asperity. If now the asperity and the flat are brought into (stationary) contact under a normal load ($(W_e)_s$), they will *at first* both deform *elastically*. This means that the spherical cap of asperity material will deform into a flat, whilst the flat will deform into a depression, with both surfaces being in contact over a *circular* region of radius (a_e) given by the following equation:

$$a_e = 1.1 \left(\frac{(W_e)_s(R)}{2} \left(\frac{1}{E_1} + \frac{1}{E_2} \right) \right)^{1/3} \tag{1.8}$$

where R is the radius of the spherical asperity. This equation can be derived from the classical *elastic* deformation equations of Hertz (see, for instance, Holm (1946) or Timoshenko (1934)). From Equation (1.8), we can deduce that the area of contact due to elastic deformation only is proportional to $[(W_e)_s]^{2/3}$, that is

$$(A_e)_s = \pi a_e^2 \propto [(W_e)_s]^{2/3} \tag{1.9}$$

where we have used the subscript 's' to denote that we are considering a *single* contact only and the subscript 'e' to denote an area formed entirely by *elastic* deformation. The mean pressure (p_m) across the single contact is given by:

$$p_m = (W_e)_s / (\pi a_e^2) \tag{1.10}$$

From Equations (1.10) and (1.9), we can readily see that

$$p_m \propto [(W_e)_s]^{1/3} \tag{1.11}$$

This equation tells us that, as we increase the load $(W_e)_s$, the mean pressure increases elastically as the third power of the load. How long can this elastic deformation continue? The answer is until the softer of the two materials in contact begins to yield. According to Timoshenko (1934), the softer material begins to deform *plastically* (i.e. permanently) at a point P, which is about $(0.6a_e)$ below the surface, when the mean pressure equals $1.1Y$, where Y is the elastic limit. For any *further* increase in load, the deformed region around the initial plastic zone continues to increase in size until it reaches the surface of the softer material (i.e. the flat in this case). The hard indenter then sinks into the softer material until the area of contact is sufficient to support the load. At this stage, Bowden and Tabor (1954) show that the mean pressure (p_m) will be about $3Y$. This is the condition for full plastic flow. Figures 1.2(*a*) and (*b*) contain diagrams of (*a*) the onset of plastic flow and (*b*) full plastic flow for the single asperity contact model. This is an idealized drawing, but it essentially shows how the area of contact increases elastically up to (πa_e^2), after which it increases plastically so that p_m becomes identical with mean flow pressure.

We notice that the mean pressure (p_m) is independent of the load $(W_p)_s$ once full plasticity is obtained. This means that any increase in the load beyond that necessary to obtain (p_m) greater than $3Y$ will only cause the area of contact to increase such that (p_m) remains constant. It is experimentally difficult to verify the predictions relating to the dependence of (p_m) upon load both during elastic and plastic deformation, mainly because plastic deformation generally increases the elastic limit through work-hardening. In fact, it is true to state that most of the basic assumptions made in the plastic deformation theory of friction all tend to be experimentally unproved. Nevertheless, the theory has survived over the last 30 years or so, due mainly to its essential correctness in so far as it tells us how the coefficient of friction depends on the shear strength and hardness of the softer material. Let us briefly list these almost *a priori* assumptions:

(a) The theory of the *single* spherical asperity on a flat *holds* for the *multiple* asperity contacts of arbitrary geometries.
(b) The material around contacting asperities is subject to stresses *well beyond* the elastic limit.
(c) The mean pressure (p_m) for full plastic flow over all the multiple contacts of real surfaces is the load divided by the sum of all the different individual areas of contact between the two surfaces. If we call that sum the real area of plastic, multiple contact $(A_p)_m$, then we have:

$$p_m = \frac{(W_p)_m}{(A_p)_m} \qquad (1.12)$$

(d) When two surfaces come into contact, the harder material sinks into the softer material (of yield pressure p_m) until the area of contact is sufficient to support the load $(W_p)_m$. Under stationary conditions, this entails the 'cold welding' of the mating asperities which are said to be broken when a shearing force equal to the static frictional force (F_{stat}) is applied to the system.
(e) Welded junctions still form when the surfaces are slid over each other. The force required to make and break the welded asperities during sliding is one component (the adhesive component) of the kinetic friction force (F_{kin}).

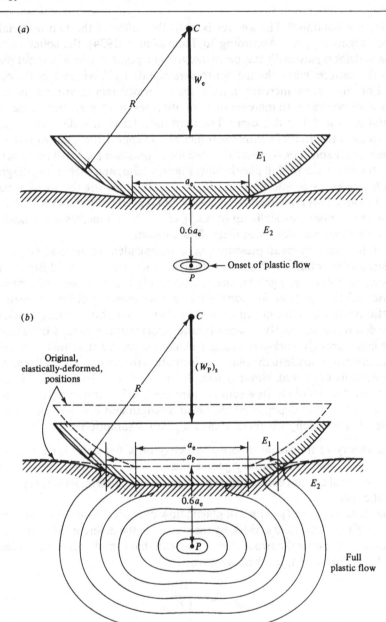

Figure 1.2*(a) Schematic diagram showing onset of plastic flow of the soft flat under a hard spherical surface at a load* $(W_e)_s$ *such that the mean pressure* (p_m) *approximately equals* $1.1Y$ *(where Y is the elastic limit of the softer material).*

(b) Schematic diagram showing full plastic flow of the soft flat at a load $(W_p)_s$ *such that the mean pressure* (p_m) *equals approximately 3 times the elastic limit* (Y)

(f) For softer materials, for example polymers, there is another component (the ploughing component) of the kinetic friction force. This component arises from the interpenetration of surface asperities as they slide tangentially to the general surface and are thereby elastically and plastically deformed, so that sliding can occur.

These assumptions have formed the basis for much tribological hypothesising over the past few years. By introducing the mean tangential shearing stress (s) – the stress required to shear the welded junctions – and ignoring the ploughing term, Bowden and Tabor (1954) produced their elegant expression for (μ_{kin}) (the kinetic coefficient of friction), namely:

$$(\mu_{kin}) = \frac{(F_{kin})}{(W_p)_m} = \frac{(A_p)_m s}{(W_p)_m} = \frac{s}{p_m} \tag{1.13}$$

By introducing a further assumption, namely that the welded junctions will tend to shear at the welded-junction/softer material interface, realizing that (p_m) was, in fact, the Brinell hardness (expressed in terms of a stress), Bowden and Tabor (1954) produced the following expression:

$$(\mu_{kin}) = \frac{\text{shear strength of the softer material}}{\text{hardness of the softer material}} \tag{1.14}$$

This equation is probably more relevant to metals sliding on metals, where any ploughing term becomes negligible after the initial running-in period. It should not be used to calculate the coefficient of kinetic friction using material properties. However, it does indicate what actions could be taken to *change* the coefficient of kinetic friction. To decrease the friction, the shear strength of the contact points must be reduced and the hardness increased. Because shear strength and indentation hardness are positively correlated, one given material is unlikely to have *both* low shear strength and high yield strength (i.e. high hardness). However, low shear strength films can be put upon hard substrates to achieve the desired results. Thus, coating a steel surface with a thin lead film, or a fluid or solid lubricant, can achieve a low shear strength junction with the steel providing a hard substrate to limit the contact area.

For polymers, and other non-metallic materials, in contact with a metallic surface (a not uncommon combination), the ploughing component appears to be the major source of the friction force during the initial stages of contact. Let us, therefore, define the interpenetration area, (A_p')$_m$, measured normal to the sliding direction which is related to the force (P_p) required to deform or plough asperities, by the relation:

$$(P_p) = (A_p')_m (p_m) \tag{1.15}$$

It seems reasonable to postulate that the resistance to ploughing will have much the same value as the hardness (p_m) so that the ploughing force (P_p) must be (A_p')$_m$ multiplied by (p_m). Let us now write the expression for (F_{kin}) in terms of the adhesive component (S) and the force (P_p) required to deform or plough asperities, that is

$$F_{kin} = S + P_p \tag{1.16}$$

that is

$$F_{kin} = [(A_p)_m]s + (A')_m(p_m)$$

Therefore

$$(\mu_{kin}) = \frac{(F_{kin})}{(W_p)_m} = \frac{(A_p)_m s}{(W_p)_m} + (A'_p)_m \frac{(p_m)}{(W_p)_m}$$

that is

$$(\mu_{kin}) = \left(\frac{s}{p_m}\right) + \frac{(A'_p)_m}{(A_p)_m} \tag{1.17}$$

Although Equation (1.17) has the merit of being relevant to all materials, Equation (1.13) has been shown to be relevant to the friction of metals, especially where these materials slide in the absence of an oxidizing environment or during the initial stages of any practical sliding experiment (i.e. during the initial running-in, when severe wear is occurring). With metals that readily form oxides during sliding (e.g. steels), it has been shown (Quinn, 1983) that the concept of plastically-deformed welded junctions may not be relevant. The relatively thick (of the order of micrometres) oxide films formed during sliding are a vivid proof that not all surface interactions occur as plastic deformation of newly-exposed metallic surfaces. Elastic deformation can also 'explain' friction forces, as shown in the next section.

1.2.3   The elastic deformation hypothesis of friction

We have seen (Equation 1.9) that $(A_e)_s$ is proportional to $(W_e)_s^{2/3}$. Remember that this relates to a single asperity. Until recently, all attempts to 'explain' the Laws of Friction were necessarily incomplete, since they assumed that every encounter between asperities had to be of a plastic nature. Then Archard (1953) produced his simple surface model, which assumes all the interactions between surfaces are *entirely elastic* (see Figure 1.3).

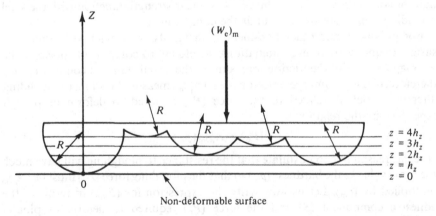

Figure 1.3 *Archard's (1953) simple model of multiple areas of elastic contact (for the very low value of $N_T = 5$, that is 5 asperities in contact)*

This model has the redeeming feature that is simple yet exhibits most of the characteristics of a real, deformable surface being loaded against a real non-deformable surface.

For light loads, our model surface would touch only at one point (O). For an increase in load $(W_e)_m$, both the *number* and the *size* of the contacts will also increase, since it is assumed that the N_T asperities in contact are evenly distributed in depth in the z-direction, that is there is only one asperity tip at each of the z-coordinates $0, h_z, 2h_z, 3h_z, ..., rh_z, ..., (N_T - 1)h_z$. Archard (1953) claimed that this was consistent with the behaviour of real surfaces, although later work by Greenwood and Williamson (1966) and Quinn (1978), indicates that the number of contacts increases with increasing load, whilst the size of the contacts remains essentially constant.

When the load is applied, the total contact area, $(A_e)_m$, resulting from the relative movement of the non-deformable (flat) surface through a distance $(N_T - 1)h_z$ (so that N_T asperities are brought into contact) is:

$$(A_e)_m = \sum_{r=0}^{r=N_T-1} \delta A_r$$

where (δA_r) is the contribution to the total contact area from the asperity in the rth level. Obviously, there is no contribution from the asperity where $r = N_T$, which is only just touching the flat. For an asperity deformed through a distance (rh_z) we have:

$$\delta A_r \approx 2\pi R(rh_z)$$

from the saggital formula. For deformation of all the $(N_T - 1)$ asperities, we have to sum the contributions from the zeroth to the $(N_T - 1)$ level. Thus:

$$(A_e)_m = 2\pi R h_z \left(\sum_{r=0}^{r=N_T-1} r \right)$$

that is

$$(A_e)_m = \pi R h_z N_T^2 \tag{1.18}$$

where R is the radius of curvature of each asperity, which we have assumed to be spherical. We have also assumed N_T to be large for a typical surface, so that the summation sign can be replaced by an integration sign and the resulting expression taken to the limit where N_T tends towards infinity, namely $(N_T^2/2)$. Since $(N_T - 1)h_z = z$, it is clear that $N_T^2 = z^2/h_z^2$ for large N_T. Hence we may write an expression for $(A_e)_m$ as follows:

$$(A_e)_m = Bz^2 \tag{1.19}$$

where $B = \pi R/h_z$, which is a constant.

If now we use Equation (1.8) to deduce the load elastically supported by (N_T) asperities, clearly we have to evaluate:

$$(W_e)_m = \sum_{r=0}^{r=N_T-1} \delta W_e \tag{1.20}$$

where δW_e is the load supported elastically at each asperity. For simplicity, let us assume $E_1 = E_2 = E'$, and hence deduce a relation between (δW_e) and a_r (the radius of contact of the asperity in the rth level), as follows:

$$\delta W_e = \left(\frac{a_r}{1.1}\right)^3 \left(\frac{E'}{R}\right)$$

Therefore

$$(W_e)_m = \frac{E'}{(1.1)^3 R} \left(\sum_{r=0}^{r=N_T-1} a_r^3\right)$$

that is

$$(W_e)_m = \frac{E'(R)^{1/2}(2h_z)^{3/2}}{(1.1)^3} \left(\sum_{r \times 0}^{r=N_T-1} r^{3/2}\right)$$

For large N_T, we can replace the summation sign by integration, write $N_T = z/h_z$, so that

$$(W_e)_m = Cz^{5/2} \qquad (1.21)$$

where $C = 0.86 ER^{1/2}/h_z$.

From Equations (1.19) and (1.21) we can get a relationship between $(A_e)_m$ and $(W_e)_m$, by substituting for z, thus:

$$(A_e)_m = \left(\frac{B^2}{C^{4/5}}\right)(W_e)_m^{4/5}$$

$$(A_e)_m \propto (W_e)_m^{0.80} \qquad (1.22)$$

Thus, by considering a simple distribution of asperity heights, Archard (1953) was able to show that the area of multiple contact of his model was proportional to the (4/5)th power of the load that could be borne elastically, which is considerably closer to unity than the (2/3)rd power of the load for a single elastic contact (see Equation 1.19). It should be recalled that the aim of friction theories is to show that the *real area of contact* (upon which the friction force *must depend*) is directly proportional to the load. Encouraged by this result, Archard (1957) carried out some experiments with a perspex facsimile of the deformable surface and found that the area of multiple contact (as viewed directly from behind the optically transparent perspex model), *and also the frictional force*, was proportional to a power of the load very close to (but below) 0.80.

The combination of theoretical analysis, backed up by experimental verification using a physical model, is, of course, the classical researcher's way of carrying out his/her research. It is not, however, giving us much information about *real* surfaces. Archard (1957) suggested that one could approximate to real surfaces by making his model more complex. He proposed that each spherical asperity in his simple model was itself covered by spherical asperities of radius very much less than R, and calculated that, for such a model, the real area of multiple *elastic* contact would be proportional to the 0.93th power of the load, an improvement of 16%. He then calculated the effect of each of the second order of magnitude spherical asperities being itself covered with spherical asperities of the third order of magnitude less than the simple model, and obtained the result that the real area of multiple *elastic* contact would be proportional to the 0.98th power of the load, an improvement of 22% upon his simple model. Archard said that *real* surfaces were even more complex than his third order asperity model, hence we might expect the real area of multiple elastic contact to be *directly* proportional to the load. Not everybody will agree with this expectation.

Although some doubts have since arisen regarding some aspects of Archard's (1953, 1957) attempts to explain the Laws of Friction entirely on the basis of elastic deformation, there is no doubt that elastic deformation is an important aspect of surface interaction, even in the interactions between surfaces exhibiting severe wear (e.g. see the work of Halliday (1957)). Recent work by Allen, Quinn and Sullivan (1985) shows that the converse is also true, namely that during mild wear, *some* plastic deformation occurs *somewhere* in the conjunction between the surfaces.

1.2.4 Discussion of the ranges of validity of the elastic and plastic deformation theories of friction

Although not explicitly stated by Archard (1953, 1957), it is clear that he assumed an equal probability of finding one particular asperity height as any other height. By assuming more plausible distributions of asperity heights, Greenwood and Williamson (1966) carried out a similar analysis to Archard (1953) to show that:

$$\frac{(W_e)_m}{(A_e)_m} = \frac{E''}{k_D}\left(\frac{\sigma}{R_S}\right)^{1/2} \tag{1.23}$$

where E'' is the *reduced* elastic modulus given by:

$$\frac{1}{E''} = \frac{1}{2}\left(\frac{1-v_1^2}{E_1} + \frac{1-v_2^2}{E_2}\right) \tag{1.24}$$

In these last two equations, v_1 and v_2 are the Poisson's ratios for the two surface materials, and E_1 and E_2 are the respective Young's moduli. k_D is a constant which depends on whether one assumes a Gaussian distribution of asperity heights (ϕ_g) or an exponential distribution (ϕ_e) where (ϕ_g) and (ϕ_e) are given by:

$$\phi_g = \left[\frac{1}{(2\pi)^{1/2}}\right]\left(\frac{1}{\sigma}\right)\left[\exp\left(\frac{-z^2}{2\sigma^2}\right)\right] \tag{1.25}$$

and

$$\phi_e = \left(\frac{1}{\sigma}\right)\left[\exp\left(\frac{-z}{\sigma}\right)\right] \tag{1.26}$$

where σ is the standard deviation of asperity heights. R_S is a parameter related to the assumed radius of curvature of the asperities.

Greenwood and Williamson (1966) compared Equations (1.12) and (1.23) and realized that:

$$\left(\frac{E''}{k_D}\right)\left(\frac{\sigma}{R_S}\right)^{1/2} = p_m \tag{1.27}$$

where p_m is the mean pressure at the onset of plastic flow, namely $p_m \approx 1.1Y$. Since k_D is expected to be about unity, Greenwood and Williamson (1966) introduced the concept of 'reduced flow pressure' (ψ_p), where ψ_p is given by:

$$\psi_p = \frac{E''}{p_m}\left(\frac{\sigma}{R_S}\right)^{1/2} \tag{1.28}$$

When ψ_p is less than unity, then $(E''\sigma^{1/2}/R_S^{1/2})$ must be less than p_m, so that elastic deformation dominates. If $\psi_p < 0.6$, the probability of plastic flow is very small indeed. If ψ_p is greater than unity, a large part of the contact will involve plastic flow.

Equation (1.28) is a distinct improvement upon other equations, since it contains *both* elastic and plastic deformation parameters. The concept is not very different from the idea that plastic flow ensues once the mean pressure across the area of contact exceeds the flow pressure. However, ψ_p *does* include the statistical parameters, σ and R_S, and hence is an advance on the original concept.

The main criticism which can be made of all the friction theories, whether they are based on plastic or elastic deformation, is that they depend on two assumptions, namely (a) that the force of friction is proportional to the real area of contact and (b) that one can extrapolate the contact mechanics of stationary contact to the moving contact situation. The first assumption might seem almost axiomatic, whereas the second assumption is taken for convenience, since we have so little direct experimental evidence about what is going on at the real areas of contact. It is relevant to mention here the work of Quinn and Winer (1987) in which the 'hot-spots' formed between a tool steel pin and a (transparent) sapphire disc were observed and photographed with decreasing exposure times. The authors found that the number of 'hot-spots' tended towards *seven*, that is an area density of *0.22 spots per mm²*. This number is an order of magnitude less than what had been expected from the only previously published work on the measurement of dynamic contacts (between electrographite and copper), Bickerstaff (1969), namely 1.9 spots per mm². Of course, one could assume that the *visible* 'hot-spots' only make up a small proportion of the actual contact spots (as revealed by Bickerstaff's electrical method). However, this explanation was not thought viable, since the total area of contact (for the seven 'hot-spots') was about equal to the real area of contact as defined by Equation (1.7), a definition which has remained accepted by tribologists since it was first proposed in the 1950s. Perhaps we should be thinking about an alternative to Equation (1.7), possibly a combination of this equation with the elastic Equation (1.23) to describe a plastic *and* elastic combined contact area?

Essentially, this discussion has shown that, provided the elastic limit is not exceeded at any contact between sliding surfaces, we should be able to explain the laws of friction in elastic terms only, possibly through the application of Equation

(1.23), to obtain $(A_e)_m$, as follows:

$$(A_e)_m = \frac{(W_e)_m k_D}{E''} \left(\frac{R_S}{\sigma}\right)^{1/2} \tag{1.29}$$

Hence

$$(F_e)_m = (A_e)_m \tau_e \tag{1.30}$$

where τ_e is the *elastic* shear stress between each surface asperity junction, a quantity that one should be able to estimate from the macroscopic stress/strain curve for the material of the member of the sliding pair for which the frictional force $(F_e)_m$ is being measured. If, however, the elastic limit is exceeded [as revealed by Equation (1.28)], then Equation (1.12) becomes operative so that:

$$(F_p)_m = (A_p)_m s \tag{1.31}$$

ignoring the ploughing term of Equation (1.16). It is not difficult to envisage a situation whereby *some* of the asperities are plastically-deformed according to Equation (1.30). Of course, when the interaction of the system *cannot* be described as being *either mainly elastic* or *mainly plastic*, then it is probably a mixture of both and can only be analysed with some difficulty. Let us now consider the basic concepts of lubrication between surfaces in relative motion.

1.3 Lubrication between surfaces in relative motion

1.3.1 Introductory remarks

Lubrication is the most highly developed of all the subjects covered by the term 'tribology'. Many of the papers at *any* tribology conference will deal with some aspect of lubrication. However, the amount of work involving the use of physical techniques in the study of basic lubrication processes is very small, being mainly concerned with the measurement of oil film thicknesses by optical and electrical methods. In this sub-section, we will review the present situation in lubrication as concisely as possible, with special emphasis being placed on those aspects of the subject most amenable to investigation by some of the more recently developed physical techniques. It will become apparent that most of these techniques are more useful for investigating the actual surfaces being lubricated, rather than the lubricant itself. Lubrication is seldom complete. If it were, then no wear would ever occur. There are always regions where breakdown of the lubricant's protective action occurs through the lubricant film when surface roughnesses meet. Let us now briefly review the basic essentials of the lubrication of surfaces in relative motion.

1.3.2 Types of lubrication

In *boundary lubrication*, the least well-known type of lubrication, the surfaces are *sometimes* protected by films of molecular thickness. These films either protect by physical adsorption (physisorption) or by chemical interaction with the surface (chemisorption). There is a less well-recognized form of boundary lubrication which involves *thicker films* being formed by *interaction* between the *metal or oxide* surface of the tribo-element *and the lubricant* (which typically has an additive comprising of a mixture of chemically-active elements, such as chlorine, iodine, sulphur and phosphorus). These films are often several micrometres thick. Sometimes, boundary films are formed by interaction with the ambient atmosphere. Oxidative wear is an example whereby oxide films of *3* or *4 μm* *thickness* are formed by interaction between the 20% partial pressure of oxygen in the air and the real areas of contact between metals. The formation of thick protective films is normally brought about by fairly heavy loads, although recent work by Sullivan and Granville (1984) indicates that thick oxide films can be produced at very light loads. When dealing with conventional lubricants, such as 'extreme-pressure' oils, however, the thick films are produced by wearing at extremely heavy loads.

The most well-documented type of lubrication is hydrodynamic lubrication. The plain journal bearing, for instance, has a macroscopic lubricant film in between the surfaces, the viscosity of which is very relevant to the load-carrying capacity of the bearing. The thickness of the lubricant film depends on the viscosity (η_0), speed (U) and the pressure (P) in the 'physical wedge' formed by the lubricant that becomes entrained between the shaft and the lower end of the journal bearing. With journal bearings, the thickness is typically 10 μm, which is much larger than the molecular films of 'conventional' boundary lubrication and still about 10 times the normal 'extreme-pressure' or oxidative wear surface film thickness. In a later part of this book, we will describe the optical and X-ray methods for measuring this thickness, as well as the more conventional electrical capacitance methods.

The third type of lubrication is known as 'mixed lubrication'. Traditionally, it was considered that under conditions of 'mixed lubrication', the shaft load (of a journal bearing) was supported by a mixture of oil pressure and asperity contacts. The terminology 'mixed lubrication', in fact, covers a lot of ignorance about situations which are clearly not hydrodynamic nor boundary lubrication. 'Elasto-hydrodynamic lubrication' (already mentioned in Section 1.1.3) is a type of lubrication that involves the elastic deformation of the sliding (and rolling) surfaces without the lubricant film actually 'breaking down' in between the contacting surfaces.

1.3.3 The Stribeck curve

In all research and development involving lubricants, it is always advisable to plot the Stribeck curve to deduce whether or not one's tribo-system is operating within the appropriate lubrication regime. This is a plot of coefficient of friction (which is clearly related to lubricant film thickness) against the bearing parameter, which is $\eta_0 U/P$ and a schematic representation of this is given in Figure 1.4. Notice that the coefficient of friction is virtually constant at about 6×10^{-2} that is 0.06, for values of $\eta_0 U/P$ from about 4×10^{-9} down to about 0.8×10^{-10}. This shows that considerable asperity/asperity interaction must occur during boundary lubrication, leading to increased friction compared with mixed and hydrodynamic lubrication regimes. The increase in friction with decreasing values of $\eta_0 U/P$ from about 10^{-8} down to about 5×10^{-9} is due to

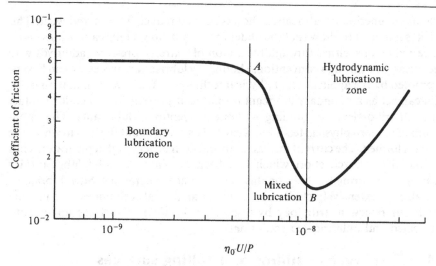

Figure 1.4 *Stribeck curve (schematic only)*

increased asperity interaction at low viscosities, low velocities and high temperatures. The lowest friction corresponds to the elasto-hydrodynamic region of hydrodynamic lubrication. Note that at *A*, the lubricant film thickness must have been virtually reduced to zero, so that the load between the shaft and the bearing is being carried entirely on asperity contacts. In the zone to the left of *A*, the coefficient of friction is almost independent of load, viscosity and shaft speed.

1.3.4 Solid lubrication

There is another form of lubrication which is difficult to classify, namely 'solid lubrication'. This involves fairly thick films of lubricant in the form of lamellar solids such as molybdenite, graphite, etc. Their action is similar to that of boundary lubricants in that they afford protection of the surfaces when starting from rest, or at any time when the fluid lubricant supply suddenly ceases. The lubricity of lamellar solids seems to depend on the ability of the individual crystallites of these solids to orient themselves during sliding so that they present 'planes of easy glide'. The mechanism of 'solid lubrication' is still not too clear, however, and in a later section, we will show how the combined use of two physical techniques (namely electron microscopy and electron diffraction) can produce extremely relevant information about how one particular tribo-system (namely the electrographite brush on copper slip ring system) successfully functions for extremely long periods, without any externally-applied lubrication.

1.3.5 Summary

In this sub-section the reader has been introduced to the fundamental aspects of lubrication, without very much of the highly complex detail normally associated with this subject. Where necessary, further details will be given in what follows.

The main function of lubrication, however, is to reduce friction and wear. This book is intended to show that the understanding of how a lubricant brings about these reductions, either through the action of 'extreme-pressure' additives with the surfaces or through conventional boundary lubrication processes, can best be promoted by the application of physical techniques. We have seen that an oxide film can act as a boundary lubricant insofar as it prevents intermetallic contact between tribo-elements running without conventional lubricants. The application of modern physical techniques can tell us much about the way in which the oxide is formed. The current interest in lubrication at very high temperatures has revived the research effort which had been put into the solid lubricant/dry bearing field during the halycon days of the space programme. Such lubricants can also be extensively studied by physical analytical techniques. Let us now finish our review of tribology by discussing the wear of surfaces under both lubricated and unlubricated conditions.

1.4 The wear of sliding and rolling surfaces

1.4.1 The wear of surfaces under lubricated conditions

At the start of an experiment, there are many solid–solid contacts through the boundary lubricant film. The protection of the surfaces under these conditions depends on the active material in the lubricant becoming physically or chemically absorbed by the surfaces. As the 'running-in' proceeds, the solid–solid contacts become less frequent as the roughness of the surfaces become reduced so that they can be accommodated within the thickness of the film. Even when it is fully 'run-in', however, a surface can still fail due to local fatigue stresses at the sites where the solid–solid contacts occur (or are about to occur) due to the lack of (or minute thickness of) the lubricant at these sites. One normally gets localized systems of cracks developing, which spread, and eventually cause flakes of material to become loosened and thereby leave pits in one or both of the surfaces in contact. This process, known as 'pitting' or 'contact fatigue', has been studied extensively. It has been shown that there is a strong negative correlation between 'the revolutions to the onset of pitting' and the surface roughnesses, when expressed in terms of the thickness of the lubricant film, that is for large values of D, the ratio of combined surface roughnesses to the minimum thickness of the lubricant film, one obtains the smallest probability of pitting.

We have been considering a tribo-system which had successfully 'run-in' before failure through pitting. However, with a new, un-tried system, 'running-in' can be disastrous. Suddenly, the conditions locally become too severe for boundary lubrication (i.e. the intrinsic lubricative properties of the lubricant) to remain effective. The surfaces become severely torn and higher friction values ensue. This process of surface modification is known as 'scuffing' and is due to local welding of the surfaces in much the same way as that envisaged by Bowden and Tabor (1954). Again, there is a dependence of 'scuffing' upon the thickness of the lubricant film. For instance, in some two-disc experiments reported by Crook (1962), failure did not start with the starting of the machine, but was delayed for several minutes, although there was no change in the external conditions. This behaviour can be explained in terms of the influence of the temperature upon the film thickness. As the surfaces become hot, through the dissipation of viscous or frictional forces, the viscosity of the oil in the contact regions falls and, eventually, so also does the thickness of the film. Solid–solid contacts increase in severity until the support of the hydrodynamic film is no longer sufficient to prevent welding. Scuffing also depends on the surfaces themselves. Changing the materials of the surfaces, pre-treating the surface with some of the exotic coatings that are currently being discovered each day, or using extreme-pressure oils (see later), will often eliminate scuffing.

Under full hydrodynamic lubrication, there is no opportunity for the opposing surface asperities to come into contact. Consequently, it is very rare for a tribo-system involving truly hydrodynamic lubrication to exhibit the continuous wear characteristic of unlubricated conditions. Typically, these tribo-systems fail catastrophically. The failure is generally due to the intrusion of foreign bodies, for example metal particles, into those regions of the contact zone where the lubricant film is at its thinnest, namely at the point where the bulk of the load is being carried. This is where efficient filtration of the lubricant being delivered to the tribo-element is of prime importance. This is also where the physical techniques of ferrography and condition monitoring play an important part in predicting the *onset* of wear, rather than analysing what has happened after the event.

Having shown that the wear of surfaces under lubricated conditions can be more readily understood and, in some cases, reduced to acceptable amounts, by the application of physical techniques forensically and as on-line detectors of imminent failure, let us now turn our attention to wear under unlubricated conditions. It was in this regime of tribological endeavour that physical techniques were originally applied. If one agrees with the view that when a lubricant breaks down at asperity/asperity contacts, the situation becomes akin to the unlubricated condition, then one can see how important it is to know how to apply physical techniques to dry wear. Not everybody will agree with this view. However, there is no doubt in the author's mind that we are going to have to live with tribo-systems that are operating at high temperatures without conventional lubrication as we know it today. The application of physical techniques to dry sliding will be an obvious way to understanding the basic principles involved in producing the machines of the future, such as the ceramic diesel.

1.4.2 The dry wear of surfaces

There had not been much research carried out into the mechanisms of wear prior to the publication of the late Ragnar Holm's classic book in 1946 (Holm, 1946) dealing with so many aspects of contacting surfaces. Holm produced the first published account of a mechanism of wear in terms of an *atomic transfer process* occurring at the real areas of contact formed by plastic deformation of the contacting asperities. He obtained an expression for the wear rate (*w*), namely:

$$w = \frac{Z_N W}{p_\mathrm{m}} \tag{1.32}$$

where Z_N is the number of atoms removed per atomic encounter, p_m is the flow pressure of the softer material and W is the load. Holm had a model in which he assumed that N_T circular regions of contact occur (which he calls 'a-spots'), each of radius a, which make up the total real area of contact (A_{real}). He further assumed that ξ_N atomic layers (of interplanar spacing, d_i) are removed from the contacting a-spots at each encounter, so that:

$$Z_N = \frac{\xi_N d_i}{2a} \qquad (1.33)$$

It is interesting to calculate what value ξ_N must take in order to satisfy experimentally measured wear rates. Let us consider the wear of stellite on tool steel as reported by Archard (1959a). He showed that the wear rate of this system, when run at 4 N applied load was 3×10^{-14} m^3/m sliding for a hardness of stellite equal to 7×10^9 N/m^2. Putting these values in Equations (1.32) and (1.33), assuming $d_i \sim 10^{-10}$ m and taking $a \approx 10^{-5}$ m (the maximum possible value, assuming all the load is borne at one spot so that $\pi a^2 = A_{real}$, where A_{real} is given by Equation 1.7), one obtains a value of $\xi_N \approx 20$. This would entail a wear fragment of about 20 or 30 Å thick. This is much too small for the sort of wear normally obtained with metals. Rabinowicz and Tabor (1951) using a combination of metals which normally exhibit a severe form of wear, showed that metallic transfer during sliding did not occur uniformly (as would be expected by the Holm hypothesis) but in a relatively small number of discrete fragments. Their work is also notable for being one of the first experiments carried out using an autoradiographic technique.

It should be noted that for materials exhibiting *mild* wear, the wear particle size is of the order of a few micrometres, which is three orders of magnitude greater than our calculated ξ_N value. About the only case of wear particles being about 20 or 30 Å thick must be the electron diffraction evidence described in Chapter 5 of this book. This evidence relates to the size of particles inferred from the diffraction line broadening in electron diffraction patterns from a carbon brush on copper slip-ring system. Even here, however, it has been shown (by electron microscopy) that these extremely small graphite particles are generally found to exist in *agglomerates*, with an average size of a few micrometres. The wear of carbon brushes is extremely mild, once the graphite crystallites have become oriented (through tribological interactions) with their low friction planes approximately parallel to the general surface. We will return to the wear of carbon brushes later in this book.

About six years after the publication of Ragnar Holm's book, two groups of tribologists published very similar hypotheses regarding wear. They were the Surface Physics Group under the leadership of Hirst in Aldermaston, England and the group led by Burwell and Strang in the United States. On the basis of many wear experiments with various combinations of materials sliding under an inert lubricant, Burwell and Strang (1952a, b) published their results, all of which indicated that the following two laws of wear would apply:

1. *Wear rate (w) is proportional to the applied load (W), that is, $w \propto W$.*
2. *Wear rate (w) is independent of the apparent area of contact.*

In fact, Burwell's and Strang's influence was a great help in my early days in the field of tribology. Their model of a surface as one in which the average size of the individual contact area is constant, whilst the number of contact areas increases with increasing load, is still considered a viable surface model (see Quinn, 1983). It is when Burwell and Strang (1952a, b) describe the various types and mechanisms of wear, that I have to take an *opposite* stand, in the interests of clarifying the position. In fact, my paper (Quinn, 1983) makes a strong case for assuming that *all wear phenomena* must be either 'mild wear' or 'severe wear', the definitions of these terms being given in the original paper by Archard and Hirst (1956).

Essentially, Quinn (1980) maintains that the Archard and Hirst (1956) classifications of mild and severe wear are the most basic and easy to apply to any wear situation, since they are entirely phenomenological. Because of the every day use of the words 'mild' and 'severe', it is not always appreciated that these classifications are based on:

(a) measurements of contact resistance (severe wear is characterized by low contact resistance, whereas mild wear provides surfaces that give predominantly high contact resistance);
(b) analysis of the wear debris both as regards size and composition (severe wear normally consists of large, about 10^{-2} mm diameter, metallic particles, whilst mild wear debris is small, about 10^{-4} mm diameter, and has been produced primarily by reaction with the ambient atmosphere or fluid); and
(c) microscopic examination of the surfaces (severe wear leaves the surfaces deeply torn and rough, whereas mild wear produces extremely smooth surfaces, often much smoother than the original surface finish).

It must be emphasized that these classifications do not specify a range of wear rates for each class of wear. Under certain circumstances, it is possible for mild wear processes to occur at a rate equal to severe wear processes for the same combination of materials. In general, however, the severe wear processes are often two orders of magnitude more effective at removing material from the sliding surfaces, especially close to the transition loads found to occur in the dry wear of steels (Welsh, 1965).

The term 'wear rate' is itself slightly misleading. As used by most tribologists, it means the volume of material removed from a surface per unit sliding distance of that surface with respect to the surface against which the relative motion is occurring. Typically, wear experiments in which wear rates can readily be measured, involve one of the pair presenting a much smaller area over which wear can occur than the other member. Hence, although the wear rate of the other member (which is normally the moving surface) may be the same as for the stationary surface, the removed volume comes from the whole of the wear track. Molgaard and Srivastava (1975) have pointed out that such asymmetrical sliding pairs will have very different heat flows from the frictional sources of heat at the interface into each member. This will affect surface temperatures and the wear. Molgaard and Srivastava (1975) suggests that all experiments should be made with symmetrical pairs, such as two annuli with the same dimensions. Experimentally, however, this geometry has been found a difficult one with which to deal. Hence, the continuing preference for the pin-on-disc geometry by most wear researchers. Provided allowance is made for the different heat flows in such geometries, it is probably as relevant to practical wearing situations as the mating annuli mentioned above. Some tribologists still use the time wear rate, that is the volume removed per unit time. This is a much less fundamental quantity than the volume removed per unit sliding distance, since it involves the speed of sliding.

1.4.3 The mechanisms of wear (with special emphasis on mild-oxidational wear)

We have already described how the mechanisms of friction are related to the plastic and elastic deformational characteristics of the materials being worn. These 'explanations' of the origin of frictional forces during sliding must surely have some relevance to the 'explanations' of wear. In other words, both friction and wear must occur by related 'mechanisms'. In view of the surface characteristics of severe wear, there seems to be adequate grounds for believing that the plastic deformation mechanism of friction needs only to be slightly amended in order to explain severe wear as well. Similarly, elastic deformation should be dominant in mild wear. It is extremely difficult to conceive of adhesion being involved in mild wear. Let us consider what are the mechanisms of *mild wear*, since this type of wear will often be referred to in what follows.

Mild wear clearly involves reaction with the environment. It is not always appreciated, even by experienced tribologists, that mild wear can occur even under wet, lubricated conditions, due to the large amount of air absorbed and entrained in all typical lubricants. Most typically, mild wear involves reactions between the surface and any ambient oxygen, that is, mild wear occurs through an oxidational wear mechanism. The understanding of this mechanism has evolved gradually over the past 20 years (see Quinn, 1962, 1967, 1968, 1978 and Sullivan, Quinn and Rowson, 1980), and is summarized in the next few paragraphs.

In the initial (severe wear or scuffing) stages, the surfaces achieve conformity, so that the real areas of contact consist of several comparatively large areas, each of which is about the size to be expected from the Bowden and Tabor (1954) plastic deformation theory, namely, of the order of (W/p_m), where W is the normal applied load and p_m is the Brinell Hardness (expressed in units of stress). At any given instant, *one* of these areas bears most of the load. That area then expands, in a manner similar to that proposed by Barber (1967), so as to become a plateau of contact and remain the only region of contact until it is removed by wear. If the sliding speed is comparatively slow, or the loads so light, that frictional heating is negligible, then the expansion of the contacting plateau will not be sufficiently large for it to become a preferred contact region. Furthermore, the rate of oxidation of the contacting plateau will not be very different from that of the remainder of the surface under these conditions.

Given sufficient frictional heating, however, the contacting plateau will oxidize preferentially with respect to the non-contacting plateaux and the remainder of the surface. It will then oxidize at a temperature (T_o) normally well in excess of the general surface temperature (T_S). The surfaces of the plateaux are extremely smooth with fine, wear tracks parallel to the direction of sliding. Typically, each of these plateaux had an area of about 10^{-2} mm^2 and a height of about 2–3 µm. The plateaux often show surface cracks perpendicular to the sliding direction, somewhat similar to the fatigue-crack systems found in fracture mechanics. It seems very probable that the intermittent heat and stress cycles suffered by the plateaux, as they come into contact with their 'opposite numbers', would bring about wear through the sort of fatigue indicated by the crack systems. The surfaces surrounding each plateau are rough and strewn with debris fragments. It would seem that these fragments were once part of a previously existing contact plateau.

The contacting plateau (the 'operative' plateau) is the site for *all* the asperity/asperity interactions between two opposing surfaces at any given moment in time. According to the oxidational wear mechanism, these asperities are the sites for oxidation at the temperature (T_o). Since oxidation occurs by diffusion of oxygen ions inwards and (sometimes) by metal ions outwards, one would expect the plateaux to grow in height from the interface between the oxide and the metal beneath each asperity contact. In the course of many passes, one would expect that the increases in height would be spread over the whole contact area of the plateau. When it reaches a critical oxide film thickness (ξ), the film becomes unstable and breaks up to form flakes and, eventually, wear debris.

The oxidational wear mechanism proposed by the author and his co-workers does not attempt to explain why the plateaux break up at a critical thickness (ξ) of a few microns. (This could be an area of investigation for future tribologists well-versed in both tribology and fatigue.) The mechanism merely states that, when the contacting plateau finally breaks up completely, then another plateau elsewhere on the surface becomes the operative one. The virgin surface beneath the original plateaux is now free to oxidize at the general temperature of the surface (T_S). Without externally-induced heating, the amount of oxidation at a T_S of 80 °C (say) is several orders of magnitude *less* than the oxide growth at $T_O = 400$ °C (say). Hence, the original plateau's sub-surface region will not oxidize significantly until that part of the surface becomes the operative area of contact once again. This mechanism is based on experimental evidence. For instance, it has been shown (Quinn, 1969), that the height of a cylindrical pin wearing its smallest dimension against a flat rotating disc often exhibits periodic, sharp decreases in height about every five minutes. Taking into account the speed and the decrease in height, it seems very likely that these periodic variations relate to the sudden removal of a plateau of thickness about 1 to 2 µm.

An interesting new development in the study of wear has been the appearance of a wear mechanism map for steels. Lim and Ashby (1987) have sub-divided the mild wear of steels into ultra-mild wear and mild-oxidational wear. They have also sub-divided severe wear into delamination wear (Suh, 1977), severe oxidational wear and melt wear. The authors plot a diagram which shows the rate and regime of dominance of each of the above mechanisms of dry wear. The diagram can be constructed empirically (i.e. from experimental data alone) and by modelling (by theoretical analysis calibrated to experiments). Lim and Ashby (1987) used the oxidational wear theory (see Section 1.4.4) together with thermal analysis of sliding contacts (see Chapter 6) to obtain the wear mechanism map for steel (as shown in Figure 1.5). In this figure, the abscissa is the normalized velocity $\tilde{U}$ given by:

$$\tilde{U} = \frac{U(R_t)}{\chi_d} \tag{1.34a}$$

where U is the speed of sliding at the pin, (R_t) is the radius of the pin and χ_d is the thermal diffusivity of the disc. The ordinate is normalized pressure $(\tilde{W})$, given by:

$$\tilde{W} = \frac{W}{(A_n)(p_m)_0} \tag{1.34b}$$

where W is the normal load, A_n is the nominal (apparent) area of contact and $(p_m)_0$ is the room-temperature hardness. The contours of constant normalized wear rate $\tilde{w}$ where $\tilde{w}$ is given by:

$$\tilde{w} = \frac{w}{(A_n)} \tag{1.34c}$$

are superimposed on the field showing the regions of dominance of the different wear mechanisms. There are discontinuities in the contours when they cross the field boundaries into the regimes of severe-oxidational wear and melt wear. The wear rates given in parentheses are the values when mild wear takes place and the shaded areas indicate a transition between mild and severe wear.

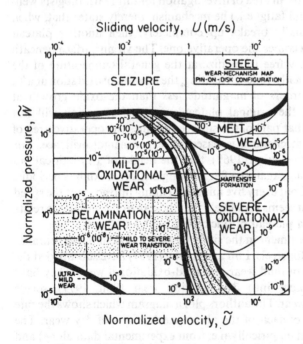

Figure 1.5 *The wear mechanism map for a steel sliding pair using the pin-on-disc configuration (Lim and Ashby, 1987)*

Although the map relates only to the unlubricated wear of steels, Lim and Ashby (1987) see no obvious obstacles to applying the concept to other materials or combinations of materials. We will return to this topic in Chapter 6. It was only introduced at this early stage in the hope that it will help the newcomer to tribology to appreciate some of the consistencies behind the apparent complexities of wear.

1.4.4 The oxidation theory of wear

(i) The Archard wear law

Most of the current theories of wear begin with Archard's (1961) interpretation of the K-factor in the expression relating wear rate (w) to the real area of contact (A_{real}), namely:

$$w = K(A_{real}) \qquad (1.35a)$$

Archard suggests that, since the units of wear rate (volume/distance slid) are the units of area, then K is dimensionless and can be interpreted as the probability of producing (on the average) a wear particle at each asperity encounter. This means that, on the average, $(1/K)$ encounters are needed to produce a wear particle. Another way of interpreting the K-factor *could* be that $(1/K)$ encounters need to occur at a plateau in order for it to build up to the critical oxide thickness (ξ). It is astonishing how much K can alter for the same operating conditions, but different material conditions. For instance, for mild steel sliding upon mild steel, without any lubricant, one obtains $K = 10^{-2}$, whereas for stellite on tool steel, one has a K-factor of 10^{-5}. The K-factor can alter by two orders of magnitude for the same material combination for just a slight change in load (or speed), as shown by Welsh (1965).

(ii) Oxidational wear in which no significant oxidation occurs at the general surface temperature (T_S)

The following expression was evolved some 20 years ago, (Quinn, 1967):

$$w_{theory} = \left[\frac{d(A_p)\exp(-Q_p/\bar{R}T_o)}{\xi \rho_o^2 f_o^2 U} \right] (A_{real}) \qquad (1.35b)$$

The expression inside the square brackets is the K-factor for mild oxidational wear, *assuming no* appreciable oxidation occurs when the wearing area is 'out-of-contact' with the opposing surface. It should be noted that the *oxidation temperature* (T_o) is generally assumed to be not very different from the temperature (T_c) of the real areas of contact during sliding, where (T_c) is related to the 'hot-spot' temperature (θ_m) by the relation:

$$T_c = T_S + \theta_m \qquad (1.36)$$

In order to derive Equation (1.35b) it was assumed that there exists a parabolic dependence of mass uptake of oxygen per unit area (i.e. $\xi \rho_o f_o$) upon the time of oxidation, that is, the time required for $(1/K)$ contacts to occur at a given asperity site on a plateau. Hence (A_p) and (Q_p) are, respectively, the *tribological* Arrhenius constant and the activation energy for *parabolic oxidation*. The other terms in

Equation (1.35) are the density of oxide in the plateaux (ρ_o), the mass fraction of oxide which is oxygen (f_o), the speed of sliding V, the gas constant $\bar{R}$ and the distance of a sliding contact between two asperities d.

(iii) The growth of oxide films in tribo-systems

This topic is basic to oxidational wear and has been the subject of several papers. Yoshimoto and Tsukizoe (1957) assumed that an oxide film grows and is removed between each asperity contact, which then becomes the source of wear debris. Quinn (1962) proposed two possible mechanisms. The first one was that the bulk of oxidation occurs at the instant the virgin metal is exposed, with further metal contact causing shear at the metal–oxide boundary. The second mechanism (upon which Equation 1.35 is based) assumes that oxidation occurs at each contact until the oxide grows to a critical film thickness when it becomes detached to become a wear particle. Tao (1969) considered two models similar to those of Quinn (1962), one in which the time for oxide growth is negligible compared with the time taken to remove the oxide layer, and an alternative model where it is assumed that the oxide grows gradually until it attains a critical thickness when it is removed immediately. Molgaard and Srivastava (1975) also postulated a critical thickness, but they suggested that this thickness is maintained by abrasive removal of the surface of the oxide film. Models which suggest the continuous removal of thin oxide films without the build-up of a film of critical thickness are not supported by the experimental evidence, which suggests that the production of relative thick oxide plateaux is a necessary prerequisite for mild wear.

Alternative explanations of mild, oxidative wear involve various models of transfer, back transfer and agglomeration and compaction of oxide and/or oxidized metal particles to form load-bearing protective films. For example, Stott and Wood (1978), Stott, Lin and Wood (1973) and Wilson, Stott and Wood (1984) report the formation of *oxide glazes* on a number of materials under reciprocated sliding conditions. They propose that oxide fragments are formed by an oxidation-scrape-oxidation mechanism, where the oxide growth both during the contact and out-of-contact is completely removed during the next few asperity interactions at the particular site. The oxide debris may then either be swept away or cause abrasive removal of the metal surface, in which case severe wear will ensue. Alternatively, the oxide debris will become fragmented, compacted and undergo plastic deformation to form a protective oxide glaze, in which case mild (oxidational) wear is the result. In other experiments, Stott, Glascott and Wood (1984) found that the rate of oxide production could not account for the rate of development of the oxide glaze film assuming, of course, that one can extrapolate from bulk oxidation studies to tribo-oxidational behaviour. They propose that *metal* debris is produced, broken up and reduced in size until the surface-to-volume ratio is such that spontaneous oxidation occurs. The running-in time would then simply be the time required to reduce the metallic particles to the required critical size. The oxidized particles would then form the metallic glaze mentioned earlier in this paragraph.

(iv) Oxidational growth rate laws and their relevance to oxides formed in tribo-systems

The first stage of growth of oxide on metal surfaces consists of adsorption of oxygen onto the surface and incorporation into some metal–oxygen structure. The second stage is the appearance of nucleation sites due possibly to defects on the surface or oxygen precipitation. The third stage consists of *lateral growth* of the oxide from the nucleation sites and the final stage, is the *vertical growth* of the oxide on the surface. The initial growth is very quick, but then rapidly slows down, often becoming negligibly slow. All of the following 'growth with time' laws have been observed experimentally:

1. *Inverse logarithmic*

$$\left(\frac{1}{X}\right) = B - [(k_i)\ln(t)]$$

2. *Direct logarithmic*

$$X = B\ln[(k_d)t + 1]$$

3. *Linear*

$$X = (k_l)t$$

(1.37)

4. *Parabolic*

$$X^2 = (k_p)t$$

5. *Cubic*

$$X^3 = (k_c)t$$

where X is the increase in thickness, volume or mass of an oxide in a time t. B and the various ks are the growth constants.

In general, the first three growth laws relate to thin films, whereas the latter two relate to thick films (typically produced at higher temperatures). The definition of 'thin' seems to be a thickness much less than 10 nm. Uniform transport of metal and/or oxygen ions is also assumed (e.g. see Cabrera and Mott (1949)). However, Davies, Evans and Agar (1954) reject this assumption insofar as iron is concerned, since they could not account for oxide growth in any but the lowest part of the thin range. Davies *et al.* (1954) assume that paths for rapid ionic transport do exist, for instance along grain boundaries, dislocations and surface pores. These paths are progressively blocked as growth proceeds. It is suggested that there is a *direct logarithmic relationship*. This is much more appropriate to the growth of oxides on the surfaces of tribo-elements, where there is substantial surface disruption due to wear. If a surface layer is non-protective and offers no resistance to the continued oxidation of that surface, the thickness of the layer has no influence on the reaction rate, leading to a *linear law* (Kubaschewski and Hopkins, 1962). The two groups of metals and alloys found to obey a linear relationship are those which produce porous or volatile oxide layers. The only possibility for linear oxidation to occur in a tribo-system is when the surface is continually being cracked and removed whilst still at the *initial*, very thin stage. Could the method of the formation of oxide glazes proposed by Stott and Wood be the result of many thin layers of oxide (which were formed in a linear manner with respect to time) becoming compacted together to produce the thick (of the order of several micrometres) oxide glazes detected by scanning electron microscopy? We will return to this question later on in this sub-section. Cubic rate laws are rare and seem to be found only in p-type semi-conductors.

It has been generally accepted that thick oxide layers (and oxides formed at higher temperatures) follow a *parabolic* rate law, due to the decreasing diffusion of metal and/or oxygen ions through the oxide film as it grows. *Most* oxidational wear models assume the parabolic law. There are circumstances, however, where it is possible that the *linear* rate law is more appropriate. This possibility is discussed in Chapter 8, which deals with the oxidational wear of stainless steels. At the time of writing, however, there is no evidence that the linear rate law is generally applicable to *all* oxidational wear. Equation (1.35b) has been obtained on the assumption that *tribo-elements* will oxidize *parabolically*.

(v) The dependence of oxidation growth rate constants upon temperature

The k-factors in Equation (1.37) are all strongly dependent upon *temperature*, through the Arrhenius equation, which is a particular form of the general Boltzmann relationship between temperature and the percentage of molecules having energies exceeding a certain threshold value. The Arrhenius equation may be written in the form:

$$k = A \exp\left(-\frac{Q}{\bar{R}T}\right) \qquad (1.38)$$

where A is the Arrhenius constant, Q the activation energy for oxidation, $\bar{R}$ the molar gas constant and T the absolute temperature of oxidation. This equation relates to the k-factors of Equation (1.37), with subscripts i, d, l, p and c used to denote the inverse logarithmic, direct logarithmic, linear, parabolic and cubic growth constants.

(vi) The importance of temperature in oxidational wear

It is because of this strong dependence of k upon temperature that so much importance is placed on determining the temperature at which the oxidation occurs during oxidational wear. There is still some controversy, however, over whether it is the general surface temperature (T_S) or the 'hot-spot'/contact temperatures which govern oxide growth. Quinn (1962) considered a number of experiments in which oxidational wear debris from sliding steel systems had been analysed, and attempted to correlate the results with calculated values of 'hot-spot' temperatures determined from the curves published by Archard (1959b). It was concluded that the constituents of the wear debris were consistent with being formed at the contact temperature, rather than the surface temperature. The contrary view was expressed by Clark, Pritchard and Midgley (1967) and is implied in the arguments of Lancaster (1957), Razavizadeh and Eyre (1983) and Sexton (1984).

The view that contact temperature rather than surface temperature is the most important factor determining oxide growth is supported by Tao (1969) and Molgaard and Srivastava (1977), although in earlier published work (1975), these authors suggested that the oxidation temperature may be somewhat lower than the contact temperature, even for oxidation nominally in the contact region. This suggestion is possibly closer to the truth, since during contact there is no free access of oxygen to the interface. Thus oxidation probably takes place immediately following contact at temperatures intermediate between contact and surface temperatures. Models assuming a *constant* oxidation temperature must, therefore, contain approximations, since the true oxidation kinetics for transient temperature changes would be almost impossible to determine.

Although it is reasonable to assume that the contact temperature (or something close to this) is the major factor in low ambient temperature experiments where relatively high loads and speeds are employed, this is not so for tribo-systems operating at high ambient temperatures and/or low speeds. Stott, Glascott and Wood (1985) have produced a quantitative expression for the relative amounts of oxide generated in and out of contact and they conclude that either mode could predominate. They tie in the number of load-bearing contacts with the excess contact temperature through the statement that, if the excess contact temperature rises are significant and the number of load-bearing contacts is small, the majority of the oxide is formed at the contact. Conversely, these authors point out that if there is a large number of contacts, particularly under low speed and high ambient temperatures, then 'out-of-contact' oxidation predominates.

(vii) Oxidational wear in which the 'out-of-contact' oxidation is an important factor in the total wear mechanism

In 1981, the author published an expression for the wear rate of a tribo-system exhibiting mild-oxidational wear, in which both oxidation at T_c and at T_s is taken into account (Quinn, 1981). A slightly modified version was also published in 1983 (Quinn, 1983). For the full derivation of this equation, the reader is advised to refer to these early publications. In fact, the wear rate (w) consists of three components as shown below:

$$w = w_\alpha + w_\beta + w_\gamma \tag{1.39}$$

where

$$w_\alpha = \frac{dW[k_p(T_c)]}{[p_m(T_S)][f_o(T_c)]^2[\rho_o(T_c)]^2\xi^2 U} \tag{1.40}$$

and

$$w_\beta = \frac{[k_p(T_S)](A_n)W}{[p_m(T_S)][f_o(T_S)]^2[\rho_o(T_S)]^2\xi^2 dUN} \tag{}$$

and

$$w_\gamma = \left\{ \frac{W}{[p_m(T_S)][f_o(T_c)][f_o(T_S)][\rho_o(T_c)][\rho_o(T_S)]\xi^2 U} \right\}$$

$$\left\{ \frac{2[k_p(T_c)][k_p(T_S)](A_n)}{N} \right\} \tag{1.42}$$

All the terms in the above equations have already been defined, apart from N, the number of contacts within the apparent (or nominal) area of contact. Clearly, the area density of contact spots is given by (N/A_n). For a pin of diameter equal to 6 mm, Quinn and Winer (1987) experimentally deduced $N \sim 7$, giving a value of 0.35 contacts/mm^2 for (N/A_n).

The terms which depend on either T_S or T_c have been indicated by the parenthesis adjoining each term. k_p is, of course, the parabolic oxidation rate constant governed by Equation (1.38) with the Arrhenius constant for parabolic oxidation (A_p) replacing A, the activation energy for parabolic oxidation (Q_p) replacing Q, and the oxidation temperature (T) assumed to be either (T_C) or (T_S) according to whether the term relates to 'in contact' or 'out of contact' oxidation. Note that ω_α is Equation (1.35b) with (W/p_m) replacing (A_{real}) and k_p replacing $A_p \exp(-Q/RT_o)]$.

(viii) Oxidational wear in which the growth law is not parabolic

We have seen that, although there is a wealth of evidence pointing to parabolic tribo-oxidation, there is also some support for the possibility that tribo-oxidation might occur *linearly* with respect to time. Based on the author's (1969) early work on a wear equation that did *not* depend on the validity of the Archard wear law, Hong, Hochman and Quinn (1987) have shown that the mild wear of AISI 316 stainless steel *could* be described mathematically in terms of linear tribo-oxidation at the real areas of contact. The equation they use has the form:

$$w = \frac{(A_{real})(k_1)}{U\rho_o f_o} \qquad (1.43)$$

Clearly, this does not take into account 'out-of-contact' oxidation. In principle, however, it should be possible to derive an equation similar to Equation (1.39) on the basis of linear oxidation only. The main advantage of Equation (1.43) is that it does not contain 'difficult' terms such as the critical oxide thickness ξ nor the number of contacts N. The real area of contact has been left in the expression, rather than equating it straight away to (W/p_m), since recent work by Quinn and Winer (1987) has indicated some possible doubts about this basic tenet of all tribologists!

1.5 Concluding remarks

This first chapter has attempted to introduce the reader to tribology, with particular emphasis on the current position in the theories and mechanisms of wear. The remainder of the book will tend to assume that the reader has, in fact, read this first chapter, with its definitions and its descriptions of modern theories of friction and wear. It will rapidly become evident that much of our current knowledge of the wear behaviour of tribo-systems arises from the application of physical analytical techniques to the study of worn surfaces and wear debris. We will, wherever possible, try to relate the analyses quantitatively to the theories and mechanisms discussed in this chapter. It is hoped that this first chapter will have been sufficiently self-contained for the reader to feel fully-equipped as far as the basic tribological principles are concerned.

2 Macroscopic physical techniques

2.1 Optical microscopy and optical interferometry

2.1.1 Introduction

The availability of powerful tools for viewing surfaces at very high magnifications through the medium of highly resolved images has often caused the optical microscope to be overlooked. As we shall see, this may be related to the fact that the depth of field of an optical micrograph varies inversely with increased magnification, whereas with electron microscopy, the depth of field is almost infinite. Nevertheless, in tribology, many of the processes occur on a scale for which the optical microscope is probably the best and easiest instrument to use, especially when used in the reflection mode. Furthermore, when the optical microscope is combined with an interferometer device, we have a surface examination tool which has a vertical resolution equal to the resolution of many electron microscopes (and at a fraction of the capital outlay). We shall see how optical interferometry has indeed been used to give information about lubricant film thickness contours within the contact between a ball and a flat. These thicknesses are of the order of micrometres. Interference techniques have been used (Tolansky, 1970) to deduce the height of 40 Å steps in cleared mica. The present author has used a home-made interferometer to deduce the thickness of vacuum-evaporated films of iron used by Halliday (1960) in his work on using the contrast of electron diffraction patterns to determine thickness. The thinnest film used in that investigation was about 250 Å.

From the previous paragraph, it is clear that optical microscopy should always be used in the initial stages of any surface examination. Let us now review the main features of the optical microscope.

2.1.2 The optical microscope

The essential feature of an optical microscope is that the lens nearest the object (the objective lens) forms a real image *inside* the front focal plane of the lens nearest to the eye (the eyepiece lens), as shown in Figure 2.1. The object (O) is placed at a distance *greater than* the focal length (f_{obj}) of the objective lens (B) but *less than* $2f_{obj}$. An intermediate image (I_1) is formed so that it lies *within the focal distance* (f_{eye}) of the eyepiece (G). The eyepiece then forms a virtual image (I_2) of the real image (I_1) as shown in Figure 2.1. In order to deduce where the second image is formed, we draw KL and GK. We then extend GK until it meets FL extended at W. All we are doing is to use the basic rules for drawing ray diagrams, namely (a) all lines passing through the centre of a thin lens are undeviated (e.g. $ABCEGF$, UBK, KGR) and (b) all rays travelling parallel to the axis are refracted through the focal point of the lens (e.g. UNC and KLF). If an eye is placed just to

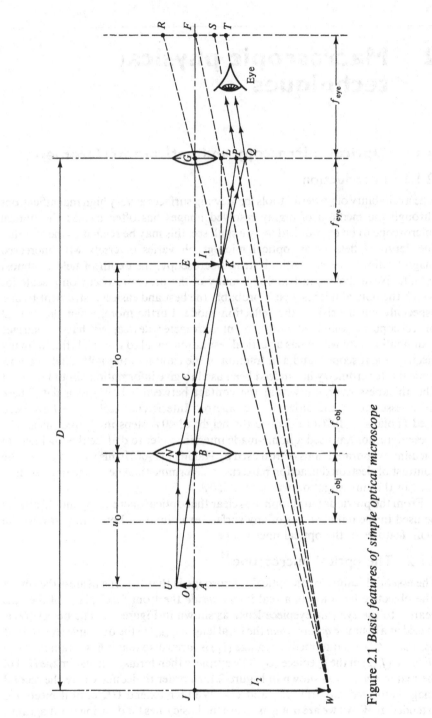

Figure 2.1 Basic features of simple optical microscope

the right of the eyepiece lens, it will see the light rays from the object ($UNCKQT$ and $UBKPS$) as if those rays diverted from the tip of the second image (I_2) at W. The magnification ($M_{v=D}$) is the magnification one obtains with the final image (I_2) at the *least distance* (D) *of distinct vision*, namely:

$$M_{v=D}= \left(1+ \frac{D}{|f_{eye}|}\right)\left(\frac{|v_O|}{|u_O|}\right) \qquad (2.1)$$

For low-power objectives, the working distance (u_O) is large enough to allow oblique illumination of the object. For high-power objectives, the working distance is too small to allow for illumination from the side. Most optical microscopes will have the facility for passing the light from a source at the side of the microscope column to be reflected down on to the object, through the objective lens, via a half-silvered mirror placed at 45° with respect to the axis of the microscope.

2.1.3 Limitations of the optical microscope as an image forming device

There are several factors influencing the effectiveness of an optical microscope. They are related to (a) aberrations in the lenses and (b) the ultimate resolving power of any optical system.

Figure 2.2 illustrates the origin of spherical aberration in a converging lens. Light rays from a point object on the axis of the lens converge to points C or D, according to whether they are travelling respectively at large or small values of the semi-aperture angle (α_{SA}) with respect to the axis. If our image plane is situated somewhere to the right of point D, the point object (O) provides an image which is a 'disc of confusion' about the *true image point* (E). We can reduce the radius (r') of this disc of confusion by using an aperture (AB), as shown in the figure. This aberration can be virtually eliminated by using a suitable chosen

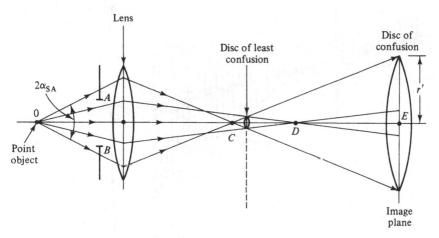

Figure 2.2 *Spherical aberration in lenses*

diverging lens in conjunction with the converging lens. It can be shown that the radius of the disc of confusion is given by:

$$r' = (C_S)(\alpha_{SA})^3 \tag{2.2}$$

where C_S is the spherical aberration coefficient. We will see (in Section 2.2) that spherical aberration is the main limiting factor in electron microscopy. Unfortunately, it cannot be so readily overcome as with optical microscopy so we have to use very small semi-aperture angles (α_{SA}) in order to get sharp electron micrographs. For electron microscopy, C_S has a value of about 0.03 cm.

Figure 2.3 illustrates what is meant by 'astigmatism'. We can see that *vertical* portions of an extended object have been shown to bring rays parallel to the axis to a focus at point (F_V), whereas *horizontal* portions of the objects give rise to light which becomes focused into point (F_H). Instead of one sharply focused image, one either has to focus so that the vertical parts of the object are in the image plane (I_V) or the horizontal parts are in the image plane (I_H). This aberration can be overcome by the use of a cylindrical lens in conjunction with the astigmatic lens. In the electron microscope, it is possible to rectify astigmatism by the use of quadrupole magnets.

Figure 2.4 illustrates how chromatic aberration can cause loss of definition of the final image of a point object, due to the blue wavelengths (λ_b) being brought to provide an image of radius Δr_i, compared with the point image which would be formed by the red wavelengths (λ_r). In order to optimize the aberration, the image plane can be arranged to coincide with the position of the disc of least confusion (C), which has a radius approximately equal to ($\frac{1}{2}\Delta r_i$). If the magnification of the lens is M, this means that any objects in the object plane passing through O, must be larger than $\Delta r = (\Delta r_i/2M)$, before they can be distinguished as two distinct objects. Chromatic aberration in optical lenses can be readily corrected by the

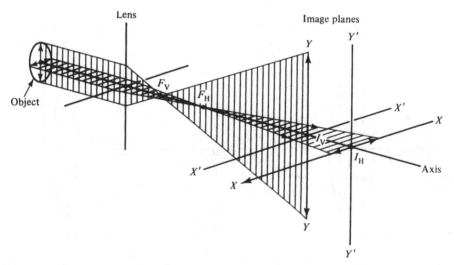

Figure 2.3 *Astigmatism*

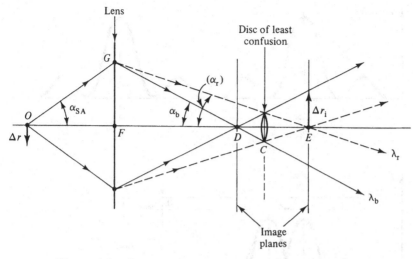

Figure 2.4 *Chromatic aberration*

use of suitable filters. In electron microscopy, where the wavelength (λ) of the electron beam is inversely proportional to the root of the voltage (V_A) through which the electrons have been accelerated, it can be shown that Δr is given by:

$$\Delta r = (C_C)(\alpha_{SA})\frac{\Delta V_A}{V_A} \tag{2.3}$$

where C_C, the chromatic aberration constant, is about 0.09 cm. Thus we see that chromatic aberration is related to the stability of the high voltage supply (i.e. $\Delta V_A/V_A$) in the electron microscope. A stability of one part in 10 000 is not too difficult to obtain, thereby showing that chromatic aberration is not much of a problem in either optical or electron microscopy.

Let us now discuss the ultimate limit of optical microscopy, that is the limit one would have to face up to even if all the lens aberrations were corrected. It is, of course, the resolving power due to the nature of the radiation being used. The resolving power is the reciprocal of the resolution limit, (d_{min}), where d_{min} is the minimum distance between two points on an object which can just be distinguished as separate points in the microscope. Rayleigh has suggested that (d_{min}) can be defined in terms of the diffraction patterns which are always formed when we restrict the light waves emanating from two point sources by the formation of images through the circular apertures of the lens. The analysis of diffraction by a circular aperture is beyond the terms of reference of this book. For the present, we shall merely say that Rayleigh's Criterion involves the suggestion that the unaided eye can just detect the 20% difference in intensity obtained when we sum the diffraction patterns from two point sources separated such that the central maximum of one pattern lies exactly over the first minimum of the second pattern. This is shown in Figure 2.5, where P_1 and P_2 are clearly resolved in Figure 2.5(a), and where the summation of the intensities of P_1 and P_2 results in a

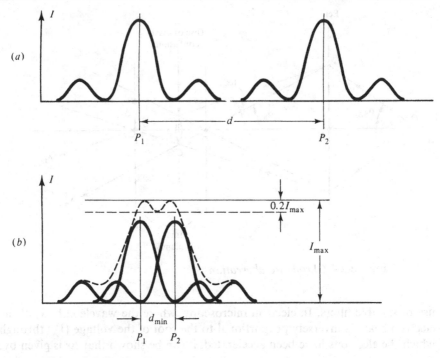

Figure 2.5 *Rayleigh criterion*

20% decrease in the combined intensities (the broken line) when separated as suggested by Rayleigh [in Figure 2.5(b)]. For the microscope, it can be shown that the Rayleigh Criterion is given by:

$$d_{\min} = K_R\left(\frac{\lambda}{NA}\right) \tag{2.4}$$

where $K_R = 0.61$, λ is the wavelength of the illuminating light and NA is the numerical aperture defined by:

$$NA = \mu\sin(\alpha_{SA}) \tag{2.5}$$

where μ is the refractive index of the medium between the object and the objective lens and α_{SA} is the semi-aperture angle (already defined). For a low-magnification objective, Equation (2.4) gives $d_{\min} = 1.1\,\mu m$. For a good, oil-immersion lens, the typical $d_{\min}$ is $0.2\,\mu m$. In a final year project carried out when I was a student, I investigated the relevance of Rayleigh's criterion insofar as the resolution of two *incoherent* point sources in a telescope was concerned. For this instrument, the Rayleigh criterion is $K_R = 1.22$. Using the unaided eye, it was shown that $K_R = 0.57$ could be obtained with the apparatus whereas, using a sensitive instrument to analyse the intensity decrease at the centre of the overlapping patterns, it was found that $K_R = 0.34$. Although some of this improvement was probably due to the occurrence of *some coherence* between the two light sources used in the experiments, the implications are that K_R should be about one-third of the Rayleigh value namely $K_R \approx 0.2$ for microscopes and $K_R \approx 0.4$ for telescopes. Putting K_R equal to 0.2 and assuming $\lambda = 5461$ Å (Hg green), together with the numerical aperture for an oil-immersion lens of about 1.35, Equations (2.4) and (2.5) give us d_{min} about 0.08 μm, that is about 800 Å, which is only a factor of 4 more than the resolution obtainable with some electron microscopes.

It has been shown that the optical microscope has an unexpectedly high resolution when one has a sensitive detector of intensity changes in overlapping diffraction patterns of pseudo-point sources. We must, of course, point out that the main *disadvantage* of the optical microscope is that the in-focus depth of field of the light microscope is only about (1/200)th that of the electron microscope, especially the scanning electron microscope (see Section 2.3). Let us now discuss the essential features of optical interferometry, since this is where one of the light optical techniques (namely multiple beam interferometry) has a better vertical resolution than that of the electron microscopy.

2.1.4 Interference between two beams of light

Interference will occur between two separate waves of light if they derive from the same essential light source (i.e. if they are 'coherent'), the combined intensity (I_x) being given by:

$$I_x = 4(a_L)^2 [\cos^2 (\Delta/2)] \tag{2.6}$$

where Δ is the phase difference between the two waves (ϕ_1 and ϕ_2) actually reaching a particular point (P) at distances (x_1) and (x_2) from the two sources. In fact, ϕ_1 and ϕ_2 are given by:

$$\phi_1 = (a_L) \sin \left[\frac{2\pi}{\lambda} (x_1 - ct) \right] \tag{2.7a}$$

$$\phi_2 = (a_L) \sin \left[\frac{2\pi}{\lambda} (x_2 - ct) \right] \tag{2.7b}$$

(a_L) is the amplitude of each of these two waves, λ is the wavelength of the illuminating light, c is the speed of light and t is the time at which the two waves reach the point P. The terms inside the square brackets are called the 'phases' of ϕ_1 and ϕ_2 so that the phase difference Δ is merely given by:

$$\Delta = \frac{2\pi}{\lambda} (x_1 - x_2) \tag{2.8}$$

Clearly, Δ can take any value according to the position of P and, if we plot I_x versus Δ from Equation (2.6), we obtain the well-known 'cosine squared' curve shown in Figure 2.6. This variation of total intensity applies to any situation in which two parts of a coherent wave front are brought together after having travelled different distances, x_1 and x_2. For instance, Young's double-slit experiment (illustrated in Figure 2.7) gives an intensity distribution of this kind,

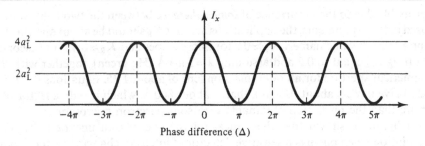

Figure 2.6 *Graph of I_x versus Δ based on Equation (2.6)*

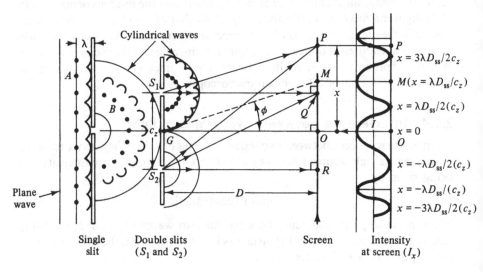

Figure 2.7 *Young's double-slit experiment*

except that now Δ is given by:

$$\Delta = \frac{2\pi}{\lambda}(S_2P - S_1P) = \frac{2\pi}{\lambda}\left[\frac{(C_z)x}{D_{ss}}\right] \tag{2.9}$$

that is

$$x = \left[\frac{\lambda D_{ss}\Delta}{2\pi(C_z)}\right] \tag{2.10}$$

Let us examine Equation (2.9). For instance, when $\Delta = \pi$, then $2(C_z)x/D = \lambda/2$ so that $x = \lambda D_{ss}/2(C_z)$. Obviously, from Equation (2.6), when $\Delta = \pi$, then $I_x = 0$, so that the first minimum occurs at $\lambda D_{ss}/2(C_z)$ (as shown in Figure 2.7). It can easily be shown that *minima always occur whenever* $\Delta = (2n + 1)\pi$, where $n = 0, 1, 2, 3, \ldots$, that is when the optical path difference is an odd number of half-wavelengths ($\lambda/2$). Maxima occur when $\Delta = 2n\pi$, where I_x is given by:

$$I_x = 4(a_L)^2\left[\cos^2\left(\frac{2n\pi}{2}\right)\right] = 4(a_L)^2 \tag{2.11}$$

and $n = 0, 1, 2, 3, 4, \ldots$, that is when the optical path difference is an even number of half-wavelengths. This is shown in Figure 2.7, where A is a typical 'Huygen Wavelet' source for a plane wave, B is a typical wavelet source for a cylindrical wave, as also are S_1 and S_2 (the interfering sources).

In Young's double-slit experiment, interference occurs between two parts of the same wavefront formed by the two slits (S_1 and S_2). It is also possible for a primary beam to interfere with its *reflection*. Let us consider the path difference between a beam that is reflected from the top surface of a plane-parallel film and a beam that is transmitted through the top surface, reflected at the bottom surface, and *then* transmitted at the top surface (in the manner shown in Figure 2.8). In this figure, we have ignored the transmitted ray at C. Clearly, however, *some* of the transmitted wave along AC is also transmitted into the lower air space. Similarly, *some* of the light impinging on B from C will be reflected and then transmitted through the lower surface, thereby giving a set of transmitted rays (similar to (1) and (2)), each one of which can interfere with its neighbours. The *optical* path difference between rays (1) and (2) is $(AC + CB)\mu - AD$. (Note that AD is in air, which has a refractive index of unity, whereas AC and BC are in the medium of the film of refractive index μ.) It can be readily shown that the optical path difference (OPD) is given by:

$$\text{OPD} = \mu\, 2d_g \cos\theta \qquad (2.12)$$

where d_g is the film thickness and θ is the angle shown in Figure 2.8. Since the phase difference Δ is given by a generalization of Equation (2.8), namely:

$$\Delta = \frac{2\pi}{\lambda}(\text{OPD}) \qquad (2.13)$$

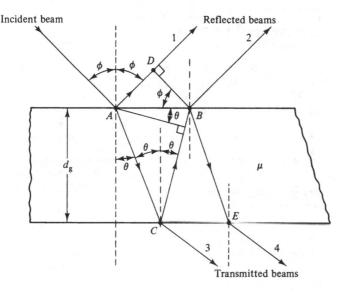

Figure 2.8 *Interference between reflected wave (1) and the transmitted, reflected and transmitted wave (2) for a plane-parallel film*

we see that the phase difference between waves (1) and (2) in Figure 2.8 is given by:

$$\Delta = \frac{2\pi}{\lambda}(\mu\,2d_g\cos\theta)$$

This is only true if there is no change of phase at A (transmission) nor at C (reflection). If there is a change (known as the Stokes' change and equal to π radians) at A, there will *not* be one at C. Equation (2.6) still holds for the interference of a primary wave with its reflection, so that, with Stokes' change, we have:

$$I_x = 4(a_L)^2\cos^2[(\Delta+\pi)/2] \qquad (2.14)$$

This gives $I_x = 0$ when $\Delta = 0$ and $I_x = 4(a_L)^2$ when $\Delta = \pi$. Thus the following conditions will hold true:

Minima occur where $2\mu d_g\cos\theta = n\lambda$

Maxima occur where $2\mu d_g\cos\theta = (n+\tfrac{1}{2})\lambda$

An interesting case of *two-beam interference*, involving the original beam and its reflection, is the *Newton's rings interferometer*. We shall see (in Chapter 5) how this interferometer can be used to study the thickness variations within the contact zone of a lubricated tribo-system. For the present, we will merely describe its basic principles, as illustrated in Figure 2.9. Essentially, we get interference between the beam reflected by the *upper* surface of the air film (i.e. the lower

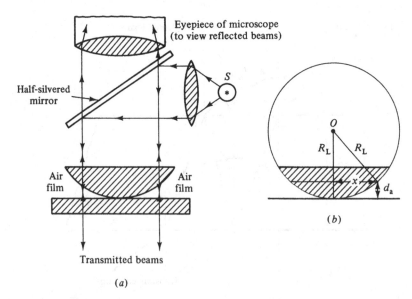

Figure 2.9 *Newton's Rings: (a) experimental details; (b) geometry of saggital formula*

surface of the plano-convex lens) and the beam reflected by the *lower* surface of the air film (i.e. the upper surface of the optical flat upon which the convex surface of the lens is placed). The incident angle θ is almost a right angle. Hence whenever $2\mu d_g$ equals $(n+\frac{1}{2})\lambda$ we get bright fringes (by reflection). Using the saggital formula we know that, for small (d_a), we have:

$$(d_a) = \frac{x^2}{2(R_L)} \tag{2.15}$$

where (d_a) is the thickness of the air film at a distance x from the centre of the system. In fact, we get concentric circles which get closer together as we increase x, thereby effectively giving us a *contour* map of the air film between the convex and the plane surfaces. We may summarize the above by saying:

(a) When viewing Newton's Rings by *reflection*, we get a central *dark spot* with *dark rings* at a distance x, whenever $x^2 = (R_L)n\lambda/\mu$ for $n = 0, 1, 2, 3, \ldots$

(b) Conversely, for viewing *by transmission*, we get a central *light* area with *bright rings* at a distance x, whenever $x^2 = (R_L)(n+\frac{1}{2})\lambda/\mu$ for $n = 0, 1, 2, 3, \ldots$

2.1.5 Multiple beam interferometry

The technique of multiple beam interferometry is a modification of the two-beam interference technique in which the primary beam interferes with its reflection. By coating one of the surfaces (AB or CD) of the plane-parallel film of Figure 2.8 with a highly-reflecting, yet still partially transmitting, thin film of silver, it has been possible to sharpen up the so-called 'cosine squared' fringes obtained with only two beams. In fact, we are using several of the reflected beams (or transmitted beams) in order to achieve this increase in fringe sharpness. Let us analyse the multiple beam reflection obtained with *strictly parallel* plane surfaces. We will then go over to the wedge geometry, since this is the most useful for surface analysis.

(i) Parallel surface analysis

The essential geometry is given in Figure 2.10. It is assumed that the incident beam (S) has unit intensity. Let a fraction (R_R) of that intensity be reflected at the face AB and a fraction (T_T) transmitted. Hence, successive *transmitted* beams have their intensities given by:

$$I_1 = (T_T)^2; \qquad I_2 = (R_R)^2(T_T)^2; \qquad I_3 = (R_R)^4(T_T)^2; \qquad \ldots \tag{2.16}$$

The optical path difference (OPD) between *successive* transmitted beams is $(EF + FG - EH)$. It can be shown that, if the film between the plates has a refractive index μ, then OPD is given by:

$$\text{OPD} = 2\mu d_a \cos \theta \tag{2.17}$$

Hence the phase difference Δ is given by:

$$\Delta = \frac{4\pi\mu d_a \cos \theta}{\lambda} \tag{2.18}$$

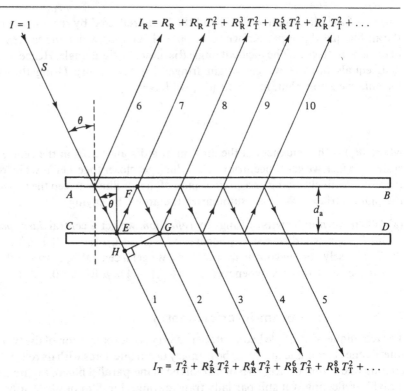

Figure 2.10 *Multiple beam interference between two parallel plates with a film of thickness (d_a) between the plates (refraction effects are ignored since they do not influence the calculations)*

and I_T, the total intensity of *all* the transmitted beams arising from the incident beam (*S*) of unit intensity, is given by:

$$I_T = \frac{(T_T)}{(1-R_R)^2 + 4(R_R)\sin^2(\Delta/2)} \tag{2.19}$$

If one uses vacuum-evaporated silver for faces *AB* and *CD* then it is very close to the truth to assume $[(T_T)+(R_R)]=1$. Hence we may write Equation (2.19) as:

$$I_T = \frac{1}{1+(F_R)\sin^2(\Delta/2)} \tag{2.20}$$

where

$$(F_R) = 4(R_R)/(1-R_R)^2 \tag{2.21}$$

It can also be shown that we could have summed all the *reflected* beams (such as beams 6, 7, 8, 9 and 10 of Figure 2.10) and obtained:

$$I_R = \frac{F_R \sin^2(\Delta/2)}{1+F_R\sin^2(\Delta/2)} \tag{2.22}$$

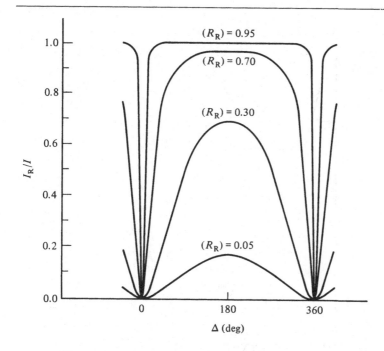

Figure 2.11 *Dependence of multiple-beam band width on surface reflectivity (R_R)*

It is interesting to see how Equation (2.22) varies with the surface reflectivity. This is shown in Figure 2.11 for values of (R_R)=0.95, 0.70, 0.30 and 0.05. We can see that, at a value of (R_R) of about 0.05 (the typical value obtained with unsilvered glass surfaces), the ratio (I_R/I) of the reflected intensity (I_R) and the incident intensity (I) is always low and varies *sinusoidally* between zero and its maximum. This shows that, for low reflectance, we need only consider two beams, and the form of the intensity curves are not very different from the 'cosine squared' form of Figure 2.6. As (R_R) increases, the ratio of (I_R/I) approaches unity everywhere except in the vicinity of the *minima* which are zero, thereby generating a pattern of *narrow dark* fringes on a *bright* field. The *complementary transmitted* pattern is given by Figure 2.11, by replacing (I_R/I) by (I_T/I) and renumbering the ordinate so that it goes from 1 to 0 in the positive direction. The very narrow fringes [for (R_R)=0.95] shown in Figure 2.11 have a half-width of only (1/60)th of the interval between successive orders.

(ii) Non-parallel surfaces

The fringes, described in the previous paragraphs relating to parallel surfaces, are formed at infinity and can only be viewed with a telescopic instrument. They are, therefore, of no help in any studies aimed at deducing surface topography. This can only be revealed by *localized* wedge fringes, as shown in Figure 2.12. In this

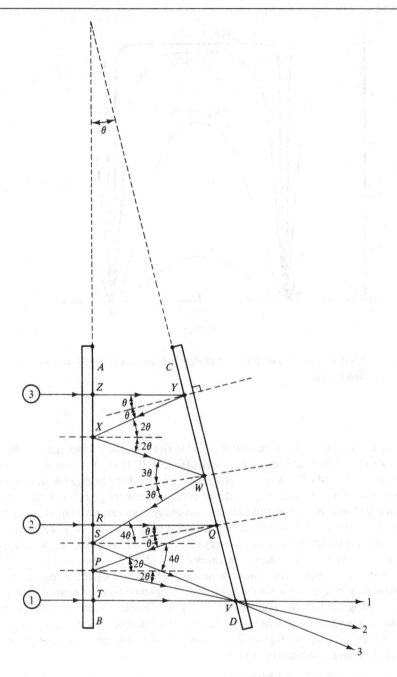

Figure 2.12 *Optical paths for three successive beams transmitted through a wedge-shaped film*

figure, AB and CD are two silvered plates at a *slight* angle θ. (Figure 2.12 has been drawn with an exaggerated angle for clarity.) For ray 1, the separation is (TV) and is equal to (d_a). The ray passes straight through (from T to V), (we are still ignoring refraction). Note also that ray 2 is reflected twice before it emerges at V, ray 3 is reflected four times before it also emerges at V, and so on. Clearly, successive path differences are not the same for each beam. The nth beam is reflected $2(n-1)$ times. It is beyond the scope of this book to show that, by the time the nth beam has been reflected $2(n-1)$ times it is, in fact, $\frac{4}{3}(n)^3\theta^2 d_a$ behind the initial beam that emerges at V. When this lag is equal to $(\lambda/2)$, we have the nth beam completely out of phase with the original beam (ray 1). Thus, it is assumed one will only get constructive interference between n rays when the following condition is approximately obeyed,

$$\left(\frac{4}{3}\right)n^3\theta^2\mu d_a \ll \frac{\lambda}{2} \tag{2.23}$$

Now n must be less than 60 for any practical contribution from the reflected beams. It is also desirable to see at least 3 or 4 fringes so that, if there are x *fringes/unit length* (say the metre) across the object, then $\theta = x(\lambda/2)$. The maximum thickness $(d_a)_{max}$ is then given by Equation (2.23) with $n=60$, namely:

$$(d_a)_{max} = \frac{1}{7.2x^2} \tag{2.24}$$

We have assumed $\lambda = 5000$ Å in order to get this expression for $(d_a)_{max}$. For Equation (2.19) to apply, we must have thickness (of the film) considerably less than $(d_a)_{max}$. For a magnification of $(\times 100)$ at the surface and 1 cm fringe separation in the image, then the real distance of fringes in the film would be about 0.01 cm. Clearly this means that $x = 100/$cm, so that $(d_a)_{max}$ is given by:

$$(d_a)_{max} = \frac{10^8}{7.2 \times 10^4}\,\text{Å} = 0.139 \times 10^4\,\text{Å}$$

that is

$$(d_a)_{max} = 1390\,\text{Å}$$

Thus, when the separation between the two non-parallel surfaces is very small, preferably of the order of a light wavelength, then, *and only then*, will constructive interference occur. From all this, we can conclude that, for the best results, one needs the reflectivity (R_R) to be high. One also needs the two surfaces to be extremely close to each other.

We have examined in some detail the fundamental aspects of optical microscopy and optical interferometry. In a later chapter, we will see how these two (very basic) techniques have been used in tribological studies. Let us now discuss transmission electron microscopy of replicas of worn surfaces.

2.2 Transmission electron microscopy of replicas of surfaces

2.2.1 Introduction

We have seen that the resolution limit with optical microscopy is about 2000 Å (according to Rayleigh's criterion). Using specially sensitive devices and optimum conditions, I have been able to halve that figure to about 1000 Å. However, in order to obtain a micrograph with features of this order of magnitude, the optical microscope would have to be operated at such a high magnification that the concomitant reduction in the field of focus would render the micrograph useless as far as surface topography is concerned. We will see that, by using electrons instead of light, we will obtain a much lower resolution limit *and* much larger depths of field and focus. We will see that, for analysis of the topography of surfaces, we can use a replica technique which, when combined with transmission electron microscopy, provides us with a resolution limit of around 200 Å. Later in this section we will discuss Scanning Electron Microscopy (SEM). In some ways this technique has replaced the examination of replicas by Transmission Electron Microscopy (TEM). We will see, however, that the replica has some advantages over SEM, especially as regards 'in situ' examination of surface topographical changes during an experiment. In this sub-section, we will discuss the essential features of the electron microscope and how one uses it, in conjunction with replicas, to provide topographical information about surfaces formed in the typical tribo-system.

2.2.2 The transmission electron microscope

Figure 2.13 is a comparison of the typical optical microscope and the essential transmission electron microscope. From this figure, we see that the electron microscope differs very little from the optical microscope in its mode of operation. There is an obvious difference in the fact that the electron beam must always be enclosed in a reasonably good vacuum (i.e. 10^{-6} torr or $133\,\mu\text{Pa}$), otherwise it will rapidly be absorbed by interaction with the molecules of air. The enclosure in a vacuum can have some restrictive effects on the types of specimen which can be examined in the electron microscope, which will be dealt with later. What, then, is the point of using electrons in place of light? The answer lies in the magnitude of the wavelength 'associated with' an electron beam. It is beyond our terms of reference to discuss the concepts of 'waves' and 'particles'. Let us accept that it has been shown that these words are just two different ways of describing how energy is transmitted. Both are equally 'correct'. One uses whichever description fits the physical phenomenon concerned. For example, the electric energy of an electron can be adequately described in terms of its kinetic energy (which is essentially a *particle* property) when it is moving freely (i.e. outside the influence of any force fields) through a vacuum, so that:

$$\tfrac{1}{2}m_e v^2 = eV_A \tag{2.25}$$

where v is the speed of the electron acting as a particle of mass (m_e) and electric

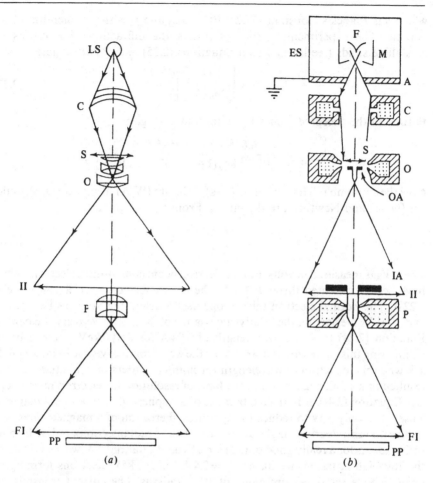

Figure 2.13 *Comparison of the optical (a) and electron (b) microscopes;*
LS, Light source; ES, electron source; C, condenser lens; F, filament; S,
specimen; M, modulator; O, objective lens; A, anode; P, projector lens;
OA, objective aperture; FI, final image; IA, intermediate aperture; PP,
photographic plate; II, intermediate image

charge (*e*) accelerated through a potential difference (V_A). When the electron
moves in the intense short-range force fields present in crystalline solids, however,
it is affected in such a way that we cannot accurately predict what will happen to
it. Nevertheless, if we consider a very large assembly of electrons, such as in an
electron beam, we can show (experimentally) that the behaviour of the electron
beam is *precisely* what would be expected of a plane wave of wavelength (λ), the
magnitude of which is given by:

$$\lambda = [h/(p_1)] \tag{2.26}$$

51

where h is Planck's Constant $(6.62 \times 10^{-34} \text{ J-s})$ and p_l is the momentum of the particle. The experimental proof of this is the diffraction of electrons by crystalline solids (see later). From Equations (2.25) and (2.26) we get:

$$\lambda = \left[\frac{h^2}{2m_e eV_A} \right]^{1/2} \tag{2.27}$$

Putting in the accepted values for h, m_e and e we get:

$$\lambda = \left[\frac{6.62^2 \times 10^{-68} \text{ J}^2 \text{ s}^2}{2(9.11 \times 10^{-31} \text{ kg})(1.60 \times 10^{-19} \text{ C}) V_A \text{ V})} \right]^{1/2}$$

Now the coulomb (C) has the units (A–s), volts are (W/A), Watts are (J/S), joules are (N–m) and Newtons are (kg–m/s²). From this we obtain:

$$\lambda = \left(\frac{150}{V_A} \right)^{1/2} \text{ Å} \tag{2.28}$$

where V_A is measured in volts. For an electron beam consisting of electrons which have been accelerated through 100 V, the wavelength of the 'electron wave' is 1.22 Å. The normal electron microscope uses accelerating voltages between 50 and 100 kV. Neglecting the relativistic change of mass with velocity, we can use Equation (2.28) to give a wavelength of 0.04 Å for a 100 keV electron beam. Comparing this value with the 4000 Å for the wavelength of visible light, we might ask why this reduction of wavelength through *five* orders of magnitude has not resulted in a similar reduction in the limit of resolution for electron microscopy (see Equation (2.4)). It is, in fact, because of the spherical aberration of magnetic lenses. The only way to reduce the spherical aberration of a magnetic lens is to reduce the semi-aperture angle (α_{SA}) to about 10^{-2} radians, compared with about $(\pi/2)$ radians for a really good optical lens. From Equation (2.4) we see that, from the Rayleigh criterion, we have $d_{min} = 2.4$ Å for 100 keV electrons forming an image with a semi-aperture angle of 10^{-2} radians. The current transmission electron microscopes have not significantly reduced the resolution limit from this value. Could this be due to the fact that very few practical applications of transmission electron microscopy need to be carried out at the huge magnifications necessary for this resolution to be effective? Certainly, if one uses a replicating technique for viewing work surfaces, then the limiting factor is the size of the evaporated particles forming the replica. This will be about 200 Å.

Although the use of paraxial rays limits the resolution of the electron microscope, it also provides an advantage over the optical microscope, since it produces large depths of field and depths of focus. In effect, we have the same conditions of focus as we would get with a 'pin-hole' camera. By using extremely small apertures, this instrument will give a sharply-focused image at virtually all possible object–lens–image distances.

The main drawback in the use of electron microscopy is the fact that it can only be used for examining extremely thin specimens (less than about 1000 Å for transmission electron microscopy). This means that one normally has to (a) make thin replicas of bulk surfaces; (b) use some thinning technique; or (c) find some way of removing the surface films formed on typical tribo-elements, before one can examine the surfaces. These operations can introduce artefacts that are not always easy to distinguish from the actual features of tribological interest. This is probably why the scanning electron microscope has tended to supercede the transmission electron microscope for examining surfaces.

2.2.3 The preparation of specimens for transmission electron microscopy (TEM)

(i) Replicas of solid surfaces

Normally, one obtains a replica by evaporating (under vacuum) a material on to the surface which can be removed and then viewed in the electron microscope (the one-stage technique). The contrast mechanism depends only upon thickness variations caused by changes in slopes of the topographical features being replicated. Typically this is often not sufficient for viewing purposes, so a shadow-casting material must be used to enhance the topographical features. For a one-stage replica, the shadowing must be performed on the actual surface before the evaporation of the replicating material. This is a destructive technique. It is mainly for this reason that the *two-stage plastic replica technique* is favoured by most electron microscopists. In this technique, a plastic replica is taken of the surface. This is then shadow-cast with a material which is more efficient at scattering than the final replicating material. The choice of materials for replicating and for shadow-casting is fairly limited but the following table (Table 2.1) gives the published values of contrast (C) per unit thickness (d_t) for some of the materials that have been tried in electron microscopy. A high value of (C/d_t) is required for a good shadowing material, to avoid geometrical distortion by 'pile-up' effects and a low value of (C/d_t) is needed for a good replicating material, so that the shadowing shall have the most effect. It is interesting to note that the values of (C/d_t) in Table 2.1 substantiate what has been found in practice; the smallest values are 0.5 for carbon and 0.7 for silicon monoxide, whilst the values for the well-known shadowing materials, chromium, palladium and platinum, are about ten times greater.

Having shadow-cast the plastic replica, one normally evaporates the replicating material (e.g. the carbon or SiO) at right angles to the general surface. If it is a two-stage technique, then the plastic replica substrate must be dissolved by a suitable reagent. This is normally the most disruptive part of the exercise. To overcome this, a thin wax layer is often placed on top of the replica in order to hold it together while the reagent dissolves the plastic. Obviously, the wax must itself be dissolved away in another reagent, the whole replica cleaned and mounted on to an electron microscope grid, before we can get an image of the surface of interest. The technique involves so many operations, it is amazing to record that electron micrographs are usually found completely free of artefacts. However, newcomers must always be alert to the possibility that their most interesting micrographs may show features induced by the preparation technique, rather than the interaction between tribo-elements.

Before leaving this discussion of replicas, we must consider the factors affecting the final image of the replica of a surface. Remember that every feature of an

Table 2.1. *Relative values of contrast per unit thickness of specimen material (from Halliday and Quinn, 1960)*

Element	Be	C	Al	Cr	Fe	Cu
C/d_t	1.3	0.5	0.9	6.5	3.3	5.9
Element	Ge	Pd	Ag	Pe	Au	Ur
C/d_t	4.6	6.6	2.9	6.3	4.0	6.0
Material	SiO					
C/d_t	0.7					

electron micrograph of the replica of a surface must be analysed in terms of (a) the scattering cross-sections for electrons when passing through the elements in the replica and (b) the thickness of the replica in a direction parallel to the transmitted electron beam. Various facets, dimples, steps, ridges, etc. of the surface will be reproduced in the replica as a combination of thickness, materials and orientation relative to the beam. Figures 2.14 and 2.15 may be used to illustrate how the shape of the shadow depends on the shape of the feature upon which the shadow falls. Figure 2.14(b) shows a sketch of a substrate which is first shadowed at 45° and then rotary-shadowed with carbon to give a more uniform layer of carbon all over the surface. The cubic hill (A) and the cubic hole (B) may be considered either as a hill and hole in the replica or as a hole and hill respectively in the surface of the specimen. If the technique used was a two-stage plastic replica and the feature casts an external 'white' shadow (as shown at A in Figure 2.14(a)) then one can be sure that the feature is a hill in the negative replica and a hole in the original surface. On the other hand, if the substrate was the actual specimen surface (as in the one-stage technique) then, of course, A would represent a hill in the original surface. Quite often, the surfaces of worn materials contain very shallow hills and holes, so that they cast no significant shadows. Figure 2.15(b) is a sketch of a convex hill (A) and a concave hole (B). Although the heights and depths are not sufficient to cast shadows, the photographic density of the print [Figure 2.15(a)] tends to be darkest where the replica is locally oriented most nearly normal to the direction of shadowing. In other words, the print of a two-stage replica of a shallow hole is darkest towards the direction of shadowing, and vice versa for a shallow hill. This also holds for shallow ridges and troughs, provided the shadowing direction is perpendicular to the features. There are various other aspects to the interpretation of electron micrographs from the replicas of solid surfaces. These are best dealt with in terms of actual tribological examples (see Chapter 8).

(ii) Thin films removed from surfaces

Although some work has been reported related to the thinning of specimens previously subjected to tribological interactions, the disruption is too great and the technique too destructive to warrant more than this brief mention. The removal of surface films from tribological surfaces has also not been very widely

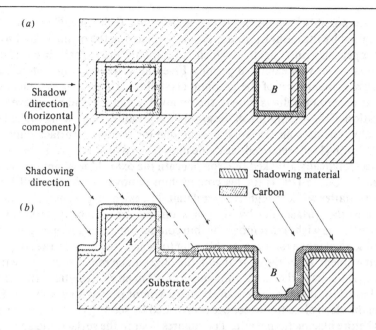

Figure 2.14 *(a) Photographic print of the electron micrograph of the model in (b); (b) model of the shadowing and carbon backing of features simplified as a cubic hill (A) and a cubic hole (B) of the substrate*

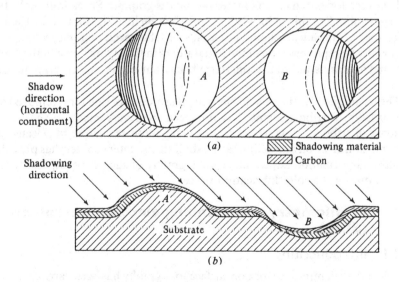

Figure 2.15 *(a) Photographic print of the electron micrograph of the model in (b); (b) model of the shadowing and carbon backing of features simplified as a convex hill (A) and a concave hole (B) of the substrate*

carried out, possibly because this is also a disruptive technique. Nevertheless, it has been used by the author (Quinn, 1964) for removing the contact film formed on a copper slip-ring, against which a graphite brush had been rubbed continuously over very extended periods. Essentially, we were only interested in the preferred orientation of the graphite crystallites. Hence, the production of unwanted reaction products (due to the removal technique) did not prevent the investigation of the orientation of the graphite crystallites from being carried out.

Typically, one uses an electrolytic technique for removing surface films. For instance, Smith (1936) removed the oxide films formed on iron, using a technique which electrolytically attacked the iron beneath the oxide film. A similar method was used by Quinn (1964) for his contact film removal work. We will briefly describe the latter work in slightly more detail. The essential part of the technique is to loosen the surface film by an attack on the sub-surface material by the electrolyte. If one wishes to remove the thin contact film from a copper slip-ring it is advisable to arrange for a removable plug to be inserted flush to the ring prior to the experiment. After the experiment, the plug can be removed from the track on the ring, and made the anode (film uppermost) in an electrolytic bath of dilute (5%) sulphuric acid and a current of about $6 \, \mu A/mm^2$ passed for about 30 s. Easy access of the electrolyte to the copper is provided by scoring the track into small squares before placing in the bath. The squares float to the surface and should be quickly transferred to a dish of distilled water. They are then caught on electron microscope grids and allowed to dry slowly.

Probably the simplest, and most effective, way of obtaining thin films which have been subjected to wearing contacts, is to rub the material under investigation *directly* onto an electron microscope grid. This method clearly can only be used for extremely soft materials such as graphite. Spreadborough (1962) did something like this in his electron microscope investigation of the frictional behaviour of graphite. It should be emphasized, however, that this technique can only really be representative of the *initial stages* in the laying down of the contact film and is *not relevant* to the long-term behaviour of the copper–graphite tribo-system.

The technique of transmission electron microscopy of thin films removed from the surfaces of tribo-elements has not been used very much in the physical analysis of tribo-systems. However, its companion technique of selected area transmission electron diffraction (as we shall see in a later chapter) has provided a wealth of information about the crystallography of such films, especially those formed during the solid lubrication process.

2.3 Profilometry and scanning electron microscopy (SEM)

2.3.1 Introduction

Much of the quantitative work on surface topography has been carried out using the Stylus Profilometer although (as we will discuss later in this sub-section) the scanning electron microscope (SEM) has also been used (in conjunction with a computer) to produce 'contour maps' of various surfaces. How do we characterize the roughness of a surface? The applied mathematician has done a disservice to tribology by using the words 'rough' and 'smooth' as shorthand for surfaces which will exhibit friction, or the absence of friction respectively, when sliding against like surfaces. What happens when one of a pair of rough surfaces slides against one of a smooth pair of surfaces? When has it ever been possible to slide one surface relative to another without *some* friction? Even with so-called frictionless air bearings, there must be traction between adjoining air layers. Leaving aside these somewhat rhetorical questions, let us return to the original question of how to characterize a surface. In order to answer that question we must first define the terms normally used in describing surfaces. We will then discuss the amplitude density function [$\rho_A(z)$], possibly the most important of all the *statistical* parameters used to characterize surfaces. Finally, we will show how the SEM has been used for obtaining the ultimate objective of surface profilometry, namely the *three-dimensional* characterization of a surface.

2.3.2 The definition of terms related to surface profilometry

The definitions used in this sub-section have been taken from the *Standard* published by the American Society of Mechanical Engineers (ASME). Some of the more commonly used terms are given below:

(a) *Surface*: The boundary which separates one object from another object, substance or space.

(b) *Surface Texture*: Repetitive or random deviations from the nominal surface.

(c) *Nominal Surface*: The intended surface contour, the shape and extent of which is usually shown and given dimensions in an engineering drawing.

(d) *Profile*: The contour of a surface in a plane perpendicular to the surface. Remember that the peaks on a *profile* are *not generally* the peaks in the three-dimensional representation of the surface.

(e) *Centre Line*: Sometimes called the 'median line', the centre line is the line about which roughness is measured and is a line parallel to the general direction of the profile, within the limits of the roughness width cut-off, such that the sums of the area contained between it and those parts of the profile which lie on either side of it, are equal. Thus the centre line in Figure 2.16 is *OX*, drawn so that the areas $(A_1 + A_2)$ are equal to the areas $(B_1 + B_2)$ for this particular surface profile.

(f) *Roughness*: The finer irregularities of the surface texture.

(g) *Roughness Width Cut-Off*: This is the greatest spacing of repetitive surface irregularities to be included in the average roughness height.

(h) *Average Roughness Height* (R_a, *AA*, *CLA*): This is the Arithmetical Average (AA) deviation from the Centre Line. It is know as the 'Centre Line Average, (CLA)' in British Standards. Considering Figure 2.16, we can define R_a as follows:

$$R_a = \frac{1}{L_{CO}} \int_{x=0}^{x=L_{CO}} |z_i| \, dx = \sum_{i=1}^{i=n_{CL}} |z_i| n_{CL} \qquad (2.29)$$

where (L_{CO}) is the length over which the average is taken. In typical profilometers, which give the average electronically, (L_{CO}) would be the

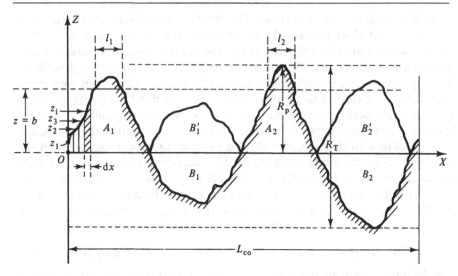

Figure 2.16 *Diagram of a given surface profile illustrating the centre line* (OX), (R_p), (R_T) *and* L_o *(see text)*

roughness width cut-off. The number (n_{CL}) of elements of area with the interval (L_{CO}) depends on the size of the interval, that is $(n_{CL} = L_{CO}/d)$. Clearly, $(R_a L_{CO})$ is equal to $(A_1 + B_1' + A_2 + B_2')$, where the primed quantities are the mirror images of B_1 and B_2.

(i) *Peak-to-Valley Roughness* (R_T): This is the *maximum* value of the peak-to-valley distance within the roughness width cut-off (see Figure 2.16).

(j) *Levelling Depth* (R_p): This is the distance between the centre-line and a line parallel to it that passes through the highest peak of the roughness profile (see Figure 2.16).

(k) *Bearing Length Ratio* (t_{bl}): If one draws a line parallel to the centre-line, but at a pre-determined distance (b) from it, then one cuts the profile in several places (provided t_{bl} is less than R_p). In Figure 2.16, the profile is cut in two places over the lengths l_1 and l_2. The bearing length ratio (t_{bl}) is defined by:

$$t_{bl} = (l_1 + l_2)/L_{CO} \qquad (2.30)$$

The above definitions are sufficient for most surface topographical descriptions. Until recently, most surface finishes were quoted in terms of their R_a or CLA values. Table 2.2 gives some typical CLA values of the surface finishes produced by common production methods. This table shows that there are five main groupings of surface finish, according to the way in which the surface has been produced. It is interesting to see that one can have a CLA value of $0.5\,\mu m$ for most methods of groups 3, 4 and 5, even though we could have used seven different methods for producing that finish. There will be significant differences (e.g. in lay and texture) in a surface that has been turned down to $0.5\,\mu m$ CLA compared with one that has been polished. There is a need for distinguishing between surface finishes by a method other than CLA, as shown in the next subsection on the Amplitude Density Function.

Table 2.2. *Typical surface roughness values*

Group no.	Production method	CLA (in μm)
1	Flame cutting	25 to 12
	Sawing	25 to 6
	Planing	25 to 2
2	Drilling	6 to 2
	Machining by electric discharge	5 to 2
	Milling	6 to 1
3	Broaching and reaming	3 to 1
	Boring and turning	6 to 0.5
4	Electrolytic grinding	1 to 0.2
	Roller burnishing	0.5 to 0.2
	Griding	2 to 0.1
	Honing	1 to 0.1
5	Polishing	0.5 to 0.1
	Lapping	0.5 to 0.05
	Superfinishing	0.2 to 0.05

2.3.3 The amplitude density function

This is the most important of the statistical parameters used for characterizing surfaces. Numerically, it is defined as the probability, per unit increase in height, of finding a profile amplitude within the slice height (Δz) above a given ordinate (z) over the length (L_{CO}) as shown in Figure 2.17. This probability is proportional to the ratio of the total projection of slice heights on to the reference line and the sampling length. Therefore

$$\text{Amplitude probability} \equiv P(z, z + \Delta z) = \sum_i \Delta l_i / L_{CO} \qquad (2.31)$$

The amplitude density [$\rho_A(z)$] is, quite simply, the amplitude probability per interval (Δz) in the z-direction and is given by:

$$\text{amplitude density } [\rho_A(z)] = P(z, z + \Delta z)/\Delta z \qquad (2.32)$$

Hence, for any profile, it is possible to generate the amplitude density function [$\rho_A(z)$], as shown in the diagram to the right of the profile in Figure 2.17. Many researchers have tried to fit analytical expressions (such as the Gaussian curve or the exponential curve) to the profile of [$\rho_A(z)$] generated from actual surfaces. This is because (as we saw in sub-section 1.2.4) they would then be able to characterize the surfaces in terms of σ (the standard deviation of asperity heights). Once the tribologist knows [$\rho_A(z)$] for a surface he can then deduce several other

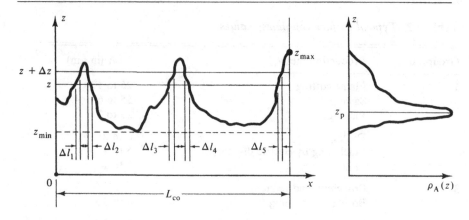

Figure 2.17 *Amplitude density function*

parameters, such as the average roughness height (R_a) and the root-mean-square roughness (RMS). Hence, from the definition of an average value, we get:

$$R_a \equiv AA = \int_{z_{min}}^{z_{max}} |z| P(z, z + \Delta z)$$

that is

$$R_a = \int_{z_{min}}^{z_{max}} |z| [\rho_A(z)] \, dz \tag{2.33}$$

and

$$(RMS)^2 = \int_{z_{min}}^{z_{max}} z^2 [\rho_A(z)] \, dz \tag{2.34}$$

These definitions of R_a and $(RMS)^2$ in terms of the amplitude density function show how important it is to be able to find a regular function [such as the Gaussian profile, described by the function $\exp(-z^2/2\sigma)$, or the exponential 'half-profile', described by the function $\exp(-z/\sigma)$] which fits the experimentally measured amplitude density function. It is clear that *all* theories of friction and/or wear *should* contain a term involving the amplitude density function. Perhaps the reason for its neglect is the fact that most surfaces change their initial topography quite significantly during 'run-in'. Hence, any analysis of the surface topography *before* the experiment may have very little relevance to the surface topography *during* and *after* the experiment.

2.3.4 Contacting techniques for studying surface characteristics

Any technique for measuring surface characteristics should be capable of measuring:
(a) *Waviness* (i.e. *long* wavelength deviations from the intended surface);
(b) *Roughness* (i.e. *short* wavelength deviations);

(c) *Form (or Lay)* (i.e. departures from the 'intended' surface due to machining processes).

The contacting (stylus) technique is in very wide use at the present time. It can measure 'waviness' and 'form' fairly well – but is insensitive to form (or lay) *differences*. There is normally a 'dedicated' computer associated with the outputs from the electronic circuits, ready to give the operator instantaneous print-out of one of the many surface characterization parameters he/she may require.

The main disadvantages of the stylus profilometer are:

(a) the stylus may damage the surface (and hence give false values of the profile);
(b) the lowest limit of the size of feature detected depends (eventually) on the stylus radius (e.g. one has a 0.1 μm radius tip on the instruments intended for research); and
(c) one normally gets presented with a distorted 'picture' of the surface when (× 20 000) magnifications are obtained perpendicular to the surface but only (× 20) parallel to the surface.

The typical worn surface is normally a gently undulating surface. The 'Talysurf' trace of such a *positive surface* gives one an entirely false picture of what it is really like. The image of two surfaces in contact being likened to Austria and Switzerland placed one on top of the other was due to Bowden and Tabor (1954). By reducing the vertical scale so that it is the same as the horizontal scale, one would find a much smoother-looking surface, more like the Berkshire Downs being inverted upon the Cotswolds! Let us now end this sub-section by describing how the scanning electron microscope has been used for surface analysis.

2.3.5 Scanning Electron Microscopy (SEM) and surface topographical analysis

Because of its great depth of focus, its wide range of magnifications of the same surface feature and its relative ease of application to the study of bulk surfaces, scanning electron microscopy (SEM) is an obvious candidate technique for surface analysis.

A simplified block diagram of the essential features of an SEM is shown in Figure 2.18. Essentially, one has a narrow beam of primary electrons scanned across the specimen in a television type raster, and the secondary electrons produced by the scanning beam are used to give an image of the specimen surface. This is done by collecting the electrons in a charged collector (EC) and amplifying this current so as to control the brightness of a cathode ray tube that is scanned synchronously with the primary beam. The magnification, which is variable from × 20 to about × 50 000 without the need to re-focus, is controlled by varying the area of scan on the specimen, while keeping the area of scan on the cathode ray tube constant.

A conventional electron gun is used as the source of electrons – these are accelerated through a potential difference between − 1 and − 30 kV, with respect to the frame of the microscope (which is at earth potential). The purpose of the lens system is to control the size of the beam so that the smallest possible 'spot' is obtained for scanning across the specimen. This determines the resolution of the instrument. Using a four-lens system, it has been possible to de-magnify the

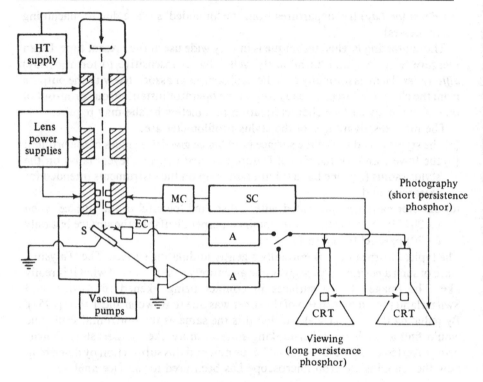

Figure 2.18 *Block diagram of the essential features of a scanning electron microscope*

electron beam width sufficiently to obtain a resolution of better than about 80 Å.

The contrast of a scanning electron micrograph is related to differences in atomic number between the various elements in the surface and the topography of that surface. For a single-element specimen, it is the relative slopes of the roughnesses that give rise to contrast, rather than their relative heights. Unfortunately, it is the latter that determines the real area of contact between the surfaces of a tribo-system. There have been many attempts to deduce relative heights from the analysis of scanning electron micrographs, especially through the use of stereoscopic pairs. The paper by Breton, Thong, and Nixon (1986) is of special interest in this respect. These authors describe how they used a real-time image processing technique to establish profiles of a surface feature observed in a stereo-microscopy system in which the electron beam is successively deflected back and forth across that feature. Breton *et al.* claim height accuracies of 100 nm at magnifications of 50 000. Apart from the work involving stereoscopic pairs mentioned above, the use of the SEM in analysing the surfaces of tribo-systems has tended to be of a qualitative nature. This is due to the great depth of field of the SEM, which precludes any measurement of relative surface heights by successively focusing on the top and bottom of the feature being observed. Another recent development, however, is one in which the use of an image-

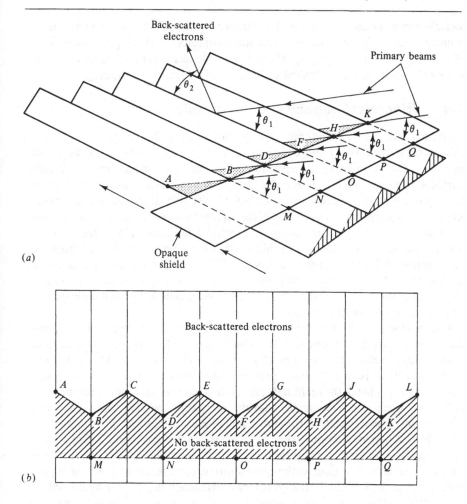

Figure 2.19 *Diagram of the principle of an image-converter to show the profile of a ridged specimen. (a) Perspective view of how a profile of a ridged specimen is generated; (b) plan view of ridged specimen showing the profile superimposed on the secondary electron image*

converting system has enabled the back-scattered primary electrons to be used to give profiles of suitably ridged specimens as shown diagrammatically in Figure 2.19. The principle is fairly simple. Effectively, an opaque shield is passed parallel to the surface of the specimen and the back-scattered primary electrons collected and superimposed on the (normal) secondary electron image. By careful choice of contrast, it is possible to only show the edge of the shadow as *ABCDEFGHJKL*, below which no back-scattered electrons can be obtained. Hence one obtains the qualitative, almost three-dimensional, image of the scanning electron microscope, with the surface contours superimposed. Since the image is an electronic signal, these contours can be digitally processed to give (in a quick and fairly straightforward fashion) quantitative data which can only be obtained by a huge number of multiple scans on the mechanical profilometer. Having discussed some of the more obviously *topographical macroscopic* techniques, let us now discuss *some other* macroscopic physical techniques of relevance to tribology.

2.4   Electrical capacitance and resistance

Electrical capacitance methods are normally used for measuring oil film thickness between sliding and/or rolling surfaces. Contact resistance methods are normally used to indicate the presence of non-metallic contact between sliding surfaces. In a paper mainly designed to demonstrate that the lubrication of rollers is essentially elastohydrodynamic, Crook (1958) also described a capacitance method for measuring the thickness of the oil film between the rollers. Essentially, the method consisted of measuring the electrical capacitance between the discs and the flat steel pads allowed to ride upon the oil films carried away from the conjunction of the discs. Crook was not satisfied with this technique and later (see Crook, 1961) he produced an improved version which allowed him to measure the *shape* of the oil film, rather than its average thickness. We will describe this improved version in more detail.

Crook (1961) took two discs, one of glass, the other of hardened steel, and rolled them together in a typical two-disc machine. The use of a non-metallic material to simulate the *geometry* of a tribo-element which would normally be made of steel (say) can only be acceptable on the grounds that it was the only way to obtain the profile of the oil film thickness. It is interesting to note that there are still some 'pockets of resistance', who do *not* accept the idea that rolling and sliding discs *actually* simulate the enmeshing of gear teeth, some thirty or so years after the first papers were published in which this claim was made. There may be some justification for this 'resistance' since, in the meshing of gear teeth, one gets sliding in *opposite directions* as the gears approach and leave the position of pure rolling at the pitch circle. In the typical two-disc machine, the sliding is fixed (at about 10% slip) and is, of course, always unidirectional. How close this simulation is to the variable sliding together with changes in sliding direction of real gears is a matter of some conjecture. Unfortunately, the use of actual gears for research (often consisting of tests to destruction) tends to be prohibitive in cost and meagre in the production of basic data about how gears are actually lubricated. Let us now see what Crook (1961) actually did.

In fact, Crook (1961) placed an evaporated chromium electrode on the glass disc, as shown in Figure 2.20(*a*). The electrical connections were made as shown in Figure 2.20(*b*). The important part of the electrode is the strip confined to the central part of the track, and the important requirement with respect to the strip is that its leading edge should lie accurately along a generator of the cylindrical cross-section of the glass disc. This requirement was checked by comparing the leading edge of the electrode with optical interference fringes formed between the glass cylinder and a suitably-placed optical flat. Crook's (1961) experiments involved an increase in the capacitance between the glass and steel discs as the leading edge of the electrode approached and passed through the conjunction.

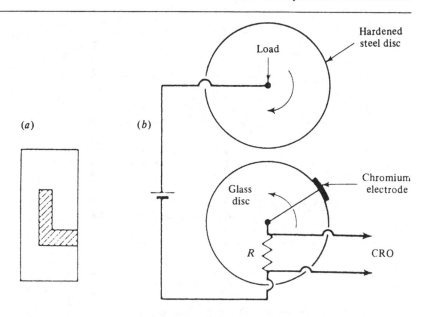

Figure 2.20 *(a) Plan of chromium electrode and (b) the electrical con-nections used in Crook's (1961) experiments*

This caused a flow of a current and a potential drop across the resistor R shown in Figure 2.20(b). Provided the resistance is small, the potential drops across R may be taken as being proportional to $(\varepsilon/h_{\text{lub}})$, where (ε) and (h_{lub}) respectively relate to the dielectric constant and the distance between the surfaces of the rollers at the instantaneous position of the leading edge of the electrode. Crook showed that the oscilloscope trace indicated a constriction in the oil film at the *exit side* of the conjunction between the discs. This constriction reduced the film thickness locally by about 10%. Otherwise, the film was almost *parallel* over the whole of the elastically deformed region. Later efforts by other workers to improve upon Crook's (1961) technique were not very successful. For instance, Kannel, Bell and Allen (1965) used a manganin *transducer* in place of the chromium electrode, in an attempt to obtain a *direct* measurement of the pressure distributions within the lubricated rolling contact. Unfortunately, their technique was insensitive to small pressure changes within the contact zone.

A typical example of the contact resistance method is given in Furey's (1961) paper. The system consisted basically of a fixed metal ball loaded against a rotating cylinder. The extent of metallic contact was determined by measuring both the *instantaneous* and the *average* electrical resistance between the two surfaces. The circuit used in the measurement of the electrical resistance is shown in Figure 2.21: Furey (1961) found that, in general, the electrical resistance oscillated rapidly between an extremely low value and infinity, suggesting that the contact was metallic but discontinuous. Alternatively, it could have indicated a mixture of metallic and oxidized contacts. The average resistance of an oil film

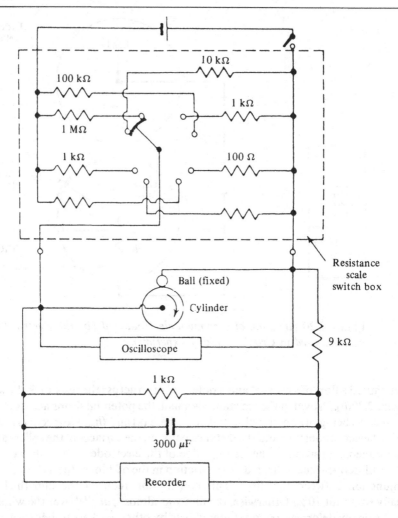

Figure 2.21 *A typical circuit used in the measurement of electrical resistance of a tribo-system (Furey, 1961)*

is, therefore, a time-average; that is, it is a measure of the proportionate time that metallic contact occurs (for severe wear, in which metallic wear prevails). It could, however, be a measure of the proportionate time that metallic or oxidized contact occurs (for most lubricants, there will always be some oxygen present in the oil to bring about oxidation of the contact area).

This has been a rather brief treatment of electrical capacitance and resistance methods. Typically, the need for making one of the tribo-elements electrically insulated from the other (apart from the point at which the resistance or capacitance is to be measured) causes great design difficulties in any attempt to measure either contact capacitance or contact resistance. For this reason, most tribometers do *not* have facilities for making these measurements.

2.5 Magnetic perturbation techniques

2.5.1 Introduction

Many failures originate from voids or inclusions in, or near, the surface. These form stress raisers and hence become the nuclei of cracks. Sometimes these cracks are propagated by hydraulic entrapment (if they are oriented towards the motion of the opposing surface). Whatever the cause of crack propagation, the presence of a crack, void or inclusion can be readily detected by a magnetic perturbation technique. Most of the techniques involve dismantling bearings and then subjecting the running surface to a magnetic field which becomes distorted in the region of an inclusion, a void or a crack. Such a *pre-testing inspection*, however, does not always lead to a significantly successful prediction rate for diagnosing possible fatigue failure (see next sub-section). There have been a few techniques that are used in situ during actual tests, where the machine is stopped as soon as the crack, inclusion or void reaches a pre-set size (related to the size of the probe signal). One such technique is described in Section 2.5.3.

2.5.2 Pre-testing magnetic perturbation techniques

Fatigue spalls in rolling-element bearings can be initiated at non-metallic inclusions in the bearing steels (Carter *et al.* 1958; Littman and Widner, 1966). Inclusions near to the rolling surface act as stress concentrations. Fatigue cracks will form at these inclusions and propagate under the repeated stresses to form a network of cracks which eventually produce a fatigue spall (i.e. a relatively large region of the surface, which suddenly detaches itself from the rolling surface in the form of a large particle). Such inclusions, however, are not the only sources of stress concentration sufficiently large to cause fatigue cracks. For example, large or unusual carbide formations, surface flaws and misalignment and edge-loading of the rollers can all lead to pitting (as contact fatigue is normally called). According to Littman and Widner (1966), however, it is the sub-surface inclusions that are the chief cause of stress concentration and, hence, fatigue cracking.

The magnetic particle technique (Betz, 1967), which will *not* be discussed here, is widely used in the bearing industry to eliminate those bearing components containing flaws. However, it is not sensitive enough to find the very small flaws, approximately 25 to 75 µm (0.001 to 0.003 in) that often cause bearing fatigue failures. The magnetic perturbation technique, to be described in the next paragraph, can detect the very small, non-metallic inclusions at the surface. It can give information on the location and size of these inclusions in ferromagnetic materials. This technique is also sensitive to the presence of cracks and voids and local metallurgical inhomogeneities (Barton and Kusenberger, 1971).

Parker (1975) produced some of the most definitive work using the magnetic perturbation technique to predict failures in rolling-element bearings. His objectives were:
(a) to inspect the inner-races of deep-groove ball bearings by magnetic pertur-
bation and document significant inclusions;

(b) to fatigue test sufficient numbers of those bearings and compare results with the inspection measurements; and

(c) to determine the relation between the magnetic perturbation results and the location and size of non-metallic inclusions as determined by metallographic sectioning.

The application of the magnetic perturbation method to the inner race of a bearing is illustrated in Figure 2.22. We see that the method consists essentially of establishing a magnetic flux in the region of the material being inspected and then scanning the surface of the material with a sensitive magnetic probe to detect anomalies in the magnetic flux. The flux lines are uniform and parallel in regions remote from the flaw or inclusion. In the region of the inclusion, the flux lines are distorted, so that they emerge from and re-enter the material surface. The intensity of the distortion or perturbation depends on the size of the flaw, its depth beneath the surface, the strength of the applied magnetic field and the overall magnetic properties of the material being examined. The signal obtained from the probe has a characteristic shape according to the nature of the flaw. For example, when viewing in the direction of the rotation of the inner race (i.e. from right to left in Figure 2.22), we see that the signal goes down first and then upwards. This indicates a void, crack or inclusion. When the signal goes up first and then downwards, it will be from a flaw *other* than a void, crack or inclusion.

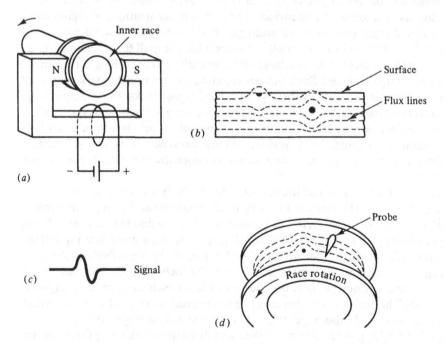

Figure 2.22 *Schematic illustration of the magnetic perturbation method applied to the inner race of a bearing. (a) Magnetized inner race; (b) perturbations in the magnetic flux at the surface on inner race; (c) signal from inclusion in surface; (d) probe scanning the surface*

Whenever one attempts to measure the fatigue lives of bearings, one must plot the results on a 'Weibull Plot'. Let us briefly examine what we mean by such a term. It arises from the Weibull distribution for the statistical analysis of failure which, in its simplest form, is given by:

$$\text{Failure rate} = a_W t^{(b_W - 1)} \tag{2.35}$$

where a_W is the scale factor, b_W is the Weibull slope and t is time in service. If $b_W < 1$, this expression indicates failures involving some initial defects, whereas if $b_W > 1$, then one is in the 'wear-out' region. A more convenient way of using the Weibull distribution is to plot the failure data on Weibull paper, examples of which are given in Figure 2.23. Component A had exhibited cumulative failure rates of 1%, 3%, 5%, 8% and 12% after 130 h, 250 h, 300 h, 400 h and 500 h respectively. When plotted on this paper, we see that it was most probable that failure of 90% of the components would have occurred after 2000 h running (see A_3 on Figure 2.23). Component B, on the other hand, exhibits a probable failure rate of only 2% after 2000 h in service (see B_3 on Figure 2.23). Note that the slope of the Weibull plot for component A is about 2, indicating a wear-out situation, whereas the Weibull slope for component B is about 0.8, suggesting initial defects (such as inclusions, cracks or voids).

Using Weibull plots for 54 identical bearings, Parker (1975) was able to characterize the results of fatigue tests run at 2750 rev/min under a radial load of 5860 N, a maximum Hertz (elastic) stress of 2.4×10^9 N/m^2 at the inner-race–ball contact and jet-lubricated with a super-refined naphthenic mineral oil. The outer-race of the bearing was held at about 70 °C. Parker (1975) took half of the bearings and, after magnetic perturbation inspection, subjected them to pre-stressing at 2750 rev/min for 25 h under a radial load producing a maximum Hertzian stress of 3.3×10^9 N/m^2 at the inner-race–ball contact. The other 27 bearings were not subjected to the pre-stress cycle. Figure 2.24 gives the Weibull plots for the pre-stressed bearings and the baseline bearings (i.e. the bearings which had *not* been subjected to pre-stressing). From this, we can see that the $B10$ life (i.e. the life which 90% of the bearings will reach or exceed without failure) is about 700×10^6 revolutions for the pre-stressed bearings, whereas it is about 280×10^6 revolutions for the baseline bearings. The Weibull slopes (b_W) of both groups of bearings are about 1.2 (pre-stressed) and about 1.6 (baseline), *indicating wear-out rather than initial defects* (see earlier paragraph regarding whether b_W was greater or less than unity).

From the fatigue tests alone, we could say that the *pre-stressing cycles* have only had a slightly *prolongation* effect on the fatigue lives. What does magnetic perturbation inspection tell us about the fatigue? Of the 27 bearings in each groups, 10 in each group showed magnetic perturbation signals which indicated the presence of inclusions, cracks or voids likely to be detrimental to bearing fatigue life. If these 10 bearings from each group are excluded from the test groups, the fatigue lives would appear as in Figure 2.25, which is the Weibull plots for the remaining 17 bearings in each of the two groups. There are only three and five inner-race failures to be considered for the baseline and pre-stressed groups, respectively. Of these eight bearings, five had definite evidence of fatigue spall

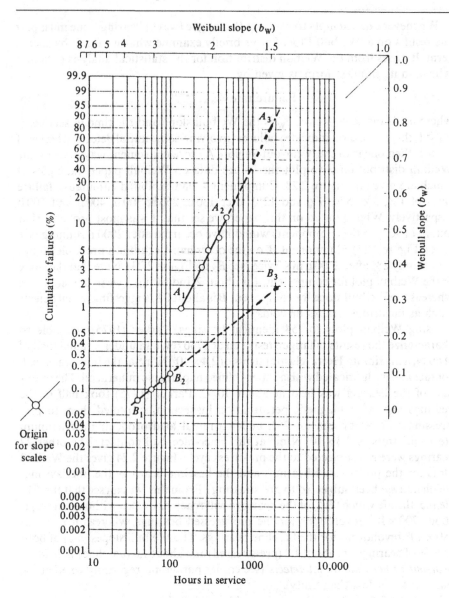

Figure 2.23 *Examples of the use of Weibull paper (from the Tribology Handbook, Butterworths, London, 1973)*

initiation at surface indentations (these are indicated by full symbols in both Figures 2.24 and 2.25). Clearly, these indentations were initiated at the surface during the fatigue tests, since the author claims that these features would have been observed at the ×30 visual inspection to which he subjected all bearings before testing. Parker (1975) suggests that these indentations were introduced during fatigue testing by foreign particles in the test apparatus or lubrication

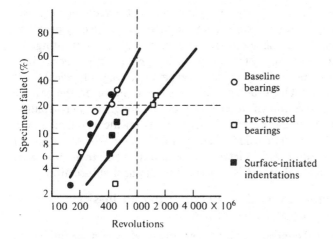

Figure 2.24 *Results of fatigue tests at a radial load of 5860 N and a shaft speed of 2750 rev/min with super-refined naphthenic mineral oil*

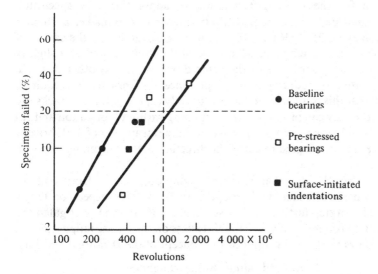

Figure 2.25 *Fatigue test results obtained by excluding the bearings which gave significant signals during magnetic perturbation inspection of their inner races*

system. The main conclusion to be drawn from the comparison of Figures 2.24 and 2.25 is that the statistical lives of the groups *without* the 'inferior' bearings (as indicated by magnetic perturbation effects) are no better than the lives of each of the groups of 27 bearings.

Although the occurrence of surface indentations had vitiated the fatigue test results, it is when Parker (1975) concentrated his attention on the non-indentation spalls that he found a very strong correlation between the magnetic perturbation signatures and the position of inclusions as determined by precision metallurgical sectioning. If some method could be found for preventing the occurrence of surface indentation by foreign particles, then one might be justified in claiming that magnetic perturbation inspection *before* testing will reveal those bearings most likely to fail. Until such a method is found, however, the best we can say is that prior inspection of inner-ball races by magnetic perturbation techniques will reveal potential fatigue nuclei such as an inclusion or a sub-surface defect. Clearly, what is needed is an *in-situ detector* which will *detect inclusions, cracks and voids* and *surface indentations* from their magnetic perturbation effects. Such a detector is described in the next sub-section.

2.5.3 Detecting failures as they occur using an in-situ magnetic perturbation technique

In this sub-section, we will concentrate on the pitting (surface contact fatigue) of sliding and rolling discs. The technique used to detect the onset of cracking at various locations on the surfaces of these discs can be modified to suit other geometries, although the details of such a modification might present a tough design problem for the tribologist wanting to adapt this very successful, continuously monitored, magnetic perturbation technique to his/her particular requirements. Way (1935, 1936) did the pioneering research into the causes of pitting. He used a disc machine in which one of the discs rotated at a slightly different speed ($\sim 10\%$ slip), but in the same direction, as the other disc. He showed that, in order to produce pitting, the presence of a lubricant was required. We now know that this is due to two factors, namely (a) the initiating cracks are worn away before they can propagate when running without lubrication and (b) propagation of cracks is encouraged by the hydraulic entrapment of the lubricant in those surface cracks oriented towards the direction of the oncoming opposite surface.

Way (1935) also showed that smooth, highly-polished surfaces had an increased life when compared with rougher surfaces. This aspect of Way's research was taken a step further by Dawson (1961, 1962), who investigated the influence of surface roughness and oil film thickness on the pitting fatigue life of a tribo-system. Dawson (1961) introduced the concept of the D-ratio, defined as:

$$D = \frac{\text{Combined } \textit{initial} \text{ surface roughness}}{\text{Lubricant film thickness}} \qquad (2.36)$$

Dawson (1962) showed that the 'life to first pit' was a function of this non-dimensional parameter (D) as shown in Figure 2.26. The straight line drawn between all the points (except points *a* and *b*, which did *not* pit) has the approximate form

$$\log_{10}(N_R) \approx -0.67(\log_{10}D) + 5.67 \qquad (2.37)$$

Dawson (1962) did not actually attempt to describe the dependence of (N_R) upon D. However, for the sake of comparison with later work (in Chapter 7), Dawson's (1962) graph has been used to obtain Equation (2.37).

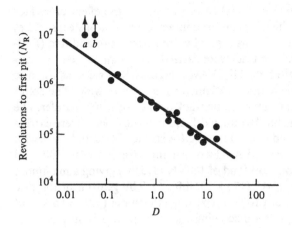

Figure 2.26 *Graph of 'revolutions to first pit' plotted against D-ratio (points a and b did not pit) (Dawson, 1962)*

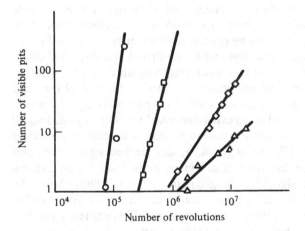

Figure 2.27 *Number of visible pits plotted against revolutions run for four typical tests (Dawson, 1963)*

The important point to note about Dawson's (1962) work is that Figure 2.26 was obtained after a very laborious extrapolation from a graph of 'number of visible pits' versus 'number of revolutions' for each D-ratio. Figure 2.27 illustrates how it was often necessary to run another million revolutions beyond the number required for the first pit before Dawson could be sure of his extrapolation. There was a clear need for a technique capable of detecting the first pit as it happened or, preferably, before it happened! In other words, there was a clear need for a crack detector. Such a detector will be described in the next few paragraphs.

Phillips and Chapman (1978) produced their crack detector in response to the needs of a research programme undertaken to investigate the effect of surface roughness and lubricant film thickness on the contact fatigue life of steel surfaces lubricated with a sulphur–phosphorus type of extreme-pressure additive (Phillips and Quinn, 1978). The fatigue tests were carried out on an Amsler disc-on-disc machine with EN26 steel (2.5% Ni–Cr–Mo, high carbon steel) as the disc material. The specimens were 40 mm in diameter and 10 mm wide. The softer (pitting) disc (260 VPN) had a 4 mm running track defined by a 30° chamfer. The harder disc (560 VPN) was the full 10 mm width. Some experiments were carried out with the disc surfaces ground to a 0.025 μm CLA surface finish and some with both discs ground to 0.25 μm CLA. Two driving disc speeds (200 and 400 rev/min) were used at a constant load of 1668 N (375 lbf) giving a maximum Hertzian stress of 1.20×10^9 N/M^2 (175×10^3 lbf/in^2). The gearing of the disc machine was such that there was a 10% slip between the upper (pitting) disc and the lower (driving) disc, the pitting disc being the slower moving surface.

Phillips and Chapman (1978) used a commercially-available tape recorder erase head as the magnetic detector. The erase head had a non-magnetic gap width of 0.127 mm and a gap length of 4.0 mm, as measured by optical microscopy. The erase head was made to track the surface of the pitting disc by applying a light load (by means of a spring) and allowing the head to slide hydraulically on the available oil adhering to the disc surface. The magnetic field was localized around the erase head by passing a direct current (1 mA) through the erase head. The detected signal was amplified, filtered and displayed on an oscilloscope. The oscilloscope was externally triggered once every revolution from another tape recorder head and magnet located at the end of the shaft supporting the pitting disc. A delay circuit enabled any part of the signal from the entire disc surface to be examined on an expanded time base. The amplified signal was also fed into two parallel discriminator-operated relay circuits providing automatic recording on a UV oscillograph, at any desired signal level, and automatic shut-down of the disc machine at a given signal level. Figure 2.28 is a schematic diagram of the signal-handling circuits. Figure 2.29(*a*) shows the recorded signal from a typical surface fatigue crack and Figure 2.29(*b*) shows the expanded time base oscilloscope trace of the same signal. The fatigue test had run for 58.67 h (1.26×10^6 cycles) before this signal was detected and the disc machine shut down automatically.

It is interesting to note that Phillips and Chapman (1978) examined the disc surface both optically and magnetically *before* the fatigue tests. There appeared to be no correlation between pre-test magnetic analysis of the surface of the pitting disc and the site of subsequent pitting failure. Signals detected prior to testing generally disappeared during the running-in period, when the whole surface signal changed continuously and, in general, no evidence of the site of pitting failure recurred until a signal developed similar to the one shown in Figure 2.29. From this, one might suggest that *any* pre-test inspection technique involving magnetic perturbation, will have its reliability invalidated by the changes in surface topography and composition during the running-in, which always occurs in any tribo-system (other than one which is lubricated by full-fluid film lubrication).

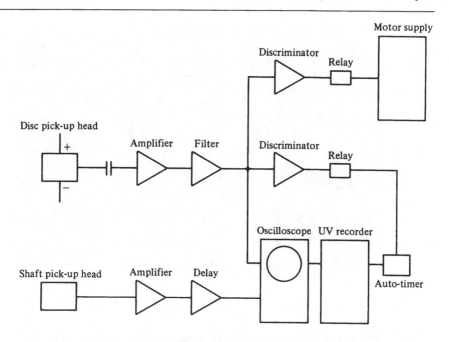

Figure 2.28 *Schematic diagram of the signal-handling circuits used in the magnetic detection system*

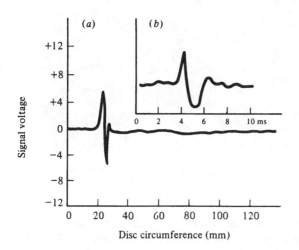

Figure 2.29 *Recorded signal from a surface crack: (a) a function of the circumference of the pitting disc; (b) the expanded-sweep trace of the same signal*

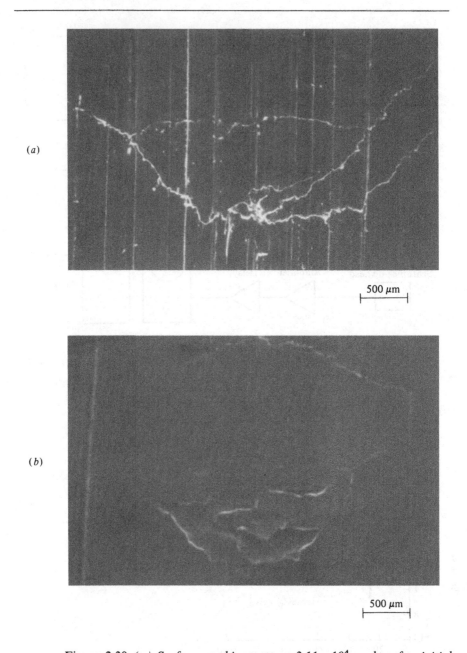

(a)

500 μm

(b)

500 μm

Figure 2.30 *(a) Surface cracking pattern 2.11×10^4 cycles after initial detection; (b) the same surface cracking pattern after a further 1.16×10^4 stress cycles (rolling direction from top to bottom)*

In order to examine the fatigue cracks as soon as they were detected, Phillips and Quinn (1978) modified the specimen stage of an SEM so that the whole of the pitting disc could be inserted into the microscope. The disc could be rotated in the microscope enabling the entire disc periphery to be examined. Figure 2.30(*a*) shows a surface cracking pattern from a fatigue test some 2.11×10^4 cycles after the initial detection as a single crack. The fatigue cracks originated from a surface flaw which, in this case, was detectable from the beginning of the test and can just be seen at the apex of the characteristic arrow-head shape. Figure 2.30(*b*) shows the development of the cracking pattern after a further 1.16×10^4 cycles, the surface cracking pattern of the trailing edge being clearly visible. It is instructive to note that, when the surface crack at the trailing edge first appeared, it was *not* connected to the surface cracking pattern of the leading edge.

Clearly the use of a crack-detecting, magnetic perturbation, continuously monitoring surface analysis techniques *in situ* in the actual tribo-system, can provide a powerful tool for investigating the precise nature of surface contact fatigue failures, since:

(a) initial surface contact fatigue cracks can be easily detected, thus enabling the origin of the cracks to be investigated;
(b) having detected the initial fatigue cracks, some measure of the propagation time leading to pitting failure can be obtained; and
(c) the progress of surface cracks towards a pitting failure can also be studied and the actual pit formation detected, thus enabling the test to be terminated and the 'life-to-first-pit' obtained.

In Section 7.5.3, in which we discuss the application of physical analytical techniques to the *study of pitting in line-contact tribo-systems*, we will see how the problem of contact fatigue in rolling and sliding tribo-systems has been investigated, with some measure of success, using the magnetic perturbation crack detector combined with other physical techniques, such as scanning electron microscopy and Auger electron spectroscopy.

2.6 Macroscopic techniques for monitoring oil-lubricated machinery (with special emphasis on 'Ferrography')

2.6.1 Introduction

To end this chapter on *macroscopic* physical techniques for tribology, we will now discuss some of the methods used for monitoring oil samples with the aim of being able to *separate* the wear debris for subsequent analysis. Clearly, that analysis can be *macroscopic* (such as measurement of the concentration of debris in the oil, particle morphology and their size distribution) or it may be *microscopic* (elemental and crystallographic). In this sub-section, we will be concerned only with the *macroscopic* aspects of separating the wear debris.

The better known methods of wear debris analysis are spectrographic oil analysis, magnetic plug inspection, filter inspection, 'Ferrography' and particle counting. The results given by the different methods depend upon the size of the particles present in the oil. For example, spectrographic oil analysis is thought to

Table 2.3. *Some properties of techniques frequently used for monitoring oil-lubricated machinery*

	Measurement of distribution concentration	Identification of type of metal	Particle morphology	Size
Spectrographic oil analysis	Excellent	Excellent	Not suitable	Not suitable
Magnetic plug inspection	Very good if automated but ferrous only	Best for ferrous	Good	Not suitable
Filter inspection	Good	Fair	Good	Not suitable
'Ferrography'	Good, if ferrous	Good	Excellent	Good
Particle counting	Good	Good	Good	Excellent

be insensitive to particles *larger* than 15 µm. It is this dependence upon particle size that has given rise to some contention regarding the relative effectiveness of the various techniques listed above. Table 2.3 lists the effectiveness of each technique insofar as it measures concentration, identifies the type of metal in the debris, provides some indication of the morphology (i.e. external shape) of the particles and gives information on particle size distribution.

Spectrographic Oil Analysis is a well-developed procedure for analysing hydraulic and lubrication systems by the removal of a known amount of oil for external analysis. The sampling and reporting of results has been standardized and is termed the Spectrographic Oil Analysis Programme (SOAP). The presence of certain elements above established concentrations (normally determined from experience) is used as an indicator of which components may be experiencing excessive wear. Magnetic plug inspection is fairly self-explanatory as also is filter inspection. As regards particle counting, there are many commercially-available automated image analysis systems, many of which can be connected to a scanning electron microscope with an X-ray microanalysis capability (KEVEX), thereby giving concentration and size distributions together with elemental identification and particle morphologies. As can be seen from Table 2.3, 'Ferrography' scores well as regards all the desirable properties of an oil-monitoring technique. For this reason, the remainder of this sub-section will be concerned with the details of this technique.

2.6.2 'Ferrography'

Lubrication is provided between moving surfaces to minimize wear, but during operation of the system many millions of minute wear particles enter the lubricating oil. These particles range in size from several micrometres to a few hundred ångströms (1 Å $= 10^{-10}$ m). Most of the particles remain in the oil and do not settle out if a sample is taken out and kept in a bottle. As well as wear particles, there will be millions of other particles that do not owe their origin to wear of the tribo-elements, but come from the air or some other contaminating source. The total number of particles from all sources can be as large as $10^{18}/m^3$. Knowledge of the quantity, and of the rate of increase in the quantity, of foreign material (particularly iron) in the lubricating oil of an engine, can give a valuable insight into the condition of the engine.

The highly-stressed wearing parts of a machine are usually made of steel. Hence, if the ferrous particles which collect in the oil are separated and examined, significant information about the condition of the machine is obtained. Particles of iron, nickel or other magnetic material are, of course, influenced by magnetic fields. The 'Ferrograph' forms the basis of a technique which permits the collection of magnetic particles from an oil sample onto a microscope slide and the sorting of the particles in such a way that the larger particles appear at one end and the smaller particles at the other end. The 'Ferrograph' is a commercial instrument – hence the use of the capital letter. However, in much the same way as the word 'Hoover' has become the generic name for vacuum cleaners, so has the word 'Ferrography' come to mean the study of magnetic particles present in used samples of lubricating oil. The paper by Seifert and Westcott (1972) was perhaps one of the first to be published on this powerful method of failure analysis. We will review this paper, bearing in mind developments since those early days.

The 'Ferrograph' consists basically of a magnet designed to develop a field having an extremely high gradient near the poles, a pump to deliver the oil sample at a slow steady rate (4.2×10^{-9} m^3/s) and a substrate on which the magnetic particles are deposited and on which they can be examined subsequently either by means of an optical densitometer or a high performance microscope. The substrate on which the particles are deposited is treated by a proprietary process so that the oil will flow down the centre of the slide and the wear particles will adhere to the surface of the slide upon removal of the oil. The slide is mounted at a slight angle to the horizontal, so as to use the small component of the gravitation force to help separate the particles according to their mass. The axis of the slide is also aligned with the gap between the polepieces of the magnet. The oil, which is first diluted with a special solvent to increase the mobility of the contaminating particles within the oil, is delivered to the higher end. Since this end is at a slightly larger distance from the pole pieces (compared with the exit end), the magnetic field is weaker at the delivery end, so that the magnetic particles are always subjected to a continually increasing force as they flow along the slide. Clearly, viscous drag also affects the particles according to their size and shape. The combination of magnetic forces and drag forces (and to a much lesser extent, gravitational forces) on the particles as they are carried by the oil, has the net effect of sorting the particles by size. The larger particles are deposited first, while the smaller particles tend to migrate some distance along the slide before being deposited.

The analysis of the 'Ferrogram', as the slide with its particles is called, can be carried out using an optical densitometer. The authors, in a later publication (Westcolt and Seifert 1973), show that the optical density (D_o) of the 'Ferrogram' is related to the ratio of the area (A_2) in the densitometer aperture covered by opaque particles divided by the area (A_1) of the densitometer aperture through the expression:

$$D_o = \log_{10}\left(\frac{1}{1-(A_2/A_1)}\right) \qquad (2.38)$$

Having obtained an experimental value of (A_2/A_1) from Equation (2.38), Westcott and Seifert (1973) attempt to deduce the total volume of metal deposited on the 'Ferrogram' through some rather obscure reasoning, although just how they use this quantity for predicting failure is not made very clear.

I have many reservations about the value of 'Ferrography' as a reliable indicator of incipient failure. As an example, let us consider the appearance of spherical metal particles. These have been claimed to be indicative of fatigue crack development (Scott and Mills, 1970). These authors suggest they form in some way from the interaction of lubricants and surface cracks that continually open and close. Hack and Feller (1970) claim that these spherical particles are associated with melting as a result of high instantaneous surface temperature during sliding. This idea is also supported by Swain and Jackson (1976): Scott and Mills (1973) have also given another possible suggestion, namely that they result from the rolling-up of flat wear particles. Rabinowicz (1977) suggests that, in order to produce spherical particles, 'chunky' wear particles must be formed as a result of a severe wear process, and that they become rounded and polished as a result of a burnishing process. Thus, we see that spherical wear particles in a 'Ferrogram' could mean different things to different investigators. The fact that these investigators were interested in different wearing mechanisms, for example Rabinowicz (1977) was interested in fretting, Hack and Feller (1970) in sliding and Scott and Mills (1970, 1973) in rolling-contact fatigue, could indicate a common factor in wear, namely that all wear originates in a fatigue process as asperities interact with each other, time and time again (in the case of rotating machinery). About all we can say if we find spherical particles is that they appear to be associated with wear of some kind. However, the severity of that wear may differ according to the tribo-system being considered. To withdraw an aircraft engine from normal operation because of the presence of a few spherical particles in a 'Ferrogram' might be considered an uneconomic way of monitoring the condition of that engine!

2.7 Concluding remarks regarding the importance of macroscopic physical techniques in tribology

In this chapter, we have been mainly interested in the topography of tribo-elements and how the *macroscopic* techniques of optical microscopy, optical interferometry, transmission electron microscopy of replicas of the surfaces of tribo-elements, profilometry using a contacting stylus, and scanning electron microscopy can all contribute in different and complementary ways to our knowledge of that topography. We have also seen how electrical and magnetic physical techniques can contribute to our knowledge of the *macroscopic* aspects of lubricated tribo-elements, such as the oil film thickness, the onset of pitting in contact fatigue situations and the study of the morphology of wear debris fragments. The actual application of these techniques to particular tribological problems will be discussed in later chapters, after we have described some of the *microscopic* physical techniques for studying atomic arrangement (Chapter 3) and for studying atomic structure (Chapter 4).

3 Microscopic physical techniques for studying atomic arrangement

3.1 Basic crystallography

3.1.1 Introduction

Most solid substances are polycrystalline, that is, they consist of many small regions of near-perfect simple crystals (crystallites) which are generally oriented at random with respect to each other. Sometimes, however, these crystallites have preferred orientations with respect to each other. The facility for taking up preferred orientations can be a distinct advantage in metal-forming operations. It is also partly responsible for the successful operation of carbon brushes (as we shall see in Chapter 5). Separating the crystallites there are regions of very poor crystallinity, that is the atoms in these regions are not arranged in the regular arrays expected of a perfect single crystal. These regions are called 'grain boundaries' and they have a strong influence upon the mechanical strength of a polycrystalline solid. It is generally assumed that polycrystalline solids are isotropic as regards mechanical strength, electrical and thermal conductivity, and other physical properties relevant to the friction and wear of such solids. We shall see, however, that this assumption is not valid for materials which take up a preferred orientation upon interacting with another surface during motion. It is also not true in respect of materials that change the chemical composition of their interacting surfaces during contact, especially those materials that exhibit mild (oxidational) wear.

In order to study the changes in crystal structures and crystallite orientations caused by tribo-elements interacting with each other during sliding, rolling or reciprocating, we need to be aware of how crystals are described; that is we need to know a minimal amount of 'Crystallography' (as the study of crystals is called). Hence the next few sub-sections.

3.1.2 The classification of crystals

A classification of crystals has evolved over the past two hundred years or so in which any given single crystal (with its own special arrangements of crystal facets) is classified on the basis of the *macroscopic* symmetry it possesses. These symmetry elements are readily revealed by rotating the single crystal about various prominent and obvious axes, and observing the angles at which reflections of light occur from a fixed source as the single crystal is rotated. The instrument for carrying out this search for the main crystallographic axes of a given single crystal is known as a 'Goniometer'. Typically, however, this instrument would not be used in studying the crystallography of tribo-elements, even if the tribo-elements themselves were single crystals. In fact, the tribologist will use the *microscopic* techniques of X-ray diffraction and electron diffraction. These techniques involve the scattering of X-rays and electrons in characteristic and symmetrical ways by the regular arrays of atoms in a crystal. However, because X-rays and, to a much lesser extent, electrons are able to penetrate below the surface of a crystal or crystallite, it turns out that X-ray and electron diffraction patterns can be used to deduce symmetry properties in addition to those revealed by the optical goniometer. These additional symmetry properties are called 'microscopic symmetry elements'. Let us first consider the *macroscopic* symmetry elements and their representation by means of a stereographic projection.

3.1.3 Macroscopic symmetry elements

There are three types of macroscopic symmetry elements. These are:
(a) rotational symmetry;
(b) inversion with respect to a point; and
(c) reflection with respect to a plane.
If a crystal has *rotational symmetry*, it is possible to find an axis about which rotation through $(360°/n)$ will enable the observer to find each of the n crystal faces parallel to that axis. It turns out that n can only take on the values 2, 3, 4 or 6. If a crystal has a *centre of inversion*, this means that every face of the crystal has its inverse face on the other side of the crystal, obtained by taking each point on the given face and plotting its complete inversion with respect to a point within the crystal (known as the centre of symmetry). If a crystal has a *plane of symmetry*, or a *mirror plane* as it is sometimes called, then every crystal face on one side of the plane has its mirror image on the other.

These three basic symmetry operations can be combined in thirty-two different ways to give the total number of crystal classes. In order to show the three-dimensional properties of each crystal class, it is usual to use a *stereographic projection*. Although we shall not need to know much about the thirty-two crystal classes (or 'point groups' as they are sometimes called, because it is always possible to find at least one point within the crystal which is unaffected by the symmetry operations), it is important for the intending tribologist to understand the stereographic projection in order to adequately represent some of the preferred orientations caused by sliding or rolling of *tribo-elements*. This is now discussed in the next sub-section.

3.1.4 The stereographic projection

In the stereographic projection of a crystal, one imagines a sphere, of radius (R_S) much larger than the crystal, with the crystal at its centre. The reason for R_S being much greater than the crystal is to allow us to assume that the normal to each crystal face (i.e. the pole of that face) can be extended inwards and meet all the other poles at the same point in the centre of the 'reference sphere' (as we call our imaginary sphere). We now imagine we have extended the poles outward until they meet the surface of the reference sphere. In order to represent the intersections of these poles with the reference sphere, we imagine a line joining that intersection with the north pole (if the intersection is in the southern hemisphere) or the south pole (if the intersection is in the northern hemisphere). Where that line passes through the central (diametral) plane (similar to the equatorial plane in geography), we have the stereographic projection of the pole of the crystal face we wish to represent. Figure 3.1 shows how we would represent a single crystal having a two-fold axis of rotation (i.e. rotation of this crystal through 180° about this axis leaves the stereographic projections of the 6 poles unaltered by the rotation).

In this figure we have imagined that the crystal has the shape of a right parallelopiped, that is the poles A and B are at right angles to the plane containing the other poles (C, D, E and F), and the unequal angles θ_1 and θ_2 are such that $\theta_1 + \theta_2 = 180° - \beta$, where β is the acute angle between the sides of the parallelopiped which, in general, is not equal to 90°. We note that, when a pole is in the lower hemisphere, its projection in the diametral plane is conventionally denoted by an open circle (e.g. E' and F'). The closed circles (D' and C') denote the

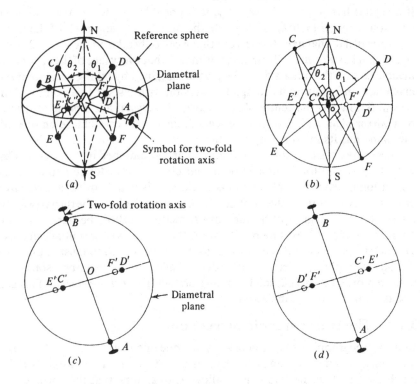

Figure 3.1 *Perspective and stereographic projections of a crystal with a two-fold axis of rotation. (a) Perspective representation; (b) section through C,E,D,F,N and S of the reference sphere; (c) stereogram of crystal shown in (a); (d) stereogram of crystal after having been rotated through 180° about axis AB.*

projection of poles in the upper hemisphere. When a pole is in the diametral plane, for example *A* or *B*, then its projection coincides with its pole. For all poles *not* in the diametral plane, however, common usage is such that we often refer to the projection in the diametral plane as if it were the actual pole. This can lead to some confusion for newcomers to the subject.

In order to project poles making a given angle (θ_1) with respect to the north pole of the reference sphere, for example, the pole *D* of Figure 3.1(*a*), we can use the relation:

$$OD' = R_S \tan(\theta_1/2) \tag{3.1}$$

It would be, however, a tedious procedure to use Equation (3.1) for plotting every pole of a crystal. There are aids to plotting stereograms, the best known being the Wulff net in which the radius of the sphere is taken to be 63.5 mm (2.5 in). We will see (in Chapter 5) how important it is to know how to use these nets when one wishes to draw *pole figures* indicating preferred orientations of crystallites due to sliding and/or rolling.

3.1.5 The unit cell and its relation to the description of crystal planes and crystallographic directions

The concept of the unit cell was first suggested by Huyghens in 1690 in his *Treatise on Optics*, when he was attempting to account for the anomalous optical behaviour of the calcite crystal. This crystal, when cleaved, breaks up into essentially rhombic shapes. In 1784, Abbe Hauy, an honorary canon of Notre Dame, suggested that the calcite crystal must be made of tiny rhombs; he showed that any face of the crystal could be obtained by stacking these rhombs in regular arrays (ignoring the small scale irregularities caused by the corners of the rhombs). This hypothesis was then extended by supposing that every crystal has its own characteristic 'building block', for example cubes, rectangular prisms, square prisms, etc. During the nineteenth century, the idea of basic blocks was replaced by the assumption that the external shapes (i.e. the morphologies) of crystals were due to the various ways in which atoms or molecules could be packed together in regular arrays of *lattices*. Thus the 'building block' could be replaced by a single point on a lattice – the shape of the lattice thereby becoming more important than the shape of the originally proposed 'building block'.

The basic repeat unit of the lattice is, of course, the unit cell. There are many choices open for this basic repeat unit. By way of illustration, consider the *two-dimensional* lattice shown in Figure 3.2. Obviously, one could take *A*, *B* or *C* as the basic repeat unit and still obtain the full lattice. Each unit has the same area. However, unit cell (*A*) has the shortest combined length of the four sides. Although it is quite valid to take *B* or *C* as the repeat unit, it is *conventional* to take *A*. This is a generalization which has many exceptions, for instance, *any* three-dimensional array of lattice point can be built up from a triclinic unit cell, such as that shown in Figure 3.3, provided one imposes certain conditions on the magnitudes of the length of the sides a_{uc}, b_{uc} and c_{uc} and the magnitudes of the angles between the sides, namely α_{uc}, β_{uc} and γ_{uc}. This unit, however, is only taken when the crystal either has no symmetry, or just a centre of inversion. Where one

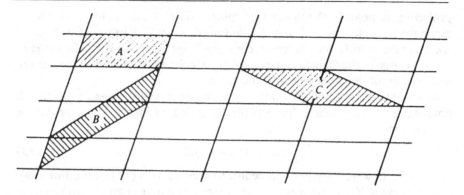

Figure 3.2 *The choice of unit cell for a two-dimensional lattice*

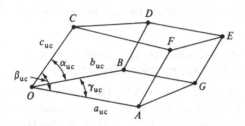

Figure 3.3 *A triclinic unit cell*

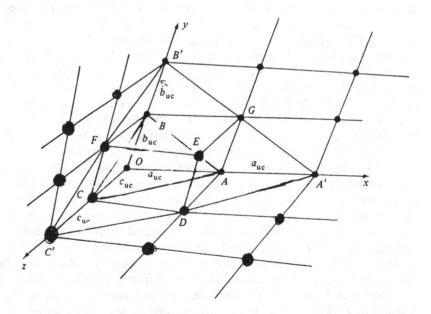

Figure 3.4 *A small part of a lattice showing the unit cell and a typical set of planes ABC and A'B'C' within that lattice*

can detect other symmetries, it is often more convenient to take a unit cell which readily reveals these symmetries.

Any three-dimensional array of point can be split into sets of parallel planes of lattice points in various ways, besides the obvious ones defined by the edges of the unit cell. For example, in Figure 3.4, we can see that ABC and $A'B'C'$ are two parallel planes in a set which cuts the lattice at equal numbers of *repeat distances* (a_{uc}, b_{uc} and c_{uc}) along the respective X-, Y- and Z-axes. Let us recall the general equation to a plane, namely:

$$Ax + By + Cz = m \qquad (3.2a)$$

This represents a plane which cuts the X-axis at (m/A), the Y-axis at (m/B) and the Z-axis at (m/C). We need to describe such a plane in terms of a_{uc}, b_{uc} and c_{uc}. Now, any plane containing lattice points must meet the crystallographic axes (defined by the edges of the unit cell – see Figure 3.4) at an integral number of repeat distances along these axes, that is

$$
\begin{aligned}
m/A &= pa_{uc} \\
m/B &= qb_{uc} \\
m/C &= rc_{uc}
\end{aligned}
\qquad (3.2b)
$$

where p, q and r are integers. Putting the values of A, B and C from Equation (3.2b) into (3.2a), we get

$$(x/pa_{uc}) + (y/qb_{uc}) + (z/rc_{uc}) = 1 \qquad (3.2c)$$

The early crystallographers used the integers p, q and r to describe crystal planes. However, this led to difficulties when describing planes parallel to crystallographic axes, since these intercept at infinity. It is now universal convention to define a set of planes by the *reciprocal* of their intercepts on the crystallographic axes, that is we set $h = (1/p)$, $k = (1/q)$ and $l = (1/r)$, so that Equation (3.2c) becomes:

$$(h/a_{uc})x + (k/b_{uc})y + (l/c_{uc})z = 1 \qquad (3.2d)$$

h, k and l are called the 'Miller indices' of the set of parallel planes. It is also conventional to describe the family of planes in terms of the intercepts made on the crystallographic axes by the plane nearest the origin.

Let us now consider a few examples of planes and their associated Miller indices. Referring to Figure 3.4 again, we can see that, of the family of planes parallel to $A'B'C'$, the plane ABC is the one nearest to the origin. For this plane, we have $p = 1$, $q = 1$ and $r = 1$. Hence h, k and l, being the reciprocals of p, q and r respectively, are 1, 1 and 1. We say that *any* plane parallel to ABC is the (111) plane. Consider now the planes illustrated in Figure 3.5. Take $A_1B_1C_1$, the plane nearest the origin, which cuts the X-axis at $2a_{uc}$, the Y-axis at $3b_{uc}$ and the Z-axis at c_{uc}. Hence $p = 2$, $q = 3$ and $r = 1$, so that $h = \frac{1}{2}$, $k = \frac{1}{3}$ and $l = 1$. It is conventional to clear all fractions from the Miller indices of the plane. Multiplying by 6, we get $h = 3$, $k = 2$ and $l = 6$, so that instead of the plane being described by:

$$(x/2a_{uc}) + (y/3b_{uc}) + (z/c_{uc}) = 1 \qquad (3.3a)$$

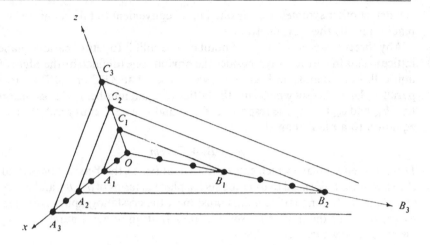

Figure 3.5 *Diagrammatic representation of the set of planes having the Miller indices of (326)*

it is now given by:

$$(3x/a_{uc}) + (2y/b_{uc}) + (6z/c_{uc}) = 6 \tag{3.3b}$$

In fact, the plane described by Equation (3.3b) does not pass through any lattice points *within the unit cell*. This does not matter, since the Miller indices are used to describe a set of parallel planes, some of which we know will go through lattice points outside of the unit cell. We describe the plane as (326).

Figure 3.6 shows the unit cell of a lattice in which $a_{uc} = b_{uc} = c_{uc}$ and the crystallographic axes are orthogonal, that is it shows the unit cell of a cubic lattice. The plane $GCDF$ cuts the Z-axis at a; hence $r = 1$, so that $l = (1/r) = 1$. However, it is parallel to the X- and Y-axes so that $p = q = \infty$, which gives us $h = (1/p) = 0 = (1/q) = k$. Hence the plane $GCDF$, *and all planes parallel to it*, are described as (001) planes. It is fairly obvious, now, that the plane $ABCD$ is the (100) and the plane $FDBE$ is the (010). Figure 3.7 illustrates the set of planes described by the Miller indices (110) and (1$\bar{1}$0) for a cubic lattice. The plane $HJBD$ cuts the X- and Y-axes at $(+a_{uc})$ and the Z axis at infinity. Hence $h = k = 1$ and $l = 0$, so that this is the (110) plane. The set of planes parallel to $EFGO$ are described as (1$\bar{1}$0), as can readily be seen by considering the plane $ABCD$. This is parallel to $EFGO$ and passes through the X-axis at $(+a_{uc})$, the Y-axis at $(-a_{uc})$ and the Z axis at infinity. Hence $h = 1$, $k = -1$ and $l = 0$.

Finally, let us define what we mean by a *crystallographic direction*. The direction $[pqr]$ – note the square brackets – is the direction of the radius vector from the origin to the point on the lattice which has the coordinate pa_{uc}, qb_{uc} and rc_{uc}, where a_{uc} is the vector of length a whose direction coincides with that of the X-axis, and b_{uc} and c_{uc} are the vectors of length b and c in the direction of the Y- and Z-axis respectively. Note that the definition does *not* require the axes of

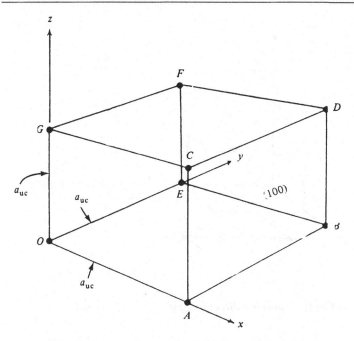

Figure 3.6 *A cubic unit cell, showing (100) (010) and (001) planes*

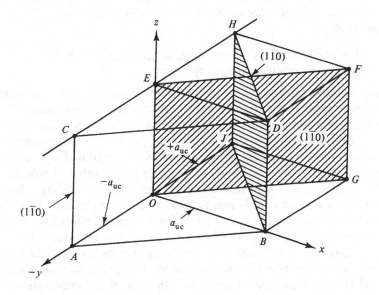

Figure 3.7 *A cubic lattice showing (110) and (1̄10)*

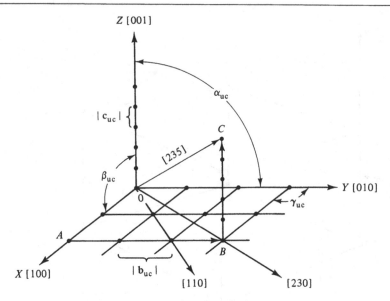

Figure 3.8 *Crystallographic directions for a triclinic lattice*

reference to be orthogonal. This is illustrated in Figure 3.8 for the directions [100], [010], [110] and [235] for the most general lattice, namely the triclinic lattice, whose unit cell sides are all different and whose angles are all different and not equal to 90°.

3.1.6 The classification of crystals according to their minimum basic symmetry

The early crystallographers, whose only instrument was the optical goniometer, brought some order to the classification of crystals, which grow many different faces in many different ways according to the conditions of crystallization. They discovered that all crystals could be classified into *seven crystal systems* according to the basic symmetry possessed by each individual crystal. These are listed, together with their appropriate basic symmetry and unit cell, in Table 3.1. It is possibly worth mentioning, in view of the 'materials' bias of tribology, that very few materials of interest to tribologists crystallize into the triclinic form. Some polymers take up the monoclinic or orthorhombic form. Most metals, are cubic, hexagonal or tetragonal. The rhombohedral unit cell is favoured by the hard oxides, for instance α-Fe_2O_3 and Al_2O_3 are both rhombohedral in their crystal structure. Most of the layer-like materials often used as solid lubricants have the hexagonal structure, for instance graphite. In view of the importance of the graphite crystal structure in so far as it facilitates low friction across planes of 'easy glide', we will briefly discuss the hexagonal classification in a little more detail.

Figure 3.9 illustrates the special features of the hexagonal unit cell. If we repeat the unit cell *OABCPQRS* in all directions, we will obtain hexagonal arrays of

Table 3.1. *The seven crystal systems*

System	Basic symmetry	Unit cell
Triclinic	None	$a_{uc} \neq b_{uc} \neq c_{uc}$ $\alpha_{uc} \neq \beta_{uc} \neq \gamma_{uc} \neq 90°$
Monoclinic	One two-fold rotation axis	$a_{uc} \neq b_{uc} \neq c_{uc}$ $\alpha_{uc} \neq \gamma_{uc} = 90° \neq \beta_{uc}$
Ortho-rhombic	Three two-fold rotation axes (at right angles to each other)	$a_{uc} \neq b_{uc} \neq c_{uc}$ $\alpha_{uc} = \beta_{uc} = \gamma_{uc} = 90°$
Rhombohedral	One three-fold rotation axis	$a_{uc} = b_{uc} = c_{uc}$ $\alpha_{uc} = \beta_{uc} = \gamma_{uc} \neq 90°$
Tetragonal	One four-fold rotation axis	$a_{uc} = b_{uc} \neq c_{uc}$ $\alpha_{uc} = \beta_{uc} = \gamma_{uc} = 90°$
Hexagonal	One six-fold rotation axis	$a_{uc} = b_{uc} \neq c_{uc}$ $\alpha_{uc} = \beta_{uc} = 90°$; $\gamma_{uc} = 120°$
Cubic	Four three-fold rotation axes (plus three *implied* two-fold axes)	$a_{uc} = b_{uc} = c_{uc}$ $\alpha_{uc} = \beta_{uc} = \gamma_{uc} = 90°$

points lying in regular positions, one above the other, as shown in plan by Figure 3.9(b). Any crystal having its atoms or molecules placed in special positions relative to this truly hexagonal lattice will possess as external form which has a six-fold (hexagonal) rotation axis. The direction of this axis is conventionally taken to be that of the c_{uc}-axis. In some cases, it is found that the crystal also has two sets of two-fold axes, namely one set along X, Y and N, and the other along X', Y' and N' (see Figure 3.9(b)). Hence we cannot fix the axes $(a_{uc})_1$, $(a_{uc})_2$ and $(a_{uc})_3$ unambiguously by symmetry alone. Whilst this ambiguity is only of academic interest to dedicated crystallographers, tribologists should be aware of the third possible axis in the hexagonal (basal) planes. If one refers to any planes, made equivalent by the presence of the six-fold rotation axis along the c_{uc}-axis, to the axes defined by $(a_{uc})_1$ and $(a_{uc})_2$ only, one sees no obvious relationship between the Miller indices of these planes. However if one introduces Miller indices containing *four* symbols, that is h, k, i and l, where i is the reciprocal of the intercept of the plane along the a_3-axis, one immediately sees the relationship. For example, see Figure 3.9(c), where the planes parallel to the sides of the hexagonal unit cell are described by the Miller indices (100), (010) and ($\bar{1}$10), whereas their four-symbol form (called 'Miller–Bravais' indices) is (10$\bar{1}$0), (01$\bar{1}$0) and ($\bar{1}$100) respectively. It can readily be shown that the extra symbol (i) is related to the normal symbols h and k by the relation:

$$h + k + i = 0 \qquad (3.4)$$

Many of the metals and alloys of interest to the practising tribologist consist of

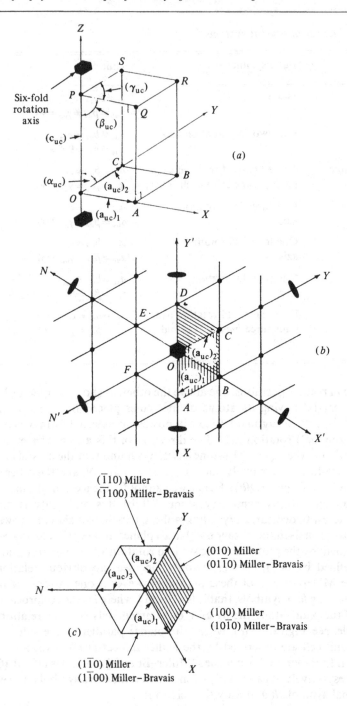

Figure 3.9 *The hexagonal unit cell. (a) Perspective view; (b) plan view of hexagonal lattice; and (c) the Miller–Bravais system*

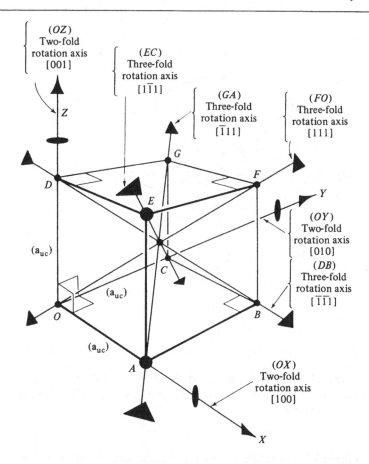

Figure 3.10 *The cubic unit cell (perspective view)*

polycrystalline materials in which the individual grains have the *cubic* crystal structure. For this reason, we should be aware of how we can represent the various possible crystal faces (or more precisely, their poles) on a stereogram, together with the various axes of symmetry associated with the cubic system. For the sake of illustration, Figure 3.10 is a perspective view of the cubic unit cell, showing its four three-fold axes along the body diagonals *OF, AG, BD* and *CE* and the three (implied) two-fold axes parallel to the cube edges. The non-crystallographer might suggest that these are actually four-fold axes and, for some crystals in the cubic system, there will be faces having four-fold axes parallel to the cube axes. In general, however, it is found that if a particular crystal has faces consistent with four three-fold axes along the body diagonals of a cube, then that crystal must also have faces consistent with three two-fold rotation axes at the appropriate angles with respect to the four three-fold axes. In essence, this means that the unit cell of a crystal in the cubic system has higher symmetry than is required for the system.

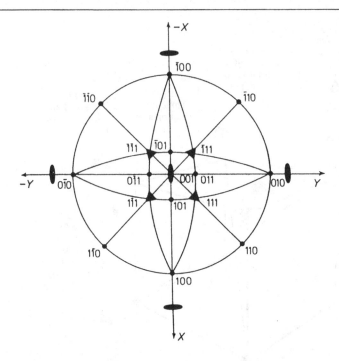

Figure 3.11 *Stereogram (standard projection on [001]) of a cubic crystal possessing only the basic cubic symmetries. Indices refer to pole in upper hemisphere only*

Clearly, Figure 3.10 is a rather complex and unsatisfactory way of representing the symmetry properties of crystalline materials with higher symmetries. This is where the stereographic projection (see Sub-section 3.1.4) becomes very important. For instance, Figure 3.11 is the stereogram of a crystal possessing only the basic symmetries of four two-fold and three two-fold rotation axes. Let us consider how we obtain this figure. First of all, we realize that, if we make the [001] direction coincide with the North pole of the reference sphere, then the (001) pole must project to a point in the centre of the diametral plane (i.e. the centre of the stereogram). Clearly, the poles of (100), (010), ($\bar{1}$00) and (0$\bar{1}$0), will be at right angles to the [001] direction and will intersect with the reference sphere at the circumference of the stereogram (as shown in Figure 3.11). These poles project as themselves. As well as the cube edge, it is possible that the crystal may have faces parallel to [001], such as the (110), ($\bar{1}$10), ($\bar{1}\bar{1}$0) and (1$\bar{1}$0). We see that these poles can also be drawn as their own projection on the circumference of the stereogram lying at 45° to the other poles already drawn. We say that the {$hk0$} planes lie in the [001] zone, that is all the planes such as ($hk0$), having various values of h and k, are parallel to the [001] zone axis. Note that the pole of ($hk0$) planes are at 90° to the projection of the [001] zone axis (as measured from Equation (3.1) with $\theta_1 = 90°$, so that $OD' = R_S \tan 45° = R_S$).

The pole of the (011) plane lies at 45° to the (001) and (010) poles. From Equation (3.1), this means we set $\theta_1 = 45°$, so that $OD' = R_S \tan(22.5°) = 0.414 R_S$. Obviously, we can more easily plot the (011) pole by using the Wulff net to find the position of 45 'stereographic degrees' between the poles (001) and (010). Clearly, the pole $(0\bar{1}1)$ will be at right angles to (011), its pole being 45 'stereographic degrees' on the other side of the line representing the Y-axis in Figure 3.11. This pole will be the zone axis for all planes with poles lying at 90° to the pole of the $(0\bar{1}1)$ plane, namely the $(\bar{1}00)$, $(\bar{1}11)$, (011), (111) and (100) poles. Note these all lie on the projection of the great circle between the $(\bar{1}00)$ and (100) poles lying at 45° to the plane of the stereogram. Similarly, the $(\bar{1}00)$, $(\bar{1}\bar{1}1)$, $(0\bar{1}1)$, $(1\bar{1}1)$ and (100) poles are plotted on the other great circle through (100) and $(\bar{1}00)$, having a zone axis of [011]. The reader is left to see for himself/herself how the other great circles in Figure 3.11 are projected onto the stereogram. Note that when a great circle passes through the North pole, it projects as a straight line, for example the [110] zone axis relates to the planes whose poles $(1\bar{1}0)$, $(1\bar{1}1)$, (001), $(\bar{1}11)$ and $(\bar{1}10)$ all lie on a diameter at right angles to that zone axis.

It will be apparent that, when a pole lies in the same zone and between adjacent poles, then the Miller indices of the middle pole is given by merely adding the Miller indices of the two outside poles. For instance, the (110) pole lies in the [001] zone between the poles (010) and (100). This property is often used in constructing stereograms and is well worth committing to memory. Another useful aid is the Weiss zone law. This states that, if a plane (hkl) is in the $[pqr]$ zone, then:

$$hp + kq + lr = 0 \qquad (3.5)$$

From this equation, we can readily check that the planes $(\bar{1}00)$, $(\bar{1}\bar{1}1)$, $(0\bar{1}1)$, $(1\bar{1}1)$ and (100) all lie in the [011] zone. Once the principles behind the construction of the stereograms are fully understood, the stereogram is the best way to represent the three-dimensional properties of crystals in two dimensions. We shall see that, when we wish to describe preferred orientation effects in polycrystalline materials, the stereogram (or more correctly, the pole figure, which is strongly related to the stereogram of the single crystal) is always used.

3.1.7 The microscopic symmetry elements, Bravais lattices and space groups

The previous sub-sections have been concerned with the macroscopic symmetries of crystals, except where we introduced the idea of the unit cell. Until the discovery of X-ray diffraction by crystals at the turn of the century, crystallographers had been unable to *definitely prove* that crystals consist of atoms oscillating about regular positions on a lattice. Nevertheless, there had been a certain amount of theoretical work done on the possible lattices that real crystals *should* possess. For example Bravais, in 1848, showed that there are only fourteen possible ways in which a lattice of points can be arranged so that the surroundings of each point are the same as any other in that lattice.

If a lattice has only one lattice point per unit cell, we say it is a 'primitive' lattice. It is, in fact, possible to describe the fourteen Bravais lattices entirely in terms of primitive lattices. However, it is much more convenient to describe them in terms of both 'primitive' and 'non-primitive' lattices, where the latter refers to lattices with more than one lattice point per unit cell. This is done by considering the unit cells of the seven crystal systems with a lattice point in the centre of each unit cell (an I-type lattice), with lattice points in the centres of each face (an F-type lattice), or with lattice points in the (100) faces (A-type), the (010) faces (B-type), or the (001) faces (C-type). The resulting fourteen Bravais lattices are shown in Figure 3.12.

Probably the most interesting Bravais lattices, from the point of view of the structures relevant to many industrial materials, are the body-centred cubic (bcc) and the face-centred cubic (fcc). Many metals and alloys have their atoms actually situated on the lattice points, for example iron has a structure in which there is an iron atom at each point of a bcc lattice. Sometimes, a structure will have more than one atom associated with each Bravais lattice point. For instance, zinc has an hexagonal crystal structure in which there are *two* atoms per lattice point, namely one at the lattice point and the other at a position having the coordinates $[(a_{uc})_1/2]$, $[2(a_{uc})_2/3]$ and $[(c_{uc})/2]$ with respect to that lattice point. This structure is called 'hexagonal close-packed', because such a structure can be built up by laying hexagonal layers of close-packed spheres (representing atoms) of equal radii on top of each other in such a way that tops of the spheres in one layer fit in the holes between the spheres in the adjacent layer. A further requirement is that each *alternate* layer should be positioned exactly over its companion layers. If we denote layer positions as being either *A* or *B*, then a hexagonal close-packed structure can be indicated by the symbol *ABABAB...*. It should be noted that the hexagonal close-packed *structure* is *not* a Bravais lattice,

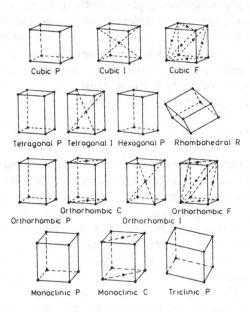

Figure 3.12 *The Bravais lattices*

since it does not satisfy the criterion that each of its *atom* positions should have the same environment.

It will be obvious that the *macroscopic* (optical) techniques cannot distinguish between the various Bravais lattices belonging to a particular crystal system. For instance, they would not be able to distinguish between simple cubic, body-centred cubic and face-centred cubic crystals. Only by using the *microscopic* techniques (such as X-ray and electron diffraction) is it possible to show the effects of different atomic dispositions within a given type of unit cell. Because of this, a more complete classification of crystals is required. The 'space group' concept has produced such a classification, since it involves *both* macroscopic and microscopic symmetry elements in a unique fashion. We shall not be dealing with the 230 different possible space groups into which crystalline materials can be uniquely classified. However, it is relevant to mention one or two of the microscopic symmetry elements which help to distinguish one space group from the next.

Normally, the *microscopic symmetry elements* are associated with some amount of *translation* through a distance equal to the lattice spacing, or some *simple* fraction of a lattice spacing. For instance, one can have a lattice which exhibits *screw axis symmetry*. This means that rotation through 180°, 120°, 90° or 30°, together with a translation along the screw axis, brings the lattice back into a position indistinguishable from its initial position. If a lattice has a (P_q) screw axis along the a-axis, this means that a P-fold rotation about that axis, followed by a translation $[(q/P)(a_{uc})]$ along that axis, leaves the lattice as it was originally. One can also have *glide plane symmetries* in which a lattice can be brought back into coincidence with a position unidentifiably different from its initial position, by a reflection in a mirror plane, together with a translation parallel to that plane through a distance equal to the lattice spacing in that direction (the *a*-, *b*- or *c*-glide planes), or through a combination of fractions of lattice spacings (the diagonal glide planes with translations such as $[(a_{uc})/2 + (b_{uc})/2]$ and the diamond glide planes, with translations such as $[(a_{uc})/4 + (b_{uc})/4]$).

By taking *each* of the fourteen Bravais lattices and operating on each one the various *macroscopic* and *microscopic* symmetry elements, one can deduce the 230 possible space groups. Although the determination of the space group is one of the main objectives in an X-ray diffraction examination of a crystal of an unknown structure, it is only important to the tribologist in so far as it affects the *relative intensities* of the diffraction maxima from a specimen containing two or more crystalline components. For instance we shall see, in Chapter 6, how proportional analysis of wear debris can only be carried out if we know the positions of the atoms in the unit cells of the constituent crystalline materials in the debris. Typically, the space group and atomic positions are known for most materials used in tribo-systems. Hence it is just a matter of some rather complex calculations to deduce what intensities to expect in the X-ray diffraction maxima from such known crystal structures. We will describe how to do these calculations in the next section (Section 3.2).

This concludes our brief review of the elements of crystallography. The main function of this section has been to introduce important concepts, such as the unit cell, Miller indices, the stereographic projection and the classification of crystals according to their macroscopic and microscopic symmetry elements, so that the applications of X-ray diffraction (and electron diffraction) to tribology can be understood with, one hopes, very little reference to other sources of information. Let us now describe those X-ray diffraction techniques of special interest to the tribologist.

3.2 X-ray diffraction techniques

3.2.1 Bragg's law

It is not really necessary, *at this stage*, for the reader to be introduced to the origin of X-rays. This topic will be covered later in Section 4.2 on X-ray spectra. By filtering out most of the unwanted wavelengths from a spectrum, it is possible to obtain a virtually monochromatic X-ray beam. If we make such a beam impinge on a set of crystal planes (hkl) at an angle (θ_{hkl}), then Bragg's Law tells us the relationship between that angle, the interplanar spacing (d_{hkl}) and the wavelength (λ) of the incident X-rays. This law is a remarkable simplification of a very complicated phenomenon, namely the interaction of the electro-magnetic field of the X-ray photon with the electron 'clouds' around atoms, vibrating about their positions of equilibrium (at all temperatures above that of absolute zero) in a three-dimensional lattice. By neglecting the atomic nature of the interaction and dealing with X-ray diffraction *as if* it arose entirely as a result of the X-ray beam being 'reflected' from parallel crystal planes, Bragg was able to predict the angles (ϕ_{hkl}) at which the X-ray beam is scattered from its original direction by the crystal, through the following relation:

$$2(d_{hkl})(\sin \theta_{hkl}) = \lambda \qquad (3.6)$$

The angle (θ_{hkl}) is called the 'Bragg angle' and is equal to $(\phi_{hkl}/2)$. The derivation of this relation can be seen from inspection of Figure 3.13. The path difference between an X-ray 'reflected' by the top plane (2) and an X-ray 'reflected' by the adjacent plane (1), is equal to $(AB + BC)$. Bragg proposed that these two parts of the X-ray wavefront would interfere *constructively* whenever $(AB + BC)$ was equal to an integral number of wavelengths, that is whenever $[2(d_{hkl}) \sin \theta_{hkl}]$ equalled an integral number of wavelengths. It is usual to assume that the second and higher orders of the diffraction maxima (i.e. the maxima occurring when the path difference equals $2\lambda, 3\lambda, 4\lambda, ..., n\lambda$) occur at the same angles as would be expected from interplanar spacings equal to $(d_{hkl}/2), (d_{hkl}/3), (d_{hkl}/4), ..., (d_{hkl}/n)$. It can readily be seen that the Miller indices corresponding to a plane with an interplanar spacing of (d_{hkl}/n) are, in fact, nh, nk and nl. In this way, the n associated with second and higher orders of diffraction can be included in the subscripts to d and θ.

It is important to understand that a plane X-ray wave being scattered by a three-dimensional crystal lattice, will give rise to scattered X-rays *in all directions*. Only in certain directions of scatter (ϕ_{hkl}), however, will the scattered X-ray waves *constructively* reinforce each other. Bragg's law enables us to predict these special angles, provided we realize that the constructively-interfering X-ray

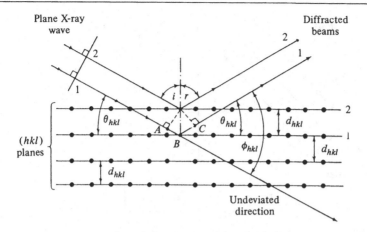

Figure 3.13 *Derivation of Bragg's law*

waves are scattered by the planes of atoms *as if* they were a series of semi-transparent mirrors 'reflecting' a parallel beam of light. This means, of course, that the incident beam, the diffracted beam and the normal to the diffracting planes must all lie in the same plane. Also, the angle of incidence i equals the angle of reflection r, where these angles are each equal to $(90 - \theta_{hkl})$ – see Figure 3.13.

3.2.2 The relationship between the crystal lattice and its diffraction pattern

We shall see, in later sections of this book, that one can explain the geometry of diffraction patterns (whether they be X-ray or electron diffraction patterns) much more readily in terms of the 'reciprocal lattice(s)' of the crystal(s) than by reference to the actual crystal lattice(s). It must be emphasized right from the start that the reciprocal lattice has no physical reality. It is merely a *convenient aid* for interpreting diffraction patterns.

Essentially, the reciprocal lattice of a crystal is a lattice of points, each point representing an infinite number of parallel atom planes within the actual crystal. Each reciprocal lattice point lies on a line through an arbitrarily chosen origin of reciprocal space, perpendicular to the corresponding planes of the real crystal, and at a distance from that origin which is proportional to the reciprocal of the spacing between the planes, as shown in Figure 3.14. From this figure, it can be seen that $(\mathbf{d}_{hkl}^{*})$, the reciprocal lattice vector of a set of parallel planes in real space, is parallel to $(\mathbf{d}_{hkl})$, the atomic interplanar spacing vector, and is given in magnitude by the relation:

$$|\mathbf{d}_{hkl}^{*}| = \frac{(K_{xe})}{|\mathbf{d}_{hkl}|} \tag{3.7}$$

where (K_{xe}) is arbitrarily taken to be λ (or 10λ) for interpreting X-ray diffraction patterns and unity for interpreting electron diffraction patterns (see Section 3.3), where λ is the wavelength of the X-radiation used to obtain the patterns.

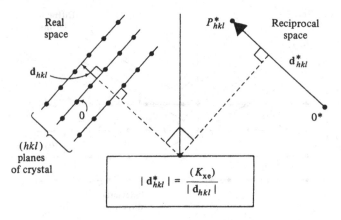

Figure 3.14 *Definition of the reciprocal lattice vector* $(\mathbf{d}_{hkl}{}^{*})$

In order to demonstrate how to construct a reciprocal lattice (for the purposes of interpreting X-ray diffraction patterns), let us consider a crystal, with an orthorhombic unit cell, in which $|\mathbf{a}_{uc}| = 1\,\text{Å}$, $|\mathbf{b}_{uc}| = 2\,\text{Å}$ and $|\mathbf{c}_{uc}| = 3\,\text{Å}$ (the ångström unit (Å) is in common usage in X-ray crystallography and is equal to $0.1\,\text{nm}$, that is $10^{-10}\,\text{m}$). Imagine we are looking along the c-axis of this crystal. The lattice would look like Figure 3.15(a), with all (hk0) planes looking, end-on, like straight lines as shown. The interplanar spacing vectors $\mathbf{d}_{100}$ and $\mathbf{d}_{010}$ lie along the X- and Y-axes respectively. Hence d_{100}^{*} and $\mathbf{d}_{010}^{*}$ also lie along directions parallel to the X- and Y-axes. As shown by the construction notes in Fig. 3.15, these reciprocal lattice vectors will have magnitudes equal to $1.54\,\text{Å}/|\mathbf{d}_{100}^{*}|$ and $1.54\,\text{Å}/|\mathbf{d}_{010}^{*}|$, that is 1.54 and 0.77 units in reciprocal space. We have assumed that (K_{xe}) has the value of the wavelength of copper X-radiation, namely $\lambda = 1.54\,\text{Å}$. We may also note that $|\mathbf{d}_{110}| = 2/(5)^{1/2} = 0.89\,\text{Å}$ from Figure 3.15(a) and that $\mathbf{d}_{100}$ makes an angle ε of $\cos^{-1}(0.89) = 26.6°$ with respect to the x-axis. If we plot, in reciprocal space, a point at a distance equal to $(1.54\,6\text{Å}/0.89\,\text{Å})$, that is at a distance of 1.72 units and at an angle of 26.6° to the X-axis, we will find that this reciprocal lattice point (which represents an infinite number of (110) planes in real space) lies exactly at the point we would expect, if the following relation were valid, namely:

$$\mathbf{d}_{hk0}^{*} = h\mathbf{a}_{uc}^{*} - k\mathbf{b}_{uc}^{*} \tag{3.8a}$$

where

$$|\mathbf{a}_{uc}^{*}| = (K_{xe})/|\mathbf{d}_{100}| \tag{3.8b}$$

and

$$|\mathbf{b}_{uc}^{*}| = (K_{xe})/|\mathbf{d}_{010}| \tag{3.8c}$$

Thus we find that all the atomic planes in the [001] zone of this orthorhombic crystal can be represented by a rectangular net of reciprocal lattice points.

If now we were to imagine we were looking along the X-axis of the real crystal, we

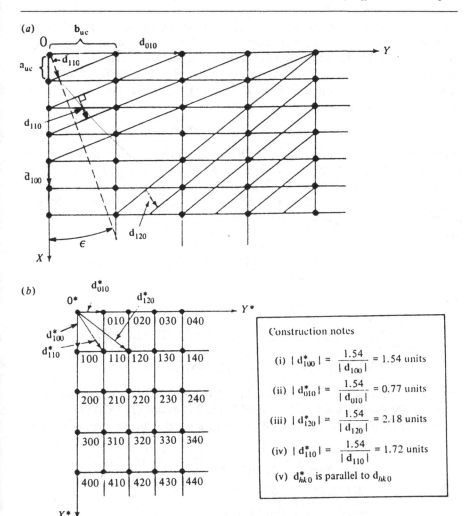

Figure 3.15 *The construction of the [001] zone of the reciprocal lattice of an orthorhombic crystal (see text). (a) Real space (hk0) planes only; (b) reciprocal space – zero layer only*

would also find that we would obtain a rectangular net in the [100] zone of the crystal such that:

$$\mathbf{a}^*_{0kl} = k\mathbf{b}^*_{uc} + l\mathbf{c}^*_{uc} \tag{3.9a}$$

where

$$|\mathbf{c}^*_{uc}| = (K_{xe})/|\mathbf{d}_{001}| \tag{3.9b}$$

and

$$|\mathbf{b}^*_{uc}| = (K_{xe})/|\mathbf{d}_{010}| \tag{3.9c}$$

101

Clearly, what we get when we apply the definition of a reciprocal lattice point (as illustrated in Figure 3.14) to every possible set of atomic planes in a *real* crystal lattice, is indeed a lattice of reciprocal points, where each point represents a whole infinity of atomic planes. Combining Equations (3.8) and (3.9), we obtain a relation giving the coordinates of *any point* in that reciprocal lattice in terms of the Miller indices of the plane, namely:

$$\mathbf{d}^*_{hkl} = h\mathbf{a}^*_{uc} + k\mathbf{b}^*_{uc} + l\mathbf{c}^*_{uc} \tag{3.10}$$

It should be emphasized that the unit cell in the reciprocal lattice, in general, has its sides ($\mathbf{a}^*_{uc}$, $\mathbf{b}^*_{uc}$ and $\mathbf{c}^*_{uc}$) lying in different directions from $\mathbf{a}_{uc}$, $\mathbf{b}_{uc}$ and $\mathbf{c}_{uc}$. This is because $\mathbf{a}^*_{uc}$ is parallel to $\mathbf{d}_{100}$, $\mathbf{b}^*_{uc}$ is parallel to $\mathbf{d}_{010}$ and $\mathbf{c}^*_{uc}$ is parallel to $\mathbf{d}^*_{001}$.

It is possible to use Equations (3.7) and (3.10) to deduce an expression for the magnitude of the interplanar spacing vector ($\mathbf{d}_{hkl}$) for *any crystal system*, in the following way:

$$\frac{(K_{xe})^2}{|\mathbf{d}_{hkl}|^2} = |h\mathbf{a}^*_{uc} + k\mathbf{b}^*_{uc} + l\mathbf{c}^*_{uc}|^2$$

that is

$$\frac{(K_{xe})^2}{|\mathbf{d}_{hkl}|^2} = h^2|\mathbf{a}^*_{uc}|^2 + k^2|\mathbf{b}^*_{uc}|^2 + l^2|\mathbf{c}^*_{uc}|^2 + 2lh|\mathbf{c}^*_{uc}||\mathbf{a}^*_{uc}|\cos\beta^*_{uc}$$
$$+ 2kl|\mathbf{b}^*_{uc}||\mathbf{c}^*_{uc}|\cos\alpha^*_{uc} + 2kh|\mathbf{a}^*_{uc}||\mathbf{b}^*_{uc}|\cos\gamma^*_{uc}$$

where α^*_{uc}, β^*_{uc} and γ^*_{uc} are the angles ($\mathbf{b}^*_{uc}, \mathbf{c}^*_{uc}$), ($\mathbf{a}^*_{uc}, \mathbf{c}^*_{uc}$) and ($\mathbf{a}^*_{uc}, \mathbf{b}^*_{uc}$) respectively. Inserting, the values for $|\mathbf{a}^*_{uc}|$, $|\mathbf{b}^*_{uc}|$ and $|\mathbf{c}^*_{uc}|$ from Equations (3.8) and (3.9), we get the most general expression for ($1/|\mathbf{d}_{hkl}|^2$), which is valid for *any* crystal system:

$$\frac{1}{|\mathbf{d}_{hkl}|^2} = \frac{h^2}{|\mathbf{d}_{100}|^2} + \frac{k^2}{|\mathbf{d}_{010}|^2} + \frac{l^2}{|\mathbf{d}_{001}|^2} + \frac{2kh\cos\gamma^*_{uc}}{|\mathbf{d}_{100}||\mathbf{d}_{010}|}$$
$$+ \frac{2kl\cos\alpha^*_{uc}}{|\mathbf{d}_{010}||\mathbf{d}_{001}|} + \frac{2lh\cos\beta^*_{uc}}{|\mathbf{d}_{001}||\mathbf{d}_{100}|} \tag{3.11}$$

For the orthorhombic system, we know $\alpha^*_{uc} = \beta^*_{uc} = \gamma^*_{uc} = 90°$, so that Equation (3.11) becomes:

$$\frac{1}{|\mathbf{d}_{hkl}|^2} = \frac{h^2}{|\mathbf{a}_{uc}|^2} + \frac{k^2}{|\mathbf{b}_{uc}|^2} + \frac{l^2}{|\mathbf{c}_{uc}|^2} \tag{3.12}$$

since $|\mathbf{d}_{100}|^2 = a^2$, $|\mathbf{d}_{010}|^2 = b^2$ and $|\mathbf{d}_{001}|^2 = c^2$ and $\cos 90°$ equals zero. Obviously, for tetragonal systems, we have:

$$\frac{1}{|\mathbf{d}_{hkl}|^2} = \frac{h^2 + k^2}{|\mathbf{a}_{uc}|^2} + \frac{l^2}{|\mathbf{c}_{uc}|^2} \tag{3.13}$$

and for cubic crystals we have:

$$\frac{1}{|\mathbf{d}_{hkl}|^2} = \frac{h^2 + k^2 + l^2}{|\mathbf{a}_{uc}|^2} \tag{3.14}$$

We will be using Equations (3.13) and (3.14) quite often in our interpretation of diffraction patterns. Another common system is the hexagonal system, in which we have $\alpha_{uc} = \beta_{uc} = 90°$, $\gamma_{uc} = 120°$ and $|(\mathbf{a}_{uc})_1| = |(\mathbf{a}_{uc})_2| \neq |\mathbf{c}_{uc}|$. Figure 3.16 shows the relationship between the real and reciprocal lattices. It should be noted that the unit cell of the reciprocal lattice is also hexagonal except that now the angle between the X^*- and Y^*-axes is 60°, as compared with 120° between the X- and Y-axes. Hence $\gamma_{uc}^* = 60°$. Obviously $\alpha_{uc}^* = \beta_{uc}^* = 90°$. Also $|\mathbf{d}_{100}^*| = |\mathbf{d}_{010}^*| = (3)^{1/2}a_{uc}/2$. Hence, for hexagonal crystals we have (from Equation (3.11)):

$$\frac{1}{|\mathbf{d}_{hkl}|^2} = \frac{4}{3}\left(\frac{h^2 + hk + k^2}{|\mathbf{a}_{uc}|^2}\right) + \frac{l^2}{(\mathbf{c}_{uc})^2} \tag{3.15}$$

So far, we have discussed the reciprocal lattice in a rather abstract manner. However, it was invented (by Bernal in 1923) for a very practial reason, namely the interpretation of complex diffraction patterns. The relationship between the reciprocal lattice of a crystal and the diffraction pattern produced when X-radiation (or an electron beam) is passed through the crystal is given by the Ewald Sphere Constructions. We will now give the essential details of this construction.

Draw a sphere of radius $(R_{ES}) = (K_{xe}/\lambda)$, where $(K_{xe}) = \lambda$ (or 10λ) for X-ray diffraction, or $(K_{xe}) = 1$ for electron diffraction (see Section 3.3). Draw the incident beam of radiation QO, where O is the centre of the crystal (see Figure 3.17). Extend QO so that it cuts the sphere at the opposite end of the diameter from Q. If we make that point of intersection the origin (O^*) of the reciprocal lattice of the crystal at O, then it can be shown that whenever a reciprocal lattice point (P_{hkl}^*) cuts the Ewald sphere surface, one has the correct conditions for diffraction of the incident radiation by the planes represented by P_{khl}, the angle of diffraction (ϕ_{hkl}) being the angle between (OP_{hkl}^*) and the line (OO^*). To prove that the construction is correct, all we need to show is that α (the glancing angle made by the incident radiation with the (hkl) planes of the crystal) is indeed the Bragg angle (θ_{hkl}) whenever P_{hkl}^* lies on the surface of the Ewald sphere. If one draws the line QP_{hkl}^*, one obtains a right-angled triangle with QO^* ($= 2R_{ES}$) as the hypotenuse. This arises from basic geometry, namely the angle subtended by a diameter of a circle at the circumference is a right angle. From this, it follows that $O^*P_{hkl}^* = 2(R_{ES})(\sin \alpha)$. But from our definition of a reciprocal lattice vector, we know that $O^*P_{hkl}^* = (K_{xe})/|\mathbf{d}_{hkl}^*|$ (see Equation 3.7). Thus

$$\sin \alpha = \frac{(K_{xe})}{2(R_{ES})|\mathbf{d}_{hkl}|} \tag{3.16}$$

From our definition of the Ewald sphere radius, we know that $(K_{xe})/(R_{ES})$ must be equal to λ, regardless of whatever arbitrary value we take for (K_{xe}). Hence $\sin \alpha = \lambda/(2|\mathbf{d}_{hkl}|)$, which is the Bragg relation (see Equation 3.6), so that $\alpha = \theta_{hkl}$.

The arbitrary choice of scale factor for our reciprocal lattice sometimes causes some confusion. Historically, the reciprocal lattice was invented before electron diffraction was discovered in 1927, so that many text books draw their reciprocal

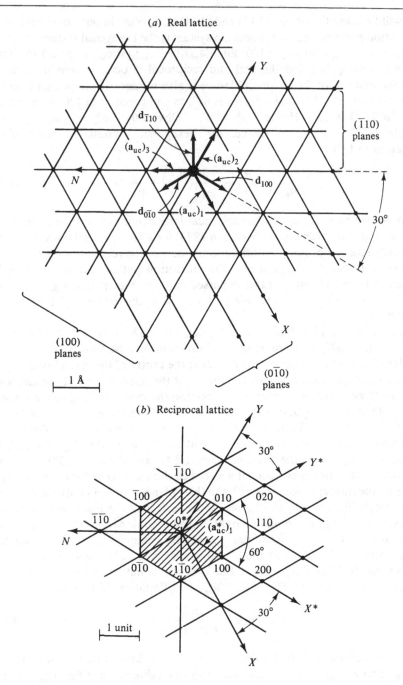

Figure 3.16 *The construction of the* [001] *zone of the reciprocal lattice of an hexagonal crystal;* $(K_{xe}) = 1.54\ Å$. *(a)* $a_1 = a_2 = 1\ Å$; $d_{01\bar{1}0} = 0.87\ Å$; $\mathbf{d}_{11\bar{2}0} = 0.5\ Å$. *(b)* $|\mathbf{d}^*_{01\bar{1}1}| = 1.54/|\mathbf{d}_{01\bar{1}0}| = 1.54/0.87 = 1.78$ *units.* $|\mathbf{d}^*_{01\bar{1}0}| = |\mathbf{d}^*_{10\bar{1}0}| = \mathbf{d}^*_{\bar{1}\bar{1}00}|$. $\mathbf{d}^*_{11\bar{2}0}|1.54/|\mathbf{d}_{11\bar{2}0}| = 1.54/0.5 = 3.08$ *units*

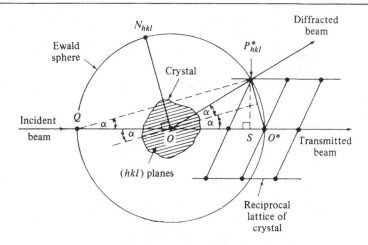

Figure 3.17 *Relation between the reciprocal lattice and the direction of the diffracted beam*

lattice to a scale in which $(K_{xe}) = \lambda$. For ease of drawing this scale factor is sometimes multiplied by ten. The units of reciprocal space are, of course, dimensionless for this scale factor. When electron diffraction was discovered, it was not easy to measure the wavelength of the electron beam – hence the use of a unity scale factor for (K_{xe}). This does, of course, mean that the units of conventional reciprocal space (as far as electron diffraction is concerned) are inverse ångström units, that is (Å^{-1}). Let us now describe some of the X-ray diffraction techniques, bearing in mind the needs of the tribologist.

3.2.3 The flat-film, X-ray diffraction camera

This technique is very useful in tribological applications, since it places very few restrictions on the size of the specimen. Essentially, a monochromatic (i.e. single wavelength) X-ray beam is passed through a crystal (or a whole host of crystals or crystallites, if the specimen is polycrystalline) and a diffraction pattern obtained on a flat photographic film placed at a suitable distance behind the specimen (as shown in Figure 3.18(a)). Alternatively, if the specimen is thicker than about 0.3 mm, one can pass the monochromatic X-ray beam through a hole in the photographic film, and obtain a diffraction pattern (on the side of the film nearest to the specimen) arising from the X-rays that are diffracted *back* from the specimen surface (see Figure 3.18(b)). Another method of obtaining crystallographic information about the surface of a bulk specimen, is to make the X-ray beam impinge on that surface at a fairly small angle (α), so that the diffracted X-ray beams are recorded on a flat photographic film placed in front of the specimen. This is the glancing-angle technique and is illustrated in Figure 3.18(c).

How does one relate the diffraction pattern obtained on the flat photographic film to the structure of the crystal (or crystals) being irradiated with the X-ray beam? Normally, in tribological applications, we have to examine polycrystalline

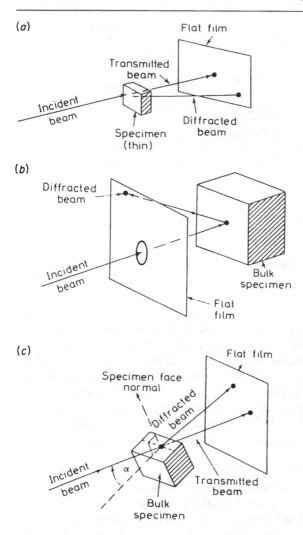

(a)

(b)

(c)

Figure 3.18 *Typical specimen/film geometries for the flat-film X-ray diffraction method: (a) Forward transmission; (b) back reflection; (c) glancing angle*

specimens. Even if we start with model single crystals, as soon as we begin sliding then wear ensues very quickly, giving us a polycrystalline surface and polycrystalline wear debris. A polycrystalline surface (as we discussed in Section 3.1.1) consists of many thousands of crystallites, each crystallite *normally* being in a *random orientation* with respect to its neighbours. The diameters of these crystallites are often around 10^{-3} mm or less. The typical collimated X-ray beam is about 0.3 mm in diameter. When it passes through a thickness of 0.3 mm of specimen, this beam will irradiate a volume of about 0.02 mm^3. If each crystallite were spherical, its volume would be about 0.5×10^{-9} mm^3. This implies that about

40 million crystallites are irradiated when the beam passes through a polycrystallic specimen. This is, of course, an over-estimate. If we assume, however, that the crystallites are cubes with sides equal to 10^{-3} mm and that the cross-section of the X-ray beam is a 0.3 mm × 0.3 mm square, then clearly the volume of specimen irradiated is about 0.03 mm^3. The volume of each cube is 10^{-9} mm^3, thereby giving us about 30 million crystallites! Either estimate indicates the scale of the process.

Figure 3.19 must be examined in the light of the last paragraph. Remember that there are millions of crystallites at O, the centre of the Ewald sphere. Remember also that the Ewald sphere has a radius of 1 unit (or 10 units) in reciprocal space [if we assume $(K_{xe}) = \lambda$ or 10λ] and that the typical specimen to plate distance is about 40 mm. Consider the point P_1^* in the zero layer of the reciprocal lattice of *one* of the millions of crystallites making up the polycrystalline specimen at O. Quite clearly, the planes $(h_1 k_1 l_1)$ represented by P_1^* cannot diffract the X-ray

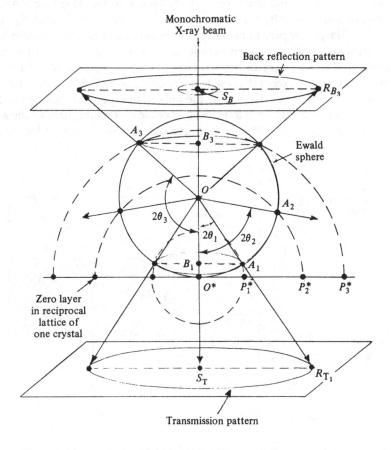

Figure 3.19 *Relation between the X-ray diffraction pattern and the reciprocal lattices of a polycrystalline specimen with randomly oriented crystallites*

beam passing along the direction (OO^*), since P_1^* does not lie on the surface of the Ewald sphere. Nevertheless, out of the millions of crystallites irradiated by the X-ray beam at O, there *will be* some which *do* have their (similar) reciprocal lattice point (P_1^*) lying on the Ewald sphere. These will be crystallites with their reciprocal lattice point (P_1^*) lying on the circle of intersection between the sphere of radius $(O^*P_1^*)$ and the Ewald sphere, centred at O. Thus we can see (from Figure 3.19) that all the diffraction maxima from a particular set of $(h_1 k_1 l_1)$ planes must arise from those crystallites at O with their corresponding reciprocal lattice points situated on a circle of radius $(B_1 A_1)$, thereby giving rise to a cone of diffraction of semi-angle $2\theta_1$. Thus the radius of the diffraction *ring* formed by transmission through the specimen at a specimen-to-film distance (D_S) is $(S_T R_{T_1})$, given by

$$(S_T R_{T_1}) = (D_S) \tan 2\theta_1 \qquad (3.17)$$

For a set of planes such as $(h_3 k_3 l_3)$, represented in reciprocal space by the point (P_3^*), we can see that the sphere of radius $(O^*P_3^*)$ cuts the Ewald sphere in a circle of intersection of radius $(B_3 A_3)$. Hence a *back-reflection* ring (of radius $S_B R_{B_3}$) is formed on a film placed perpendicularly in front of the specimen (see Figure 3.19).

The glancing-angle technique is one that is very useful in tribological applications. The geometry of the technique is similar to that of the flat-film transmission method, except that the forward cones of radiation cutting the film in concentric circles are obstructed by the bulk of the specimen, giving rise to a diffraction pattern consisting of less than half of the diffraction rings, as shown in Figure 3.20. In this figure, (OS_T) is the original direction of the X-radiation and

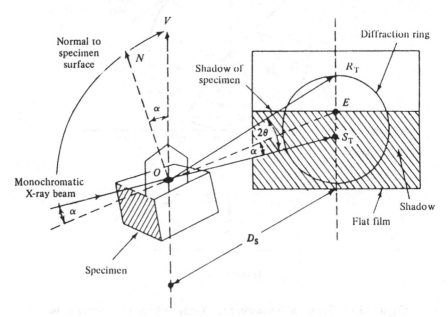

Figure 3.20 *Formation of a glancing-angle X-ray diffraction flat-film pattern from a polycrystalline bulk specimen*

$(S_T R_T)$ is the radius of the intersection of the diffraction cone (of semi-angle 2θ) with the flat-film placed perpendicularly to the beam. Note that more than half of the ring is obscured by the bulk of the specimen, which forms a shadow edge at a distance ($DS \tan \alpha$) from the centre of the diffraction ring (at S_T). It should also be noted that the glancing-angle, flat-film X-ray diffraction technique leads to *broadened* diffraction rings, since the whole of the surface being irradiated by the collimated X-ray beam contributes to each diffraction maximum. This broadening can be overcome by setting the specimen so that the X-ray beam impinges on the *edge* of the specimen, instead of the *centre*. We will describe this technique in the next sub-section, which discusses the various forms of cylindrical X-ray diffraction cameras.

3.2.4 Cylindrical X-ray diffraction cameras

The most usual geometry of an X-ray diffraction camera is cylindrical. If the specimen is polycrystalline, cylindrical in shape with its axis coinciding with the axis of the cylindrical film cassette, then we have what is colloquially called 'the X-ray powder camera'. It is essentially a transmission technique, the sharpness of the diffraction maxima being limited by the apertures used to collimate the incident X-ray beam. The technique is very useful for investigating the structures of powders – hence, it is used to identify the constituents of wear debris. It is possible to use the cylindrical powder camera to examine the crystallography of bulk surfaces provided that the X-ray beam is made to impinge at one edge of a fairly flat surface. We shall see, later, how powerful both of these techniques can be in obtaining important information about the mechanisms of wear. Although not as sensitive to minority constituents as the above film techniques, X-ray diffractometry is probably the most accurate and satisfactory way of analysing *both* the *positions* and the *intensities* of the diffraction maxima. Let us now discuss each technique in turn.

(i) The cylindrical X-ray diffraction powder camera

A diagram of the typical powder camera is shown in Figure 3.21. This figure shows how the forward-reflection diffraction lines from a polycrystalline specimen are formed by the intersection of the diffraction cone (of semi-angle 2θ) with the cylindrical film. The reader is referred to Figure 3.19 regarding the origin of this diffraction cone. The only difference from the flat-film geometry is the way in which the maxima are recorded as circles in the flat-film systems but as hyperbolas in the cylindrical film systems. The back-reflection lines are formed by the intersection of diffraction cones of semi-angle ($180° - 2\theta$) with the cylindrical film. The lines for 2θ greater than about 160° are split into doublets, due to the fact that the filtered X-ray beam contains two wavelengths (called K_{α_1} and K_{α_2}). These wavelengths differ by only a very small amount, so that only the large angle diffraction maxima are affected.

The cylindrical film is held in position by a cylindrical cassette. Normally, the manufacturers arrange for the radius (i.e. the distance from the centre of the camera to a point halfway between the inner and outer surfaces of the photographic film) to be 57.3 mm. This means that the distance between the

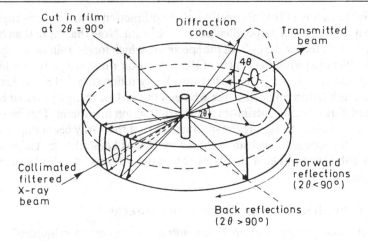

Figure 3.21 *Diagram showing how the X-ray powder pattern arises using a cylindrical camera*

centre of the forward cones and the centre of the back-reflection cones, should be 180 mm. Hence, one can measure the angle 2θ directly in degrees by measuring the distance (in millimetres) from the centres of the forward cones of diffraction. The reason for cutting the film at $2\theta = 90°$ (measuring in an anti-clockwise direction) is so that any film shrinkages (or elongations) caused by the photographic processing can be allowed for.

One of the uses of the X-ray diffraction powder camera is for *identification* purposes. This is effected through a measurement of $(2\theta_{hkl})$ and thence via a simple calculation (the Bragg relation) to give the interplanar spacing (d_{hkl}). Once we have obtained a set of interplanar spacings for a particular specimen, we can compare against the interplanar spacings noted in the 'X-ray powder data' file, issued by the American Society for Testing and Materials (ASTM). This file consists of several thousand cards which list the interplanar spacings of the diffraction lines obtained by previous investigators, the relative intensities of the diffraction line (see later regarding X-ray diffraction intensities), and (where possible) the relevant Miller indices. The cards are indexed in 'Hanawalt order' (after the originator of the idea), by means of the interplanar spacing values of the three strongest (i.e. most intense) lines in X-ray diffraction pattern. Each substance is also listed in name and 'Hanawalt' order in index books covering organic and inorganic materials. Most tribologists will be mainly interested in the *inorganic* materials. However, the additives and polymers used in tribo-systems will probably be found in the *organic* index.

Each crystalline substance has its own characteristic X-ray diffraction pattern, so it is possible to use this technique to identify the constituents of that substance, without knowing very much about the crystallography of the substance. We shall see, however, that identification is only one small part of the many applications of X-ray (and electron) diffraction to tribo-systems. An example of a typical X-ray powder data file card is given in Table 3.2.

Table 3.2. *An X-ray powder data file card for haematite* (α-Fe_2O_3)

13-534

d	2.69	1.69	2.51	3.66	α-$Fe_2\overset{o}{3}$	Haematite
I/I_1	100	60	50	25	Alpha iron oxide	

	d(Å)	I/I_1	hkl	d(Å)	I/I_1	hkl
Rad^n:CoK$_a$; λ:1.790; Filter; Dia: 11.46cm						
	3.66	25	012	1.484	35	214
Cut off:——I/I_1: Recording Microphotometer	2.69	100	104	1.452	35	300
Reference: Aravindakshan and Ali, etc.						
System: Rhombohedral SG:$R\bar{3}_c$	2.51	50	110	1.349	4	208
$a_0 = 5.4228$, b_0: c_0: A: c:	2.285	2	006	1.310	20	191
$\alpha = 55°17'$, β: γ: z: D_x:						10, 10
	2.201	30	113	1.258	8	220
Reference: As above						
E_a: nw β: $\varepsilon\gamma$: sign:	2.070	2	202	1.226	2	036
2V: D: mp: Colour:	1.838	40	024	1.213	4	223
Reference:	1.690	60	116	1.189	8	128
Indexing based on hexagonal cell	1.634	4	211	1.162	10	02, 10
$a_H = 5.0317$	1.596	16	018	1.141	12	134
$c_H = 13.737$						

The top-left hand corner of the ASTM card gives the interplanar spacings and relative intensities of the three strongest lines in the X-ray diffraction patterns obtained by Aravindakshan and Ali (and then submitted for inclusion in the ASTM Powder Data File). The reliability of *some* of the data included in the file should be treated with some scepticism. Quite often, the investigator is better off using standard materials (if available) for comparison of actual diffraction patterns. ASTM are aware of this unevenness in reliability from card to card and they have commissioned an X-ray crystallographer to go through the index in a critical way. Hence, whenever one sees a large star printed on the right hand top position of the card, one can be fairly sure of the data on that card. The information on the rest of the card is fairly self-explanatory. If one had approximate correspondence between one's unknown d_{hkl} values and the twenty or so d_{hkl} values given in this card, then one would be confident it was α-Fe_2O_3.

(ii) The edge-irradiated, glancing-angle X-ray diffraction technique

A diagram of the geometry of this technique is given in Figure 3.22. The single crystal rotation camera is most readily adapted to provide powder patterns from the surfaces of bulk specimens. This type of camera has a small radius (30 mm), which means the technique is not so accurate as it could be with a suitably modified powder camera (with its typical radius of 57.3 mm). Nevertheless, it does have a goniometer stage together with a telescope for accurate alignment of the specimen relative to the collimator and the axis of rotation. This is most important if one is to be sure that the X-ray beam impinges on the *edge* of the surface (as at *C* in Figure 3.22) and *not* over the whole surface. When accurately aligned, the edge of the specimen will obscure half of the circular collimator aperture for all rotations of the specimen about the camera axis. Typically, the best glancing angle (α) value is about 30°. Quinn (1970) has shown that this technique will provide very sharp maxima from any thin surface layers (such as a few micrometres of oxide) and fairly broad maxima from the bulk of the material. Quinn (1970) discusses the transmitted and reflected components of the diffraction

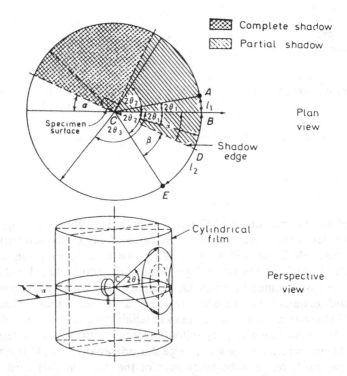

Figure 3.22 *Schematic representation of specimen alignment for the edge-irradiated, glancing-angle film technique (Isherwood and Quinn, 1967; Quinn 1970)*

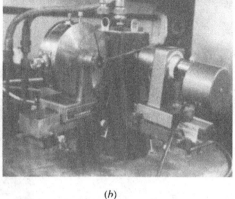

(a)   (b)

Figure 3.23 Whitaker's (1967) modification to the conventional X-ray powder camera to accommodate small bulk specimens: (a) specimen holder; (b) specimen orientation and oscillation mechanisms

pattern and concludes that this technique is very suitable for accurate determination of the structure of oxide films of the order of thickness of a few micrometres. Since the oxide films formed on the surfaces of steel tribo-elements during sliding have been shown to be of this thickness, it is clear that this is a powerful tool for examining tribological specimens.

The technique is at its best when diffraction maxima are also obtained in the partial shadow on the opposite side of the direct beam from the shadow-edge, such as at A in Figure 3.22. The centres of the forward diffraction cones give us the undeviated direction (CB). All diffraction maxima *without* their comparison maxima on the other side of the undeviated direction, are then measured with respect to the point B. For example, a diffraction maximum at ($2\theta_2$) intersects the film at E, a distance l_2 along the film from B. Obviously, we have:

$$2(R_{RC})(\theta_2) = l_2 \qquad (3.18)$$

where (R_{RC}) is the radius of the single crystal rotation camera and θ_2 is the Bragg angle (in radians). Since θ_2 must satisfy the Bragg relation (Equation (3.6)), we can deduce the interplanar spacings d from measurements of l_2 (for a given wavelength of X-radiation).

If no diffraction maxima are obtained on the shadow side of the undeviated direction, we cannot readily ascertain the actual position of the centres of the forward diffraction cones with the method used by Isherwood and Quinn (1967). It has been suggested that one could overcome this problem by taking all diffraction line positions in terms of their distance from a sharp prominent line of a *known* constituent (Quinn, 1971). The position of the centre for the various diffraction cones from the known constituent is readily obtained from the Bragg angles (θ) relevant to that constituent. This suggestion has been used fairly successfully by Isherwood and Quinn (1967). However, it does depend on being

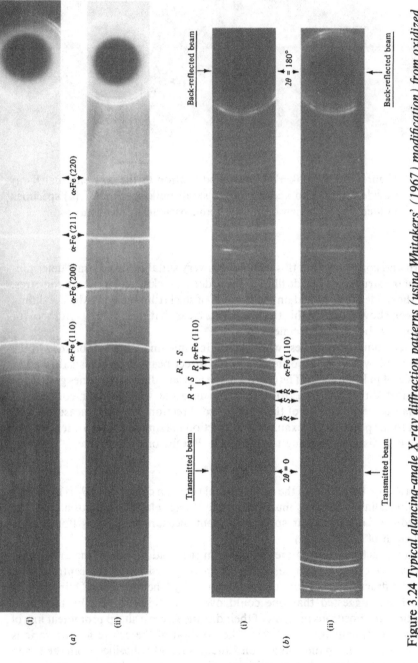

Figure 3.24 Typical glancing-angle X-ray diffraction patterns (using Whitakers' (1967) modification) from oxidized specimens of: (a) 12% Cr, 0.29% C steel (i) half-pattern and (ii) double-exposure pattern (b) 0.01% Cr, 0.30% C steel (i) half-pattern and (ii) double-exposure pattern

able to identify unambiguously the lines of the known constituent, for example, one must be sure that the matrix material in steel is either α-Fe (ferrite) or γ-Fe (austenite) before using this technique when only half-patterns can be obtained from the surface.

An interesting modification to the conventional, 57.3 mm radius, cylindrical X-ray diffraction powder camera has been suggested by Whitaker (1967). This also has the advantage in that it enables maxima to be obtained *on both sides* of the undeviated direction. The modification consists of two parts, one to hold the specimen and the other to provide and measure its oscillation. The specimen holder consists of a disk of diameter 3 cm and thickness 1 mm. On one face around the edge is a flange 1 mm high, while in the centre of the other face is a small shaft of diameter 3 mm and length 6.5 mm. This shaft fits into the jaws of the specimen holder of the camera. The specimen is held by placing it in a bed of plasticine which, in turn, is held by the flange.

The oscillation of the specimen is accomplished by fastening a collar over the drive shaft (see Figure 3.23(*b*)). To the collar is fitted an arm, the other end of which rests on a steel pin which rotates about an axis parallel to its length, the diameter of rotation of the pin and the distance of the arm-pin pivot from the centre of the driving shaft determining the angle of oscillation, set at $\pm 7\frac{1}{2}°$ for the patterns shown in Figure 3.24. The angle of oscillation is readily measured by directly observing the movement of a pointer (screwed onto the end of the specimen driving shaft) against a scale which is clamped to the hub of the camera. The vertical alignment of the specimen or, more correctly, the line-up of the specimen surface relative to the axis of specimen rotation, is not so readily accomplished as with the single crystal camera used by Isherwood and Quinn (1967). In fact, this is the disadvantage of Whitaker's (1967) modification. It is not really suited to the *edge-irradiated* geometry shown in Figure 3.22. In fact, Figures 3.24 were all obtained using $\pm 7\frac{1}{2}°$ oscillation about a mean glancing angle of a few degrees (say 10°), the incident X-ray beam impingeing on the *centre* of the specimen surface. The radiation used was Cobalt K_α ($\lambda = 1.79$ Å) and the diameter of the diffraction camera was 57.3 mm.

These diffraction patterns have been reproduced in this book in order () to show what an X-ray powder pattern actually looks like, (*b*) to demonstrate the importance of being able to measure both sides of the undeviated direction, and (*c*) to demonstrate the sharpness of the innermost lines (i.e. lines with 2θ less than about 40°), even though the geometry used was not the edge-irradiated geometry. It can be shown that the first three lines of Figure 3.24(*b*) are due to diffraction of the Cobalt K_α X-rays by the (110) planes of the rhombohedral (*R*) oxide, the (220) planes of the spinel (*S*)-oxide and the (211) planes of the rhombohedral oxide. Since these oxides must lie on top of the α-Fe bulk material, we see a slight self-collimation effect must occur, where parts of the X-ray beam impinge on the oxidized surface near to the edge of the specimen. The sharpness of the oxide lines rapidly falls off with 2θ greater than 50°, whereas with the edge-irradiated technique of Isherwood and Quinn (1967), sharp oxide lines are obtained over values of 2θ from very low angles up to 120° or more. Figure 3.24(*a*)(i) and (ii) are good examples of wide α-Fe lines in the same diffraction patterns as sharp oxide lines. Note that there are no oxide lines visible at 2θ greater than $\sim 78°$ (the position of the 200 line of α-Fe). Let us now consider a method for obtaining sharply focused maxima at all values of 2θ, namely the X-ray diffractometer.

3.2.5 The X-ray diffractometer

Although the X-ray diffractometer has a cylindrical geometry, it is *not* a camera. In fact, the X-rays are detected by a scintillation counter, thereby giving much more *direct* information on the position and the intensity of X-ray diffraction maxima than with the film cameras already discussed. In the X-ray diffractometer, we arrange for the X-ray detector to make an angle (2α) with the incident beam, when the specimen surface makes an angle α (for all rotations of the specimen). Typically, the specimen is a polycrystalline powder set in a holder about $1\,cm \times 2\,cm \times 0.2\,cm$ deep, or it can be a bulk specimen. The diverging beam from the line source of X-rays will be made to converge at the detector slit by the action of the diffracting crystallites in the specimen surface. This is shown in Figure 3.25 for a particular value of α, namely a value which is equal to the Bragg angle (θ) for the given set of planes of interplanar distance d. In order to ensure that this angular condition holds for all Bragg 'reflections', one must rotate the specimen about an axis perpendicular to the plane of the diffractometer circle at an angular velocity (ω_0) equal to half the angular velocity of the detector ($2\omega_0$) about the same axis (see Figure 3.26). Quite clearly, the line SOE is fixed. The specimen surface and the detector slit must lie along this line at $2\theta = 0$. It should be noted that perfect focusing will only occur when the specimen surface is curved so that it coincides with the 'focusing circle', that is a circle that passes through S, O and D. Since the radius of the focusing circle varies according to the angle α, there is no real advantage in making our specimen surface slightly curved.

It may be relevant to mention 'focusing X-ray diffraction film cameras' at this point. These are made so that the film is coincident with the focusing circle, the source lying on the focusing circle and the specimen consisting of a powder adhering to a specimen holder curved to fit the focusing circle. Since perfect

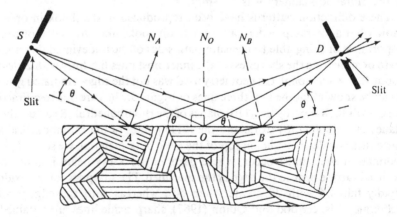

Figure 3.25 *Schematic diagram of the focusing action of a polycrystalline specimen examined in the X-ray diffractometer*

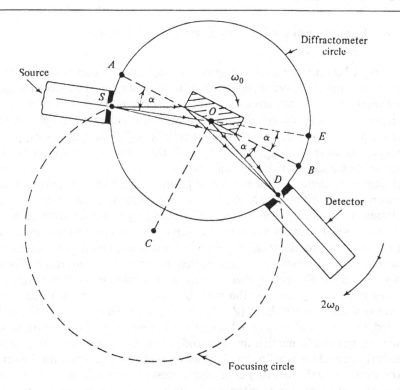

Figure 3.26 *The Bragg–Brentano focusing system*

focusing will only occur over a small range of diffraction angles about a given value of (2α), it is necessary to have three or four focusing cameras of different radii in order to cover a sufficient range of Bragg angles. This emphasizes the usefulness of the edge-irradiated, glancing-angle X-ray diffraction technique mentioned earlier, since it requires no special cameras and yet provides sharply focused maxima over a wide range of Bragg angles.

The main advantage of using the X-ray diffractometer over the film cameras is that the detector of the diffracted beams provides an output that is directly connected with the intensity of those beams. This output can be counted accurately by sophisticated counter techniques, or it can be fed to a pen recorder, thereby giving the line profile as a function of the diffraction angle (2θ). This record can be used for determining both line position and its relative integrated intensity in a very direct method. With the X-ray diffraction *film* techniques, the intensity of the X-rays can only be found after a rather tedious operation involving the characteristic curve of the photographic film as regards its response to X-radiation. We will see, however, that both counter and film techniques complement each other when dealing with specimens obtained from tribo-systems. To complete this concise review of X-ray diffraction technique, we will briefly outline the essential theory of X-ray scattering by polycrystalline materials.

3.2.6 Basic theory of X-ray scattering by polycrystalline materials

An X-ray photon is an electromagnetic wave, that is it consists of an electric vector and a magnetic vector, oriented at 90° to each other. This electromagnetic wave interacts with each electron in the electron 'cloud' around each atom, in each crystallite of a polycrystalline specimen. It can be shown that the interaction with the positively-charged nucleus of the atom is negligible. X-rays are essentially scattered by their exchange of energy with the extra-nuclear electrons. These negatively-charged particles are alternately *accelerated* and *decelerated* in the oscillating magnetic field of the X-ray photon. Basic physics tells us that an electron accelerated in a magnetic field gives off radiation. Since the forcing X-radiation is oscillating, so also will be the secondary radiation given off by each electron. Thus a plane X-ray wave impinging on an electron will give rise to secondary X-rays (of the same frequency as the incident radiation) but emanating now from a point source (the extra-nuclear electron) *in all directions*. For each atom, there are Z electrons [where Z is the atomic number, i.e. the charge on the nucleus is $(+Ze)$, where e is the basic unit of electric charge]. If the atom is *positively* ionized, it will have $(Z-1)$ extra-nuclear electrons. If it is *negatively* ionized, the atom will have $(Z+1)$ extra-nuclear electrons. *For very small angles of scatter*, we can assume that the amplitude of the X-rays scattered by an atom is merely the sum of the amplitudes of the waves scattered by each of the Z electrons in the atom. Since the intensity of any wave must be proportional to the square of the amplitude, we may assume that the intensity (I_a) scattered by an atom is given by:

$$I_a = (Z^2)(I_e) \text{ (for small scattering angles)} \qquad (3.19)$$

where I_e is the intensity scattered by a single electron. *For large angles of scattering*, we would expect I_a to be less than $(Z^2)(I_e)$. Let us therefore re-write Equation (3.19) in terms of an atomic scattering factor (f_X), where $F_X \to Z$ at low angles of scattering.

$$I_a = f_X^2 (I_e) \qquad (3.20)$$

Hence (f_X) is the ratio of the amplitude of the X-ray wave scattered by the atom compared with the amplitude scattered by an electron.

In order to obtain a value for (f_X) physicists have computed the wave-mechanical probabilities of finding electrons in particular orbits around the nucleus of the atom. Most estimates of the atomic scattering factor are based on the Hartree approximation (for light elements) and the Thomas–Fermi model of the atom (for the heavier elements). Tables of atomic scattering factors, as functions of $(\sin \theta)/\lambda$, are given in most books dealing with X-ray crystallography (for instance, see the book by Peisser, Rooksby and Wilson, 1955).

The intensity scattered by a three-dimensional lattice of atoms is proportional to the intensity of X-rays scattered by each unit cell. The amount scattered by each unit cell will, of course, depend on the positions of each atom (or ion) within that unit cell. Very few materials consist of only one atom per unit cell. For instance, any body-centred cubic element will have *two* atoms per unit cell, namely one at the origin (000) and one at the point $(\frac{1}{2},\frac{1}{2},\frac{1}{2})$ measured in terms of the repeat distance of the cubic lattice. Compounds are even more unlikely to have their component atoms placed at only the corners of their unit cell. Let us denote the coordinates of the ith atom (or ion) in the unit cell by the symbols x_i, y_i, z_i, it being understood that these coordinates are measured in terms of the fraction of the repeat distances of the crystal lattice in the X-, Y- and Z-directions. The phases of the X-ray waves scattered by the atoms in a unit cell will depend on their position within that cell. It can be shown that the phase difference (Φ_i) between an X-ray wave scattered (in the direction $2\theta_{hkl}$) by an atom at the origin and that same wave when scattered by an atom at $(x_i y_i z_i)$ within the unit cell, is given by:

$$\Phi_i = 2\pi(hx_i + ky_i + lz_i) \tag{3.21}$$

The amplitude $(F_X)_{hkl}$ of the resultant of all the X-ray waves scattered by the n atoms in the unit cell will be the *vector* sum of all the amplitudes of each individual X-ray wave. Since we are still only considering the unit cell, we have to divide this amplitude by the unit cell volume (v_{uc}) to get the amplitude scattered per unit volume of crystal.

It is relevant, at this stage, to introduce the concept of 'primary extinction' and 'secondary extinction' of X-rays, since this can have an effect on the relationship between the amplitude factor $[(F_X)_{hkl}]$ and the intensity of the *hkl* diffraction maximum. Within a *perfect crystal*, if the Bragg condition is met for the *primary* ray, it is also satisfied for the 'reflected' ray. The twice-'reflected' wave (180° out of phase for two 'reflections') travels along with, and cancels part of, the primary wave. This is *primary* extinction. When the upper portions of a crystallite reflect the X-rays completely out of the primary beam, they are effectively acting as an absorbing layer as far as the lower portions are concerned. Since this is a very special form of absorption of X-rays (see later), it is called *secondary* extinction. In tribology, we rarely have perfect crystals. Hence, we will not be concerned with either effect at this stage. However, just for completeness, it will be pointed out that if one does have a *perfect* crystal, then these effects will cause the *intensity scattered per unit volume* to be proportional to the structure amplitude divided by the volume of the unit cell. Hence we have:

$$(I_{hkl})_{\text{perfect crystal}} \propto \{[(F_X)_{hkl}]/[v_{uc}]\} \tag{3.22}$$

For a small *ideally imperfect* crystal, that is a crystal made up of small blocks of perfect crystal that are slightly misaligned with respect to each other (the 'mosaic' model of a crystal), it can be shown that the intensity of X-rays scattered per unit volume of crystal is proportional to the *square* of the structure amplitude for X-rays for each unit cell, that is:

$$(I_{hkl})_{\text{real crystal}} \propto \left\{ \frac{[(F_X)_{hkl}]}{[v_{uc}]} \right\}^2 \tag{3.23}$$

We shall see (in Section 3.3.8) that a similar situation occurs in the diffraction of electron waves, except that the *dynamical effects* of re-diffraction into the beam and diffraction out of the beam are much more common.

It can readily be shown that $(F_X)_{hkl}$ is given by:

$$[(F_X)_{hkl}]^2 = \left\{ \sum_{i=0}^{i=n-1} [(f_X)_i \cos 2\pi(hx_i + ky_i + lz_i)] \right\}^2$$
$$+ \left\{ \sum_{i=0}^{i=n-1} [f_X)_i \sin 2\pi(hx_i + ky_i + lz_i)] \right\}^2 \qquad (3.24)$$

To obtain some idea of where this equation comes from, consider Figure 3.27, in which we show the situation for a unit cell with atoms at 000, $x_1y_1z_1$, $x_2y_2z_2$ and $x_3y_3z_3$. Clearly, the phase at the origin (Φ_0) is equal to zero. The phases at $x_1y_1z_1$ and $x_2y_2z_2$ are less than 90°. The phase at $x_3y_3z_3$, however, is greater than 90°, leading to a resultant which is less than would have been obtained if $(f_X)_3$ were at an angle less than 90°. From simple trigonometry, we can see that:

$$[(f_X)_{hkl}]^2 = (OE)^2 + (DE)^2 \qquad (3.25a)$$

where

$$OE = (f_X)_0 + (f_X)_1 \cos \Phi_1 + (f_X)_2 \cos \Phi_2 + (f_X)_3 \cos \Phi_3 \qquad (3.25b)$$

and

$$DE = (f_X)_1 \sin \Phi_1 + (f_X)_2 \sin \Phi_2 + (f_X)_3 \sin \Phi_3 \qquad (3.25c)$$

Thus, Equations (3.25a, b, c) are what Equation (3.24) gives for $n=4$.

To allow for the fact that the total volume (V) giving rise to diffraction (at $2\theta_{hkl}$) will only be a fraction of the crystallites traversed by the scattered X-ray beams on their way through the specimen, it is usual to introduce an absorption factor. This

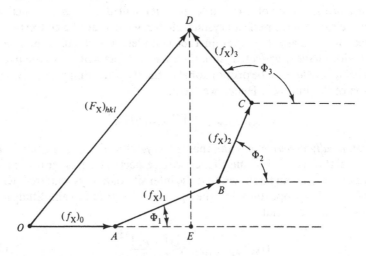

Figure 3.27 *The relationship between $(F_X)_{hkl}$ and the atomic scattering factors (f_X) for a unit cell with atoms at 000, $x_1y_1z_1$, $x_2y_2z_2$ and $x_3y_3z_3$*

will be smallest at low angles and will increase in an undetermined way with increasing angle. Let us denote the fraction by the symbol $A(\theta)$, so that, in order to make Equation (3.23b) suitable for a polycrystalline specimen we have to multiply the right-hand side by $[VA(\theta)]$. Here θ is written as an abbreviated form of θ_{hkl}. Hence:

$$(I_{hkl}) \propto \left[\frac{(F_X)_{hkl}}{(v_{uc})}\right]^2 [VA(\theta)] \tag{3.26}$$

We should also allow for the fact that, for the crystals of higher symmetry, there are several equivalent (hkl) planes in each crystallite. For instance, there are *six* equivalent $(h00)$ planes and twenty-four equivalent $(hk0)$ planes in a crystallite belonging to the cubic system. Obviously, there are four times as many chances for a crystallite to be oriented just right for the $(hk0)$ compared with the orientation for the diffraction from the $(h00)$ planes. To allow for this 'multiplicity' of crystal planes, we introduce the factor (m), as given in Table 3.3, so that Equation (3.26) can now be written:

$$(I_{hkl}) \propto \frac{[(F_X)_{hkl}]^2 \, mVA(\theta)}{v_{uc}^2} \tag{3.27}$$

There are two other factors affecting the intensity (I_{hkl}) of the scattered X-rays that are normally considered together. One is the Lorentz Factor, which is a correlation factor which takes into account the fact that a crystallite will not only diffract a monochromatic beam of X-rays at the *exact* Bragg angle (θ_{hkl}) but will also diffract, with smaller intensities, at angles deviating by some seconds of arc from the exact Bragg value. Since we are *always* only interested in the *integrated intensity*, that is the total area under the profile of intensity versus angle, this factor must be taken into account. The other factor, the *polarization factor*, takes into account the fact that the incident X-ray beam normally consists of *unpolarized* electromagnetic waves, that is the electric and magnetic vectors are not aligned in any particular plane. This means that any electron which scatters the incident X-radiation will oscillate in *any plane* containing the incident direction, thereby giving rise to unpolarized secondary X-radiation. Both the Lorentz and the polarization factors have been combined into the Lorentz-polarization (LP) factor, namely:

$$(LP) = \frac{(1 + \cos^2 2\theta)}{(\sin^2 \theta)(\cos \theta)} \tag{3.28a}$$

For tables of (LP) factors, the reader is referred to the book by Henry, Lipson and Wooster (1951).

The last factor which need concern us is the Debye–Waller temperature factor (B_{DW}). This allows for the fact that the atoms in a crystallite are not stationary, but are oscillating about their positions of equilibrium in the crystal lattice, at an amplitude depending on the temperature of the crystallite. Since this factor also depends on θ, it must be allowed for when comparing diffraction maxima at low θ with those at high θ. The Debye–Waller factor is associated with the atomic

Table 3.3. Multiplicity factors (m) for powder photographs

Cubic		Hexagonal		Rhombohedral		Tetragonal		Orthorhombic		Monoclinic		Triclinic	
Indices	m	Indices	m	Indices	m	Indices	m	Indices	m	Indices	m	Indices	m
h00	6	00l	2	h00	6	00l	2	h00	2	h0l	2	hkl	2
hhh	8	h00	6	hh0	6	h00	4	hk0	4	hkl	4		
hh0	12	hh0	6	hk0	12	hh0	4	hkl	8				
hhl	24	h0l	12	hhl	6	h0l	8						
hk0	24	hhl	12	hkl	12	hhl	8						
hkl	48	hk0	12			hk0	8						
		hkl	24			hkl	16						

scattering factor f_X through the equation:

$$(f_X)_T = (f_X)\left[\exp\left(\frac{-B_{DW}\sin^2\theta}{\lambda^2}\right)\right] \qquad (3.28b)$$

where $(f_X)_T$ is the atomic scattering for X-rays at temperature T. It will replace (f_X) in Equation (3.24). B is equal to $(B_0 + B_T)$, where B_0 is derived from the zero point energy of the atom and B_T derived from the thermal vibrations. Values of B_0 and B_T are given for various elements in Volume III of the *International Tables for Crystallography* (see also Cullity, 1962). Assuming that the thermal vibrations affect every atom in the unit cell equally, it is possible to take the *exponential* part of $(f_X)_T$ outside the summation of Equation (3.24). We can also replace the proportional sign in Equation (3.27) by an equality (since we are only interested in *relative* integrated intensities) and obtain:

$$(I_{hkl})_w = \frac{[(F_X)_{hkl}]^2 (m)(LP)\left\{\exp\left[\frac{-2(B_{DW})\sin^2\theta}{\lambda^2}\right]\right\}}{(v_{uc})^2} \, [VA(\theta)] \quad (3.29)$$

where $[(F_X)_{hkl}]^2$ is given by Equation (3.24), m is found from Table 3.3, (LP) is given by Equation (3.28a), $\exp(-2B_{DW}\sin^2\theta/\lambda^2)$ is found from references (such as Cullity, 1962) and v_{uc} is computed from solid geometry. The atom positions within the unit cell, so necessary for deducing $(F_X)_{hkl}$, can be obtained from the appropriate volume of Wyckoff's *Crystal Structures* (an indispensable part of *any* X-ray crystallography laboratory). We will discuss how we deduce $A(\theta)$ later in the book, when we describe how to carry out proportional analysis of mixtures of different crystals, that is where we are measuring the relative volume percentages $(V_1, V_2, ..., V_n)$ of an n-component mixture. Let us now go on to discuss another form of diffraction, namely electron diffraction.

3.3   Electron diffraction methods

3.3.1   Introduction

In Chapter 2, we saw that an electron beam may be considered to be equivalent to a plane wave of wavelength (λ) given by Equation (2.28). As far as electron microscopy is concerned, the wavelength of the electron beam is only important insofar as it affects the ultimate resolution of the electron microscope. For most purposes of image formation, we can visualize the electron beam as consisting of an extremely large number of particles, namely electrons, that are acted upon by the magnetic fields of the electron lenses exactly as if they were *particles*. In this sub-section, however, we must visualize the electron beam as a plane wave *when it is interacting with crystalline materials*. We have already seen that the wavelength of a beam of electrons accelerated through a potential difference of 100 kV is about 0.04 Å (where 1 Å = 0.1 nm). We should ask ourselves, what sort of Bragg angle (θ_{hkl}) do we expect for the diffraction of this electron beam from crystal planes 2 Å apart? Since the Bragg relation (Equation (3.6)) is valid for the diffraction of *any* form of radiation of wavelength (λ), we see that $\sin(\theta_{hkl}) = 0.01$.

Hence $\theta_{hkl} = 0.57°$ so that the angle of diffraction $(2\theta_{hkl}) \approx 1°$. This angle is typical of electron diffraction, or to be more correct, typical of High-Energy Electron Diffraction (HEED). We shall see, in the following sub-sections, how this small value of $(2\theta_{hkl})$ for the diffraction of high-energy electrons by normal crystalline materials leads to a *very simple relationship* between the reciprocal lattice of a crystal and its electron diffraction pattern. For completeness, some mention will also be made of Low-Energy Electron Diffraction (LEED), since this technique has found some applications in surface analysis related to the tribological properties of surfaces. Finally, in this review of electron diffraction, we will discuss the basic ideas of electron scattering by atoms in crystalline arrangements. Although the application of electron diffraction to the proportional analysis of the surfaces (or thin surface layers) of tribo-elements is still in its very early stages, I am convinced that the mechanisms of (for example) the anti-wear additive protection of tribo-elements may eventually be revealed by a surface-sensitive technique such as electron diffraction. Let us start by discussing the reciprocal lattice of a thin crystal.

3.3.2 The reciprocal lattice of a thin crystal and its relation to the transmission electron diffraction pattern

In Section 3.2.2, we saw how to construct the reciprocal lattice of a crystal. In that construction, it was assumed that the real crystal lattice was of infinite extent in all directions. Let us consider the effects of (a) limiting the extent of the crystal in a direction parallel to the incident direction of the electron beam and (b) using a beam of high energy electrons instead of a collimated X-ray beam upon the construction of the reciprocal lattice of a crystal.

In order to allow for the effect of limited crystallite size, we use the formula for the diameter (b) of a reciprocal lattice point representing a typical set of (hkl) planes within a spherical crystal of diameter (ε), namely:

$$b = \frac{2(K_{xe})}{(\varepsilon)} \tag{3.30}$$

where (K_{xe}) is (arbitrarily) taken to be unity for electron diffraction purposes. Thus for an infinite crystal, b is zero. In other words, the reciprocal lattice points representing crystal planes in a perfect, infinite crystal are merely mathematical points. In order that electrons may pass through a specimen in the electron diffraction camera it must be extremely thin in a direction parallel to the electron beam, compared with the dimensions of the crystal in directions perpendicular to the beam. For instance, iron must be less than about 1000 Å thick (i.e. 100 nm) if it is going to give rise to a diffraction pattern, when a 100keV electron beam passes through a thin evaporated film of this metal. If one uses the glancing-angle geometry, known as 'reflection electron diffraction', the angles used are so small that one finds diffraction can only occur from crystal planes that lie within the top 20 Å or so of a perfectly flat crystal. Effectively one has a crystal of limited extent in a direction perpendicular to the surface. This limit to the extent of the crystal in both transmission and reflection electron diffraction examination of crystals is what makes these techniques so very different (in a *complementary way*) from X-ray diffraction.

For the sake of simplicity, let us assume that we are passing an electron beam (of infinitesimally narrow beam width) through a crystalline specimen with a crystal size (ε_p) equal to 10 Å parallel to the beam. Assume the beam passes along the [100] zone axis of a cubic crystal whose unit cell dimension $|\mathbf{a}|$ is 1 Å. Typically, the thickness (ε_p) will be more than 10 Å and the unit cell dimension more than 1 Å. These values have been chosen in order to show the principles, rather than to be an accurate representation of a real system. From Equation (3.30), we know that $b = 0.2$ Å^{-1} along the [100] direction and is effectively zero in all directions perpendicular to the [100] zone axis. From Equation (3.7), we know that $|\mathbf{a}_{uc}^*| = |\mathbf{d}_{001}^*| = |\mathbf{d}_{010}^*| = |\mathbf{d}_{100}^*| = 1$ Å^{-1}, setting $(K_{xe}) = 1$ (for electron diffraction) in both of these equations. Figure 3.28 is a perspective view of the reciprocal lattice unit cell. Note that the reciprocal lattice points are elongated into 'spikes' in the direct parallel to [100].

What effect will this 'spiky' reciprocal lattice have on the electron diffraction pattern? When we draw the Ewald sphere of radius 25 Å^{-1} (we have set $K_{xe} = 1$ and assumed our electron beam has a wavelength of 0.04 Å), we find that the surface of that sphere will intersect several of the 'spikes' in the $(hk0)$ layer of the reciprocal lattice (i.e. those spikes representing the planes in the [001] zone of the crystal), as shown in Figure 3.29. This is because the radius of the Ewald sphere is 25 times as large as $|\mathbf{d}_{001}^*|$, so that its surface approximates to that of a plane. Thus, one has *several* beams *simultaneously* satisfying the Bragg condition, that is there

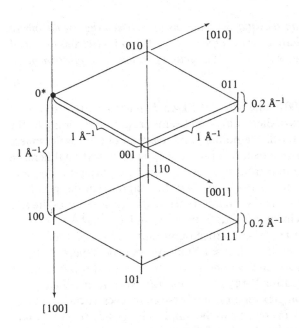

Figure 3.28 *The reciprocal lattice unit cell of a cubic crystal 10 Å thick and of infinite extent perpendicular to the* [100] *direction*

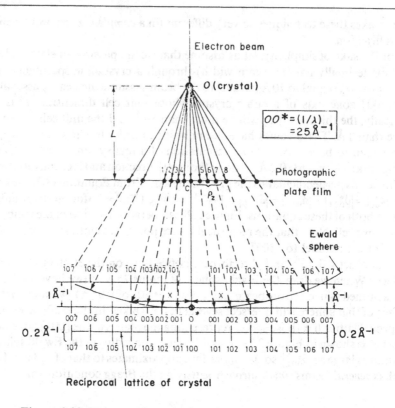

Figure 3.29 *Diagram showing the relationship between the Ewald sphere and the reciprocal lattice for the diffraction of 100 keV electrons ($\lambda \approx 0.04$ Å) by a cubic crystal with its [100] direction parallel to the electron beam (not to scale)*

are several places in the [100] zone where the Ewald sphere cuts the reciprocal lattice points. The directions of these diffracted beams are given by joining O (the origin of the crystal lattice) with the points of intersection of the Ewald sphere with the corresponding spikes in reciprocal space. This is an important difference from X-ray diffraction. The penetration of X-rays through a crystal is so great (compared with the penetration of electrons) that the reciprocal lattice point is a *sphere* of radius $(2\lambda/\varepsilon)$, where ε is several micrometres (the typical crystallite size (ε) of a polycrystalline specimen is about 2 or 3 µm) and λ is 1.54 Å (for CuK$_\alpha$ radiation), thereby giving a reciprocal lattice point diameter (b) of about 10^{-4} units. Since $|\mathbf{d}^*_{001}| = \lambda/|\mathbf{d}_{001}|$, we have $\mathbf{d}^*_{001} \approx 1.5$ units in reciprocal space. Thus we cannot easily find intersections between the surface of the Ewald sphere and the reciprocal lattice constructed for X-ray diffraction *without rotating the crystal*. This leads to a somewhat complex relationship between the reciprocal lattice and the X-ray diffraction pattern, in contrast to the simple relationship that obtains in electron diffraction. This can be stated as follows:

The electron diffraction pattern from a thin single crystal, with a zone axis along which the electron beam is passing, is directly proportional to the section in the reciprocal lattice of that crystal appropriate to that zone.

As an example, we see (from Figure 3.29) that the diffracted beams '1' to '8' impinge on the photographic plate at distances from the centre spot (C) that are (approximately) proportional to the distances of the reciprocal lattice spikes from the origin of reciprocal space (O^*). For instance, we can see that the distance between the centre-spot (C) and the diffraction spot numbered '5', that is ($C5$), is given by:

$$OC/(C5) = (1/\lambda)/|\mathbf{d}_{001}^*| \qquad (3.31)$$

If we denote the distance of any diffraction spot from C by the symbol (R_{ED}), and letting (OC) be denoted by L (the crystal-to-photographic plate distance), then Equation (3.31) can be written more generally as:

$$(R_{ED})(d_{hkl}) = \lambda L \qquad (3.32)$$

where (d_{hkl}) has been written for ($1/|\mathbf{d}_{hkl}^*|$), that is Equation (3.7), with (K_{xe}) set equal to unity.

Since a crystal is three-dimensional, its reciprocal lattice is also three-dimensional. Thus, all reciprocal lattice spikes lying inside a circle of radius (x) will intersect the Ewald sphere and this gives rise to diffraction. Hence, the diffraction pattern will appear as a square array of spots all lying within a circle of radius ρ_z, where ρ_z is given by $\rho_z = xL\lambda$ (see Figure 3.30). But, from the sagittal formula, we know that $x^2 \approx (2/\lambda)(1/\varepsilon_P)$, that is the diameter of the Ewald sphere multiplied by half the height of the reciprocal lattice spike. From these two equations, we can substitute for x and obtain the following relation:

$$\rho_z = (L)[2\lambda/(\varepsilon_P)]^{1/2} \qquad (3.33)$$

Hence, if the 'spot pattern' extends over a circle of radius (ρ_2) equal to 20 mm, on a photographic plate positioned at a distance (L) equal to 750 mm from the single crystal irradiated with a monochromatic electron beam of wavelength equal to 0.04 Å, then this would imply a crystal thickness (ε_p) of about 112 Å in a direction normal to the flat surface of the thin crystal. This calculation assumes that the crystal is perfectly flat (which may be a slight over-simplification). It will be noted that (in Figure 3.30) there is indicated a circle of spots, known as 'Laue Zones', arising from where the Ewald sphere surface intersects the spikes of the ($\bar{1}kl$) layer in the reciprocal lattice. This phenomenon is rarely found in normal electron diffraction patterns. Laue zones have been observed in patterns obtained from very thin layers of mica. The limited extent of the spot pattern arising from planes in the zone whose axis coincides with the direction of the beam is, however, quite commonly observed and provides us with a useful estimate of the crystal thickness (ε_p) through the application of Equation (3.33) above, especially when the spot pattern is obtained by 'Selected Area Electron Diffraction' (see Section 3.3.6).

So far we have only been considering thin perfect crystals being irradiated by an infinitesimally narrow electron beam. In practice, the electron beam width is a

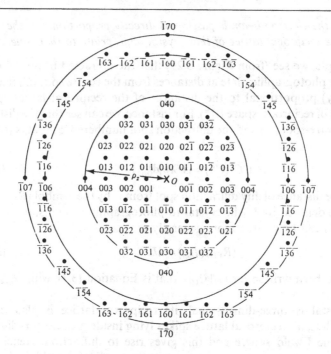

Figure 3.30 *Diagrammatic representation of the type of pattern that could be obtained from a thin cubic crystal with its [100] zone axis parallel to the electron beam (see Figure 3.29)*

few micrometres, although there are some instruments which can now reach beam widths less than 0.1 μm. Furthermore, our single crystal may be bent or strained in some way. Almost certainly, the crystal will consist of 'mosaic-like' crystallites, each crystallite being slightly misaligned with respect to its neighbours. All these factors will cause the diffraction maxima to broaden, even in directions parallel to the most extensive regions of the crystal. We can best represent these effects by suitably modifying the shape of the reciprocal lattice spike to one that is more like an 'oblate spheroid' than a spike, as shown in Figure 3.31 (for the special case of an hexagonal crystal with its thinnest dimension parallel to [001] direction). We will be returning to Figure 3.31 in the next sub-section, which discusses the tribologically-important variation of electron diffraction, namely 'Reflection High-Energy Electron Diffraction' (RHEED).

3.3.3 Reflection high-energy electron diffraction

In the previous sub-section (Section 3.3.2) we mentioned the limited extent of the crystal normal to the surface which contributes to the typical reflection electron diffraction pattern. If the surface was perfectly flat, the penetration would only be about 20 Å beneath that surface. *No* surfaces are perfectly flat and what we normally consider to be 'reflection' electron diffraction patterns are probably electron diffraction patterns arising from *transmission* through thin regions of the

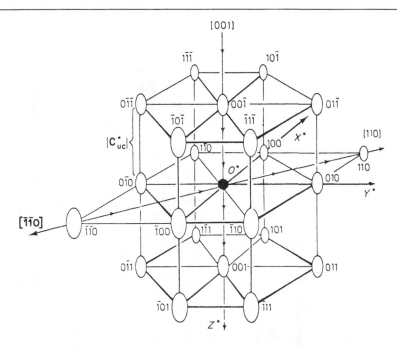

Figure 3.31 *Reciprocal lattice of an hexagonal crystal with its thinnest dimension parallel to* [001], *showing the effects of slight misorientation and strain upon the reciprocal lattice points*

higher regions of that surface. Nevertheless, there is limited penetration of the electron beam in a direction perpendicular to the surface of the specimen and this will have a pronounced effect on the reflection electron diffraction pattern from that surface. Let us refer back to Figure 3.31. Imagine we are irradiating an hexagonal single crystal with the electron beam along the [100] direction, at grazing incidence to the surface of the specimen, which has its [001] direction perpendicular to that surface. Now the Ewald sphere (with its centre O at a distance $(1/\lambda)$ back along the $[\bar{1}\bar{1}0]$ direction) cuts the reciprocal lattice 'points' in the [110] zone, for example the 'points' relating to (001), $(\bar{1}11)$, $(\bar{1}10)$, $(\bar{1}1\bar{1})$, $(00\bar{1})$, $(1\bar{1}\bar{1})$, $(1\bar{1}0)$ and $(1\bar{1}1)$ planes, along their longest dimension, thereby giving rise to broad 'streaks' in the reflection electron diffraction pattern, as shown in Figure 3.32.

This figure is fairly self-explanatory, except perhaps for the shadow-edge. This arises whenever one uses glancing-angle techniques, whether it be glancing-angle X-ray diffraction or reflection electron diffraction. The optimum angle of incidence for glancing-angle X-ray diffraction is around 30°. For reflection *electron* diffraction, however, we try to make the electron beam graze the surface of the specimen at as low an angle as possible. Typically that angle will be less than 1°, so that the shadow edge (due to the flat surface of the bulk of the specimen obscuring any diffracted beams below the level of that surface) will pass through the streaks arising from the $(11\bar{1})$ and $(\bar{1}10)$ planes in Figure 3.32. If any maxima

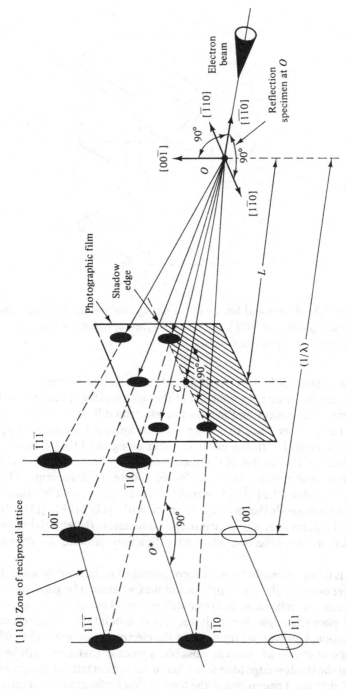

Figure 3.32 Diagram showing how 'streaks' arise in the reflection electron diffraction pattern from the hexagonal crystal.shown in Figure 3.31 when the beam passes along [110], parallel to the (001) plane

appear below the shadow edge, this is a good indication that the diffraction is arising from transmission through upstanding regions on the surface. The appearance of 'streaks' instead of spots is a good indication that the surface is truly flat and smooth and that we really do have reflection electron diffraction. Let us now discuss the electron diffraction patterns from polycrystalline specimens.

3.3.4 Electron diffraction patterns arising from polycrystalline materials

The formation of ring patterns, when a polycrystalline sheet of material is used to diffract a monochromatic beam of X-rays onto a flat photographic film placed perpendicularly to the beam on the other side of the sheet from the X-ray source, has already been described in some detail (see Section 3.2.3). It turns out that this is also the typical geometry used in the transmission electron diffraction camera. Hence we will use Figure 3.19, modified to suit electron radiation. This means that the distance (OO^*) is now $(1/\lambda)$ and, since $(2\theta_{hkl})$ is so small for electron diffraction, A_1 approximately coincides with P_1^*. Furthermore, all points such as $P_1^*, P_2^*, P_3^*, \ldots, P_n^*$ will typically lie at distances that are multiples of $(1/d_{001})$ from O^*, the origin of the reciprocal lattice, that is at distances such as $0.3, 0.6, 0.96 A^{-1}$ (for $d_{001} = 3 \text{ Å}$). Such distances are less than $(1/25)$th of the typical Ewald sphere radius of 25 Å^{-1} for 100 keV electrons. Hence, in general, for a polycrystalline thin film we have $(O^*P_n^*)$ coinciding (to a good approximation) with (A_nB_n) for electron diffraction. In other words, one obtains an electron diffraction maximum whenever the sphere of radius $(O^*P_n^*)$ cuts the 'plane' drawn through O^* perpendicular to the electron beam. The word 'plane' has been put in inverted commas because it is in fact the Ewald sphere surface for a very large radius. Thus we see, in Figure 3.33, that one obtains a circular maximum (i.e. ring) for each set of reciprocal lattice points having the magnitude $|O^*P_n^*|$, in which the radius (R_{ED}) of the diffraction ring will be given by:

$$\frac{L}{(1/\lambda)} = \frac{CP_n}{|O^*P_n^*|} = (R_{ED})(d_{hkl})$$

that is

$$(R_{ED})(d_{hkl}) = \lambda L$$

which is the same as Equation (3.32).

If the crystallites in the polycrystalline specimen are randomly oriented with respect to each other, the rings will be of uniform intensity around their complete circumference. If there is any preferred orientation within the specimen, this will generally appear as a strengthening of these rings into arcs at certain azimuths about the electron beam. Sometimes, however, there are no arcs to indicate preferred orientation, for example if there is a fibre axis parallel to the beam. Under these circumstances, the relative intensities of the uniform diffraction rings will be completely different from those expected of a randomly oriented polycrystalline specimen, due to the multiplicity factors of equivalent (hkl) planes now being quite different from those for a host of randomly oriented crystallites.

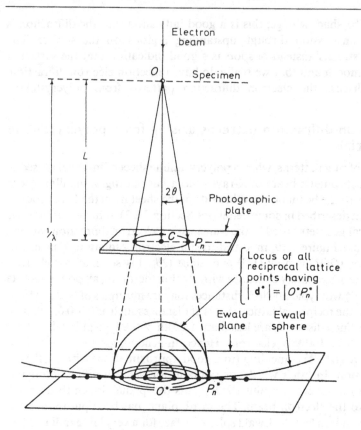

Figure 3.33 *Relation between the reciprocal lattices of a polycrystalline specimen and its transmission electron diffraction pattern*

We shall return to preferred orientation in electron diffraction patterns later in the book (Chapter 5).

3.3.5 The essential features of a high-energy electron diffraction camera

We have already seen that a beam of energetic electrons (acting as if it were a plane wave) may be diffracted by a regular array of atoms in a crystal. Let us now consider the technical details of how this diffraction is accomplished in practice.

Essentially, we need:

(a) an electron gun (the source of the electrons);

(b) a high voltage generator (50 to 100 kV, to accelerate the electrons);

(c) a weak magnetic lens (to obtain sharply focused diffraction patterns);

(d) an evacuated enclosure ($\sim 10^{-5}$ torr, i.e. 1.33×10^{-3} Pa, to enable the electrons to travel sufficient distances to obtain a resolvable pattern);

(e) a specimen loading and manipulator stage (for quick insertion and flexible manipulation of the specimen in the electron beam); and

(f) a means of detecting and recording the diffracted electrons (for example, a screen that fluoresces when it absorbs electrons or a photographic film or (more rarely) a radiation detector).

In the electron gun, there is a hair-pin filament through which passes a current of 2 to 3 A to produce electronic emission. Normally, the filament is connected to the high voltage generator. There is generally a cylindrical modulator cap, with a hole in the centre of the closed lower surface of the cap through which the point of the filament just protrudes. The stainless steel modulator is slightly more negative than the filament, for example it is at $-50\,500$ volts compared with $-50\,000$ volts at the filament. This serves as a crude electrostatic lens, in order to produce a 'cross-over' of the electron beams, thereby providing an effective source as far as the electron diffraction camera is concerned. The electrons, after emerging from the cross-over are accelerated towards the anode plate (A) (which is at earth potential), and then pass through a weak magnetic (condenser) lens through the specimen, on to the fluorescent screen (or photographic plate/film), as shown in Figure 3.34.

By arranging for an image of the cross-over (CO) to be obtained on the photographic plate (P) at the centre-spot (C), we shall, in fact, be in the correct focusing condition for electron diffraction. Remember that the beam of electrons acts only as a plane wave *when passing through the specimen at* S. Thus, the directions ($2\theta_{hkl}$) through which the electrons are scattered depends only on the wave nature of the beam, whilst the focusing of these diffracted beams depends on the effect of the magnetic lens on each electron, as if it were merely a point charge of electricity. Although it may be *conceptually difficult* to see why an electron beam is diffracted by a crystalline specimen, it is *practically* very *easy* to obtain. All the operator needs to do is to alter the current in the condenser lens in order to obtain the minimum size for the centre-spot, as he/she sees it on the fluorescent screen (FS). The angles of electron diffraction (at high energies) are so small, that the slight differences in the specimen-fluorescent screen and the specimen-photographic plate distances have no significant effect on the focused nature of the electron diffraction pattern obtained in the photograph, compared with that seen on the fluorescent screen.

In the caption for Figure 3.34 we described the apparatus as being suitable for 'general area' high-energy electron diffraction. This was to distinguish the technique from the selected area electron diffraction technique discussed in the next sub-section (Section 3.3.6). Some of the electron microscopes available commercially are capable of being converted into general area instruments. This is done by placing a 'general area electron diffraction' attachment *below* the final projector lens of the electron microscope, and turning off the appropriate lenses (e.g. turn off one of the projector lenses, the objective lens and one of the condenser lenses, for a normal two-condenser, two-projector instrument). One then uses the electron microscope exactly as if it were the instrument described above. Let us now describe how to use an electron microscope for electron diffraction, without any changes in specimen position from its normal position for electron microscopy.

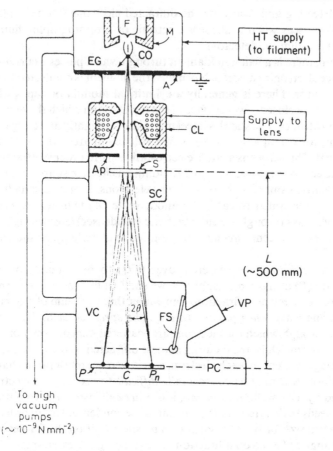

Figure 3.34 *A typical general-area, high-energy electron diffraction camera. P_n, diffraction maximum; A_p, limiting aperture; C, centre spot; SC, specimen chamber; F, filament; S, specimen; M, modulator; VC, viewing chamber; EG, electron gun; VP, viewing port; A, anode plate; FS, fluorescent screen; CO, 'cross-over'; P, photographic plate/film; CL, condenser lens; PC, plate chamber*

3 3.6 Selected-area, transmission electron diffraction

The selected-area, transmission, high-energy electron diffraction technique, although not so accurate as the general area mode of operation, is nevertheless one of the most powerful aids to microanalysis yet discovered. With this technique, it is possible to use the high-resolution electron microscope image of a (crystalline) transmission specimen in order to select a very small area for analysis, and then obtain an electron diffraction pattern from that small selected area of the specimen. Quite often, the area selected will be less than the grain size of the polycrystalline thin foil being examined, so that the electron diffraction is that expected from a single crystal, i.e. a spot pattern.

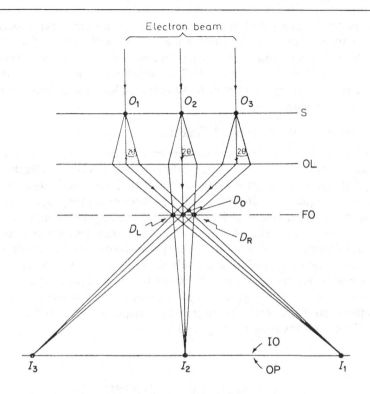

Figure 3.35 *Ray diagram showing how the electron diffraction pattern of a crystalline specimen is formed in the focal plane of the objective lens. S, crystalline specimen; IO, image plane of objective lens; OL, objective lens; OP, objective plane of projector lens; FO, focal plane of objective lens*

Figure 3.35 shows how the image of a small part of a crystalline object (of which O_1, O_2 and O_3 are typical points) is formed in the image plane (IO) of the objective lens, which also coincides with the object plane (OP) of the projector lens, *for correct operation as an electron microscope*. Note that all the electrons diffracted through ($\pm 2\theta_{hkl}$) upon passing through the specimen, will be brought to a focus at D_L and D_R in the focal plane of the objective (FO). All undeviated beams will be brought to a focus at D_O. If we wish to view the diffraction pattern formed in the focal plane of the objective (FO), all we need to do is to reduce the current through the projector lens (in the case of a two-projector instrument, this will be called 'projector 1' or 'the intermediate lens') until the object plane of the projector lens (OP) coincides with (FO), the focal plane of the objective lens. The diffraction pattern will relate *only* to the circular part of the area for which (O_1O_3) is the diameter, if we arrange for a circular aperture (in the object plane (OP) of the projector lens) to *just* accept the extreme rays I_1 and I_3. In most electron microscopes, all one needs to do is to remove the objective aperture (typically of such a radius that it will only accept electrons scattered through angles less than

10^{-3} rad), switch the projector lens current to a pre-set value, and then make fine adjustments to that current so as to obtain a well-focused electron diffraction pattern. The reason for removing the objective aperture is so that the diffracted beams (i.e. those scattered through angles, $2\theta_{hkl}$, which are an order of magnitude larger than 10^{-3} rad) may pass through the objective lens and be brought to a focus at the object plane.

3.3.7 Low-energy electron diffraction

So far, in this section, we have been concerned with electrons that have been accelerated through voltages greater than 50 kV. In fact, it was the diffraction of low-energy electrons by single crystals of nickel that first established the de Broglie formula (Equation (2.26)) as more than a theoretical idea proposed in the PhD thesis of a member of a French noble family! The original experiments are described in almost any textbook on modern physics and were carried out by Davisson and Germer. The modern low-energy electron diffraction camera owes its current important status as an extremely surface-sensitive technique to the relative ease with which one is now able to obtain the necessary high vacuum ($\sim 10^{-9}$ torr, i.e. 1.33×10^{-7} Pa). The high vacuum is to allow the low-energy electrons (between 5 and 500 volts) to travel the required distances from their source to the crystal (as shown in Figure 3.36).

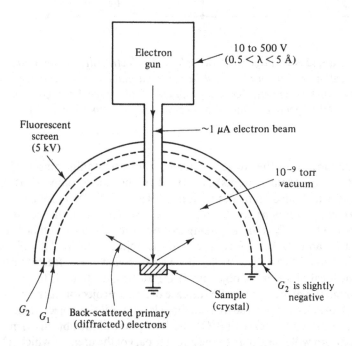

Figure 3.36 *Schematic diagram of low-energy electron diffraction (LEED) camera (the fluorescent screen is maintained at 5 kV relative to the sample, thereby enabling the low-energy diffracted electrons to be accelerated on to the screen)*

Essentially a beam of low energy electrons (say about 100 eV) is made to impinge on the surface of a crystal. Because of the low energy of the electrons, the penetration d_p beneath the surface is limited to (say) 3 Å for electrons which will emerge without any loss of energy (i.e. will have suffered an elastic collision, one of the prerequisites for diffraction). This limited penetration normal to the surface of the crystal means that the width b of the reciprocal lattice point is $(\frac{2}{3})$ Å^{-1} in that direction. Assuming a cubic crystal of lattice constant equal to 2 Å, then we have $|\mathbf{d}^*_{001}| = (\frac{1}{2})$Å$^{-1} = |\mathbf{d}^*_{100}| = \mathbf{d}^*_{010}|$. The Ewald sphere radius (for 100 eV electron beam) is $(1/1.22$ Å$) = 0.82$ Å^{-1}. Because the width of a reciprocal lattice point is greater than the distance between the $(00l)$ layer and the $(\bar{1}0l)$ layer in reciprocal space, we have a continuous spike at (001) (as shown in Figure 3.37) which cuts the Ewald sphere at B, as well as at A. This spike relates to all reciprocal lattice points such as $(\bar{3}01)$ in '*back* reflection' and will give rise to a diffraction spot at S on the spherical fluorescent screen. It is not possible to obtain a diffraction maximum in the *forward* direction OA, where the electron beam *would have cut* the reciprocal lattice spike due to (001), because of obscuration by the surface of the crystal.

Effectively, this shows that the Bragg condition is relaxed for very thin crystals (or for limited penetration) of the order of a few ångström units, so that we still obtain diffraction at the angle defined by (O^*OA), even though this is significantly greater than $2\theta_{001}$. We do not see this diffraction maximum because of the bulk of the crystal, but we can see the maximum (at S) which arises from the reciprocal lattice spike at $\bar{3}01$ cutting the Ewald sphere at B. The angle defined by (O^*OB) in Figure 3.37 differs only slightly from the obtuse angle $2\theta_{301}$. Hence we will get a square pattern on the surface of the fluorescent sphere, where all the points have their k and l indices defined by the [100] zone in reciprocal space, but where the h index cannot really be assigned. This is because it could also have arisen from the $(2kl)$, the $(\bar{1}kl)$ or even the $(0kl)$ spikes for a crystal where $d_p = 3$ Å, that is $b = 2/3$ Å $= 0.67$ Å^{-1} (for an extremely thin layer of a crystal lying on top of a bulk crystal). This is why the patterns are sometimes referred to as arising from a *two-dimensional lattice*.

We need not concern ourselves with the technical details of the LEED apparatus, since it is most unlikely that such an apparatus exists in the typical materials analysis laboratory. Essentially, however, the electrons are accelerated through a few thousand volts, *after* being diffracted at the crystal, so as to obtain sufficiently intense spots on the spherical screen. The pattern is photographed through a glass window placed behind the crystal. (The crystal is in the very high vacuum of the apparatus.) The LEED pattern will contain information about chemisorption, physisorption and surface topography. As an example of its power for tribological investigations, it has been found that LEED is capable of detecting the effect on a surface of as minute a quantity as 5×10^{-14} kg/mm^2 of adsorbate. However, apart from some work regarding the effects of adsorbed layers on metal surfaces as regards their catalytic action, very little research seems to have been done of practical significance to tribologists. This is unfortunate for tribology, since, for example, it is possible that the early stages of the mechanism whereby 'anti-wear' additives provide their protection could usefully be in-

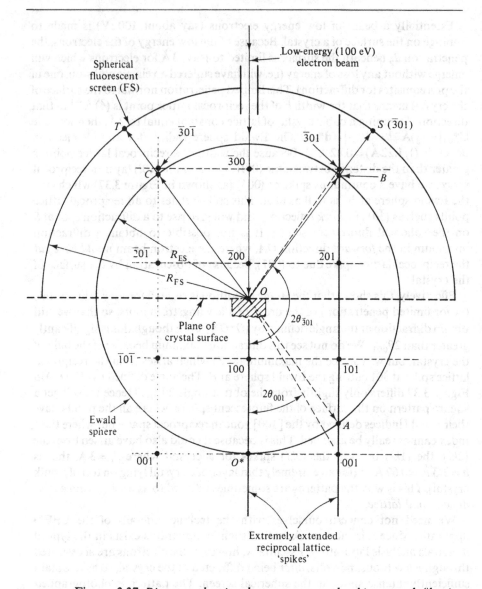

Figure 3.37 *Diagram showing how an extremely thin crystal film (or limited penetration of electron beam into a bulk crystal) gives rise to the 'back reflection' of 100 eV electrons.* $\lambda = 1.22$ Å; $R_{ES} = 1/\lambda = 0.82$ Å^{-1}; $d_p = 3$ Å; $b = 2/d_p = 0.66$ Å^{-1} $|d^*_{\bar{1}00}| = 2$ Å^{-1}; $|d^*_{\bar{1}10}| = 1/d_{\bar{1}00}|$; *therefore* $|d^*_{\bar{1}00}|1 = 0.5$ w^{-1}

vestigated with this technique, combined with (say) Auger electron spectroscopy. For further information on this technique, interested tribologists are advised to consult the original papers of Ehrenberg (1934), Dillon, Schlier and Farnsworth (1959), Germer and Hartman (1960) and MacRae (1963). Let us now end this review of electron diffraction by discussing the basic ideas behind the scattering of electrons by crystalline solids, that is the theory of electron diffraction intensities.

3.3.8 Electron diffraction intensities

When an electron beam interacts with a crystalline solid, *some* of the constituent electrons are scattered with a significant loss of energy, that is they are scattered *inelastically*, and some are scattered with virtually no exchange of energy, that is they are *elastically* scattered. Most of the *inelastic* scattering occurs when an electron in the electron beam interacts with the *extra-nuclear electrons* in an atom. Not all the electrons in the beam interact with the extra-nuclear electrons in the atoms. Some manage to pass through the 'electron cloud' without any loss of energy. Once inside the sphere of influence of the positively-charged nucleus, they are scattered, that is they have their trajectory changed from their original direction, by amounts that are related to the 'nearest distance of approach' between the incident electron and the nucleus. Due to the comparatively massive nature of the nucleus, it does not move under the influence of the electrostatic force of attraction between it and the passing electron. Hence the electron is *elastically scattered by the positive nucleus*. This scattering can occur in any direction. When all the nuclei are placed at the positions of equilibrium in a crystal lattice, it is clear that, if we regard the electrons in the incident beam as plane waves travelling in the direction of the beam, then the scattering of these electrons can be regarded also as spherical waves originating from within each atom. Hence, only in certain directions ($2\theta_{hkl}$) from the original direction will the elastically scattered electron waves constructively interfere (where θ_{hkl} is the Bragg angle for electron diffraction). Remember that, in this direction, we will also have electrons which have been inelastically scattered by the extra-nuclear electrons. These will contribute to the general background of the electron diffraction pattern, whereas the elastically scattered electrons contribute to the diffraction maxima at ($2\theta_{hkl}$). It can be shown, by wave mechanical calculations beyond the scope of this book, that the atomic scattering amplitude (factor) for electrons ($f_e(\theta)$) is given by:

$$f_e(\theta) = \left[\frac{(Z - f_X)e}{16\pi V_A \sin^2 \theta} \right] \tag{3.34}$$

where the first term (Z) represents the elastic scattering of the electron wave by the ($+Ze$) point charge of the nucleus. The second term (f_X) represents the inelastic scattering by the Z extra-nuclear electrons in every atom. It is, in fact, the atomic scattering amplitude factor for X-rays which, we may recall, was strongly connected with the probability of finding an electron in a particular energy level within the atom. Of course, there will be elastically-scattered electrons in other directions than $(2\theta_{hkl})$. Hence the angle θ in Equation (3.34) can relate to elastically scattered electrons travelling in *any* direction (2θ) with respect to the incident beam direction. These electrons will also contribute to the general background of the electron diffraction pattern. The other terms in Equation (3.34) are V_A, the accelerating voltage and e, the electronic charge. To be strictly correct, this expression should also contain (ε_0), the permittivity of vacuo, thereby making Equation (3.34) consistent as regards SI units. Table 3.4 is based on the above equation and is generally accepted as the best available values for $f_e(\theta)$.

The values included in Table 3.4 are reproduced by courtesy of Dr J. A. Ibers and *Acta Crystallographica*. They are based on the rest mass of the electron. The reference denoted by 'E' means that the values have been obtained by computing the *exact* wave function for hydrogen. The reference 'HF' stands for the fact that the Hartree–Fock wave function was used to determine the probability of finding an electron in a given atomic orbit. This wave function is one in which the atom is considered to be hydrogen-like, with the nucleus $(+Ze)$ and all the extra-nuclear electrons save one $[-(Z-1)e]$ acting as the central positive charge. Each of the extra-nuclear electrons are taken, in turn, as the orbiting electron and the resulting wave function is thereby refined to become approximately correct for any multi-electron atom with atomic number Z less than about 35. For Z greater than 35, one has to use the Thomas–Fermi–Dirac (TFD) potential function in the Schrödinger wave-equation. Where the reference is quoted as 'I', this means they relate to *new* Hartree–Fock wave functions deduced by Dr Ibers in 1961 and are more valid than those given in the *International Tables for X-ray Crystallography* (Volume 3).

It is interesting to note how the atomic scattering amplitude for electrons is so dependent upon the validity of the same quantity for X-rays. Any unreliabilities in the 'accepted' values for f_X will obviously lead to similar unreliabilities in the 'accepted' values for f_e. In the references for Chapter 2, we mentioned the paper by Halliday and Quinn (1960). It is interesting to note that, according to these authors, the *amorphous contrast of electron micrographs* of crystalline materials depends almost entirely on the *inelastic scattering cross-sections* (at the very low aperture angles used in electron microscopy). All attempts to relate the experimentally-measured contrast with theoretically-expected values were unsuccessful. The authors suggested that these discrepancies could be due to errors in theoretical scattering cross-sections, but were unable to substantiate their suggestions with any revised wave functions for f_X. The reader may have noticed some discontinuities or divergence from expected values when one has a new Hartree–Fock wave function, for example for tungsten (W), compared with the adjacent values of f_e for Ta and Re. This indicates the reliability of most of the TFD values, namely about 10%. The HF and I values are fairly reliable up to $Z \sim 35$. In this respect, the reader should also be aware that the values in parentheses in Table 3.4 are either interpolations or extrapolations. We will be using this table later in the book. However it should be emphasized that the basis of electron scattering cross-sections is still not so reliably known as for X-ray scattering which is, effectively, elastic (since the secondary X-rays emitted from

Table 3.4. *Atomic scattering amplitudes* $[f_e(\theta)]$ *for electrons (from Dr J. A. Ibers)*

Element	Z	Reference	$(\sin\theta)/\lambda$											
			0.00	0.05	0.10	0.15	0.20	0.25	0.30	0.35	0.40	0.50	0.60	0.70
H	1	E	0.529	0.508	0.453	0.382	0.311	0.249	0.199	0.160	0.130	0.089	0.064	0.04
He	2	I	(0.445)	0.431	0.403	0.368	0.328	0.288	0.250	0.216	0.188	0.142	0.109	0.086
Li	3	HF	3.31	2.78	1.88	1.17	0.75	0.53	0.40	0.34	0.26	0.19	0.14	0.11
Be	4	HF	3.09	2.81	2.23	1.63	1.16	0.83	0.61	0.47	0.37	0.25	0.19	0.14
B	5	HF	2.82	2.62	2.23	1.78	1.37	1.04	0.80	0.62	0.49	0.33	0.24	0.18
C	6	HF	2.45	2.26	2.06	1.74	1.43	1.15	0.92	0.74	0.60	0.41	0.30	0.22
N	7	HF	2.20	2.09	1.91	1.68	1.44	1.20	1.00	0.83	0.69	0.48	0.35	0.27
O	8	HF	2.01	1.95	1.80	1.61	1.42	1.22	1.04	0.88	0.75	0.54	0.40	0.31
F	9	HF	2.12	2.01	1.90	1.71	1.50	1.29	1.11	0.95	0.81	0.60	0.45	0.35
Ne	10	I	(1.66)	1.59	1.53	1.43	1.30	1.11	1.04	0.92	0.80	0.62	0.48	0.38
Na	11	HF	4.89	4.21	2.97	2.11	1.59	1.29	1.09	0.95	0.83	0.64	0.51	0.40
Mg	12	HF	5.01	4.60	3.59	2.63	1.94	1.50	1.21	1.01	0.87	0.67	0,53	0.43
Al	13	I	(6.1)	5.36	4.24	3.13	2.30	1.73	1.36	1.11	0.93	0.70	0.55	0.45
Si	14	I	(6.0)	5.26	4.40	3.41	2.59	1.97	1.54	1.23	1.02	0.74	0.58	0.47
P	15	I	(5.4)	5.07	4.38	3.55	2.79	2.17	1.70	1.36	1.12	0.80	0.61	0.49
S	16	I	(5.2)	4.88	4.36	3.63	2.93	2.33	1.85	1.49	1.22	0.86	0.65	0.51
Cl	17	I	(5.0)	4.69	4.24	3.62	3.00	2.44	1.98	1.61	1.32	0.93	0.69	0.54
Ar	18	HF	4.71	4.40	4.07	3.56	3.03	2.52	2.07	1.71	1.42	1.00	0.74	0.57
K	19	I	(9.0)	(7.0)	5.43	(4.10)	3.15	(2.60)	2.14	(1.90)	1.49	1.07	0.79	0.61
Ca	20	HF	10.46	8.71	6.40	4.54	3.40	2.69	2.20	1.84	1.55	1.12	0.84	0.65
Sc	21	I	(9.7)	(8.35)	(6.30)	(4.63)	(3.50)	(2.75)	(2.29)	(1.92)	(1.62)	(1.18)	(0.89)	(0.69)
Ti	22	I	8.9	7.95	6.20	4.63	3.55	2.84	2.34	1.97	1.67	1.23	0.93	0.72
V	23	I	8.4	7.60	6.06	4.60	3.57	2.88	2.39	2.02	1.72	1.28	0.97	0.76

Table 3.4 (*cont.*)

Element	Z	Reference	(Sin θ)/λ											
			0.00	0.05	0.10	0.15	0.20	0.25	0.30	0.35	0.40	0.50	0.60	0.70
Cr	24	I	(8.0)	(7.26)	(5.86)	(4.55)	(3.58)	(2.89)	(2.42)	(2.06)	(1.76)	(1.32)	(1.01)	(0.80)
Mn	25	I	7.7	7.00	5.72	4.48	3.55	2.91	2.44	2.08	1.79	1.36	1.04	0.83
Fe	26	I	7.4	6.70	5.55	4.41	3.54	2.91	2.45	2.11	1.82	1.39	1.08	0.86
Co	27	I	7.1	6.41	5.41	4.34	3.51	2.91	2.46	2.12	1.84	1.42	1.11	0.89
Ni	28	I	6.8	6.22	5.27	4.27	3.48	2.90	2.47	2.13	1.86	1.46	1.14	0.92
Cu	20	I	(6.5)	(6.00)	(5.11)	(4.19)	(3.44)	(2.88)	(2.46)	(2.13)	(1.87)	(1.47)	(1.16)	(0.95)
Zn	30	I	(6.5)	5.48	5.00	4.14	3.42	2.88	2.47	2.14	1.88	1.48	1.10	0.96
Ga	31	I	(7.5)	6.70	5.62	4.51	3.64	3.00	2.53	2.18	1.91	1.50	1.20	0.98
Ge	32	I	(7.8)	6.89	5.93	4.81	3.87	3.16	2.63	2.24	1.94	1.51	1.22	0.99
As	33	I	(7.8)	6.99	6.05	5.01	4.07	3.32	2.74	2.31	1.99	1.54	1.23	1.01
Se	34	I	(7.7)	6.99	6.15	5.18	4.24	3.47	2.86	2.40	2.05	1.57	1.23	1.02
Br	35	I	(7.3)	6.80	6.15	5.25	4.37	3.60	2.97	2.49	2.12	1.60	1.27	1.04
Kr	36	TFD	7.9	7.56	6.80	5.74	4.71	3.85	3.19	2.69	2.13	1.73	1.35	1.08
Rb	37	TFD	8.0	7.75	6.92	5.85	4.80	3.93	3.26	2.75	2.35	1.77	1.38	1.10
Sr	38	TFD	8.2	7.85	7.04	5.96	4.89	4.00	3.32	2.80	2.40	1.80	1.41	1.13
Y	39	TFD	8.3	8.04	7.16	6.06	4.98	4.07	3.38	2.86	2.45	1.84	1.44	1.15
Zr	40	TFD	8.5	8.14	7.28	6.16	5.06	4.15	3.45	2.91	2.51	1.88	1.47	1.17
Nb	41	TFD	8.6	8.23	7.40	6.27	5.15	4.22	3.51	2.97	2.54	1.92	1.50	1.20
Mo	42	TFD	8.7	8.42	7.52	6.36	5.24	4.29	3.57	3.02	2.59	1.95	1.53	1.22
Tc	43	TFD	8.9	8.52	7.63	6.47	5.31	4.36	3.63	3.08	2.64	1.99	1.56	1.25
Ru	44	TFD	9.0	8.62	7.75	6.56	5.40	4.43	3.69	3.13	2.68	2.03	1.58	1.27
Rh	45	TFD	9.1	8.81	7.85	6.66	5.48	4.50	3.75	3.18	2.73	2.06	1.61	1.30
Pd	46	TFD	9.3	8.90	7.97	6.75	5.56	4.57	3.81	3.23	2.77	2.10	1.64	1.32
Ag	47	I	(8.8)	8.24	7.47	6.51	5.58	4.75	4.05	3.46	2.97	2.22	1.70	1.35
Cd	48	TFD	9.5	9.19	8.19	6.95	5.72	4.71	3.93	3.34	2.86	2.17	1.71	1.37
In	49	TFD	9.6	9.25	8.31	7.03	5.80	4.78	3.99	3.39	2.91	2.20	1.73	1.39

Sn	50	TFD	9.8	9.38	8.40	7.13	5.88	4.84	4.05	3.44	2.95	2.24	1.76	1.41
Sb	51	TFD	9.9	9.48	8.50	7.22	5.95	4.91	4.10	3.49	3.00	2.27	1.79	1.44
Te	52	TFD	10.0	9.57	8.62	7.31	6.03	4.97	4.16	3.54	3.04	2.31	1.81	1.46
I	53	TFD	10.1	9.77	8.71	7.39	6.11	5.04	4.22	3.59	3.08	2.34	1.84	1.48
Xe	54	TFD	10.2	9.86	8.81	7.49	6.19	5.10	4.27	3.64	3.13	2.38	1.87	1.51
Cs	55	TFD	10.4	9.96	8.93	7.57	6.26	5.17	4.33	3.68	3.17	2.41	1.90	1.53
Ba	56	TFD	10.5	10.05	9.02	7.66	6.34	5.23	4.39	3.73	3.21	2.45	1.93	1.55
La	57	TFD	10.6	10.15	9.12	7.75	6.40	5.30	4.44	3.78	3.26	2.48	1.95	1.57
Ce	58	TFD	10.7	10.24	9.21	7.84	6.49	5.36	4.50	3.83	3.30	2.51	1.98	1.60
Pr	59	TFD	10.8	10.44	9.31	7.92	6.56	5.42	4.55	3.88	3.34	2.55	2.01	1.62
Nd	60	TFD	10.9	10.53	9.41	8.01	6.63	5.48	4.60	3.93	3.38	2.58	2.03	1.64
Pm	61	TFD	11.0	10.63	9.53	8.10	6.70	5.55	4.66	3.97	3.43	2.61	2.06	1.66
Sm	62	TFD	11.1	10.72	9.62	8.17	6.77	5.61	4.71	4.02	3.47	2.65	2.09	1.69
Eu	63	TFD	11.2	10.82	9.72	8.25	6.85	5.67	4.77	4.07	3.51	2.68	2.11	1.71
Gd	64	TFD	11.4	10.91	9.79	8.34	6.91	5.73	4.82	4.11	3.55	2.71	2.14	1.73
Tb	65	TFD	11.5	11.01	9.88	8.42	6.98	5.79	4.87	4.16	3.59	2.74	2.17	1.75
Dy	66	TFD	11.6	11.11	9.98	8.50	7.05	5.85	4.92	4.20	3.63	2.78	2.19	1.77
Ho	67	TFD	11.7	11.20	10.08	8.58	7.12	5.91	4.98	4.25	3.67	2.81	2.22	1.80
Er	68	TFD	11.8	11.30	10.17	8.66	7.19	5.97	5.03	4.30	3.71	2.84	2.25	1.82
Tm	69	TFD	11.9	11.49	10.27	8.74	7.26	6.03	5.08	4.34	3.75	2.87	2.27	1.84
Yb	70	TFD	12.0	11.58	10.36	8.82	7.33	6.09	5.13	4.39	3.79	2.19	2.30	1.86
Lu	71	TFD	12.1	11.68	10.44	8.90	7.40	6.15	5.18	4.43	3.83	2.94	2.32	1.88
Hf	72	TFD	12.2	11.78	10.53	8.98	7.46	6.20	5.23	4.48	3.87	2.97	2.35	1.90
Ta	73	TFD	12.3	11.87	10.63	9.05	7.53	6.26	5.28	4.52	3.91	3.00	2.38	1.93
W	74	I	(14)	—	11.80	—	7.43	—	5.16	—	3.85	2.99	2.39	1.96
Re	75	TFD	12.5	12.06	10.79	9.21	7.66	6.38	5.38	4.61	3.99	3.06	2.43	1.97
Os	76	TFD	12.6	12.16	10.89	9.29	7.72	6.43	5.43	4.65	4.03	3.19	2.45	1.99
Ir	77	TFD	12.7	12.25	10.96	9.36	7.79	6.49	5.48	4.70	4.06	3.12	2.50	2.01
Pt	78	TFD	12.8	12.35	11.06	9.44	7.86	6.54	5.53	4.74	4.19	3.16	2.50	2.03
Au	79	TFD	12.9	12.45	11.13	9.51	7.92	6.60	5.58	4.78	4.14	3.19	2.53	2.05
Hg	**80**	**I**	**(13.3)**	**12.26**	**10.82**	**9.18**	**7.70**	**6.48**	**5.50**	**4.72**	**4.09**	**3.16**	**2.51**	**2.05**

the extra-nuclear electrons moving in the field of an X-ray photon, have the same wavelength as the incident X-rays and hence, the same energy).

Having determined the atomic scattering for electrons $[f_e(\theta)]$, the calculation of the relative integrated intensity of the diffraction maxima would be fairly straightforward, if one could be sure that the only electron waves contributing to the maxima were the *original* primary wave (i.e. the incident electron beam) and the *singly-diffracted* electron wave in the direction $(2\theta_{hkl})$. This is called the 'two-beam approach', in which it is assumed that the amplitude of the primary wave (and hence, its intensity) is the same as it falls, in turn, on each atom within each crystallite. This is not strictly true, since some electrons will find themselves re-diffracted (by the same crystal planes) back into the beam. In fact, the situation is more complex than this, since the 'wave' we are talking about is merely a 'probability wave'. But when we are dealing with the countless millions of electrons in an electron beam, these probabilities of finding an electron going in a certain direction become much more well-determined. Effectively then, we find that the primary beam will *lose* electrons through elastic scattering (in the direction $2\theta_{hkl}$) by previously-traversed atoms, and through *inelastic* collisions (but this latter phenomenon is allowed for in $[f_e(\theta)]$). The primary beam will also *gain* electrons from the electron 'waves' that are diffracted back into the original direction by previously-traversed atoms. The two-beam model is called the 'kinematical theory'. The multiple-beam approach, when an equilibrium eventually occurs between these gains and losses of electrons, is called the 'dynamical theory'.

Rymer (1970) has shown that, for the kinematical theory, the intensity diffracted into a complete diffraction ring (i.e. the whole circumference at the radius R_{ED}) is given by $(I'_{hkl})_{ED}$ where:

$$\frac{(I'_{hkl})_{ED}}{I_0} = \frac{[(F_e)_{hkl}]^2 (\lambda)^2 (d_p)(d_{hkl})}{2(v_{uc})^2} \tag{3.35}$$

where I_0 is the incident intensity of the electron beam, λ is the wavelength, (d_p) is the thickness of the crystalline specimen and (v_{uc}) is the volume of the unit cell within the crystallites making up the specimen. The term $[(F_e)_{hkl}]$ is the structure factor for electron diffraction, and is completely analogous to the same factor for X-ray diffraction, except we substitute $[f_e(\theta)]$ for $[f_X(\theta)]$.

Equation (3.35) should be modified to allow for the Debye–Waller factor $[\exp(-D)]$, the multiplicity factor (m) and to allow for the fact that one normally measures *across* the profile of the electron diffraction ring with a slit placed tangentially to the ring. Such a slit (of length $\imath$) will admit a fraction $(l/2\pi R_{ED})$ of the electrons diffracted into the complete ring. But we know (from Equation (3.32)) that this fraction is $l(d_{hkl})/2\pi\lambda L$, so that the intensity diffracted per unit length of a diffraction ring can be written (from Equation (3.35)) as $(I_{hkl})_{ED}$ where:

$$\frac{(I_{hkl})_{ED}}{I_0} = \frac{[(F_e)_{hkl}]^2 (l)(d_p)(d_{hkl})^2(\lambda)}{4\pi L(v_{uc})^2} \tag{3.36}$$

Allowing for thermal vibration of the atoms (the Debye–Waller temperature factor) and the multiplicity, we have the *intensity per unit length of an electron diffraction ring* given by:

$$\frac{(I_{hkl})_{\text{ED}}}{I_0} = \frac{(\lambda)\{(F_e)_{hkl}[\exp(-D)]\}^2(d_p)(d_{hkl})^2(m)}{4\pi(v_{\text{uc}})^2} \tag{3.37}$$

This is the *relative* integrated intensity from ring to ring of the same pattern (where it can be assumed that the camera length L and the slit length l do not change). *Remember, this is the kinematical theory of electron diffraction.*

Under the conditions where the dynamical theory applies, a dynamical equilibrium occurs between the original beam intensity and that of the diffracted beam (as already discussed). Provided the crystal is thick enough, the ratio of the amplitudes of the original and diffracted beams is independent of its thickness. Rymer (1970) shows that, under dynamical theory conditions, the intensity for a complete ring (I'_{hkl}) is given by:

$$\frac{(I'_{hkl})_{\text{ED}}}{I_0} = \frac{[(F_e)_{hkl}](\lambda)(d_{hkl})}{4(v_{\text{uc}})} \tag{3.38}$$

Comparison between Equations (3.38) and (3.35) reveals the critical thickness $(d_p)_c$ when $[(I'_{hkl})_{\text{ED}}/I_0]$ are made equal to each other for kinematical and dynamical conditions, namely:

$$(d_p)_c = \frac{(v_{\text{uc}})}{2[(F_e)_{hkl}]\lambda} \tag{3.39}$$

It has been suggested that, for $(d_p) > (d_p)_c$, the dynamical theory will operate, and vice versa for $(d_p) < (d_p)_c$. There is no general agreement about the criterion to be used, but Rymer (1970) suggests that (d_p) should not be greater than the 'extinction distance' (Δ) for using the kinematical theory of electron diffraction, where Δ is given by:

$$\Delta = \frac{\pi(v_{\text{uc}})}{[(F_e)_{hkl}]\lambda} \tag{3.40}$$

The 'extinction distance' is a concept taken over from the interpretation of electron micrographs of crystalline thin films and need not concern us at this stage. Clearly, $(d_p)_c$ and Δ only differ by a factor of $\pi/2$, which is trivial, when one is really only after an 'order of magnitude' estimate.

It can be seen that $(d_p)_c$ (or Δ) depends inversely upon $(F_e)_{hkl}$ and λ. To ensure that one can assume kinematical conditions *for a given maximum* in a diffraction pattern, we can increase $(d_p)_c$ by decreasing λ, that is by increasing the accelerating voltage for the incident electron beam. For a given λ, however, it is possible for the *inner rings* of an electron diffraction (polycrystalline) pattern to have a different $(d_p)_c$ from the *outer rings*, due to differences in the $[(F_e)_{hkl}]$ values. Strictly, we should also include the Debye–Waller temperature factor $[\exp(-D)]$ in with the atomic structure factor for electrons. This also depends upon the Bragg angle (θ_{hkl}). Hence, it is difficult to predict which theory we should use in order to use electron diffraction *intensities quantitatively*. There are great possibilities in using high-energy, transmission electron diffraction for the proportional analysis of thin surface films, *provided* we can be sure that *either* the kinematical *or* the dynamical theory of electron diffraction applies and *not*, as we seem to so often find, *a mixture* of both theories, as we compare the intensities of various different maxima in a given electron diffraction pattern.

To complete this sub-section on electron diffraction intensities, we must (as with the kinematical approach), allow for the multiplicity (m) of the (hkl) planes, the temperature factor $[\exp(-D)]$ and the slit method for measuring intensities $(l d_{hkl}/2\pi\lambda L)$ to get:

$$\frac{(I_{hkl})_{\text{ED}}}{I_0} = \frac{[(F_e)_{hkl}][\exp(-D)](l)(d_{hkl})^2(m)}{8\pi L(v_{\text{uc}})} \tag{3.41}$$

This equation does not explicitly involve the thickness of the specimen. Rymer (1970) states that this equation (actually he refers to Equation (3.38)) relates to a 'single layer' of crystals. We must assume he meant a layer of sufficient thickness (d_p) to ensure that the dynamical conditions occur. Now if the specimen has a crystallite size (ε) that is very much less than $(d_p)_c$, then the kinematical theory (Equation (3.34)) will apply. If, however, the crystallite size ε is much greater than $(d_p)_c$, then the dynamical theory for each single layer within the specimen thickness (d_p) will contribute to the final diffracted intensity. Hence, the single layer expression (Equation (3.41)) must be multiplied by the number of such layers within the thickness (d_p), that is we must multiply by a factor (d_p/ε). The final expression for the *relative* integrated intensity (per unit length of diffraction ring) *for the dynamical theory* is given by:

$$\frac{(I_{hkl})_{\text{ED}}}{I_0} = \frac{[(F_e)_{hkl}][\exp(-D)](d_p)(d_{hkl})^2 m}{8\pi(v_{\text{uc}})\varepsilon} \tag{3.42}$$

As with Equation (3.37), we have assumed that Equation (3.42) is for a given l and L, for a given pattern. We can write *both* these equations in the form:

$$\frac{(I_{hkl})_{\text{ED}}}{I_0} = \left\{\frac{(F_e)_{hkl}[\exp(-D)]}{(v_{\text{uc}})}\right\}^n \left[\frac{\lambda^{(n-1)}(d_p)(d_{hkl})^2 m}{4\pi(2\varepsilon)^{(2-n)}}\right] \tag{3.43}$$

where $n=1$ for the dynamical theory and $n=2$ for the kinematical theory.

3.4 Concluding remarks regarding the importance of microscope techniques for studying the atomic arrangements in tribo-elements

In my opinion, this is the most important of the introductory chapters dealing with the techniques which are used in tribological investigations. The next chapter, which describes the microscopic techniques for studying the *atomic structures* of materials, can only give information about the various *elements* present in a tribo-system. This is very important information, especially as many of the techniques to be described in Chapter 4 can be used to give precise *locations* of those elements with respect to the surface topography, and precise determinations of *proportional amounts* of each element in a tribo-element. Most tribo-systems, however, are comprised of tribo-elements containing alloys of several metals. These alloys form different phases during sliding, due to the mechanical and thermal interactions always involved in this activity. Also, the surfaces of these tribo-elements often form oxides or some other surface film due to interaction with the ambient atmosphere, whether it be oxygen in the air or an additive in the lubricant. Knowledge of how the atoms and nuclei are changed as regards this arrangement in crystal lattices can be more important to the tribologists than merely the detection of the elements of which the crystals are composed.

X-ray diffraction is the more accurate of the two techniques discussed in this chapter. It does suffer, however, from requiring a relatively large volume of material in order to get a usable X-ray diffraction pattern. We shall see how the requirement is readily satisfied when we use wear-debris as the specimen. For thin films, however, the technique has a lower limit of about 0.5 μm, below which even the edge-irradiated, glancing-angle X-ray diffraction technique cannot be used. Electron diffraction suffers from being less accurate than X-ray diffraction. It is, however, a very surface-sensitive technique and, consequently, gives information that cannot be provided by any other technique. One must be prepared to forego accuracy for the sake of the *uniqueness* of the information. Obviously, the tribologist should be prepared to use *both* techniques, since tribological interactions will always involve both the immediate surface, and the sub-surface, layers of the tribo-elements. Low-energy electron diffraction (LEED) and selected area electron diffraction of thin foils clearly have potential for tribological studies, but their limited ranges of applicability may hamper their development as applied to tribology. The *as yet unrealized potential of proportional analysis of thin films by electron diffraction* should really bring this technique to the forefront of the physical techniques for tribology.

4 Microscopic techniques for studying atomic structure

4.1 Optical spectra

4.1.1 Description of the technique

In the simplest form of optical emission spectrometer, an electrical discharge is produced between two electrodes, at least one of which consists of, or contains, the sample to be analysed. A high voltage accelerates electrons from the cathode (the negative electrode) onto the anode, where they produce heat and the release of positive ions. Hence, some portion of the sample is vaporized by the discharge, the vaporized species become dissociated, and the resultant atoms (or ions) are excited either thermally or through collisions with each other. An excited atom or ion emits light, which is dispersed into its component wavelengths (λ) by means of a *diffraction grating*. In Section 2.1.4 we saw that if we place *two* slits in front of the sample slit source, then maxima appeared at distances from the undeviated beam that were multiples of ($\lambda D_{ss}/c_z$), where the reader may recall that D_{ss} was the distance from the slits to the screen and c_z was the distance between the slits (see Figure 2.7). We can write this in terms of diffraction angle ϕ, since in ΔGOM of that figure, we see $\tan \phi = (\lambda/c_z)$ for the first maximum. If the light source behind the single slit is *not* monochromatic but contains several wavelengths, then we would obtain several overlapping fringe patterns from each of the wavelengths present. Notice that the longer wavelengths would have their first maxima at larger angles ϕ from the undeviated direction. Imagine now that we have several thousand slits (obtained by ruling opaque lines on a glass flat). The slit distance ($2c_z$) becomes very small and the angle of the first maxima very large. For instance, for $\lambda = 500$ nm and $c_z = 1 \, \mu$m (i.e. 24 000 lines/in) then $\tan \phi = 0.5$, so that $\phi = 26.6°$. It also turns out that, for multiple slits, only the central (zero order) maximum and the first order maximum have any significant intensity. Hence, we may transmit the slit of light from the excited atoms (or ions) through the ruled diffraction grating and obtain the first maximum of all the colours (i.e. wavelengths) present in the light, in the form of coloured images of that single slit source. We could also use a prism to disperse the wavelengths, but this does not provide such sharp images of the slit source. The spectrum (as we call the collection of different coloured images of the slit source) is detected either with a photographic plate or with a photomultiplier readout system, that is a direct reader.

4.1.2 The Bohr theory of hydrogen spectra

The simplest spectrum is that obtained with excited hydrogen atoms/ions as the source of light. In about 1910, Niels Bohr, a Danish physicist, was working (under Lord Rutherford at Manchester University) on how to understand optical spectra in terms of the (then) existing theories about the atom. J. J. Thompson, the discoverer of the electron in the latter part of the nineteenth century, was convinced that the atoms of a substance were large spheres (about 1 Å in diameter) of positive charge ($+Ze$) in which were embedded Z electrons, of total charge ($-Ze$). This model has sometimes been given the humorous name of the 'currant bun model'. Lord Rutherford's team at Manchester were very interested in α-particle scattering, that is the scattering of particles emitted *spontaneously* from radioactive sources. They found that most of the α-particles were only scattered through very small angles. A significant proportion, however, were scattered through large angles. Some were even scattered through angles greater than 90°. The Thomson model of the atom was not in accord with these experimental scattering angles, especially as regards 'back-scattered' α-particles. Rutherford's experiments were very adequately explained (on the basis of classical physics) by assuming the nuclear model of the atom. This is a model whereby one assumes that almost all the atomic mass is concentrated in the positive nucleus (which has a diameter about 10^{-14} m and a charge $+Ze$) whilst the Z electrons (necessary for the atom to be neutral) moved in orbits around the nucleus at a distance of about 10^{-10} m from the nucleus. Although we are now used to the nuclear model of the atom, it must be pointed out that its acceptance by the community of physicists was not immediate.

The 'acceptability' of Bohr's ideas met with even more resistance, mainly because his basic assumptions ran contrary to 'accepted' ideas at that time. His most revolutionary suggestion was that of 'quantized stationary orbits'. These were orbits in which the electron moved around the nucleus in constantly accelerating circular motion (due to electrostatic attraction) *without giving off any radiation.* ('Classical' physics predicted that a charged particle being accelerated in this way would give off radiation and hence would continuously lose energy and spiral into the nucleus.) No other orbits were allowed. A stationary orbit occurs whenever the angular momentum equals an integral number n of Planck's constant h divided by 2π. This arbitrary choice of 2π was very lucky (although one suspects Bohr realized that multiplying the angular momentum by about 0.16 would make the final answer come right!). There is nothing wrong with this heuristic approach to research. It is the fault of the way in which we present our research results which gives the impression that everything was 'cut and dried' before we carried out our experiments. In tribology, where there is a shortage of definitive theories, such an approach is the only feasible one, as we shall see many times in this book.

We will not give the details of Bohr's theory of hydrogen spectra in this book. Essentially, he deduced the radius (r_n) of any orbit which satisfied his proposed quantization rule for angular momentum (p_ϕ), namely:

$$p_\phi = (m_e)v\,r_n = n\left(\frac{h}{2\pi}\right) \tag{4.1}$$

where v is the linear velocity of the electron as it travels around a circular path about the central nucleus (i.e. the proton in the case of hydrogen) and (m_e) is the mass of the electron. This radius also has to satisfy 'classical' (i.e. Newtonian) physics insofar as the total energy of an orbit (E_n) is the sum of the kinetic energy of the electron in that orbit and the potential energy of that orbit. Bohr's Theory gives us an expression for the energy (E_n) of each quantized orbit, namely:

$$(E_n) = -\frac{(R_B)hcZ^2}{n^2} \tag{4.2}$$

where Z is the atomic number of the nucleus, c is the velocity of light in vacuo and n is the 'quantum number' of the orbit. Clearly, $Z = 1$ for a hydrogen nucleus. However, it has since been found that ionized helium $(Z = 2)$ has the same form of spectrum as hydrogen, as also does doubly-ionized lithium $(Z = 3)$. Clearly, this is because these ions are 'hydrogen-like', that is, they are two-body problems. (R_B) is a collection of terms involving atomic constants, to which we will refer later in this sub-section. The negative sign indicates that (E_n) relates to a binding energy.

The next revolutionary idea proposed by Bohr was an application of Planck's Quantum Hypothesis regarding the radiation given off when the electron moves from one quantized orbit to another. Bohr said we could not know anything about the position or state of the electron whilst in transition. All we could say is that the difference in energies between the two stationary orbits must be equal to Planck's constant (h) times the frequency of the radiation given off. Hence:

$$E_m - E_n = h(v_{mn}) \tag{4.3}$$

when m and n are the quantum numbers associated with the two orbits concerned and (v_{mn}) is the frequency of the radiation. From Equations (4.3) and (4.2), we get:

$$E_m - E_n = (R_B)hc(1/n^2 - 1/m^2) = h(v_{nm}) \tag{4.4}$$

Since spectroscopists normally talk in terms of wave number $(\bar{v}_{nm})$, that is the number of waves in 1 m in vacuo, we can write Equation (4.4) as:

$$(\bar{v}_{nm}) = \frac{(v_{nm})}{c} = (R_B)\left(\frac{1}{n^2} - \frac{1}{m^2}\right) \tag{4.5}$$

How does Equation (4.5) compare with the actual experimentally-obtained wave numbers? It turns out that the hydrogen series in the normal visual range of frequencies (the Balmer series) is given by:

$$(\bar{v}_{expt})_H = 1096.78\left(\frac{1}{2^2} - \frac{1}{m^2}\right) \tag{4.6}$$

where m can have the value 3, 4, 5, ... and $\bar{v}_{expt}$ is measured in units of m^{-1}. If we assume an infinite nuclear mass for the nucleus–electron system rotating about its centre of gravity, then Bohr's theory gives $(R_B) = 1097.37 \, m^{-1}$, a value within 0.05% of the experimental value. This extremely close correlation was one of the factors which eventually forced the acceptance of Bohr's ideas. The ultra-violet series (called the Lyman series) of hydrogen was also closely predicted by Equation (4.5) with $n = 1$ and $m = 2, 3, 4, \ldots$. The infra-red (Paschen series) was given by $n = 3$ and $m = 4, 5, 6, \ldots$. It is interesting to note that, for singly-ionized helium (where $Z = 2$), we have:

$$(\bar{\nu}_{expt})_{He^-} = (4)\,(1097.22)\left(\frac{1}{n^2} - \frac{1}{m^2}\right) \qquad (4.7)$$

where $n = 2$ and $m = 3, 4, 5, \ldots$, for the Lyman series. This gives an experimental (Rydberg) constant that is even closer (0.04%) to the theoretical (R_B) value for infinite nuclear mass.

Bohr's theory of hydrogen spectra gave the first clue to the *electronic structure of the atom*. Later improvements, called 'Wave Mechanics', revealed that, as well as the main quantum number n for each orbit, there must be *three* other numbers to account for the fine and hyperfine structure of atomic spectra. Thus we have the concept of the electron shell (for each quantum number n) and several sub-shells (within that shell) due to slight differences in energy. The four quantum numbers (n, m, l and j) are inter-related (at least, n, l and j are related and we need not worry about the m quantum number). For each n value, wave mechanics tells us we have various l values from 0 to $(n - 1)$ and for each l value we can have just two values of j, namely $(l + \frac{1}{2})$ and $(l - \frac{1}{2})$, except when $l = 0$, in which case $j = \frac{1}{2}$ only. For $n = 1$, we have the K-shell containing just one energy level, since $l = 0$ and $j = \frac{1}{2}$. For the L-shell, we have $n = 2$ so that $l = 0$ (hence $j = \frac{1}{2}$) and $l = 1$ (with $j = \frac{3}{2}$ and $\frac{1}{2}$). Thus, there are *three* energy levels in the L-shell. The M-shell has $n = 3$; it therefore has *five energy levels*; and so on for the N- and O-shells.

The Balmer series of hydrogen arises from electronic transitions between the L-shells (with $n = 2$) and the M-, N- and O-shells, where $n = 3$, 4 and 5, etc. The Lyman series of hydrogen arises from electrons 'jumping' between the K-shell (with $n = 1$) and the L-, M-, N- and O-shells (where $n = 2$, 3, 4 and 5). We will return to the subject of hydrogen spectra when we describe X-ray spectra (in Section 4.2). It turns out that there is a strong resemblance between these two types of spectra, the reason for which will be explained in Section 4.2.

4.1.3 Surface analysis using optical emission spectroscopy

In tribology, we find three types of surface films, namely: (a) alien surface 'contamination', (b) intentional surface 'contamination' and (c) sputtered films. The thickness of any of these films may range from a few ångströms to several hundred micrometres. The alien and intentional surface films may generally be expected to be quite variably non-uniform, while sputtered films should be fairly uniform. In any case, the analyst's problem may assume several distinct aspects. In some cases, the analyst is merely expected to establish the identity of the principal components of the surface film. Sometimes, however, the analyst must establish the variation in composition of the film in the X-, Y- and Z-directions, or determine only the variations in the Z-dimension of thickness. Differentiation between the constituents of the surface film and those of the substrate is almost invariably required.

Although emission methods have been used extensively for the solution of a wide variety of analytical problems, their use for surface analysis has been somewhat limited. Possibly the best techniques for this specialized type of application are (i) spark methods and (ii) laser vaporization methods.

(i) Spark methods

High-voltage spark vaporization–excitation applied to a metallic surface will vaporize a milligram of material when sustained for the fairly normal exposure period of 1 min. By adjustment of the spark parameters, the area sampled may be varied from $1 \, mm^2$ to $1 \, cm^2$. Typically the depth of sampling will be less than 1 mm. According to the voltage applied and the inductance in the discharge circuit, it has been shown (Strasheim and Blum, 1971) that a medium-voltage spark applied to an aluminium surface produces sampling depths ranging from 40 to 120 μm. By systematically varying the inductance, the radius of the spark channel could be changed from approximately 0.1 to 2.0 mm. Thus spark methods offer *moderate* spatial resolution coupled with the multi-element sensitivity of emission spectroscopy. Quantification of the results in terms of concentration at various locations in the sparked surface is, however, a very complex problem. Among the various techniques for surface analysis, these problems are not unique and, in fact, it turns out that *relative* concentration profiles are frequently sufficient for the solution of most tribological problems.

(ii) Laser vaporization techniques

The use of laser vaporization and excitation techniques in emission spectroscopy has been reviewed by Baldwin (1968). The most widely used system focuses a laser beam on to the sample surface through a microscope, the area of analysis thereby obtained being approximately 5–10 μm in diameter. A pulsed laser, with a pulse duration of tens of nanoseconds (10^{-8} s) can be used to inject peak powers of 10^7 W into these small areas. Hence, most materials can be vaporized by this technique. Some characteristic radiation is emitted by the vaporized material, but its intensity is normally too low to detect anything but the major constituents of the material. Therefore, pointed graphite electrodes are placed in opposition across the path of the vapour plume. These electrodes are charged to a high potential *just below* that necessary for the gap to be conductive *without* the vapour plume. When the vapour plume is formed, the electrode gap becomes conductive and the energy dissipated in a spark pulse across the gap serves to excite the atoms in the plume. This secondary means of excitation greatly enhances the sensitivity of this surface spectrographic technique.

The future role of emission spectroscopy in surface analysis is not too clear especially in the resolution of tribological problems. Both the spark and laser methods are *destructive*. They are, however, capable of multi-element analysis at a very high absolute sensitivity (10^{-8} kg) for a diverse range of elements. Quantification of spectra in terms of either concentration or spatial terms is difficult, but no more so than many other surface analytical techniques. There are very few examples of the application of emission spectroscopy to tribological problems. The author, together with his ex-colleagues at The Georgia Institute of Technology (Professors Winer and O'Shea), used a modified emission spectrometer to study the glow from the track of a pin-on-disc tribometer, the disc being made of single crystal sapphire and the pin of tool steel. The analysis of the spectra was inconclusive as regards the origin of the glow. Nevertheless, it does show that, when flashes of light are emitted at 'hot-spots' between sliding surfaces, some interesting information about the interactions going on at the location of the 'hot-spot' could be gained by analysing that light. Obviously, the use of infrared detectors will tell us about the temperatures of the hot-spots and we will deal with this technique in Chapter 6. Let us now discuss X-ray spectra, since this subject is basic to Electron Probe Microanalysis, X-ray Photoelectron Spectroscopy and Auger Electron Spectroscopy (see Sections 4.3, 4.4 and 4.5).

4.2 X-ray spectra

4.2.1 The origin of X-rays

We have discussed (in Section 3.2) the diffraction of X-rays, without really knowing how X-rays are obtained in the first place. Essentially, X-rays are produced in *two ways*, firstly by the interaction of charged particles (in particular electrons) with the positively-charged nucleus and secondly, by interaction of these particles (or *photons*) with the electrons in the extra-nuclear electron shells around the nucleus of the atom. The first method produces a *continuous* spectrum of X-rays and the second produces a *characteristic* spectrum of X-rays. When the X-rays are produced from an energetic beam of charged particles, they are called 'primary X-rays'. When they are produced as a result of photons (i.e. primary X-rays) interacting with the electrons in the electron shells, we have what is called 'secondary X-rays' or 'X-ray fluorescence'.

Primary X-rays are produced by suddenly arresting a beam of electrons (say, one of energy equal to about 30 keV) by a water-cooled target. The X-rays, consisting of *both continuous and characteristic spectra*, emerge through aluminium 'windows', set at about 7° to the plane of the target, upon which the electrons impinge at right angles. Secondary or fluorescent X-rays are produced when the above-mentioned primary X-rays knock out the extra-nuclear electrons in the innermost shells, leaving vacancies for other electrons in the outer shells to drop into and hence give off their excess kinetic energy in the form of *characteristic X-radiation*.

4.2.2 The crystal spectrometer

Invariably, X-ray spectra are studied using *primary X-rays* to cause the unknown specimen to fluoresce, that is give off characteristic secondary X-radiation. The flat-crystal spectrometer used for the analysis of these spectra is an adaptation of the Powder Diffractometer described in Section 3.2.5. It is called a 'wavelength dispersion device' because the wavelength being diffracted by the device is a function of the measurable Bragg angle. A crystal spectrometer can measure any X-ray wavelength, provided that a crystal with a suitable interplanar spacing (d_{hkl}) can be found.

In the typical X-ray flat-crystal spectrometer, an X-ray tube excites a few square centimetres of the sample. The specimen is located close to the window of the X-ray tube in order to obtain as high an incident flux as possible. The emitted radiation is passed through a set of slits in order to collimate the beam before it impinges onto the analysing crystal, where it is diffracted to the detector, set at an appropriate angle. As with the diffractometer, the detector and the crystal rotation are coupled in such a way that the detector moves at twice the angular speed of the crystal. Thus, with the crystal set at the angle α, the detector will be at 2α to detect any diffracted intensity.

The spectrometer may be fixed to measure the intensity of a single X-ray line. It may be a scanning model which can be driven by a motor (or manually) through a range of Bragg angles, to measure a range of wavelengths. Alternatively, it may be programmed to move at a rapid speed from one angle to another, stopping long enough at each position to measure the intensity. Research instruments, however, are usually the manual or motor-driven type.

Most modern spectrometers are capable of being operated with an air atmosphere, a helium or other controlled gas atmosphere, or in a vacuum. In an air path, the analyst is limited to measuring X-ray wavelengths shorter than about 2.5 Å, because longer wavelengths are severely attenuated in the air. For softer X-rays, a vacuum, hydrogen or helium path is required. The vacuum path is the easiest to attain since it involves merely the use of a simple mechanical vacuum pump.

One of the important parameters of any spectrometer is its *resolution*. For a flat-crystal X-ray spectrometer, the ability to separate two X-ray lines close to one another in wavelength can be calculated from differentiation of the Bragg relation Equation (3.6), namely:

$$(\Delta\lambda) = 2(d_{hkl})(\cos\theta_{hkl})(\Delta\theta_{hkl}) \qquad (4.8a)$$

If we define dispersion as $(\Delta\theta_{hkl})/(\Delta\lambda)$, we see that Equation (4.8a) gives us:

$$[(\Delta\theta_{hkl})/(\Delta\lambda)] = [2(d_{hkl})(\cos\theta_{hkl})]^{-1} \qquad (4.8b)$$

Hence we get maximum dispersion when $(\cos\theta_{hkl})$ tends towards zero, that is when θ_{hkl} tends towards 90°. This means that maximum dispersion occurs in the 'back reflection' region, when the diffraction angle ϕ ($=2\theta$) tends towards 180°. We also notice that the dispersion is also inversely dependent upon the interplanar spacing (d_{hkl}).

It is interesting to calculate ($\Delta\lambda$) from Equation (4.8a) as a function of (d_{hkl}), taking ($\Delta\theta_{hkl}$) as the geometric mean of the collimator divergence (say 3.4 minutes of arc, for a collimator 10 cm long with blades spaced 0.01 cm apart) and the divergence caused by the mosaic nature of the diffracting crystal (say, 2 minutes of arc for an abraded and etched LiF crystal). The geometrical mean for such a crystal, with the irradiated face having an interplanar spacing of 2.015 Å is $(\Delta\theta_{hkl}) = (2^2 + 3.4^2)^{\frac{1}{2}} = 3.9$ minutes of arc, that is 0.0011 radians. If the fraction to be analysed is of fairly small wavelength, for example (BrK_{α}) at $\lambda = 1.04$ Å, then the resolution ($\Delta\lambda$) is 0.0043 Å, that is ($\Delta\lambda/\lambda$) is 0.41%. For (CrK_{α}) (with $\lambda = 2.29$ Å), then $\Delta\lambda$ is 0.0033 Å, and for (KK_{α}) it can be shown that $\Delta\lambda$ is 0.0017 Å (giving respectively $\Delta\lambda/\lambda = 0.14\%$ and 0.05%).

For the longer wavelengths, for example (SiK_{α}) $= 7.11$ Å, (NaK_{α}) $= 11.9$ Å and (OK_{α}) $= 23.06$ Å one normally uses cleaved potassium acid phthalate (KAP). This has a large unit cell. Typically, the crystal face irradiated by the beam to be analysed has an interplanar spacing of 13.3 Å. The dispersion of such a crystal will be about 10 *seconds* of arc. This is so small that it can be neglected,

Table 4.1. *X-ray fluorescence analysing crystals*

Crystal	(*hkl*)	$(d_{hkl})/2$ (Å)	Fluorescing elements	Integral reflection coefficient (rad)	λ_{opt} (Å)
LiF	(220)	1.42	—	1×10^{-4}	1.66
LiF	(200)	2.01	—	3×10^{-4}	2.10
Graphite	(002)	3.35	—	2×10^{-3}	2.75
PET[a]	(002)	4.37	—	2×10^{-4}	3.74
EddT[b]	(020)	4.40	—	1×10^{-4}	5.37
ADP[c]	(101)	5.3	P	5×10^{-5}	6.16
Gypsum	(020)	7.6	Ca, S	7×10^{-5}	6.16
Mica	(002)	9.9	K, Al, Si	2×10^{-5}	8.34
KAP[d]	(100)	13.3	K	7×10^{-5}	9.89
OHM[e]	—	31.7	C	1×10^{-4}	11.9
Pb St[f]	—	50	C, Pb	2×10^{-4}	18.3

a, Pentaerythritol; b, ethylene diamine ditartrate; c, ammonium dihydrogen phosphate; d, potassium acid phthalate; e, octadecyl hydrogen maleate; f, lead stearate, a multi-layer soap 'pseudo-crystal'.

compared with the collimator dispersion of 3.4 minutes of arc. Using $(\Delta\theta_{hkl})$ equal to 3.4 minutes of arc (0.0010 rad) one obtains $\Delta\lambda = 0.0257$ Å for (SiK_α) and $\Delta\lambda = 0.0133$ Å for (OK_α). These values of $\Delta\lambda$ eventually give 0.36%, 0.20% and 0.06% for $(\Delta\lambda/\lambda)$ respectively.

If the source of X-rays is small, much higher intensities can be obtained by focusing the radiation into the detector. If the *diffracting planes* of the crystal are cylindrically curved to a *radius* equal to the *diameter* of the focusing circle, and the *surface of the crystal* is ground to a *radius* equal to the *radius* of the focusing circle, it is possible to diffract a point source to a line image. This type of X-ray optics is used in electron-probe microanalysis (see Section 4.3).

As regards the choice of crystals for X-ray fluorescence, the analyst is constrained by the limited number of available materials. On the one hand, the interplanar spacing (d_{hkl}) should be large enough for the wavelength of interest to be diffracted. The longest wavelength which can be diffracted is, in theory, equal to $(d_{hkl})/2$, since $(\sin\theta_{hkl})$ in the Bragg equation cannot have a value larger than unity. In practice, it is impossible for a spectrometer to go beyond a (θ_{hkl}) value of about 70°. The longest wavelength which can be measured with a particular crystal is then equal to about 0.94 times $(2(d_{hkl}))$. On the other hand, if the analyst is really concerned about the resolution of his/her instrument, then the smallest spacing crystal should be used, consistent with the above criterion. If possible, the crystal should not have any elements in it which can fluoresce in a measurable wavelength region, because this merely increases the background. The crystal should have a high internal reflection coefficient so as to diffract with a high intensity. Table 4.1 lists some of the crystals commonly used, along with the relevant parameters mentioned above. The column headed 'λ_{opt}' relates to the optimum wavelength which can be analysed with the crystal, the optimization being a compromise between good resolution and having an interplanar spacing close enough to the wavelength being analysed. It will be noticed that, for interplanar spacings less than about 10 Å, the optimum wavelength gives rise to diffraction angles ($2\theta_{hkl}$) between 40° and 75°. For very large wavelengths, the OHM and PbSt crystals are best used, at lower ($2\theta_{hkl}$) values, namely about 20°.

Having discussed the technical details related to X-ray spectrometry, let us now describe the two types of spectra, namely (a) continuous (or white) spectra and (b) discrete (or characteristic) spectra.

4.2.3 Continuous X-ray spectra

When a beam of electrons impinges on a target, interactions occur between the electric field of each electron and the electric field of the nucleus of the atom through which the electron passes. The strength of the interaction depends on how close the path of the electron lies with respect to the position of the nucleus. If all the energy of the electron (eV_A) is lost in the interaction, then the energy (E_{max}) of the X-rays produced is simply given by:

$$(E_{max}) = eV_A = h(v_{max}) = hc/(\lambda_{min}) \qquad (4.9a)$$

where e is the electronic charge $= 1.60 \times 10^{-19}$ C, V_A is the voltage through which the electron has been accelerated, h is Planck's constant $= 6.626 \times 10^{-34}$ J s, v_{max} is the maximum frequency of the X-radiation produced, c is the velocity of propagation of electromagnetic radiation equal to 3×10^8 m/s and λ_{min} is the well-known 'short-wavelength-cut-off' experimentally observed in continuous X-ray spectra.

In addition to the maximum frequency (v_{max}), there will be a whole (continuous) spectrum of lower frequencies emitted after the encounter of an electron (which has lost part of its energy due to a previous deflection by another nucleus) with a target atom. Obviously, an electron will suffer many collisions before being brought to rest, and at each collision, some of the initial energy (eV) is lost, in the form of quanta of (hv), where v is less than v_{max}. Hence, we get a continuous spectrum, which is sometimes called 'Bremsstrahlung'. However, this particular term is usually reserved for radiation due to the 'braking', or sudden arrest, of very high energy particles (such as mesons) rather than 30 keV electrons. A typical set of X-ray continuous spectra is shown in Figure 4.1 (for a tungsten target).

It should be noted that the short-wavelength-cut-off decreases with increasing electron accelerating voltage. It is also apparent that the position of the maximum intensity (see arrows in Figure 4.1) also occurs at shorter wavelengths as the accelerating voltage increases, as one might expect from Equation (4.9a). This is quite a triumph for the quantum theory. In fact, the short-wavelength-cut-off is often used for determining Planck's constant! It turns out that the total energy of the emitted X-radiation (i.e. the area under one of the curves in Figure

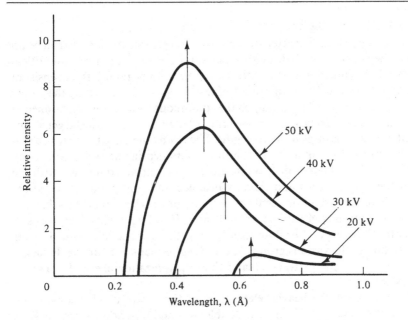

Figure 4.1 *The graph of relative intensity versus wavelength (λ) for continuous X-radiation from a tungsten target*

4.1) is proportional to (ZV_A^2), where Z is the atomic number of the target atoms. Hence, if we wish to increase the intensity of white X-radiation, we should use a higher voltage and a 'heavy' element (such as tungsten).

Equation (4.9a) is often written in terms of the actual quantities involved, namely:

$$\lambda_{min} = \left[\frac{(6.62 \times 10^{-34}\,\text{J s})(3 \times 10^8\,\text{m/s})}{1.60 \times 10^{-19}\,\text{C}} \right]\left(\frac{1}{V_A}\right) \tag{4.9b}$$

Since 1 volt equals 1 J/C, we see that this equation leads to:

$$\lambda_{min} = \left[\frac{12.4 \times 10^{-7}\,\text{m}}{V_A} \right]$$

that is

$$\lambda_{min} = \left(\frac{12\,400}{V_A}\right)\text{Å} \tag{4.10}$$

where V_A is measured in volts. Let us now discuss discrete (or characteristic) X-ray spectra.

4.2.4 Characteristic X-ray spectra

There are two types of characteristic X-ray spectra, namely *emission* and *absorption* spectra.

(i) Characteristic emission X-ray spectra

One method for exciting characteristic emission spectra is to use an electron beam which has been accelerated through a potential greater than a certain critical voltage V_c. The primary X-rays excited by the electron beam will then consist of discrete high intensity 'lines' (in groups of wavelengths designated by the letters K, L, M, N, O, etc.), *superimposed* upon the continuous background mentioned in Section 4.2.1. *Alternatively*, one can use primary X-rays to excite the target atoms to give off the same groups of wavelengths. The main difference between the two spectra lies in the low intensity background obtained in the secondary X-ray spectrum. This is, of course, due to the *absence* of electrons to interact with the nuclei of the target atoms and thereby produce any continuous X-radiation. Hence the secondary (or fluorescent) X-ray spectrum has better contrast than the corresponding primary X-ray spectrum from a given target. However, it is still necessary for the energy of the fluorescent X-rays to exceed certain critical values in order to obtain the various groups of wavelengths mentioned above. In fact, it is necessary for the energy to be greater than the ionization energy for a given electron shell.

The K-series of all the heavier elements (say, those with $Z > 13$) consist of 4 principle lines, namely K_γ, K_β and the doublet $K_{\alpha 1}$ and $K_{\alpha 2}$. Sometimes, the K_γ is called $K_{\beta 2}$ and K_β is called $K_{\beta 1}$. A typical K-series is shown in Figure 4.2. We note that the intensity of the $K_{\alpha 1}$ is almost double that of $K_{\alpha 2}$ and that $K_{\alpha 2}$ is about double that of K_β. Translating crystal table angles into wavelengths one obtains the following typical values for K-emission spectra:

$$
\left.\begin{array}{ll}
\text{Cu}K_{\alpha 1}; & \lambda = 1.5405\,\text{Å} \\
\text{Cu}K_{\alpha 2}; & \lambda = 1.5443\,\text{Å} \\
\text{Cu}K_{\beta}; & \lambda = 1.3922\,\text{Å} \\
\text{Cu}K_{\gamma}; & \lambda = 1.3810\,\text{Å}
\end{array}\right\} \text{Copper K-spectra}
$$

$$
\left.\begin{array}{ll}
\text{Co}K_{\alpha 1}; & \lambda = 1.7880\,\text{Å} \\
\text{Co}K_{\alpha 2}; & \lambda = 1.7928\,\text{Å} \\
\text{Co}K_{\beta}; & \lambda = 1.6234\,\text{Å} \\
\text{Co}K_{\gamma}; & \lambda = 1.6207\,\text{Å}
\end{array}\right\} \text{Cobalt K-spectra}
$$

In X-ray diffraction, we normally filter out the K_β and K_γ wavelengths (see later), leaving us with $K_{\alpha 1}$ and $K_{\alpha 2}$ for diffraction by our unknown crystal. At low diffraction angles, the separation of the $K_{\alpha 1}$ and $K_{\alpha 2}$ wavelengths is such that the crystal will diffract them as if they were a weighted mean wavelength (λ_{mean}) where $\lambda_{\text{mean}} = \frac{1}{3}[\lambda(K_{\alpha 2}) + 2\lambda(K_{\alpha 1})]$. For Cu X-radiation, the λ_{mean} of $K_{\alpha 1}$ and $K_{\alpha 2}$ is 1.5414 Å. For cobalt, the weighted mean is 1.7902 Å.

Moseley was one of the first scientists to investigate X-ray spectra. He was not able to resolve the $K_{\alpha 1}$ and $K_{\alpha 2}$ doublet, nor was he aware that his K_β was, in fact, K_β and K_γ. Nevertheless, he was able to show that, if one plots $(\bar{\nu}/R_B)^{\frac{1}{2}}$ against Z for a wide range of elements, one gets an approximately straight line for the K_α and K_β, and a 'not-so-straight' line for the values arising from the L- and M-series, as shown in Figure 4.3. As far as the K-series is concerned, Moseley showed that:

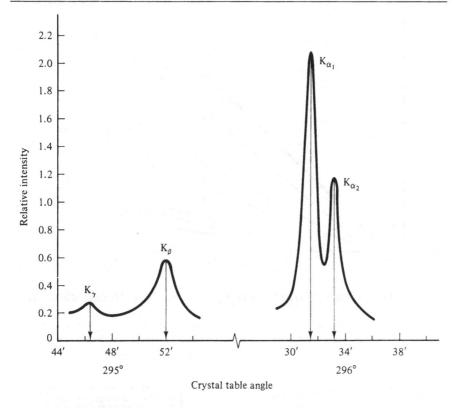

Figure 4.2 *Relative intensity versus crystal table angle for a typical K-series of X-ray wavelengths*

$$(\bar{\nu}/R_B)^{\frac{1}{2}} = 0.874(Z - 1.13) \qquad (4.11)$$

We will return to this equation a little later on. It is interesting to note that for any given Z (say $Z = Z_1$, as shown in Figure 4.3), as we increase the energy of the electron beam producing the primary X-rays, we first obtain the M-series. then the L-series and finally the K-series. One never gets just one line of a series. The energy of the impinging electrons is, of course, related to the accelerating voltage (V_A). Once we have exceeded the critical voltage (V_C) for a particular series, the intensity (I) of the X-rays increases with V_A according to the empirical relation:

$$I = (c_x)(V_A - V_C)^{1.75} \qquad (4.12)$$

where c_x is a constant.

(ii) Characteristic absorption spectra

Now X-rays can interact with matter in many ways, as shown schematically in Figure 4.4. For low energies, absorption occurs mainly by transformation into fluorescent X-rays, photoelectrons and heat. Without analysing each method of

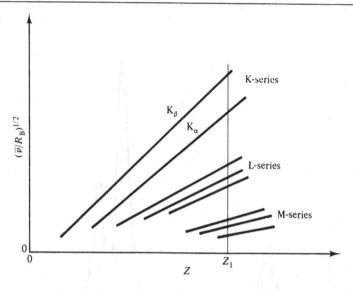

Figure 4.3 *Schematic plot of* $(\bar{v}/R_B)^{1/2}$ *versus Z (a Moseley diagram)*

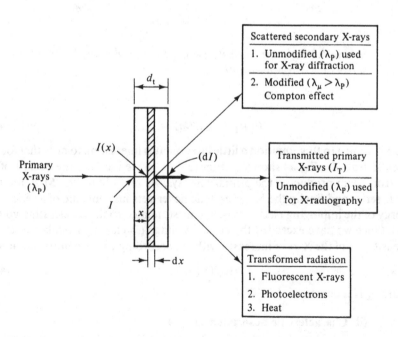

Figure 4.4 *Schematic diagram showing the interaction of X-rays with matter*

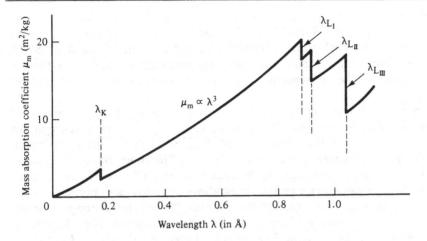

Figure 4.5 *Variation of mass absorption coefficient (μ_m) of platinum ($Z = 78$) as a function of incident X-ray wavelength (λ)*

losing energy in detail, let us consider the combined effect of all these phenomena together, that is let us study the absorption of X-rays by matter.

Imagine a beam of X-rays of intensity (I) and wavelength (λ) impinging upon a slab of material of thickness (d_t). As it passes through the material, the beam becomes attenuated by the various interactions indicated in Figure 4.5. If we consider a very thin slice of that slab at a distance x from the incident face and of thickness (dx), then we can say that the intensity of X-rays scattered by that thin slice will be d(I), where d(I) is proportional to $I(x)$ and (dx), hence:

$$d(I) = -(\mu_1)I(x)(dx)$$

where (μ_1) is the linear absorption coefficient for X-rays and the negative sign is to indicate that d(I) decreases with increasing (x). Rearranging this equation, we obtain:

$$\frac{d(I)}{I(x)} = -(\mu_1)(dx) \tag{4.13}$$

Integrating this equation, remembering that $I(x) = I$ at $x = 0$, we obtain (I_T), the total intensity of X-rays emerging from a slab of thickness (d_t), namely:

$$I_T = I \exp[-(\mu_1)(d_t)] \tag{4.14}$$

Clearly, for a given thickness, I_T depends inversely upon (μ_1). If (μ_1) is large, then I_T will be small. Values of (μ_1) have been obtained experimentally from absorption experiments involving the plotting of $\ln(I_T/I)$ versus (d_t). The slope of this line will be ($-\mu_1$). It is more usual, however, to quote the mass coefficient of absorption (μ_m) where (μ_m) = (μ_1)/(ρ). In this equation, (ρ) is the density, so that Equation (4.14) becomes:

$$I_T = I \exp[-(\mu_m)(\rho)(d_t)] \tag{4.15}$$

161

Values of (μ_m) as a function of wavelength (λ) are given in Volume 3 of the *International Tables of X-Ray Crystallography*. The main reason for quoting the *mass* absorption coefficient is because it is independent of the state of the absorber. For example, diamond and graphite have the same (μ_m), but (μ_l) will be different due to the difference in structure between these two types of carbon crystals!

A remarkable characteristic of the sets of (μ_m) values is the appearance of *absorption discontinuities* (or *edges*) as one varies λ (the X-ray wavelength). In Figure 4.5, we show the appearance of these edges in the mass absorption coefficient versus wavelength curve for platinum. The nomenclature assigned to the K and L absorption edges will be explained later. Clearly, however, we see that an X-ray beam with a wavelength just above the (L_{III}) absorption edge will be absorbed much less than a beam with a wavelength just below the edge (at about 1.0 Å). In fact, μ_m increases from about $10\,\text{m}^2/\text{kg}$ up to about $18\,\text{m}^2/\text{kg}$ for this slight decrease in wavelength. For the typical specimen thickness (d_t) of about $30\,\mu\text{m}$, this means that the intensity transmitted (I_T) is reduced by a factor of 170 (assuming $\rho = 21.4 \times 10^3\,\text{kg/m}^3$ for Pt).

These absorption edges are very useful for selectively *filtering* out the unwanted K_β and K_γ radiation in the K-spectrum and thereby virtually obtaining a monochromatic beam for X-ray diffraction purposes (in fact the beam will still consist of the $K_{\alpha1}$ and $K_{\alpha2}$ doublet). This is shown in Figure 4.6 for the special case of a Ni filter of thickness equal to $15\,\mu\text{m}$. Assuming nickel has a density of $8.9 \times 10^3\,\text{kg/m}^3$, we can deduce that at $\lambda = 1.54\,\text{Å}$ (the wavelength of the Cu K_α

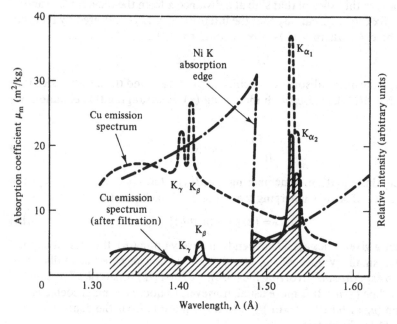

Figure 4.6 *The effect of a nickel filter on the K-emission X-ray spectrum of copper*

emission radiation), the reduction in intensity will be a factor of 50%, whereas at $\lambda = 1.40$ Å (the approximate wavelength of the Cu K_β and Cu K_γ radiation) the reduction in intensity will be 93%, or more precisely $(I_T/I) = 0.07$. Hence, although we have halved the (wanted) Cu K_α intensity with the nickel filter, we have reduced the (unwanted) Cu K_β and Cu K_γ intensities to (7/100) of their original intensities, thereby monochromatizing the Cu X-ray beam.

Having discussed the practical aspects of X-ray emission and absorption spectra, let us now consider the simple theory of X-ray spectra. This theory will also be useful when we deal with X-ray Photoelectron Spectroscopy (Section 4.4) and Auger Electron Spectroscopy (Section 4.5), as well as the obvious relevance to Electron Probe Microanalysis (Section 4.3).

4.2.5 The simple theory of X-ray spectra

X-ray energies are very much greater than optical energies. It turns out that the optical spectra of elements are derived from electron transitions between the empty shells available in the outermost regions of the atoms of those elements. One might expect, therefore, to find that X-ray spectra arise from transitions between the innermost (i.e. the most closely-bound) shells of the atom. For a multi-particle atom, Pauli's exclusion principle means that, in the ground state, the innermost shells are closed shells, that is each shell will have its K-shell filled with 2 electrons, its L-shell with 8 electrons and so on. In fact, each atom can only have $(2n^2)$ electrons in any shell with a quantum number n. Hence, in order to get a transition between the innermost shells, we need to have a vacancy in one of the shells. For example, if we remove an electron from the K-shell (either by using an energetic electron beam or a primary X-ray), then its place will be taken immediately by an electron from the L, M, N,..., shells. These shells have electrons with greater kinetic energy than that required for the stationary K-state. The excess energy can be given off as X-radiation. The essential difference between optical and X-ray spectra lies in the fact that the optical levels are created by *an electron being in* an excited atomic state, whereas the X-ray levels are due to the *absence of an electron* from a given energy state.

In order to remove an electron from the K-shell of an atom, we need radiation (i.e. an electron or an X-ray photon) of enough energy to overcome the binding energy of the K-shell. When a vacancy exists in the K-shell, the most likely transition will occur between the L- and the K-shells. However, some transitions may occur between the M- or N-shells and the K-shell. Clearly, if an electron leaves the L-shell to fill a vacancy in the K-shell, we can then obtain the L-series (since a vacancy now exists in the L-shell, ready for an electron to 'jump' from the M- or N-shells).

The early investigators of X-ray spectra were aware of the strong similarity between their X-ray spectra and the optical spectra of hydrogen, namely, they all consisted of very simple sets of wavelengths, with similar relationships between these wavelengths. Perhaps an electron dropping into a vacancy in the K-shell 'sees' the positive nucleus (of charge $+Ze$) and the other electron in the K-shell (of charge $-e$) as a positive composite nucleus of charge $(Z-1)e$? If this is so, then we can apply Bohr's theory of hydrogen optical spectra to our X-ray spectra, and deduce the energy (E_n) of the nth orbit to be given by:

$$E_n = - \left[\frac{(R_B)hc}{n^2} \right] (Z-1)^2 \qquad (4.16)$$

This equation is the same as Equation (4.2), except $(Z-1)^2$ has now replaced Z^2.

Let us deduce the wave number ($\bar{v}_{LK}$) for transition from the L-level ($n=2$) to the K-level ($n=1$). Assuming Planck's quantum condition applies (as given by Equation (4.3) for optical spectra), we obtain:

$$E_{KL} = E_L - E_K = (R_B)(h)(c)(Z-1)^2 \left(-\frac{1}{2^2} + \frac{1}{1^2} \right) = h(v_{KL})$$

hence

$$\bar{v}_{KL} = \frac{v_{KL}}{c} = (R_B)(Z-1)^2 \left(\frac{3}{4} \right)$$

that is

$$\left(\frac{\bar{v}_{KL}}{(R_B)} \right)^{\frac{1}{2}} = 0.866(Z-1) \qquad (4.17)$$

This equation should be compared with the experimental equation obtained by Moseley (see Equation (4.11)) where the constant before the bracketed term is 0.874 (i.e. 1% less than in Equation (4.17)) and the constant within the bracketed term is 1.13 instead of unity (i.e. 13% more than in Equation (4.17)).

This agreement is such as to indicate the essential correctness of the theory, at least as far as the K-series X-ray spectra are concerned. There are, however, a number of inexplicable facts such as:

(a) there are *three* absorption edges for the L-series, 5 for the M-series, and so on. In other words, there exists a fine structure in X-ray spectra not readily explained by a simple modification of the simple Bohr theory; and

(b) the Moseley diagram is only a straight line for the K-series. The L, M, N, O,..., series are all represented by non-linear curves in the Moseley diagram.

In order to explain the fine structure of X-ray absorption spectra, we have to use the same quantum mechanical ideas already formulated (in Section 4.1.2) with the aim of explaining the optical emission spectra of hydrogen. We did not, *at that stage*, introduce the concept of 'selection rules' regarding possible transitions between energy levels. These rules say that only *odd* levels can combine with *even, and vice versa*, in order to produce a transition, i.e. Δl must be ± 1 and Δj must be 0 or ± 1. Perhaps this is better understood as follows:

(a) the quantum number n *must* change;

(b) l *must* change by *one unit*; and

(c) j *may* change by *one unit*, or remain unchanged.

With these rules in mind, Figure 4.7 has been drawn (diagramatically, and not necessarily to scale). For instance, we see that the only transitions allowed from the L-shell to the K-shell are those giving rise to the $K_{\alpha 1}$ and $K_{\alpha 2}$ wavelengths

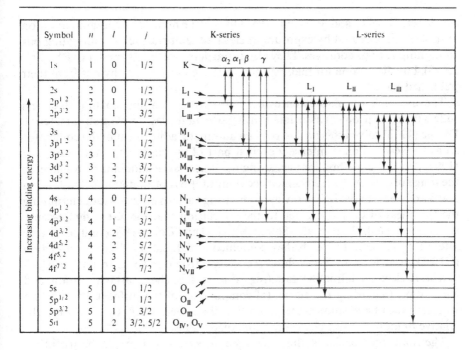

Figure 4.7 *The origin of the K- and L-series in the X-ray spectrum of a typical atom*

(from the L_{III} and L_{II} sub-shells respectively). This is because l must change by one unit (which would *not occur* for the L_I shell). The change of n and the unchanging nature of j is allowed, but all three rules must be obeyed. However, the change from L_{II} to K and the change from L_{III} to K both satisfy all three rules.

The probability of transitions to the K-level from the outer energy levels (i.e. N, O and higher) is so small that the resulting intensity of the emitted X-radiation is undetectable by normal instruments. Hence, we have only shown how the experimentally observed spectra (e.g. the $K_{\alpha1}$, $K_{\alpha2}$, K_β and K_γ lines) are obtained. We will need to refer to Figure 4.7 later in the book. Before we go on to a particular form of *primary X-ray* spectra suited to the surface analysis of tribological specimens, namely electron probe microanalysis, let us briefly describe some applications of X-ray fluorescence to surface studies.

4.2.6 Applications of X-ray fluorescence to surface analysis

One of the first studies of the surfaces of metallurgical materials using X-ray fluorescence was carried out by Koh and Caugherty (1952). They did a theoretical examination of the maximum thickness of the surface which emitted an X-ray intensity equal to a bulk sample. Their calculations led them to conclude that it should be possible to analyse corrosion layers *in situ*. Rhodin (1955), on the other hand, stripped off the surface layer and analysed it as a thin film, showing that as small an amount as $3.7 \times 10^{-7} \, \text{kg/m}^2$ could be detected for nickel.

Schrieber, Ottolini and Johnson (1963) have determined the contamination on bulk steel discs caused by exposure at elevated temperatures to lubricating oils containing various additives. They also claim detection limits similar to Rhodin's (1955), but the criteria for making this claim are not readily obvious from their published paper.

The use of X-ray fluorescence to determine the 'on-line' coating thickness suffers from the problem associated with flutter in the fast-moving strip material. Dunne (1963) solved this problem by an automatic adjustment of the collimator so as to view the most intensely excited part of the sheet. Cline and Schwartz (1967) propose a method for measuring the thickness of aluminium on silicon in the range of $0-4\,\mu m$, with a claimed accuracy of $\pm 6\,\text{Å}$ of aluminium. Smuts, Plug and Van Niekirk (1967), in a review comparing various methods for determining the thickness of tin plate, came to the conclusion that the X-ray fluorescence technique gave the best results. Sewell, Mitchell and Cohen (1969) did some very interesting research in which they combined a high-energy electron diffraction (HEED) unit with an X-ray fluorescence spectrometer so as to obtain *simultaneously* electron diffraction data (regarding surface chemical composition) and X-ray data (regarding the elemental composition of the surface). Francis and Jutson (1968) have studied the oxidation of stainless steels by measuring the relative amounts of five elements in the oxide films.

The main drawbacks of this technique, in so far as it might be useful for tribological analysis, are that (a) it is an elemental technique, that is, it is not easy to obtain information about the way in which the elements are combined in an irradiated compound; and (b) it requires a fairly large area to be irradiated in order to get sufficient intensity of the characteristic X-rays for the elements to be identified in an unambiguous manner. We will see that, when we use an electron beam to excite the primary X-rays from the specimen, we obtain elemental information about very small areas of the specimen. This is the main advantage of electron probe microanalysis, which we will now discuss in some detail.

4.3   Electron probe microanalysis

4.3.1   Introduction

The X-ray fluorescence method for chemically analysing an unknown substance developed much earlier than the 'emission method' that is the use of an electron beam to cause direct emission of X-rays. The late development of the emission method was entirely due to the primitive nature of electron-optical techniques in the 1930s (the time when the X-ray fluorescence method began to be widely used). Developments in these techniques in the 1950s made it possible to focus a beam of electrons into a probe of $1\,\mu m$ diameter or less. This enabled analysis of a correspondingly small region of the specimen surface, which is still very difficult to obtain by the fluorescence method. The emission method thereby finds its main application in the *microanalysis* of surfaces, and can be applied not only to metals but also to biological and mineralogical specimens. In combination with the scanning technique to be described, it gives a rapid means of obtaining both the concentration of a given element at a selected point and its distribution over the surface. In this short review of electron probe microanalysis (as we now call this particular method of X-ray emission), we will discuss (a) the basic requirements of any electron probe microanalyser, (b) the factors involved in quantitative electron probe microanalysis, and (c) the range of applications of electron probe microanalysis to the study of surfaces. Note that we will *not* discuss the tribological applications in detail at this stage, since these will be described in later chapters.

4.3.2 The basic features of a typical electron probe microanalyser

The essential features of instruments completely dedicated to electron probe microanalysis are as follows:

(a) *An electron optical system* of high stability to focus a beam of electrons on to the specimen, the energy of the electrons being variable in steps from 1 to 50 keV.

(b) *A magnetic beam-sweeping system* for two-dimensional X-ray scans of the specimen surface. This sub-system is also required for back-scattered or secondary electron scanning of the surface (see later).

(c) *A specimen airlock*, a stage with X, Y and Z motion and a means for positioning the desired area of the specimen under the electron beam.

(d) An energy (or wavelength) spectrometer to disperse the X-rays, so that the characteristic lines can be assigned to specific elements.

(e) An X-ray detector, and associated electronics for measuring X-ray diffraction line intensities.

(f) Read outs and recording electronics to display and record the characteristic X-ray intensities as functions of energy, wavelength and/or specimen position, together with a computer for data reduction.

A schematic diagram of the typical electron probe microanalyser, which incorporates all six required features is given in Figure 4.8. The electron probe is focused by two magnetic lenses on to the specimen at the bottom of the column, and scanned over the surface by deflection coils between the lenses. X-rays are thereby generated at the surface of the specimen, and characteristic radiation of a particular wavelength may be selected with either of the crystal spectrometers shown in the diagram (wavelength dispersion). Although not shown in Figure 4.8, it is possible to use a lithium-drifted silicon detector, or some other device to obtain dispersion of the X-rays according to their characteristic energies (energy dispersion). The amplified signal modulates the beam of the cathode ray tube scanned in synchronism with the probe, which thus displays the surface distribution of the element giving rise to the characteristic X-rays.

Alternatively (or concurrently, in some instruments) back-scattered or secondary electrons can be collected by the electron multiplier (scintillation counter) to form an image of the surface topography of the same area, as in the scanning microscope (except in the latter, one normally collects secondary electrons emitted when the primary electrons impinge upon the surface of the specimen). The resolution of the X-ray image is limited by electron penetration to about 1 μm, whilst that of the electron image can be 100 Å or better. The maximum area of scan is about a 400 μm square.

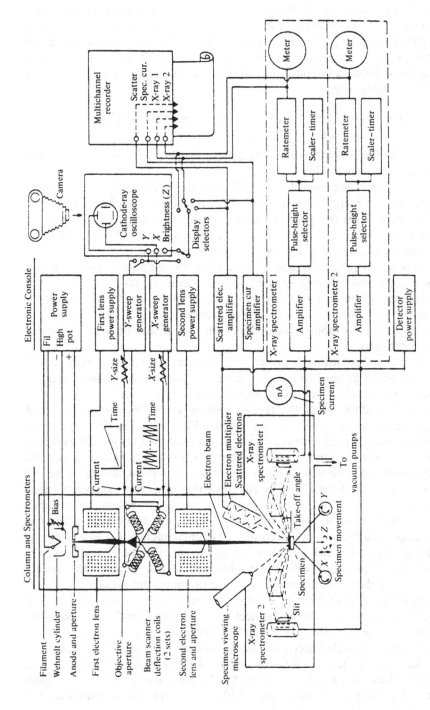

Figure 4.8 *Schematic diagram of a typical electron probe microanalyser* [*Bertin (1971)*]

Most electron probe microanalysis is now carried out in a scanning electron microscope with facilities for energy dispersive or wavelength dispersive spectrometers.

4.3.3 Quantitative electron probe microanalysis

The X-ray generation, absorption and fluorescence processes taking place in electron probe microanalysis are extremely complex. Figure 4.9 is an attempt to give (diagramatically) a qualitative explanation of how X-rays are generated in and beneath the surface of a specimen being bombarded with energetic electrons. We can see that, if we place our X-ray detector at angle α with respect to the surface of the specimen, X-rays will be detected from four different sources (denoted by the letters W, P, F and B in Figure 4.9). Let us follow the path of one of the millions of electrons impinging on the specimen surface, namely the one

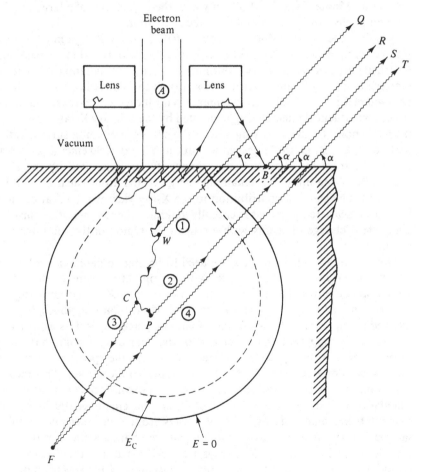

Figure 4.9 *Diagram illustrating how X-rays are generated in the surface of a specimen in the electron probe microanalyser $Q \equiv White \ X$-rays; $R \equiv Back$-scattered X-rays; $S \equiv Primary \ X$-rays; $T = Fluorescent \ X$-rays*

marked '*A*' in the diagram. As it travels *into* the specimen, the electron will lose small decrements of its energy, due to inelastic scattering by the atoms in its path. At each scattering incident, some continuous ('white') radiation will be created (such as at *W*). Clearly, some of this white radiation will find its way into the X-ray detector. Continuing with our original electron (which has now lost a considerable amount of its original energy (eV_A), where (V_A) is the accelerating voltage), we now see it interact at *C* to produce primary X-rays (in the way already described in Section 4.2.4) characteristic of the atoms present in the small volume of specimen at *C*. This will only occur, however, provided the energy of the electron (when it arrives at *C*) still has an energy greater than E_c, the critical excitation energy for the atoms at *C*. Typically, it is possible to use the computer to deduce the limits of the region (denoted by the broken line in Figure 4.9), beyond which it is unlikely for primary X-rays to be produced. The method is called the 'Monte Carlo Method', in which the trajectories of a large number of individual electrons are simulated in the computer.

It is fairly safe to say that all *characteristic* primary X-rays must arise from interactions between the incident electrons and all atoms within the pear-shaped (hatched) area. The *white* (or continuous) X-rays (originating from inelastic scattering of the incident electrons) arise from atoms anywhere within the outer pear-shaped region (indicated in Figure 4.9). Fluorescent X-rays, characteristic of the constituent elements being irradiated by the primary X-rays, can originate from any point (such as *F*) lying *outside* the pear-shaped outer surface indicated in the diagram. By far the largest contribution to X-ray yield emerging at an angle α must be from the characteristic primary X-rays generated by the electron eventually giving up all its remaining energy to the atom (such as at *P* in the diagram). The smallest contribution to the X-ray yield will be that originating from back-scattered electrons eventually making either an inelastic collision or a complete exchange of energy with atoms in the region of the specimen surface around *B*.

Now for quantitative analysis, we need to be able to define the volume from which the X-rays originate. Hence, X-rays originating from *B* (due to *back-scattering* of electrons) and from *F* (due to *fluorescent X-rays* originating within the pear-shaped volume calculated from the Monte Carlo method), have to be allowed for in any reliable estimates of volume percentages. It is also clear that X-rays originating near the lower end of the pear-drop volume have to pass through more of the specimen before emerging at the specimen surface, than those X-rays originating in those regions closest to the surface. This means we must take into account the *absorption of X-rays*. Another factor affecting our calculations must be the excitation (of *fluorescent* characteristic X-rays of the element being analysed) by *primary* X-rays from other elements within the sample. These primary X-rays could be white or characteristic. For instance, in the (Cu–Au) alloy, some Cu K_α X-rays are produced as a result of absorbing Au L_α and Au L_β X-rays. Due to the stronger continuous X-ray spectrum produced by the Au atoms compared with the Cu atoms, one can get a false idea of the percentage of Cu (since the continuous Au X-rays cause an enhancement of the Cu K_α characteristic radiation).

Most quantitative analysis involves comparison with respect to a *standard*. Quite often, that standard is a compound or alloy, rather than the element itself. There is, therefore, an *atomic number* effect due to the differences in electron scattering and energy losses between the sample and the standard. For instance, for Cu–Au alloy, the presence of Au increases the elastic scattering and reduces the stopping power $(\overline{SP})$, which is defined as:

$$\overline{SP} = \frac{Z}{A}\ln\left(\frac{1.17\overline{E}}{J}\right) \tag{4.18}$$

where Z is the atomic number and A is the atomic weight of the atoms constituting the element in question. J is the mean ionization potential and $\overline{E}$ is the mean energy of the electrons, given by:

$$\overline{E} = \tfrac{1}{2}(E_p + E_c) \tag{4.19}$$

where E_p is the energy of the primary beam and E_c is the energy required for K-line excitation.

The atomic number also affects the proportion of the incident electron beam that is back-scattered, and is thus unavailable for generating X-rays (apart from the insignificantly small amount occurring when the back-scattered electrons are themselves scattered back into the specimen surface). For example, more energy is lost due to back-scattering in the Cu–Au alloy than in Cu, and this causes a decrease in the number of Cu K_α X-rays generated in the alloy. Hence, it is usual to include the back-scattering coefficient (R_{BS}) in the atomic number correction, where (R_{BS}) is defined as the ratio of the X-ray intensity actually generated to that which would be generated if all of the incident electrons remained within the specimen. Table 4.2 gives a list of (R_{BS}) values.

Quantitative analysis by electron-probe microanalysis is generally carried out assuming the validity of the following equation:

$$\frac{I_A}{I_{AB}} = C_A\left[\frac{(R_{BS})_{AB}(\overline{SP})_A}{(R_{BS})_A(\overline{SP})_B}\right]\left[\frac{f(\chi)_{AB}}{f(\chi)_A}\right][1 + \gamma_F] \tag{4.20}$$

This equation is known as the 'Philibert–Duncumb–Reed' (PDR) expression, and it relates the *true* concentration (C_A) of constituent A to the experimental ratio of intensities between X-rays characteristic of constituent A from the actual specimen compared with the X-rays characteristic of (A) from the standard specimen consisting of a compound or alloy (AB). It is a combined equation extracted from the published work of Philibert (1962), Duncumb, Shields-Mason and da Casa (1969), and Reed (1965).

The terms in the first square bracket in Equation (4.20) comprise the Atomic Number Correction. We have already discussed the back-scattering coefficient (R_{BS}). The ratio $((R_{BS})_{AB}/(R_{BS})_A)$ is found quite readily from Table 4.2, knowing the atomic number of the standard (AB), together with E_p (the energy of the primary beam) and E_c (the energy required for K-line excitation). The ratio of the stopping powers for A and AB can be deduced from Equations (4.18) and (4.19), provided one knows the mean ionization potential J (see Table 4.3).

Table 4.2 *Values of back-scattering coefficient (R_{BS}) as a function of $(1/U_0)$ and Z (Duncumb and Reed, 1968), where $U_0 = (E_p)/(E_c)$*

	$Z1/U_0$										
	0.01	0.10	0.20	0.30	0.40	0.50	0.60	0.70	0.80	0.90	1.00
0	1.000	1.000	1.000	1.000	1.000	1.000	1.000	1.000	1.000	1.000	1.000
10	0.934	0.944	0.953	0.961	0.968	0.975	0.981	0.988	0.993	0.997	1.000
20	0.856	0.873	0.888	0.903	0.917	0.933	0.948	0.963	0.977	0.990	1.000
30	0.786	0.808	0.828	0.847	0.867	0.888	0.911	0.935	0.959	0.981	1.000
40	0.735	0.760	0.782	0.804	0.827	0.851	0.878	0.907	0.938	0.970	1.000
50	0.693	0.718	0.741	0.764	0.789	0.817	0.847	0.881	0.919	0.959	1.000
60	0.662	0.668	0.713	0.737	0.764	0.793	0.825	0.862	0.904	0.950	1.000
70	0.635	0.663	0.687	0.713	0.740	0.770	0.805	0.844	0.889	0.941	1.000
80	0.611	0.639	0.665	0.691	0718	0.750	0.785	0.826	0.874	0.932	1.000
90	0.592	0.613	0.639	0.665	0.695	0.730	0.767	0.811	0.862	0.924	1.000
99	0.578	0.606	0.634	0.661	0.691	0.725	0.763	0.806	0.858	0.921	1.000

The second term in square brackets in Equation (4.20) is known as the 'X-ray absorption correction'. The absorption function ($f(\chi)$) is defined as the ratio of the intensity of X-rays actually measured and the intensity that *would be* measured in the *absence* of absorption. Duncumb, Shields-Mason and da Casa (1969) improved the earlier model of Philibert (1962) and Heinrich (1967) improved on the results of Duncumb et al. (1969), to give the following expression for $f(\chi)$:

$$f(\chi) = \frac{(1+H)}{(1+\chi/\sigma_c)[1+H(1+\chi/\sigma_c)]} \qquad (4.21)$$

with

$$H = 1.2(A_w)/Z^2 \qquad (4.22a)$$

$$\sigma_c = \frac{4.5 \times 10^5}{(E_p)^{1.65} - (E_c)^{1.65}} \qquad (4.22b)$$

In Equation (4.22b), E_p and E_c are measured in keV. σ_c is sometimes known as the 'Modified Lenard Coefficient'.

The fourth term in square brackets in Equation (4.20) is the Fluoresence Correction. It ignores the very small contribution to the fluorescence by the white (or continuous) radiation. In fact, Hutchins (1974) tells us that the omission from Equation (4.20) is likely to result in significant error ($> 1\%$) only if the measured radiation is of high energy and a large spread of atomic numbers are present in the specimen. For instance, in (40 at % Cu–60 at % Au) at 20 keV, the correction factors are approximately 0.995 and 0.987 for X-ray take-off angles of 15.5° and 52.5° respectively.

Table 4.3. *Mean Ionization Potential J determined empirically as a function of atomic number* Z *(Dumcumb and Reed, 1968)*

Z	J(eV)	Z	J(eV)	Z	J(eV)	Z	J(eV)
6	146	29	377	52	706	75	1017
7	135	30	392	53	720	76	1031
8	127	31	407	54	734	77	1044
9	123	32	422	55	747	78	1057
10	123	33	437	56	761	79	1071
11	126	34	451	57	775	80	1084
12	133	35	466	58	788	81	1097
13	142	36	481	59	802	82	1111
14	154	37	495	60	815	83	1124
15	166	38	510	61	829	84	1137
16	180	39	524	62	843	85	1151
17	194	40	538	63	856	86	1164
18	209	41	553	64	870	87	1177
19	224	42	567	65	883	88	1191
20	239	43	581	66	897	89	1204
21	255	44	595	67	910	90	1217
22	270	45	609	68	923	91	1231
23	286	46	623	69	937	92	1244
24	301	47	637	70	950	93	1257
25	316	48	651	71	964	94	1270
26	332	49	665	77	992		
27	347	50	679	73	991		
28	362	51	692	74	1004		

To correct for fluoresence by characteristic X-rays, it is necessary to calculate the ratio:

$$v_F = \frac{I_F}{I_{px}} \qquad (4.23)$$

where I_F is the intensity of X-rays produced by characteristic-line fluorescence and I_{px} is the intensity due to primary excitation by electrons. Both of these factors depend on many imponderables beyond the scope of this introductory section. A very lengthy equation has been derived in an applicable form by Reed (1965). Let us just say that the characteristic fluoresence correction factor $(1 + \gamma 7_F)$ takes on values of 0.966 (for a take-off angle of 15.5°) and 0.954 (for a take-off angle of 52.5°) for Cu K_α in (40 at % Cu–60 at % Au) for an electron accelerating voltage of 20 keV (Hutchins, 1974). Clearly, these values indicate that fluorescence has contributed to an error in analysing proportional amounts of the order of $\pm 4\%$, compared with an analysis that does not allow for this additional source of X-rays. It must be pointed out, however, that fluorescence is only significant when the radiation from the fluorescing element is strongly absorbed by the specimen. Depending on the accuracy desired, a fluorescence correction is usually not necessary in oxides, nor in samples that contain large amounts of many different elements.

Much of the quantitative theory and tables given in this subsection need not concern the non-specialist, since there are now many computer programmes available to carry out the relevant iterative calculations involved in accurate quantitative analysis. Nevertheless, if tribologists are to obtain the most benefit from electron probe microanalysis they should be aware of what the software is actually doing for them.

4.3.4 Applications of electron probe microanalysis to the study of surfaces

Electron probe microanalysis provides us with information about surfaces in the following ways:

(a) the *analysis of secondary phases and the determination* (or confirmation) of *equilibrium phase diagrams.*

(b) the measurement of diffusion gradients *directly* on a *sectioned specimen.*

(c) *Homogeneity testing on a micrometre scale.* The detection of differences in concentration (< 1%) is a statistical problem and requires a stable instrument and an accumulation of many counts on a number of specimen areas. A variation of $\pm 5\%$ in a major constituent can be detected with a line scan.

(d) The *identification of surface contamination* that, possibly, had been introduced during the manufacturing process, is readily carried out using the spectrometer illustrated in Figure 4.8. This can be done quickly by a line-scan or a two-dimensional scan.

(e) The measurement of the thickness and the compositional analysis of *thin films.* This is an accurate technique and can even be applied to the *substrates* upon which these thin films are laid.

Most applications of electron probe microanalysis to the study of surfaces involve the topography of those surfaces in some way. Topography can be a severe problem on rough surfaces, particularly if a *quantitative* analysis is required. As the irregular surface is traversed, a number of negative and positive tilt angles relative to the horizontal plane of a flat specimen will be encountered. X-ray measurements will thereby be vitiated in two ways, namely (a) the electron beam incidence is no longer normal to the surface and (b) the X-ray absorption path length becomes either shorter or longer because of the change in *the effective value* of the 'take off' angle ψ. The larger or smaller effective ψ *primarily affects the absorption correction.* Provided the slopes of asperities on the (rough) surfaces are less than $\pm 10°$ there should be *no need* for making this *absorption correction.*

We will be returning to the tribological application of electron probe microanalysis in later chapters. Now is the time for us to discuss a technique that involves the excited atoms returning to their ground state by the emission of photoelectrons, namely X-ray Photoelectron Spectroscopy (XPS) – also known as Electron Spectroscopy for Chemical Analysis (ESCA). It will also serve as an introduction to the phenomenon of Auger electron emission.

4.4   X-ray photoelectron spectroscopy

4.4.1   The principles of the technique

Surface analysis by X-ray photoelectron spectroscopy (XPS) is achieved by irradiating a sample with mono-energetic *soft* X-rays. MgK_α ($hv = 1253.7$ eV) or AlK_α($hv = 1486.6$ eV) are ordinarily used. These low energy photons can penetrate approximately 1 to 10 μm in a solid. In fact, they interact with atoms in the surface region by the photoelectron effect, thereby causing photoelectrons to be emitted. The emitted electrons have *kinetic energies* (*KE*) given by:

$$(KE) = hv - (BE) - \phi_{sp} \qquad (4.24)$$

where (hv) is the energy of the photon, (*BE*) is the binding energy of the *atomic shell* from which the electron originates, and (ϕ_{sp}) is the spectrometer work function (i.e. the work necessary for the photoelectron to leave the influence of the surface atoms).

The binding energy may be regarded as an ionization energy of the atom for the particular shell involved. Since there are many ways in which an atom may be ionized, there is a corresponding variety in the kinetic energies of the emitted photoelectrons (see Figure 4.10(*a*) for the emission process involving a photoelectron originating from the (1s) level of the atom).

In addition, to the photoelectrons emitted in the photoelectric process, Auger electrons (see Section 4.5) are emitted due to the excited ions returning to their ground states *after* the photoelectric event. This *associated Auger electron emission* occurs some 10^{-14} s *after* photoelectric emission. In the associated Auger process shown in Figure 4.10(*b*), an outer electron falls into the vacancy in the (1s) level of the atom, caused by the emission of the photoelectron (in Figure 4.10(*a*)). In order to bring the atom back to its ground state, *a second* (*Auger*) *electron* is emitted, carrying off the excess energy. This Auger electron possesses a kinetic energy equal to the difference between the energy of the initial ion (i.e. the atom *immediately* after the photoelectron has left the atom) and the energy of the final (doubly-charged) ion. This kinetic energy does *not* depend on the mode of the initial ionization. Thus, *photoionization normally leads to two emitted electrons*, a photoelectron and an Auger electron.

Although the path length of the *photons* causing ionization is of the order of several micrometres, the path length of the *electron* is considerably less, being of the order of a few nanometres (i.e. 10^{-9} m compared with 10^{-6} m for the photons). Hence the need for a good vacuum (less than 10^{-7} torr). The very short path length of the ejected photoelectrons and Auger electrons means that only those atoms in the immediate surface of the specimen can give rise to those electrons which are detected by the instrument. The surface specific nature of the technique of XPS makes it ideal for use in our attempts to understand the role of surface films under anti-wear conditions. Let us now give some brief details of the instrumentation.

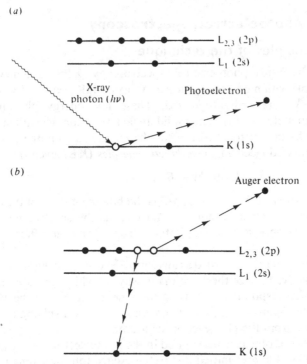

Figure 4.10 *(a) Diagram of the photoelectron process; (b) diagram of the associated Auger process*

4.4.2 Instrumentation

The major components of any XPS system are (i) the X-ray source, (ii) the specimen, (iii) the electron energy analyser, (iv) the ion gun (used for sputter-etching of the specimen surface), (v) the detector and (vi) the recorder. Let us discuss each component in turn:

(i) The X-ray source

Typically, targets are made of magnesium or aluminium. These elements give rise to $(K_{\alpha_{1,2}})$ X-ray photon energies of 1253.7 and 1486.6 eV respectively. This is enough energy to excite the 1s and valence levels of the atoms in the first row ($Z = 3$ to 10) of the periodic table, the 2s and 2p and valence levels of second row elements and so on. For an Al or Mg target, the operating conditions are typically 10 kV accelerating voltage with 60 mA beam current (i.e. 600 W). Copper (K_{α}) and chromium (K_{α}) X-ray sources can be used, but these more energetic X-rays (8047.8 and 5414.17 eV respectively) have much larger 'line widths' than the 0.7 eV of Mg and 1.0 eV of Al, thereby causing a drop in resolution.

(ii) The specimen

In most XPS applications, sample preparation and mounting are not critical. Typically, the sample can be mechanically attached to the specimen mount and analysis begun, with the sample in the 'as-received' condition. Sample preparation is even discouraged in many cases, especially when the natural surface is of interest. Typically, in tribological applications, the surface will contain volatile material. This is normally removed by long-term pumping in a separate vacuum system or by washing in a suitable solvent. If, under exceptional circumstances, the volatile surface layers are of interest, the specimen may be cooled for analysis using a 'cold finger'. The volatile components of the lubricant are typically not involved in providing the protective surface films *per se*. However, they are *indirectly* involved through reaction with the surface layers of the specimen to form oxides and/or anti-wear films.

Ion sputter-etching can also be used to remove surface contaminants. Argon or xenon ion etching is often used to obtain information on composition as a function of depth into the specimen (known as 'depth profiling'). We should always be aware, however, that the use of this method for surface removal is likely to change the chemical nature of the surface.

The specimen mounting angle is not very critical in XPS, but it does have *some* effect on the spectra. The use of a *low* electron take-off angle accentuates the spectrum of any layer segregated *on the surface, whereas a sample mounting angle* normal to the analyser axis minimizes the contribution from such a layer. We will see that the highly-favoured double-pass cylindrical-mirror type analyser (CMA) effectively integrates over a large range of electron take-off angles when used in the normal configuration. This reduces the effects of variations in sample geometry and mounting angle to an insignificant level in most cases.

(iii) The electron energy analyser

In order to be effective, the energy analyser must have a resolution at least as good as 1 part in 10^4. It was this requirement for high precision that held back development of the technique for many years after discovery. There are two types of analyser, namely magnetic and electrostatic analysers, but most attention is now focused on electrostatic analysers. The earlier electrostatic analysers were the hemispherical, double-focusing design shown diagrammatically in Figure 4.11. Note the retarding lens inserted just in front of the analyser. The purpose of this retarding lens is to slow down the electrons before they enter the analyser, thereby enabling the analyser to be made smaller (and hence, less expensive). The more recent instruments use the cylindrical mirror geometry. A typical set-up is shown in Figure 4.12. The particular geometry shown here is a double-pass, cylindrical-mirror type analyser (CMA) with a retarding grid input stage. The retarding grids are used to scan the energy spectrum of the ejected photoelectrons and Auger electrons from the specimen, while the CMA is operated at constant pass energy. This gives a *constant* energy resolution (ΔE) across the entire energy spectrum. The size of the area being analysed is defined by the size of the circular apertures at the output of the CMA stages. The energy resolution of the analyser ($\Delta E/E$) is also determined by these apertures. Typically, there will be about three aperture sizes available. Use of the largest aperture provides the user with the maximum signal intensity, a circular area of analysis on the actual sample of about *5 mm* in diameter, and an energy resolution of about 2% of the pass energy (25 eV),

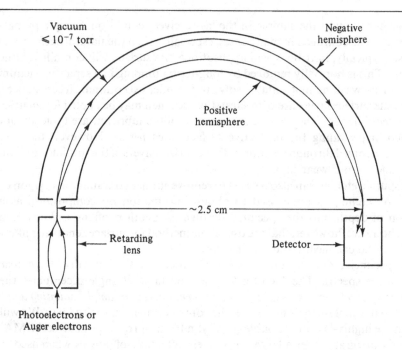

Figure 4.11 *Diagrammatical representation of a hemispherical electrostatic analyser*

namely 0.5 eV. For tribological applications this area might be considered a little too large to provide the locational ability required. Some instruments are emerging with a much smaller area of analysis, without the expected loss in intensity.

(iv) The ion gun

Most surface analysis involves 'depth profiling', that is, the controlled erosion of the surface by ion sputtering. Table 4.4 details some data on sputtering rates as a general guide (see page 180).

Chemical states can often be changed by the sputter technique, but useful information on elemental distribution can still be obtained.

(v) The detector

The very minute electron currents involved in XPS means that single and/or continuous channel electron multipliers have to be used.

(vi) The recorder

Typically, when faced with an unknown sample, a broad (survey) scan spectrum should be obtained first to identify the elements present. Once the elemental composition has been determined, narrower detailed scans of selected peaks can be used for a more comprehensive picture of the chemical composition. Before these scans can be done, however, the analyser has to be calibrated by reference to known binding energies (see Equation (4.24)). For instance, in order to deduce ϕ_{sp} (the work function of the spectrometer) one can assume that the binding

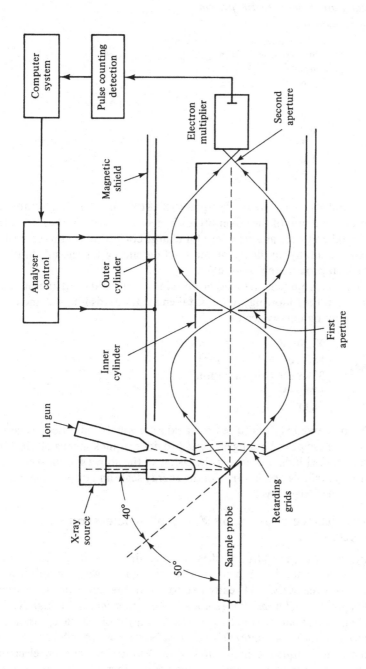

Figure 4.12 *Diagrammatical representation of a typical cylindrical-mirror analyser*

Table 4.4. *Representative sputter rates*
(2 keV argon ion beam with 100 $\mu A/cm^2$
impinging on sample)

Target	Sputter rate (Å/min) $\pm 20\%$
Ta_2O_5	100
Si	90
SiO_2	85
Pt	220
Cr	140
Al	95
Au	410

energy of the gold $4f_{7/2}$ peak is exactly equal to 83.8 eV. On insulating samples, a high resolution spectrum of the adventitious hydrocarbon on the surface of the sample can be taken to use as a reference for charge correction. In this respect, it has been experimentally verified that the binding energy for the adventitious carbon peak is approximately 284.6 eV.

There is no general agreement on *accurate* values of any standard line energies. It is suggested that the following values be taken as standards for gold and clean copper (on a binding energy scale):

Cu $2p_{3/2}$ 932.4 eV

Cu $(L_3M_5M_5)$ $\begin{cases} 567.9 \text{ eV (Al radiation)} \\ 334.9 \text{ eV (Mg radiation)} \end{cases}$

Cu $3p_{3/2}$ 74.9 eV

Au $4f_{7/2}$ 83.8 eV

Since the Cu $2p_{3/2}$ and $3p_{3/2}$ photoelectron peaks are widely separated in energy, measurement of these peak binding energies provides a quick means of checking the magnitude of the binding energy scale. Using all the above standard energies establishes the *magnitude* and *linearity* of the energy scale and its position, that is the location of the Fermi level.

4.4.3 Quantitative analysis by X-ray photoelectron spectroscopy

In tribology, we are generally satisfied, in the first instance, with merely identifying a surface species. When we wish to deduce wear or lubricating mechanisms, however, we would really like to know *how much* of each surface constituent is present *and where it is relative to the topography*. Although XPS is not well known for its ability to analyse microscopic areas of a surface, it is indeed able to give a *quantitative estimate of the various elements*. Where XPS scores over other elemental techniques is in its ability to distinguish between electrons coming from atoms of the same element combined in different ways, for example it can tell us whether the electron came from Fe^{2+} or Fe^{3+} in an oxidized surface, thereby showing that the oxide is either FeO (for Fe^{2+}) or α-Fe_2O_3 (for Fe^{3+}).

Methods for quantifying XPS measurements have been developed using 'Peak Area' and 'Peak Height' Sensitivity Factors. It is generally agreed that peak area sensitivity factors are more accurate. This is simply given (for most elements other than the transition metals) by the vertical peak height to baseline times the width parallel to the baseline at half the height. The baseline is tangential to the background at the points where the 'shoulders' of the peak begin. For a sample that is homogeneous in the analysis volume, the number of photoelectrons per second in a specific spectral peak (denoted by the symbol I) is given by:

$$I = (N_A)(f)(\sigma_{PE})(\theta_E)(y_E)(\lambda_T)(A_{PE})(T_D) \qquad (4.25)$$

where N_A is the number of atoms of the element per cm^3 of sample, f is the X-ray flux in photons/(cm^2 s), σ_{PE} is the photoelectric cross-section for the atomic orbital (or shell) of interest (in units of cm^2), θ_E is the angular efficiency factor for the given instrumental arrangement based on the angle between the photon path and the detected electron, y_E is the efficiency of the photoelectric process for the formation of photoelectrons, λ_T is the mean free path of the electrons in the sample, A_{PE} is the area of the sample from which photoelectrons are detected, and (T_D) is the detection efficiency for electrons emitted from the sample. From Equation (4.25) we can write:

$$N_A = \frac{I}{(f)(\sigma_{PE})(\theta_E)(y_E)(\lambda_T)(A_{PE})(T_D)} \qquad (4.26)$$

The factors in the divisor of Equation (4.26) are usually combined into what is called the Atomic Sensitivity Factor (S_S). If we consider a *strong* line from each of two elements, then:

$$\frac{(N_A)_1}{(N_A)_2} = \frac{I_1/(S_S)_1}{I_2/(S_S)_2} \qquad (4.27)$$

If the ratio $(S_S)_1/(S_S)_2$ is independent of the matrix for all materials, then one can develop a set of relative values of S_S for all the elements. The values of S_S given in Table 4.5 are based on peak area intensities relative to the (F 1s) intensity, which has been used as a standard unit (i.e. 1.00) intensity. These are relevant to a particular instrument and are only presented here as an illustration of the effect of the atomic sensitivity factors upon calculations based on peak area intensities alone.

A generalized expression for the determination of the atomic fraction (C_x) of any constituent in an n-component sample can be written as an extension of Equation (4.27), namely:

$$C_x = \frac{(N_A)_x}{\sum\limits_{i=1}^{i=n} (N_A)_i} = \frac{I_x/(S_S)_x}{\sum\limits_{i=1}^{i=n} I_i/(S_S)_i} \qquad (4.28)$$

There is much more that we could discuss about quantitative analysis by XPS. We will find, however, that many of the tribological applications of this technique rely mainly on using it as an *identification* technique. Let us end this sub-section

Table 4.5. *Atomic sensitivity factors for XPS peak area intensities*

Z	Element	Line	S_S	Z	Element	Line	S_S
3	Li	1s	0.012	48	Cd	$3d_{5/2}$	2.25
4	Be	1s	0.039	49	In	$3d_{5/2}$	2.85
5	B	1s	0.088	50	Sn	$3d_{5/2}$	3.2
6	C	1s	0.205	51	Sb	$3d_{5/2}$	3.55
7	N	1s	0.38	52	Te	$3d_{5/2}$	4.0
8	O	1s	0.63	53	I	$3d_{5/2}$	4.4
9	F	1s	1.00	54	Xe	$3d_{5/2}$	4.9
10	Ne	1s	1.54	55	Cs	$3d_{5/2}$	5.5
11	Na	1s	2.51	56	Ba	$3d_{5/2}$	6.1
12	Mg	2p	0.11	57	La	$3d_{5/2}$	6.7
13	Al	2p	0.11			4d	1.22
14	Si	2p	0.17	58	Ce	3d	6.7
15	P	2p	0.25			4d	1.22
16	S	2p	0.35	59	Pr	3d	12.5
17	Cl	2p	0.48			4d	1.48
18	Ar	$2p_{3/2}$	0.42	60	Nd	3d	15.7
19	K	$2p_{3/2}$	0.55			4d	1.48
20	Ca	$2p_{3/2}$	0.71	61	Pm	3d	17.6
21	Sc	$2p_{3/2}$	0.90			4d	1.57
22	Ti	$2p_{3/2}$	1.1	62	Sm	3d	20.3
23	V	$2p_{3/2}$	1.4			4d	1.66
24	Cr	$2p_{3/2}$	1.7	63	Eu	3d	23.8
25	Mn	$2p_{3/2}$	2.1			4d	1.76
26	Fe	2p	3.8	64	Gd	3d	29.4
27	Co	2p	4.5			4d	1.84
28	Ni	2p	5.4	65	Tb	4d	2.03
29	Cu	$2p_{3/2}$	4.3	66	Dy	4d	1.93
30	Zn	$2p_{3/2}$	5.3	67	Ho	4d	2.12
31	Ga	$2p_{3/2}$	9.2	68	Er	4d	2.19
32	Ge	$2p_{3/2}$	9.2	69	Tm	4d	2.28
		3d	0.30	70	Yb	4d	2.36
33	As	3d	0.38	71	Lu	4d	2.45
34	Se	3d	0.48	72	Hf	4f	1.55
35	Br	3d	0.59	73	Ta	4f	1.75
36	Kr	3d	0.72	74	W	4f	2.0
37	Rb	3d	0.88	75	Re	$4f_{7/2}$	1.25
38	Sr	3d	1.05	76	OS	$4f_{7/2}$	1.4
39	Y	3d	1.25	77	Ir	$4f_{7/2}$	1.55
40	Zr	$3d_{5/2}$	0.87	78	Pt	$4f_{7/2}$	1.75
41	Nb	$3d_{5/2}$	1.00	79	Au	$4f_{7/2}$	1.9
42	Mo	$3d_{5/2}$	1.2	80	Hg	$4f_{7/2}$	2.1
43	Tc	$3d_{5/2}$	1.35	81	Tl	$4f_{7/2}$	2.3
44	Ru	$3d_{5/2}$	1.55	82	Pb	$4f_{7/2}$	2.55
45	Rh	$3d_{5/2}$	1.75	83	Bi	$4f_{7/2}$	2.8
46	Pd	$3d_{5/2}$	2.0	90	Th	$4f_{7/2}$	4.8
47	Ag	$3d_{5/2}$	2.25	92	U	$4f_{7/2}$	5.6

(i) This table is based on $S_S = 1.00$ for (F 1s).

(ii) It is only valid for a particular double-pass, cylindrical-mirror type analyser supplied by Physical Electronics.

(iii) Data are for Mg X-rays. However, S_S for Al X-rays agree with those for Mg X-rays within 10%.

on X-ray Photoelectron Spectroscopy by discussing *line identification by X-ray photoelectron spectroscopy*. We will leave over, until a later chapter, the many applications of XPS to tribology and surface analysis.

4.4.4 Line identification by X-ray photoelectron spectroscopy

(i) Introduction

The spectrum is typically displayed as a plot of electron binding energy against the number of electrons in a small, fixed interval of energy. The position on the kinetic energy scale equal to the photon energy ($h\nu$) minus the spectrometer work function (ϕ_{sp}) corresponds to a binding energy of zero with reference to the Fermi level (see Equation (4.24)). Therefore, a binding energy scale beginning at that point and increasing to the left is normally used.

It is beyond the scope of this book to explain much about the Fermi level, except to say it relates to the various electron energy states available when we take into account the effect of the crystalline structure upon the electron energy levels of each individual atom within that structure. Instead of electron 'shells' one obtains various 'bands'. A conducting material has its conduction band contiguous with its valence band, whereas an insulating material has its conduction band significantly separated from its valence band. The Fermi level is related to the energy necessary to completely remove an electron from one of the ground states (1s, 2p, 3d, 4d, etc.), namely it is the free electron energy level minus the work function (ϕ_{sp}). It lies between the conduction and valence bands. When two materials are in electrical contact, for example the conducting spectrometer material and the (often) insulating sample, they have common Fermi levels.

A typical spectrum will consist of well-defined peaks due to electrons that have not lost energy in emerging from the specimen. Electrons that have lost energy contribute the raised background at bonding energies *higher* than the peaks. Several kinds of peaks are observed in XPS spectra. Some are always observed (being fundamental to the technique) whilst others depend upon the precise chemical and physical nature of the specimen. They can be grouped under the following eight headings: (a) Photoelectron lines, (b) Auger lines, (c) X-ray satellites, (d) X-ray 'ghosts', (e) 'Shake-up' lines, (f) Multiplet splitting, (g) Energy loss lines and (h) Valence lines and bands. For our purposes, it will be sufficient to discuss only the first two types of peaks, leaving the remaining six to be considered only when the identification is not clear.

(a) *Photoelectron lines.* The most intense of the photoelectron lines are normally almost symmetrical and are typically the narrowest lines observed in the spectrum. Photoelectron lines of pure metals can, however, show significant asymmetry due to coupling with the conduction electrons. The actual peak width is a combination of the natural line width, the width of the X-ray line (causing the ejection of the electron) and the instrumental broadening.

(b) *Auger lines.* These are groups of lines in somewhat complex patterns. There are *four* main Auger series observable in XPS. They are the KLL, LMM, MNN and NOO series, identified by specifying the initial and final vacancies in the Auger transition. For example, the KLL series includes those processes

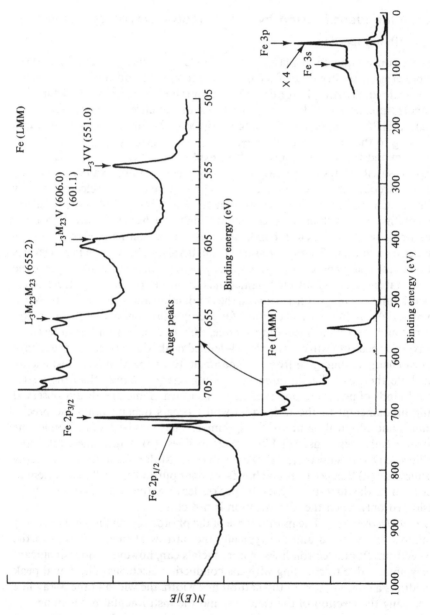

Figure 4.13 *The XPS spectra obtained from pure iron (using Mg K_α X-rays)*

with an initial vacancy in the K-shell and a final (double) vacancy in the L-shell. The symbol (V), for example KVV, indicates that the final vacancies are in the valence levels. Theoretically, the KLL series has nine lines, and the other series have still more. Since Auger lines have kinetic energies that are independent of the ionizing radiation, they appear (on a binding energy plot) to be in different positions when ionizing photons of different energy (i.e. different X-ray sources) are used. Core-type Auger lines (with final vacancies deeper than the valence levels) usually have at least one component of intensity and width similar to the most intense photoelectron lines.

(ii) Typical X-ray photoelectron spectra

Figure 4.13 gives an example of the typical XPS spectrum one obtains from an element, in this case iron. We can see that the sharp (Fe $2p_{3/2}$) and (Fe $2p_{1/2}$) peaks lie between binding energies of about 700 and 730 eV. An enlarged version of the spectra (Figure 4.14(a)) shows that, in fact, the binding energy of the $2p_{3/2}$

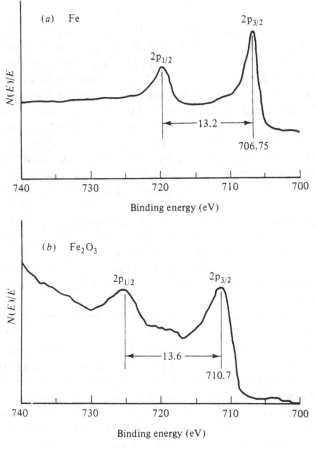

Figure 4.14 *(a) The $2p_{3/2}$ and $2p_{1/2}$ photoelectron peaks for pure iron; (b) the same peaks for iron combined with oxygen to form Fe_2O_3*

Table 4.6. *Core level electron binding energies for iron and iron oxides (after Sullivan)*

	Fe 2p$_{3/2}$ (± 0.2 eV)	Fe 2p$_{1/2}$ (± 0.2 eV)	Fe 3p (± 0.2 eV)	Fe 3s (± 0.2 eV)	O 1s (± 0.2 eV)	C 1s
Pure Fe	706.5	719.7	52.9	91.3	530.1 532.4	284.9
FeO	709.6	722.9	54.9	93.3 ± 0.5	530.1	284.8
Fe$_3$O$_4$	710.3	723.5	56.0 and 55.2	93.6 ± 0.5	530.1	284.8
α-Fe$_2$O$_3$	710.8	724.4	56.0	93.4	530.2	284.8
Hydroxide	711.6	725.1	56.5	93.9	531.4	284.8

line is 706.15 eV, with the (2p$_{1/2}$) peak lying at 13.2 eV below that peak. Note that, when the sample is Fe$_2$O$_3$ (see Figure 4.14(*b*)) then the (2p$_{3/2}$) line has a binding energy of 710.7 eV and the (2p$_{1/2}$) peak now lies at 13.6 eV below this energy. This immediately illustrates the power of XPS for chemical analysis. In view of my interest in the wear of steels, it is worth showing at this stage the differences one sees between the (Fe 2p$_{3/2}$) and (Fe 2p$_{1/2}$) for Fe, FeO, Fe$_3$O$_4$, α-Fe$_2$O$_3$ and iron hydroxide (see Table 4.6). It is interesting to note that there is not *exact* agreement between these energies (which were obtained by my colleague, Dr J. L. Sullivan) and the energies in Figures 4.13 and 4.14 (which were extracted from the *Handbook of X-ray Photoelectron Spectroscopy* from Physical Electronics). The binding energies for (Fe 2p$_{3/2}$) and (Fe 2p$_{1/2}$) for pure iron are only about 0.3% different between Table 4.6 and Figure 4.14(*a*). This shows the typical instrumental difference obtained with XPS data when attempting to give *absolute values*. Note that the *difference between the binding energies* of (Fe 2p$_{3/2}$) and (Fe 2p$_{1/2}$) for any given instrument is accurately the same (at 13.2 eV for pure iron and 13.6 eV for Fe$_2$O$_3$). This shows that we should place more reliance on the *relative differences* when attempting to determine how a given ion is combined with another to form a compound.

It should be noted that the lines in the row called 'hydroxide' are due to the adsorbed layer on oxidized iron surfaces, which was assumed to be a combination of various hydroxides of iron. Also, the lines due to Fe$_3$O$_4$ are not true binding energies since Fe$_3$O$_4$ contains *both* Fe^{2+} and Fe^{3+} ions. In fact these lines are the measured maxima of the photoelectron peaks. The column headed (C 1s) are the electron binding energies of carbon present as a contaminant on the oxide sample surfaces. The (O 1s) column gives the binding energies relating to oxides found on unetched metal surfaces. Note that the oxygen peaks are relatively unaffected by whether or not the oxygen ions are combined to form FeO or α-Fe$_2$O$_3$.

The reader may have noted the presence of (LMM) Auger lines between about 500 and 700 eV for the iron XPS spectrum obtained with Mg K$_\alpha$ X-rays. There do

Table 4.7. *Auger and photoelectron energies for iron and selected iron compounds*

Material	$(2p_{3/2})$	(L_3VV)	α'
Fe	706.9	702.4	1409.3
Fe_2O_3	711.0	703.1	1414.1
$FeWO_4$	711.5	703.1	1414.6
$Fe_2(WO_4)_3$	711.1	702.5	1413.6
FeS	710.4	703.2	1413.6
FeS_2	707.4	702.7	1410.1

Table 4.8. *Line positions from Mg X-rays, in numerical order*

17	Hf 4f$_7$	100	Hg 4f$_7$	191	B 1s	335	Th 4f$_5$
23	O 2s	101	La 4d$_5$	191	P 2s	336	Au 4d$_5$
25	Ta 4f$_7$	102	Si 2p$_3$	197	Lu 4d$_5$	337	Pd 3d$_3$
30	F 2s	105	Ga 3p$_3$	199	Cl 2p$_3$	337	Cu (A)
31	Ge 3d$_s$	108	Ce 4d$_5$	206	Nb 3d$_5$	342	Yb 4p$_3$
34	W 4f$_7$	110	Rb 3d$_5$	208	Kr 3p$_3$	347	Ca 2p$_3$
40	V 3p	113	Be 1s	213	HF 4d$_5$	359	Lu 4p$_3$
41	Ne 2s	113	Ge (A)	229	S 2s	359	Hg 4d$_5$
43	Re 4f$_7$	114	Pr 4d	229	Ta 4d$_5$	362	Gd (A)
44	As 3d$_5$	118	Tl 4f$_7$	230	Mo 3d$_5$	364	Nb 3p$_5$
45	Cr 3p$_3$	119	Al 2s	238	Rb 3p$_3$	368	Ag 3d$_5$
48	Nn 3ps	120	Nd 4d	241	Ar 2p$_3$	378	K 2s
50	I 4d$_5$	124	Ge 3p$_3$	245	W 4d$_5$	380	U 4f$_7$
51	Mg 3p	132	Sm 4d$_5$	263	Re 4d$_5$	385	Tl 4d$_5$
52	Os 4f$_7$	133	P 2p$_3$	264	Na (A)	396	Mo 3p$_3$
55	Fe 3p$_3$	133	Sr 3d$_5$	265	Zn (A)	402	N 1s
56	Li 1s	136	Eu 4d	269	Sr 3p$_3$	402	Eu (A)
57	Se 3d$_5$	138	Pb 4f$_7$	270	Cl 2s	402	Sc 2p$_3$
61	Co 3p$_3$	143	As 3p$_3$	279	Os 4d$_5$	405	Cd 3d$_5$
62	Ir 4f$_7$	150	Tb 4d	282	Ru 3d$_5$	410	Ni (A)
63	Xe 4d$_5$	153	Si 2s	284	Tb 4p$_3$	413	Pb-4d$_5$
64	Na 2s	154	Dy 4d	287	C 1s	435	Ne (A)
67	Ni 3p$_3$	158	Y 3d$_5$	293	Dy 4p$_3$	439	Ca 2s
69	Br 3d$_5$	159	Bi 4f$_7$	293	K 2p$_3$	440	Sm (A)
73	Pt 4f$_7$	161	Ho 4d	297	Ir 4d$_5$	443	Bi 4d$_5$
74	Al 2p	163	Se 3p$_3$	301	Y 3p$_3$	445	In 3d$_5$
75	Cs 4d$_5$	165	S 2p$_3$	306	Ho 4p$_3$	458	Ti 2p$_3$
77	Cu 3p$_3$	169	Rr 4d	309	Rh 3d$_3$	463	Ru 3p$_3$
85	Au 4f$_7$	180	Tm 4d	316	Pt 4d$_5$	483	Co (A)
87	Zn 3p$_3$	181	Zr 3d$_5$	319	Ar 2s	486	Sn 3d$_5$
88	Kr 3d$_5$	182	Br 3p$_3$	320	Er 4p$_3$	498	Rh 3p$_3$
90	Ba 4d$_5$	185	Yb 4d$_3$	331	Zr 3p$_3$	501	Sc 2s
90	Mg 3s	189	Ga (A)	333	Tm 4p$_3$	515	V 2p$_3$

Table 4.8 (*cont.*)

519	Nd (A)	677	Th $4d_5$	834	Fe (A)	1003	K (A)
530	Sb $3d_5$	684	Cs (A)	839	Pr (A)	1005	Th (A)
531	O 1s	686	F 1s	846	Ti 2s	1022	Zn $2p_3$
534	Pd $3p_3$	710	Fe $2p_3$	855	Ag $3p_3$	1035	Ar (A)
553	Fe (A)	711	Xe (A)	863	Te $3d_5$	1071	Cl (A)
555	Fe (A)	715	Sn $3p_3$	872	Cr $2p_3$	1072	Na 1s
565	Ti 2s	724	Cs $3d_5$	875	Ce (A)	1082	B (A)
573	Ag $3p_3$	729	Cr (A)	882	F (A)	1083	Sm $3d_5$
575	Te $3d_5$	737	I (A)	897	Cd $3p_3$	1088	Nb (A)
577	Cr $2p_3$	739	U $4d_5$	920	I $3d_5$	1103	S (A)
594	Ce (A)	743	O (A)	928	La (A)	1117	Ga $2p_3$
599	F (A)	765	Te (A)	930	Pr $3d_5$	1136	Eu $3d_5$
618	Cd $3p_3$	768	Sb $3d_5$	934	Cu $2p_3$	1155	Bi (A)
619	1 $3d_5$	780	Ba $3d_5$	954	Rh (A)	1162	Pb (A)
632	La (A)	781	Co $2p_3$	961	Ca (A)	1169	Tl (A)
641	Mn $2p_3$	784	V (A)	970	U (A)	1176	Hg (A)
657	Ba (A)	794	Sb (A)	980	Nd $3d_5$	1184	Au (A)
666	In $3p_3$	819	Sn (A)	981	Ru (A)	1186	Gd $3d_5$
670	Mn (A)	822	Te $3p_3$	993	C (A)	1192	Pt (A)
672	Xe $3d_5$						

exist tables of the *kinetic energies* of Auger electrons excited by X-rays, but for most practical purposes one uses the binding energy positions from the XPS spectra obtained from standard specimens to reinforce the interpretation based on the photoelectron peaks. Remembering that Auger electrons have the same kinetic energy no matter how the atoms were originally excited, one can deduce the kinetic energies of the Auger electrons from the modified *Auger parameter* (α') (which is the difference in the Auger and photoelectron line energies plus the energy of the exciting radiation) or, more simply,

$$\alpha' = \text{KE (Auger)} - \text{KE (Photoelectron)} + h\nu$$
$$= \text{KE (Auger)} + \text{BE (Photoelectron)} \qquad (4.29)$$

where the initials (KE) and (BE) denote 'Kinetic Energy' and 'Binding Energy' respectively. Table 4.7 gives some of the photoelectron spectra (the $2p_{3/2}$) and the Auger spectra (L_3VV) obtained from selected iron and iron compounds, from which we see that the Auger Parameter (α') is relatively constant at 1412.6 ± 2.2 eV. The values of binding energy for ($2p_{3/2}$) are in electron volts and were obtained from Appendix 4 of the book entitled *Practical Surface Analysis by Auger and Electron Spectroscopy* (Briggs and Seah, 1983). Note the slight differences between Tables 4.7 and 4.6 as regards the photoelectron levels of pure iron and iron in the form of Fe_2O_3. The values under the heading (L_3VV) are the kinetic energies (in eV) of the Auger electrons. If we were to plot the (L_3VV) of pure iron on a *binding energy scale* it would appear to have an energy of 553 eV (using Mg K_α exciting radiation) and 786 eV (using Al K_α radiation). Note that the first number is slightly different from the value of 551.0 eV given in Figure 4.13, thereby giving us an indication of the sort of accuracy we can expect from Auger energy analysis in XPS spectra.

We have concentrated mainly on *iron* XPS spectra so far in this section. In order to use XPS spectra for identification purposes, we need a comprehensive list of all previously published Auger and photoelectron energies (similar to Table 4.7) for the most intense photoelectron and Auger lines for *all the common elements and simpler compounds involving those elements*. Briggs and Seah (1983) have attempted this immense task. Possibly the best source of X-ray photoelectron spectra from many of the commoner elements and their simpler compounds is the *Handbook of X-ray Photoelectron Spectroscopy* (Wagner *et al.*, 1979). Some of these data were obtained before the present generation of X-ray photoelectron spectrometers appeared on the scene and may need to be weighted accordingly. Nevertheless, this handbook contains the XPS spectra of many elements plotted against binding energy (with enlargements for the Auger spectra) for both Mg K_{α} and Al K_{α} exciting radiation (the bulk of the spectra are those obtained with Mg K_{α}). Each spectrum is also accompanied by an enlarged version of the most intense photoelectron peak (such as we saw in Figure 4.14 for iron) and a list of the binding energies for that peak when the element is present in one of several compounds. In fact, Table 4.7 is part of such a list given in the handbook (for iron).

Any identification must involve an initial scan of line positions from given elements in order of increasing binding energy. Such a list is given in Table 4.8 for line positions from Mg X-rays (taken from Briggs and Seah, 1983). In this table, the 'A' in parenthesis denotes an Auger line. The sharpest Auger line and the *two* most intense photoelectron lines per element are included in the table. For brevity, $2p_3$ equals $2p_{3/2}$, $3d_5$ equals $3d_{5/2}$, etc. All lines are on the binding energy scale.

(iii) The tribologist as a user of physical techniques

From the foregoing sections, it should be possible for the tribologist with no previous experience of X-ray photoelectron spectroscopy to understand the applications of this wide-ranging technique to his particular subject. The actual applications will be left over until a later chapter. Although it has been mentioned before, it still bears repeating that the use of *any* physical technique (and especially X-ray photoelectron spectroscopy) should be undertaken by the tribologist in a knowledgeable way. He (or she) must not be deterred by the perfectionism of the expert in the particular technique from attempting to use that technique in any given problem. I well remember when one of my PhD students was working for a short period in an industrial research laboratory. The expert in X-ray diffraction at that laboratory could not identify *any* of the oxides of iron we knew would be present in a particular worn specimen. His mind could not make the adjustments often necessary with tribological specimens with their rough surfaces and the special chemistry involved at the high temperatures and pressures occurring at the real areas of contact between loaded, moving surfaces. The tribologist must know enough about his/her chosen techniques to be able to talk to the expert, and work with him/her towards the understanding of surface interaction during sliding and rolling through the medium of that technique. Let us now deal with another important physical technique, namely Auger electron spectroscopy.

4.5   Auger electron spectroscopy

4.5.1   The Meitner/Auger effect

In 1925, Pierre Auger came across tertiary radiation in Wilson cloud chamber photographs when observing incident X-radiation (he called it primary radiation) ionizing atoms, which then ejected electrons (secondary β-radiation, in his terminology). These were, in fact, what later became known as Auger electrons (Auger, 1925). It is interesting to note that the Austrian woman physicist, Lise Meitner, published papers in 1922 and 1923 which, according to Sietmann (1988) pre-date the independent discovery of the Auger effect by some two years (Meitner, 1922, 1923). He suggests that, apart from obvious discrimination against women in physics in the 1920s, her papers may have been overlooked for technical reasons. Meitner's interests were mainly in what happened in the nucleus rather than the atomic shell structure. Thus, her description and interpretation of the autoionization process is only part of a larger paper on the β-decay spectrum of thorium. By contrast, Auger's papers are short, concentrate on one phenomenon, and provide easy reference. Perhaps there is a moral here for the new researcher, especially in tribology, with its many facets. Learn to communicate as efficiently as possible or your (possibly) important findings will easily be overlooked! Let us now discuss the basic aspects of the Meitner/Auger Effect.

4.5.2   The Auger process

Having made the point that Auger electron spectroscopy depends on an effect which should be jointly attributed to Meitner and Auger, let us (for the sake of convenience) refer to the process and the electron spectroscopy by its commonly known term of 'Auger'. The Auger process involves the ionization of free or surface-bound atoms by *electron bombardment* (and not by X-ray photons as in XPS). In the initial ionization process of a core state of the atom, let us say that one of the K-electrons is completely removed from that atom after collision with the energetic primary electron. This is the secondary electron shown in Figure 4.15(*a*). The original primary electron carries on its journey (having lost a little of its energy in the above-mentioned collision), ready to 'knock out' an electron in the core state of another atom (probably a state with slightly less binding energy than the K-electron of the first atom). Hence we have an ion, that has a vacancy in its K-shell (see position 4 in Figure 4.15(*a*)). This vacancy is filled (say) by an electron from the nearby (L_1) level. We now have an excited atom, with one of its K-electrons having *too much energy* for the K-shell. De-excitation occurs, either by the emission of an X-ray photon of energy $E = (E_K - E_{L_1})$, *or* by the transfer of that excess energy to another electron, the Auger electron, which is then emitted from the atom (say) from the ($L_{2,3}$) level (see Figure 4.15(*b*)).

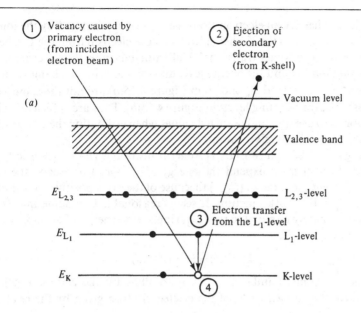

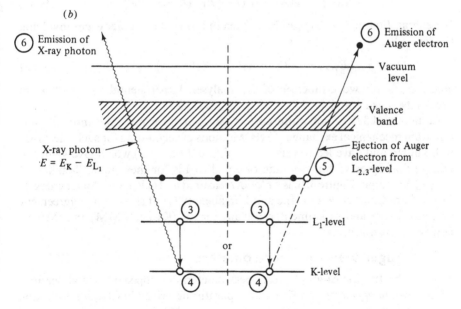

Figure 4.15*(a) Initial ionization of a core state (K-shell) followed by electron transfer from the L_1-level; (b) de-excitation of the ionized atom by emission of an X-ray photon or a $(KL_1L_{2,3})$ Auger electron*

The process whereby an electron from the $L_{2,3}$ level is emitted by the ionized atom with its vacancy in the K-shell filled by a core electron from the L_1-shell, is called the '($KL_1L_{2,3}$) Auger transition'. Unfortunately, after Auger ejection, the atom is *doubly-ionized* and the energy level diagram, in general, changes. Hence, the values of E_{L_1}, $E_{L_{2,3}}$ and E_K shown in Figure 4.15(a) are not the same as the energies of the corresponding levels in Figure 4.15(b). They are sufficiently close, however, for us to use the singly-ionized values when estimating the energy of the Auger electron from any given atom.

The energy released in a ($KL_1L_{2,3}$) Auger transition is $(E_K - E_{L_1})$ exactly, but the ejected electron must expend the energy $(E'_{L_{2,3}} + \phi_{sp})$ to escape the atom where $(E'_{L_{2,3}})$ is *not* the same as $(E_{L_{2,3}})$ (because of the extra positive charge of the atom) and ϕ_{sp} is the work function. It seems reasonable to assume that $(E'_{L_{2,3}})$ should approximately be equal to the $(L_{2,3})$ ionization energy of the next heavier element, that is:

$$E'_{L_{2,3}}(Z) = E_{L_{2,3}}(Z + \Delta) \qquad (4.30)$$

where Z is the atomic number and $\Delta \approx 1$ (to allow for the extra charge). We therefore have the energy, $E(Z)$, of the ejected electron given by the relation:

$$E(Z) = E_K(Z) - E_{L_1}(Z) - E_{L_{2,3}}(Z + \Delta) - \phi_{sp} \qquad (4.31)$$

In general, Equation (4.31) can be written in terms of an Auger event involving the levels W, X and Y, that is:

$$E_{WXY} = E_W(Z) - E_X(Z) - E_Y(Z + \Delta) - \phi_{sp} \qquad (4.32)$$

where ϕ_{sp} is the work function of the analyser. Experimental values of Δ are usually between $\frac{1}{2}$ and $\frac{3}{2}$.

Equation (4.32) is not quite correct, since it does not take into account quantum mechanical exchange effects. Methods of accounting for such effects are well known and have led to refinments of Equation (4.32) which give values of (E_{WXY}) that agree within less than or equal to 1% for the simpler 'families' of Auger transitions. Figure 4.16 is a plot of measured and calculated Auger energies (taken from Chapter 20 of Kane and Larrabee, 1974). It shows good agreement between theory and experiment, especially for the (KLL), (LMM) and (MNN) families of transitions.

4.5.3 Auger Electron Spectrometers

Historically, the first Auger electron spectrometers developed from modifications of the low-energy electron diffraction apparatus described in Chapter 3 (Section 3.3.7). These have now been largely superceded by dual-purpose, X-ray photoelectron spectrometers combined with the facility for Auger electron spectroscopy, which use either single-pass or double-pass cylindrical mirror analysers. The rate of development of new modifications of Auger and X-ray photoelectron spectrometers makes this author hesitant to go into *any* details of current instrumentation. Essentially, however, all modern instruments have facilities for scanning the surface of the specimen, giving the distribution of surface species in much the same way as electron probe microanalysis, except, of

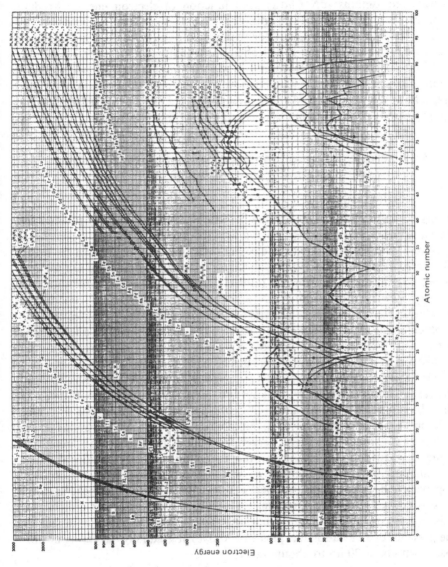

Figure 4.16 *Auger energies (E_{WXY}) versus atomic number Z*

course, we arrange to analyse the characteristic X-rays with EPMA, whereas we need to detect the characteristic Auger electrons with Auger electron spectroscopy (AES). Due to the low intensities of Auger electrons on a high background, the output from the detectors are electronically differentiated so that, in place of $N(E)$ (the number of Auger electrons of a given energy) being plotted against energy, one obtains $(dN(E)/dE)$ as a function of electron energy. Modern instruments are capable of providing beam widths less than a micron which, when combined with the facility for scanning across the surface, makes Auger electron spectroscopy a very good technique for analysing topographical features in terms of their chemical composition. We will be dealing with several applications of Auger electron spectroscopy in later chapters in this book. At this stage, however, we need to describe how the technique is used to calculate the approximate percentage atomic concentration, since this is required in order to make sense of any depth profiles obtained by ion sputtering of the surface.

4.5.4 The calculation of approximate percentage atomic concentration in depth profiles

(i) Introduction

Many depth profiles reported in the literature have been done by plotting the Auger peak-to-peak height (ppht) as a function of depth removed. In this form, only a *qualitative* assessment of element variation with depth can be made. Furthermore, an observer has to constantly remind himself/herself not to draw *quantitative* conclusions from the results. For instance, in the profile shown in Figure 7.19(a) (which is a plot of $(dN(E)/dE)$ against electron energy, obtained by Auger electron spectroscopic analysis of a steel surface which has pitted through contact fatigue in the presence of an extreme-pressure lubricant), we see that the peak-to-peak of the sulphur is about the same as the lowest energy iron peak ($\sim 590 \, eV$). We will see, however, that in fact the correct atomic percentage iron is about 12% and the correct atomic percentage sulphur is about 3%. Even in Figure 7.19(b) (which relates to the surface analysis with base oil only), we find that the sulphur is still only about 3%, whilst the iron is now about 7%. Such differences between peak-to-peak heights and actual concentrations cause confusion to the inexperienced observer. It is well worth the effort to calculate the actual concentration, especially now that computer software is readily available for this task. In this sub-section, we will give a brief outline of the *absolute* approach to deducing elemental concentration. We will then describe the relative sensitivity factor approach and illustrate it with an example from a particular tribo-system, namely aluminium–bronze sliding against a steel counterface (in the presence of an aviation fuel plus an additive) at gradually increasing normal load from about 20 up to about 300 N.

(ii) The absolute approach

According to Meyer and Vrakking (1972), the measured Auger current (I_A) from element A can be expressed as:

$$I_A = [S_A(E_p, E_i)] (1 - w_X)(n_A)(T_{AE})(I)(\operatorname{cosec} \theta_p) \left[\sum_i (N_i)(r_i)(q_i) \right] (kE_A) \quad (4.33)$$

The last term (E_A) has been added to the expression given by Meyer and Vrakking (1972) to account for the dependence of the Auger current upon the energy at which the Auger transition is measured using a cylindrical or hemi-cylindrical mirror analyser (as recommended by Bouwman, 1973). The constant k preserves the dimensional stability of the equation. The other terms in Equation (4.33) are:

$[S_A(E_p, E_i)]$ The *electron impact ionization cross-section* of element A, which is a function of both the primary beam energy (E_p) and the energy (E_i) of the inner shell (K, L, M, etc.) initially ionized.

$(1 - w_X)$ A correction for the *X-ray fluorescence* yield w_X (where X = K, L, M, etc.).

n_A The *solid angle* of detection. (This is the fraction of the Auger electrons emitted from the surface that are actually collected by the analyser).

(T_{AE}) The transparency of the analyser.

(θ_p) The angle between the incident electron beam and the specimen surface (typically 15°).

(I) The incident primary beam current.

(N_i) The number of atoms of element A per unit area on site i of the surface.

(r_i) The corresponding back-scattering factor for element A on site i.

(q_i) The corresponding screening factor for site i that tends to reduce the yield.

It is beyond the terms of reference of this book to elaborate very much more than this on the absolute approach. Clearly N_i is what we are after. Not all the other terms are always available to us. Some of them can be neglected under certain circumstances. We need to measure I_A, the Auger electron current from the expression:

$$I_A = \int_{E_1}^{E_2} N(E)dE = \int\int_{E_1}^{E_2} \frac{dN(E)}{dE}dE \tag{4.34}$$

In practical terms, this corresponds to measuring either the actual current detected ($\sim 10^{-10}$ A) or the areas under the peaks in the $[N(E)]$ or $[dN(E)/dE]$ energy distributions. If one assumes that $[N(E)]$ is a Gaussian distribution over the energy range E_1 to E_2, then the approximate area under the $[dN(E)/dE]$ Auger peak is given by Taylor (1969) as:

$$I_A = (K_i)(P_A)(W_A)^2 \tag{4.35}$$

where (P_A) is the peak-to-peak height (ppht), (K_i) is a constant dependent upon the amplitude of the modulating voltage and (W_A) is the half-width or twice the standard deviation. (W_A) can be taken as the separation of the positive and negative peaks in the $[dN(E)/dE]$ spectrum.

Bouwmann (1973) suggests that we can set $(1 - w_X)$ and (r_i) equal to unity in Equation (4.33). It has also been suggested that we replace (q_i) by $[\lambda_i(E_A)]$, which is a measure of the mean free path for inelastic scattering. This will change (N_i) from the number of atoms per unit area to (N_A), *the number of atoms of element A per unit volume*. Hence, from Equations (4.33) and (4.35) we get:

$$(P_A)(W_A)^2(K_i) = (N_A)[S_A(E_p E_i)](E_A)[\lambda_i(E_A)][G] \qquad (4.36)$$

where $G = k(n_A)(T_{AE})(I)\operatorname{cosec}\theta_p$. Rearrangement of Equation (4.36) gives the number of atoms per unit volume (N_A) as:

$$N_A = \frac{(K_2)(P_A)}{[S_A(E_p, E_i)][E_A][\lambda_i(E_A)][W_A]^{-2}(G)} \qquad (4.37)$$

where $K_2 = (K_i)^{-1}$. Similarly, for element B we get:

$$N_B = \frac{K_2(P_B)}{[S_B(E_p, E_i)](E_B)[\lambda_i(E_B)][W_B]^{-2}(G)} \qquad (4.38)$$

The approximate percentage atomic concentration of element (A) on the surface is therefore given by:

$$\% N_A = \frac{N_A}{(N_A + N_B + N_C + \ldots)} \qquad (4.39)$$

Writing this out in terms of its constituent parts, we get:

$$\%(N_A) = \frac{P_A/\{[S_A(E_p, E_i)](E_A)[\lambda_i(E_A)][W_A]^{-2}\}}{P_A/\{S_A(E_p; E_i)](E_A)[\lambda_i(E_A)](W_A)^{-2}\} + P_B/\{[S_B(E_p, E_i)](E_B)[\lambda_i(E_B)](W_B)^{-2} + \ldots}$$

$$(4.40)$$

where we see that K_2 and G have cancelled from the expression.

In order to use Equation (4.40), we can obtain (P_A) and (W_A) from the $(\mathrm{d}N(E)/\mathrm{d}E)$ versus electron energy curves. The other items require some sophisticated explanation beyond the scope of this book. For more details, the reader is referred to the papers by Vrakking and Meyer (1974, 1975). Let us now deal with the somewhat *simpler* approach via *relative sensitivity factors*.

(iii) Relative sensitivity factors

Several authors, for example Chang (1975), Moratibo (1975) and Palmberg (1976), have proposed relative sensitivity factors as a useful approach to quantitative AES. In essence, the method involves the compilation of elemental sensitivity factors referenced to *one arbitrarily chosen element* and empirically derived from elemental standards.

Following the method outlined by Chang (1975), for the case where all the elements on a surface are distributed uniformly and homogeneously, the percentage atomic concentration (X_i) of element i, in terms of the measured Auger current (I_i), can be expressed as:

$$X_i = \frac{(a_i)(I_i)}{\sum_j (a_j)(I_j)} \qquad (4.41)$$

where (a_i) is the *inverse* sensitivity factor for element i. Chang assumes that (a_i) can be replaced by (a_i^0), which is the relative inverse sensitivity factor between the standard element (s) and the pure element (i) *under the same conditions*, namely:

$$(a_i^0) = \left(\frac{I_s^0}{I_i^0}\right) \tag{4.42}$$

where (I_s^0) is the Auger current from a pure, arbitrarily chosen elemental standard and (I_i^0) is the Auger current from pure element i.

As Chang points out, the assumption that $(a_i) = (a_i^0)$ neglects any effects associated with the surface environments that are different to those on the elemental standards. These 'matrix effects' include differences in cross-sections, escape depths and back-scattering factors that can arise when the element of interest is part of a host material, for example an alloy.

Sometimes the spectra of the pure elements of interest to the tribologist are not readily available for the particular spectrometer being used for the analysis. Chang (1975) has suggested that one can use the *calibrated* spectra in the *Handbook of Auger Electron Spectroscopy* (Palmberg et al., 1972) as source. Essentially, this means that instead of assuming (a_i) is the same as (a_i^0), we now assume (a_i) is not very different from (a_i^H), where (a_i^H) is given by:

$$(a_i^H) = \frac{(I_s^H)}{(I_i^H)} \tag{4.43}$$

In fact, values of $[1/(a_i^H)]$ have been plotted for the (KLL), (LMM) and (MNN) Auger spectra for elements of atomic numbers from 4 to 80 in the above mentioned handbook. Provided one's own spectra were obtained with the same primary electron beam energy as that used by the handbook people (*namely 3 keV*), one can just read off the appropriate value of $(1/a_i)$ direct from that graph, a copy of which is shown in Figure 4.17. (I_i) can be assumed to be the peak-to-peak height of element i. Hence the atomic concentration (X_i) can be derived from Equation (4.41) provided *all* the other elements in the surface have been identified (so that the (I_j)s can be measured and the (a_j)s ascertained from Figure 4.17).

As an *example* of the use of Figure 4.17 in deducing concentrations, let us consider the $(dN(E)/dE)$ versus Auger electron energy (E_i) plot obtained (with a primary electron beam energy of 3 keV) from a pin of aluminium–bronze that had been sliding against a steel surface, at gradually increasing loads from 20 up to 300 N, under a lubricant consisting of an aviation fuel plus an additive designed to increase the lubricity of the fuel. This is shown in Figure 4.18. It was kindly supplied by my former colleague, Dr J. L. Sullivan. Since this was the first of a series of Auger spectra obtained from this specimen at various depths, the main interest was in how the carbon, oxygen, iron, copper and aluminium components varied with depth. Most of the carbon (and the nitrogen and sulphur) present in the surface in the 'as-received' condition is removed in the early stages of the depth profiling, indicating their contaminant nature. Table 4.9 gives the relevant values of peak-to-peak height (which will be assumed to be proportional to I_i),

197

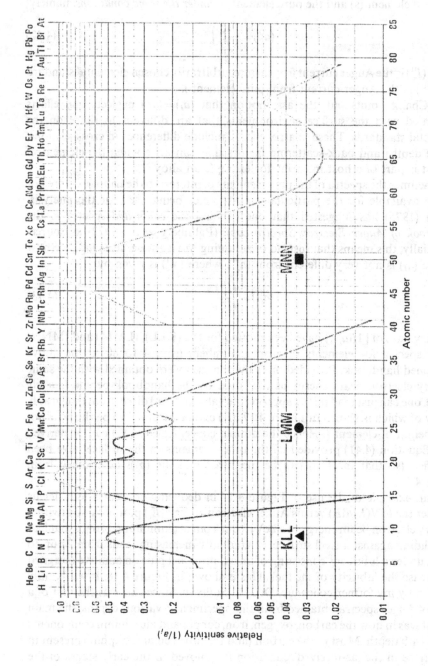

Figure 4.17 Relative Auger sensitivities $(1/a_i)$ of the elements for primary electron energies (E_p) equal to 3 keV (from Palmberg, 1972)

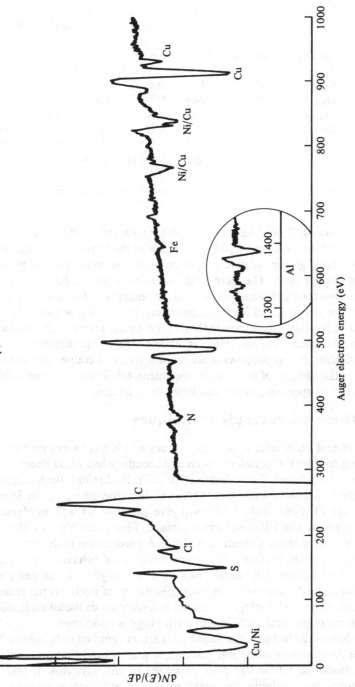

Figure 4.18 *Typical Auger spectrum from an aluminium-bronze surface worn in the presence of an aviation fuel plus lubricity additive*

Table 4.9. *Analysis of the Auger spectrum of Figure* 4.18

Auger peak		ppht		$(1/a_i)$	a_i	$(a_i)(I_i)$	X_i
Energy (eV)	Identity	Reading	%				
265	C_{KLL}	60	52.4	0.20	5.0	300	52.3%
510	O_{KLL}	28	24.4	0.51	1.96	54.9	9.6%
648	Fe_{LMM}	1.5	1.3	0.20	5.0	7.5	1.3%
920	Cu_{LMM}	19	16.6	0.21	4.8	91.2	15.9%
1390	Al_{KLL}	6	5.3	0.05	20.0	120	20.9%
Total		114.5	100%			$\sum (a_i)(I_i) = 573.5$	100%

$(1/a_i)$ from Figure 4.17 (and hence a_i), together with the identification of each Auger peak, and the product $[(a_i)(I_i)]$, for each of the important components.

Note the column giving the percentage compositions on the basis of peak-to-peak height (ppht) only. The percentage *aluminium* would have been considerably *under-estimated* if we have not used the relative inverse sensitivity (a_i) to provide the 'correct' value of 20.9%. Conversely, the *oxygen* would have been *over-estimated*. It is interesting to note that, as the surface layers of carbon on this specimen were etched away, an oxide layer of considerable thickness became revealed with constant proportions of aluminium, iron and copper. We will often find oxides in the surfaces of worn metal specimens. Obviously, the iron oxide in the surface of this specimen originated from the steel disc.

4.6   Other microscopic techniques

There are several other microscopic techniques which have some potential for investigating tribological problems, such as Secondary Ion Mass Spectroscopy (SIMS), Proton-Induced X-ray Analysis (PIXE), Rutherford Back-Scattering (RBS), Nuclear Reaction Analysis (NRA) and Electron Energy Loss Spectroscopy (EELS). Of these, only SIMS will give *chemical*, as well as elemental, information about the tribo-element's surface. This probably has the most promise of all the techniques mentioned above. A good review of the relevance of these techniques to the surface analysis and tribological behaviour of the most important of all industrial materials, namely steels, is given by Singer (1985). Although his interest is mainly in the ion implantation of steels, his paper makes interesting reading for a tribologist prepared to look towards the future insofar as the application of physical techniques to tribology is concerned.

Finally, there is the technique of radioactive tracers applied to tribology. Very early in the development of nuclear reactors, it was realized that one could use tracer techniques to follow the wear processes of tribo-systems. It was particularly useful for studying the wear of actual automotive engines. Unfortunately, the health hazards associated with radioactive tracer techniques have tended to restrict their use. The same can also be said about nuclear reaction analysis. The typical tribology laboratory is not really the place to have dangerous radiations present. Not very much new work has been done in radioactive tracers applied to tribology since 1964, when Lancaster wrote his review (Lancaster, 1964). The reader is referred to that review if he/she is interested in the possibilities of this technique for his/her particular problem. The judicious use of the techniques covered in this, and the previous two, chapters should furnish all the information provided by radioactive tracers, without the problems of providing protection from the ionizing radiations.

4.7 Concluding remarks

In this chapter we have covered three of the most important of all the microscopic techniques for studying atomic structure, at least as far as tribological problems are concerned. Electron probe microanalysis has probably been used for tribological investigations more than the other two, but it is only a matter of time before the techniques of X-ray photoelectron spectroscopy and Auger electron spectroscopy are being used as a matter of routine in all basic investigations into the *chemical* aspects of tribo-systems. It cannot be stressed enough how important it is for the tribologist to have sufficient knowledge of his/her chosen physical technique to be able to speak meaningfully with the expert who is actually 'driving' the instrument. The aim of these last three chapters has been to give the reader some ideas regarding the basic theory and practical use of three sets of techniques, namely those techniques relevant to (a) the *macroscopic properties* of tribo-systems, (b) the *microscopic properties* as regards *atomic arrangements* of the tribo-elements and (c) the *microscopic properties* of the atoms which go to make up those arrangements.

I am especially aware of how difficult it is to *predict* the tribological behaviour of any given tribo-system when there are no guiding equations to help the engineer. There has, however, been an upsurge in the modelling of friction and wear processes (see Lim and Ashby, 1987) which could easily be the start of a new era in the prediction of wear. This is the most urgent problem in tribology and the combination of physical analysis with wear modelling, along the lines suggested by Lim and Ashby, is the proper way to go about getting to that happy position where an engineer can feed his/her running conditions, materials and all other relevant input data into the model and thereby extract the required prediction!

5   The analysis of lubricant films

5.1   The analysis of extreme-pressure lubricant films formed in the presence of typical disulphide additives

5.1.1   Introduction

In this section of our chapter on the analysis of lubricant films, we will be concerned with the use of physical methods of analysis to investigate the mechanisms whereby some selected disulphides provide protection under conditions of sliding which would otherwise have resulted in the breakdown or seizure of the tribo-system if the disulphides were not present as 'extreme-pressure' additives in the lubricant. There are, in fact, two types of extreme-pressure additives, one that prevents catastrophic failure when a tribo-system is *suddenly* subjected to *unexpectedly* high loads (such an additive is truly called an 'extreme-pressure' additive), and another that reduces the wear to an 'acceptable' level in a system designed to work close to the limits of lubrication, for example a hypoid gear system. We call the latter type of lubricant additive an 'anti-wear' additive, for obvious reasons. Some tribologists suggest we should drop the word 'pressure' in the generic word for the whole family of lubricant additives and merely call them 'extreme-temperature' additives. As we shall see, the temperature between the wearing interfaces is indeed an important factor in the mechanisms whereby extreme-pressure lubricant additives provide protection of those interfaces from the incidence of severe way.

The study of extreme-pressure additives has tended to be approached in two very different ways. The practicing engineer will tend to only believe the results obtained in actual service tests, for example, in the gear boxes or differentials of actual test automobiles driven for thousands of miles running under real conditions. The experimental chemist, physicist or metallurgist will, however, maintain that the most economically-effective way of comparing newly-formulated additives, is to run them through standard tests in the 'Sliding Four-Ball Machine'. This is a device in which one ball is rotated against a 'nest' of three identically-shaped balls, clamped so that they may not rotate. There is variety of this machine which is called the 'Rolling Four-Ball Machine', in which the three lower balls *are* allowed to rotate and roll (see Figure 7.1). Either machine is fitted with balls of $\frac{1}{2}$ in (i.e. 127 mm) diameter.

In the sliding four-ball test, the upper ball produces a wear scar on each of the three lower balls. Obviously, the upper ball exhibits a continuous wear track around a circle of latitude about 55° South (assuming the driving ball lies immediately north of the centre of gravity of the clamped three-ball system). The plot of the mean wear scar diameter on the three lower balls *at the end of one-*

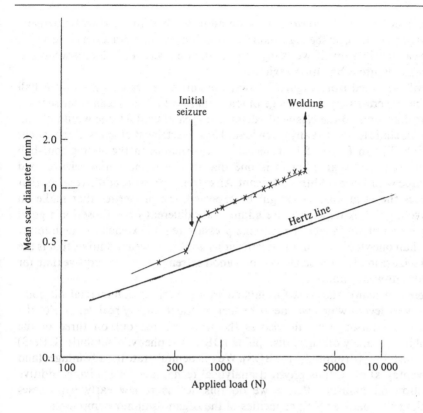

Figure 5.1 *Typical log-log plot of mean wear scar diameter (mm) versus the applied load (N); (the sliding four-ball machine)*

minute of running at a considerable load is normally drawn on logarithmic graph paper, a typical example of which is found in Figure 5.1.

It has been suggested, (Coy and Quinn, 1975) that the region before the initial seizure load (about 600 N for Figure 5.1) represents the region of effective operation of an anti-wear additive, whereas the region between the initial seizure and weld loads represents the extreme-pressure region. It must be pointed out that not all practicing lubrication engineers would necessarily agree with this definition. They would tend to say that there is no proof that the sliding four-ball wear tester can be used to predict service performance. It has been shown, however, by many investigations (too numerous to mention here) that the four-ball test will rank various additives in the same way as their service performance. Thus, we see that the anti-wear region is one in which acceptably small amounts of wear occur due to the action of the additive (or some parts of the additive molecules) with the sliding surface. Of course, there may also be (a) reaction between the air trapped within the additive and the surface and (b) between the base oil in which the additive is placed and the surface. It is the aim of most *current research* into additive mechanisms to find out the relative importance of each possible reaction with the surface, under conditions involving *anti-wear additives*. Much of the *previous work* on additives has involved truly *extreme-pressure lubrication*, on the reasonable assumption that one can learn more about anti-wear mechanisms if we study the boundaries between the anti-wear and extreme-pressure lubrication regions.

It will be noted from Figure 5.1 that the anti-wear region (in the four-ball machine experiments) is the range of loads that give rise to mean wear scars of similar diameters to the elastically-deformed area of contact one would obtain under the static loading at any given load. These Hertzian diameters are shown as a straight line in Figure 5.1. In terms of performance in the sliding four-ball machine, an anti-wear additive is one that increases the initial seizure load compared with a non-additive lubricant. An extreme-pressure additive is one that increases the weld load. As might be expected, the properties that make an additive a good extreme-pressure additive are different from those for a good anti-wear additive. In fact, the extreme pressure region extends to much lower loads than one would normally obtain with a good anti-wear additive, that is the initial seizure loads are much lower for a good extreme-pressure additive than for a good anti-wear additive.

There are many additives formulated by the various commercial oil companies, very few of which will describe their products in any real detail. For the sake of illustration, we will discuss the work carried out on three of the disulphides, namely dibenzyl disulphide (DBDS), diphenyl disulphide (DPDS) and di–tert–butyl disulphide (DTBDS). We will demonstrate the advantages (and disadvantages) of various physical analytical techniques for studying additive protection mechanisms. Before we do this, let us review early hypotheses regarding the load-carrying capacities of the organo-sulphur compounds.

5.1.2 Early hypotheses related to the load-carrying capacity of organo-sulphur compounds

The most commonly accepted hypothesis of the load-carrying capacities of organo-sulphur additives is that these compounds react with the metal, newly-exposed by the wear process, to form mixed surface layers of organic and inorganic sulphides or oxides, which tend to prevent further metal-to-metal contact and seizure. Greenhill (1948) suggested that sulphide films were formed when using sulphur-containing compounds. His suggestion was based on metallographic interpretation of the optical micrographs obtained from taper sections through wear scars.

Davey and Edwards (1957) were among the first researchers to try to explain the influence of chemical structure on the load-carrying properties of organo-sulphur compounds. They proposed that the reaction mechanism, taking place on the steel surface of the balls in a ball-bearing test rig, consisted of two or three steps, according to whether the additive used was a monosulphide or a disulphide respectively. *If one starts with a monosulphide*, the first step is:

$$R-S-R + Fe \rightarrow Fe:S \begin{matrix} R \\ \diagup \\ \diagdown \\ R \end{matrix} \qquad (5.1)$$

The second step would involve the formation of a mixed layer of ferrous sulphide and an organic compound, thus:

$$\text{Fe:S} \begin{matrix} R \\ \diagup \\ \diagdown \\ R \end{matrix} \rightarrow \text{FeS} + \text{R—R} \tag{5.2}$$

They suggest that the FeS layer is only effective at low loads. Davey and Edwards (1957) proposed that the *disulphides under low loads would also act as do the monosulphides*, namely:

$$\text{R—S—S—R} + \text{Fe} \rightarrow \text{Fe:S} \begin{matrix} R \\ \diagup \\ \diagdown \\ \text{S—R} \end{matrix} \tag{5.3}$$

The second step would be the formation of an iron mercaptide, namely:

$$\text{Fe:S} \begin{matrix} R \\ \diagup \\ \diagdown \\ R \end{matrix} \rightarrow \text{Fe:S} \begin{matrix} \text{S—R} \\ \diagup \\ \diagdown \\ \text{S—R} \end{matrix} \tag{5.4}$$

The iron mercaptide film would behave as a soap film formed from a fatty acid under anti-wear lubrication conditions. Under very severe conditions, that is extreme-pressure lubrication, the iron mercaptide film would break down to form iron sulphide and organic sulphide layers, as shown below:

$$\text{Fe} \begin{matrix} \text{S—R} \\ \diagup \\ \diagdown \\ \text{S—R} \end{matrix} \rightarrow \text{FeS} + \text{R—S—R} \tag{5.5}$$

Godfrey (1962) carried out sliding four-ball machine tests using sulphurized mineral oils, and proposed that the extreme-pressure mechanism consisted of the formation of a layer from 0.5 to 1.0 μm deep within the steel, consisting mainly of iron oxide (Fe_3O_4), with iron sulphide just a minor constituent of the worn surface. His work is interesting in that he was one of the first people to suggest that the presence of sulphur in the mineral oil promoted the oxidation of steel under extreme-pressure conditions. Thus, Godfrey (1962) was pioneering the way for later attempts to find a connection between mild-oxidational wear and extreme-pressure lubrication, using the term 'extreme-pressure' in its broadest sense. He does not, however, relegate the iron sulphide to a *minor role* in the mechanism, even though only small amounts were detected.

From this brief review of hypotheses related to the load-carrying capacities of organo-sulphur compounds, it is clear that *identification* of the surface compounds, as well as their distribution across the wearing surface and just beneath the wearing surface, is an important factor in making these hypotheses viable. Let us now review the relevant techniques that will provide the identification.

5.1.3 Physical methods of analysis relevant to the analysis of extreme-pressure lubricant films

The reactions that occur between the molecules of the extreme-pressure lubricant additives and the heavily-stressed surfaces occurring in the truly extreme-pressure region, involve much larger volumes of surface material than those involved under conditions where anti-wear additives are normally used. Clearly the techniques to be used for analysing extreme-pressure additive mechanisms involve much deeper penetration beneath the surface that those to be used for analysing anti-wear mechanisms. In this section we will be concentrating mainly on those additives most effective in the extreme-pressure region, namely the disulphides. We will leave over (until Section 5.4), the use of physical methods more suitable for the anti-wear region of activity.

Glancing-angle X-ray diffraction (GAXRD) and electron probe microanalysis (EPMA) are obviously relevant to the analysis of extreme-pressure additive mechanisms. The topography of the worn surfaces is best examined by scanning electron microscopy (SEM). The analysis with respect to depth (at the large depths involved in truly extreme-pressure mechanisms) is best done using optical microscopy of tapered sections combined with micro-hardness (OM + H) measurements. Table 5.1 is a summary of the information available from each of these four techniques.

This table illustrates the need for using *several* physical techniques when analysing tribo-systems. Note the tendency for the 'yes' answers to lie on a diagonal across the table, indicating that a more complete 'picture' can be better obtained by using these four techniques on the same tribo-system.

5.1.4 The use of glancing-angle X-ray diffraction for studying extreme-pressure additive mechanisms of organo-sulphur compounds

Only a limited amount of research has been carried out using glancing-angle X-ray diffraction for analysing worn surfaces. This must be due to lack of familiarity with the technique rather than to its irrelevance. Together with my colleagues I have long been using the glancing-angle X-ray diffraction technique described in Section 3.2.4 for examining the crystallography of many different tribo-systems. Remember that it is the irradiation of the *edge* of the specimen that provides the narrow diffraction lines so necessary for positive identification. Most investigators have dismissed the glancing-angle X-ray diffraction technique as being too inaccurate. This dismissal is inappropriate for edge-irradiated conditions, where the transmitted component is clearly visible in the X-ray diffraction pattern.

In view of the lack of other references, there is no choice but to describe the work carried out in my own laboratory as being illustrative of the value of the glancing-angle, edge-irradiated X-ray diffraction technique for identifying surface structures formed under extreme-pressure additives. Coy and Quinn (1975)

Table 5.1. *Information available from typical physical techniques used for analysing extreme-pressure mechanisms*

Features observed	Analytical techniques used			
	GAXRD	EPMA	SEM	OM+H
Positive compound identification	Yes	No	No	No
Element identification	No	Yes	No	No
Spatial distribution of elements	No	Yes	No	No
Film thickness	Yes	Yes	Yes	Yes
Surface topographies	No	Yes	Yes	Yes
Substrate structures	Yes	No	No	Yes
Hardness	No	No	No	Yes

Table 5.2. *Load-carrying capacities of the organo-sulphur compounds studied by Coy and Quinn (1975)*

	Initial seizure load (N)				Weld load (N)			
	White oil		HVI oil		White oil		HVI oil	
Additive	Balls	Flats	Balls	Flats	Balls	Flats	Balls	Flats
None	350	450	450	450	700	700	700	1000
DBDS	550	550	450	550	2600	2600	2600	2200
DTBDS	450	450	450	450	2600	2400	2200	2400
DPDS	550	650	550	450	1800	1800	1800	1800

used dibenzyl disulphide (DBDS), di–tert–butyl disulphide (DTBDS) and diphenyl disulphide (DPDS), all at 0.26 wt% concentrations of sulphur. They obtained their blends by mixing with (a) a highly refined white oil containing many saturated cyclic hydrocarbons with side chains and less than 100 ppm sulphur and (b) a high viscosity index (HVI) solvent-refined oil consisting mainly of saturated cyclic hydrocarbons with side chains, iso-paraffins, some aromatic and polar compounds, and approximately 1 wt% sulphur. The results of some of Coy and Quinn (1975) standard four-ball tests, together with the results of similar 'One-Ball-on-Three-Flats' tests, are summarized in Table 5.2.

One interesting feature of Table 5.2 is that the nature of the base oil has very little effect on the initial seizure and weld loads, suggesting that the original 1 wt% sulphur in the HVI oil is relatively inert and plays little part in the activity of the lubricant. Another feature is that there is no marked difference in the weld loads obtained with the four-ball and the one-ball-on-three flats geometry. This is rather satisfying, since the latter geometry was introduced entirely so that the scars were formed on specimens suitable for inserting into the electron probe microanalyser and the scanning electron microscope. In fact, they were 0.25 in diameter roller bearings, of the same material as the upper (driving) ball, placed so that the *top flat surface* of each roller coincided with the plane of contact between the upper ball and its lower companion. It is so much easier to view a flat surface than a round surface that one should always try to arrange for this optimum geometry to be present in one's tribo-system.

Although Table 5.2 relates to the particular oils and additives used by Coy and Quinn (1975), it does possess the character of being typical of the behaviour exhibited by many other additive blends when tested in the four-ball machine. It shows that DBDS and DTBDS additives provide *good* extreme-pressure activity (i.e. high weld loads), whereas DPDS gives a much lower weld load. It is interesting to see that DPDS provides much the same initial seizure loads as the DBDS, DTBDS and the plain oil, thereby indicating that its anti-wear activity is not particularly outstanding either! On the whole, however, the results were almost what *was expected* in the light of previous work on these materials. Let us now see what the glancing-angle, edge-irradiated, X-ray diffraction techniques can tell us about the extreme-pressure mechanisms involved when these disulphides are used as additives.

Table 5.3 is a good example of the way in which possible identification of surface structures is made using glancing-angle, X-ray diffraction analysis of wear scars formed under extreme-pressure conditions. It should be examined in conjunction with Table 5.2, since the choice of worn specimens was made on the basis of their proximity to initial seizure and final weld loads. The table is mainly concerned with the wear scars formed in the extreme-pressure region. The abbreviations HVI and RIS relate to the High-Viscosity Index oil and the Risella (white oil) respectively.

It will be noted there is no column headed 'α-Fe'. This is because the four lines that appear at $d = 2.027$, $d = 1.433$, $d = 1.170$ and $d = 1.0134$ Å, have all been assumed to arise from diffraction by the (110), (200), (211) and (220) planes of the α-Fe crystallites (Isherwood and Quinn, 1967). Note also that the columns headed 'Etched Ball' and 'Unworn Ball' were respectively calculated from the glancing-angle X-ray diffraction patterns from the flat ground surface of ball material (after etching) and from the curved (unetched) surface of an unworn ball. The interplanar spacings of both patterns are virtually identical to each other, showing that the bulk material and the unworn surface were comprised of the same constituents before wear occurs. Comparison of these two columns with the column headed '6-0688, Fe_3C, I/I_1' reveals that, for d-values greater than 1.80 Å, only the low intensity line at $d = 2.54$ Å is absent from those expected from the X-Ray Powder Data File Card 6-0688 for Fe_3C. The symbol 'I/I_1' stands for the intensity of a particular line relative to the strongest, which is at $d = 2.01$ Å (according to Card 6-0688). Of course, this line has a d-value so close to that of the (110) planes of α-Fe that it is masked by the broader line of that element. It should be clear from this table that the X-ray diffraction patterns from the surfaces worn under the 'HVI Oil' at 450 N and 'HVI oil plus DBDS' at 550 N (the initial seizure load for these lubricants) are almost identical to those of the bulk material, that is α-Fe and cementite (Fe_3C). This indicates that there was either no surface film produced during the period of anti-wear behaviour or that the film was non-crystalline. A further possible reason could be that the wear scars are so small during anti-wear behaviour that there just is not enough material in the wear scar to provide any diffraction. This third possible explanation is not so convincing as the one which I favour in which anti-wear films are much too thin for X-ray diffraction techniques to be of very much use in their analysis. The extreme-pressure behaviour of the disulphides used by Coy and Quinn (1975) is much more amenable to investigation by glancing-angle X-ray diffraction techniques. Consider, for instance, the last two columns relating to 'HVI plus DPDS' at 1400 and 1600 N, both loads being definitely in the extreme-pressure region (between 550 and 1800 N). Apart from $d = 2.10$ Å, all the lines of α-Fe and Fe_3C were present for d-values greater than 1.85 Å. The non-appearance of this medium intensity line is not very significant, since the preferred orientation of the cementite (Fe_3C) crystallites could cause prominent lines for randomly oriented crystallites to become less intense (i.e. at $d = 2.10$ Å) and low intensity lines (e.g. at $d = 1.76$ Å) to become more noticeable. By directly comparing relative intensities, Coy and Quinn found an apparent increase in the proportion of Fe_3C to α-Fe as the loads increased in the extreme-pressure region with the DPDS. With DBDS and DTBDS, however, considerably less Fe_3C was indicated in the extreme-pressure region, showing that bad extreme-pressure additives are associated with the production of more Fe_3C, an extremely hard constituent of the original bulk material. This conclusion should be treated as tentative, since it has been shown (Quinn 1971) that proportional analysis by X-ray diffraction is a much more complex procedure than merely comparing relative intensities.

The really significant result of the X-ray diffraction analysis was the iron sulphide (FeS) which was found in the wear scar with DBDS and DTBDS for loads above 1000 N in amounts which seemed to increase with increasing loads. Also, very small amounts of the rhombohedral and spinel iron oxides (i.e. α-Fe_2O_3, haematite, and Fe_3O_4, magnetite) were detected. It turns out that over 30 lines other than the α-Fe lines were obtained in the glancing-angle X-ray diffraction patterns from most of the specimens worn under extreme-pressure conditions with DBDS and DTBDS. They show that large amounts of FeS together with small amounts of iron oxides and Fe_3C are formed under extreme-pressure conditions with these good extreme-pressure additives but not with a bad extreme-pressure additive such as DPDS. Again, no differences due to the base oil could be detected.

The problem with the glancing-angle technique is that the X-rays often penetrate to depths between 1 and 10 µm. Hence it gives crystallographic information that is more relevant to the substrate material than to the actual surfaces. Clearly, the lack of evidence related to the presence of FeS in the extreme-pressure scars obtained with DPDS could be due to the inability of this additive to produce effective thicknesses of the extreme-pressure film. Azouz

Table 5.3. *The interplanar spacings from glancing-angle X-ray diffraction patterns of EN31 steel specimens worn under extreme pressure conditions*

Standards (columns 13-534, 11-614, 4-832, 11-151, 6-0688)

13-534 α-Fe$_2$O$_3$ I_1/I_1	11-614 Fe$_3$O$_4$ I_1/I_1	4-832 FeS I_1/I_1	11-151 FeS I_1/I_1	6-0688 Fe$_3$C I_1/I_1	FeS specimen	Etched ball	Unworn ball	HVI 450N	HVI+ DBDS 550N	RIS° DBDS 900N	RIS+ DBDS 1400N	HVC° DBDS 2000N
			5.4 20									
	4.85 40		4.74 10									
3.68 25												
		2.98 90	2.98 40		2.98						3.00	2.99
	2.97 70	2.93 50			2.91							
					2.86							
2.69 100		2.67 90	2.66 60								2.67	2'66
2.51 50	2.53 100	2.52 30	2.52 10	2.54 5	2.54						2.52	2.54
	2.42 10				2.48							2.46
				2.38 65		2.383	2.38	2.38	2.38	2.38	2.39	2.38
					2.29							
				2.26 25		2.257	2.26	2.26	2.26	2.26	2.27	2.26
2.20 30				2.20 25		2.203	2.20	2.20		2.20	2.22	2.212
		2.15 50			2.15							2.165
		2.14 50	2.14 10									
	2.096 70	2.09 100	2.09 100	2.10 60	2.097	2.10	2.10	2.10	2.10	2.10	2.11	2.093
2.07 2				2.06 70		2.063	2.06				2.07	
				2.02 60	2.027	2.027	2.027	2.027	2.027	2.027	2.027	2.027
				2.01 100								
		1.95 50		1.97 55	1.97	1.964	1.97	1.97	1.97	1.97	1.987	1.97
		1.92 60	1.923 30		1.94							
												1.91
					1.89							
1.837 40				1.87 30		1.878	1.87	1.87	1.87	1.87	1.875	1.87
				1.85 40		1.85	1.85	1.85	1.85	1.85		
						1.797						

1.691 60	1.712 60	1.72 90	1.72 50	1.68 15	1.73				1.433	1.433	1.686	1.72	1.68
1.634 4		1.64 60 1.62 40	1.634 30		1.64				1.34		1.65	1.64	
1.596 16	1.634 85	1.60 40	1.595 5 1.501 5	1.61 7 1.58 20	1.606	1.268	1.27		1.266	1.59 1.52	1.59 1.51	
		1.50 50 1.49 50 1.49 50 1.47 60			1.51 1.48 1.48 1.46							
1.484	1.483 85		1.469 30 1.445 30 1.422 5			1.215	1.22		1.214	1.46 1.433	1.433	
1.452 35		1.46 50 1.44 60 1.42 50			1.43	1.433	1.433	1.433				
1.349 4	1.327 20	1.34 40 1.33 60 1.32 50	1.331 40 1.320 10		1.34 1.31 1.298				1.34		1.40 1.35	
1.310 20				1.33 25								
1.258 8	1.279 30 1.264 10	1.38 40 1.27 40	1.284 5 1.271 5		1.277				1.277	1.219	1.277	
1.213 4	1.211 20	1.23 40	1.224 20		1.230				1.214		1.22	
1.167 8		1.19 40	1.188 10 1.179 10		1.190	1.170	1.170	1.170 1.170	1.170	1.170	1.170 1.146 1.127 1.11	
1.160 10 1.141 12	1.121·4 30	1.18 40	1.135 20 1.119 40 1.108 30		1.170 1.14 1.122 1.11	1.125	1.125		1.127	1.127 1.107		
1.100 14			1.091 10		1.096	1.081 1.057 1.037	1.08		1.082			
1.053 18	1.092 60 1.049 40		1.054 30		1.06							
			1.024 10		1.032 1.018 1.000	1.0134	1.0134	1.0134 1.0134	1.0134	1.0134	1.0134 1.0134	
0.989 10	0.989 10		0.994 20		0.991			0.991	0.989		0.991	

Table 5.3 (cont.)

RIS+ DBDS 2000N	RIS+ DBDS 2000N	RIS+ DBDS 2000N	HVI+ DBDS 2200N	RIS+ DBDS 2200N	RIS+ DTBDS 1000N	RIS+ DTBDS 1800N	RIS+ DTBDS 2400N	HVI+ DTBDS 1800N	HVI+ DTBDS 2000N	HVI+ DPDS 1600N	RIS+ DPDS 1400N	RIS+ DPDS 1600N	HVI+ DPDS 1400N
2.99	2.98	3.05	2.99	2.98	3.00	2.99			2.97				
2.89		2.96		2.91									
2.66	2.67	2.67	2.65	2.66	2.67	2.66	2.63	2.63	2.65				
		2.61											
2.54	2.54	2.54	2.55	2.52		2.53	2.54	2.53	2.53				
			2.48				2.49						
2.38	2.38	2.39	2.38	2.39	2.39	2.39	2.38	2.38		2.37	2.38	2.39	2.39
2.30								2.38	2.30	2.30		2.30	
2.24	2.36	2.27	2.26	2.27	2.27	2.26	2.25	2.24	2.26	2.26	2.27	2.27	2.24
2.21	2.21	2.22	2.22	2.20	2.22	2.21	2.20	2.26	2.20		2.22		2.21
		2.17				2.17							
							2.14						
2.10	2.10	2.11			2.11	2.12							
	2.097	2.09	2.09	2.09			2.09	2.09	2.09	2.09			
	2.082										2.08		
2.06	2.055	2.066	2.06		2.07	2.07	2.06	2.06	2.07	2.06			2.04
2.027	2.027	2.027	2.027	2.027	2.027	2.027	2.027	2.027	2.027	2.027	2.027	2.027	2.027
1.97	1.97	1.98	1.98	1.97	1.98	1.98	1.97	1.97	1.98	1.98	1.97	1.97	1.97
		1.95		1.94									
	1.92		1.93	1.91									
1.87	1.87	1.88	1.87	1.86	1.87	1.86	1.87	1.88	1.87	1.86	1.87	1.87	1.87
	1.85	1.86	1.85	1.84			1.84	1.85	1.85	1.85	1.85	1.85	1.85
			1.80	1.81	1.81	1.81	1.80	1.80	1.81	1.80	1.80	1.80	1.80
1.76	1.76	1.76	1.76	1.75	1.76	1.76	1.76	1.76	1.76	1.76	1.75	1.76	1.76
1.72	1.73	1.73	1.72	1.73	1.72	1.72	1.72	1.73	1.72				
							1.68						
1.68	1.68	1.69	1.68	1.68	1.69	1.69	1.67	1.67	1.68		1.68	1.69	1.68
					1.65								

1.64	1.64	1.63		1.64	1.63	1.63	1.62	1.63	1.64	1.64	1.63		1.63
1.59	1.59	1.59	1.58	1.59	1.52	1.51	1.50	1.58	1.59	1.50	1.58	1.58	1.58
1.51	1.51	1.51	1.51			1.45	1.46	1.51	1.50		1.51		1.51
1.45		1.46				1.433	1.433						
1.4332	1.433	1.433	1.433	1.433	1.433	1.433	1.428	1.433	1.433	1.433	1.433	1.433	1.433
1.41	1.40					1.41			1.41				
1.33	1.34	1.33	1.33	1.33	1.35	1.34	1.33	1.32	1.33	1.36		1.275	1.325
1.29													
1.27		1.28	1.275	1.278			1.26	1.26		1.27	1.27		1.27
1.22	1.22	1.22	1.25	1.22	1.22	1.22	1.22	1.22	1.22	1.20	1.22	1.22	1.22
			1.21				1.21						
1.170	1.19	1.19	1.19	1.170	1.170	1.19	1.170	1.19	1.170	1.170	1.170	1.170	1.170
1.15	1.170	1.170	1.170	1.13	1.13	1.170	1.15	1.170	1.15	1.13	1.12	1.15	1.12
1.126	1.15	1.14	1.125	1.11	1.10	1.15	1.13	1.10	1.13	1.10	1.09	1.13	1.10
1.10	1.13	1.13	1.104	1.09		1.13	1.11		1.11	1.09		1.11	1.09
	1.11	1.12	1.09			1.11				1.06		1.09	1.06
1.06	1.06		1.039				1.06	1.045		1.04	1.04	1.04	1.04
										1.03			
1.0134	1.0134	1.0134	1.0134	1.0134	1.0134	1.0134	1.0134	1.0134	1.0134	1.0134	1.0134	1.0134	1.0134
0.986	0.993	0.992				0.996	0.982		0.992				0.992

Table 5.4. *Comparison of interplanar spacings obtained by X-ray diffraction analysis of wear debris obtained after 20 min tests under extreme-pressure conditions* (*X-ray powder photographs*)

Wear debris spacings (Å)				Possible constituents		
	Additive used					
S	DBDS	DPDS	FeS (I/I_1)	α-Fe (I/I_1)	α-Fe$_2$O$_3$ (I/I_1)	
	3.58				3.68 (70)	
2.98 (s)	2.98	2.98 (s)	2.98 (40)			
2.65 (s)	2.62 (s)	2.62 (s)	2.66 (60)		2.69 (100)	
2.50			2.52 (10)		2.51 (80)	
2.06 (vs)	2.04 (vs)	2.06 (vs)	2.09 (100)	2.03 (100)	2.07 (10)	
1.71	1.71 (s)	1.70 (vs)	1.72 (50)		1.69 (80)	
1.61	1.62	1.60	1.63 (30)		1.63 (10)	
1.42	1.44	1.43	1.46 (20)	1.43 (19)	1.45 (80)	
1.32	1.32	1.32	1.33 (40)		1.31 (40)	
1.17	1.17	1.17	1.18 (70)	1.17 (30)	1.17 (30)	
1.10	1.11	1.10	1.11 (30)		1.10 (40)	

(1982) attempted to overcome this problem by running his extreme-pressure experiments for 20 min, for the main purpose of generating sufficient wear debris to be able to obtain *X-ray Powder Photographs*. The main advantage of this technique is that it relates to the wear occurring at the real areas of contact (Quinn, 1982). Table 5.4 gives the interplanar spacings obtained by Azouz (1982) in his analysis of X-ray powder photographs obtained from wear debris formed under extreme-pressure conditions with elemental sulphur (S), dibenzyl disulphide (DBDS) and diphenyl disulphide (DPDS). From these we see that α-Fe and FeS are present in the wear debris from all three (extended duration) tests. The main difference getween the DPDS debris and that from the S and DBDS experiments is the indication that the rhombohedral iron oxide (α-Fe$_2$O$_3$) is also present with the good extreme-pressure additive.

The combined use of glancing-angle X-ray diffraction of worn surfaces and the powder camera analysis of wear debris can give the tribologist a much more complete picture of the crystallographic changes that might occur with good and bad extreme-pressure additives. It is clear that large thicknesses of FeS are formed by good extreme-pressure additives, whereas much smaller amounts of this material seem to be produced at the real areas of contact by poor extreme-pressure additives. As the protection of the solid lubricant (FeS) decreases, Fe$_3$C (cementite) is formed beneath the surface, making a system lubricated with a poor extreme-pressure additive much more likely to exhibit severe wear (when loaded in the extreme-pressure region) than a system lubricated with a good extreme-pressure additive. Let us now see how electron probe microanalysis has added to the picture derived from X-ray diffraction analysis.

5.1.5 The use of electron probe microanalysis for studying extreme-pressure additive mechanisms of the disulphides

At the time Davey and Edwards (1957) and Godfrey (1962) produced their ideas on the mechanism of extreme-pressure lubricant protection (i.e. 1957–62), electron probe microanalysers were not readily available on a commercial basis. With the advent of this sensitive microanalytical technique, however, the way was opened up for Allum and Forbes (1968) to carry out their pioneer work on the mechanisms of extreme-pressure activity, based on electron probe microanalysis of surfaces formed in conventional sliding four-ball tests with organic sulphur compounds. Unfortunately, however, these authors concentrated entirely on the use of this single surface analytical technique. It must be emphasized that the use of just one surface analytical technique often gives a biased view of what has happened at the surface. For instance, electron probe microanalysis will only give *elemental* information about the top 10 μm or so of the surface. Whilst the depth of penetration is suited to the study of extreme-pressure additive reactions, it is much too deep for detecting the reactions of anti-wear additives (which occur at or near (~ 10 nm) the surface).

Once again, we are going to refer to research carried out at my laboratory in order to show how one uses electron probe microanalysis to elucidate the mechanisms of extreme-pressure lubricant protection, namely the paper by Coy and Quinn (1975). There are two reasons for choosing this work: (a) very little *definitive work* has been done on extreme-pressure additive mechanisms as revealed by electron probe microanalysis since the publication of this paper, and (b) the details are very familiar to me. Let us start by considering the area analysis of wear scars.

Figure 5.2 is an example of how one can use the electron probe for microanalysis of the rough surfaces so typical of extreme-pressure lubrication. Such surfaces do not readily lend themselves to accurate determinations of relative elemental proportions. Hence, instead of attempting the more conventional 'spot analysis', small areas of 150 μm × 150 μm were scanned, thereby minimizing the effects of surface roughness, and average concentrations of the surface elements were found. Figure 5.2 shows the quantitative area analysis of wear scars, formed over a wide range of loads, for (*a*) 1 wt% DBDS; (*b*) 0.72 wt% DTBDS; and (*c*) 0.89 wt% DPDS in white oil, the various percentages being chosen to give 0.26 wt% *sulphur* in each blend.

Before initial seizure, all three graphs show similar sulphur distributions, the amount of sulphur increasing slightly with increasing load until, at initial seizure, it drops to a minimum where there is very little sulphur present in the scars. After initial seizure, with the DBDS and DTBDS additives, the amount of sulphur increases to around 30 wt% just before welding occurs, whereas with the DPDS additive, the sulphur content increases only slowly to a level similar to that before initial seizure. The sum of the oxygen and carbon content in the scars does not alter appreciably throughout the whole load range. From the X-ray diffraction evidence regarding the increasing amounts of Fe_3C in the scars produced in the

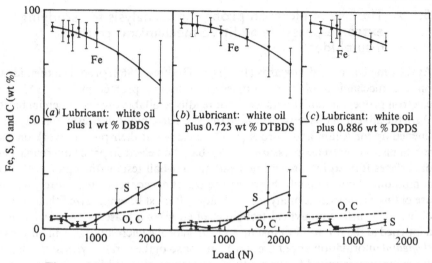

Figure 5.2 *Variation of amount of Fe, S, O+C, with load for (a) 1 wt%
DBDS, (b) 0.72 wt% DTBDS and (c) 0.89 wt% DPDS in white oil*

extreme-pressure region with DPDS, as well as the general presence of Fe_3C in all
scars in the anti-wear region, it would seem that the carbon content probably
dominates the $(O+C)\%$ which is merely the difference between 100% and the
combined percentages of S and Fe.

In the extreme-pressure region, when DBDS and DTBDS were used, the
amount of sulphur in the wear scars increases to a maximum around 30 wt%
(compared with 36.5 wt% sulphur in FeS), which gives some credence to the
detection of FeS by X-ray diffraction. Figures 5.3(a), (b) and (c) show the electron
image, and the iron and sulphur X-ray images, respectively, of part of a wear scar
formed at a load of 2200 N with DTBDS as the additive. One can readily see that
there is so much sulphur in the surface that it significantly affects the iron image.
The sulphur is concentrated in the smooth-tracked regions that make up the real
areas of contact. There are also concentrations of sulphur in the rough areas
where the film has been removed; this appears to consist of debris formed when
the FeS film is broken up. The lateral cracks across the direction of sliding show
that the extreme-pressure film is continually cracking up and, presumably, then
being renewed when that area next becomes a contacting region.

The electron probe microanalysis of wear scars formed under *anti-wear
conditions* provides *minimal* information about the nature of the surface films
taking part in the anti-wear protection mechanism, as already mentioned at the
start of this sub-section. For instance, Coy and Quinn (1975) reported that all the
scars appeared to give the same distribution of sulphur, regardless of whether
DPDS, DBDS or DTBDS was used as the additive. They also found no signs of
the extensive film cracking (so noticeable under extreme-pressure conditions).
Although EPMA showed that low amounts of sulphur are present in the surfaces
worn under anti-wear conditions, it gave no new insight into the mechanism of
anti-wear protection.

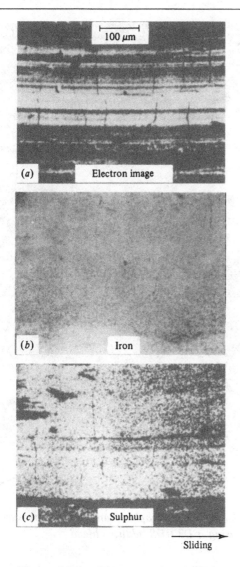

Figure 5.3 *(a) Electron image and (b), (c) elemental distributions, in the wear scar formed in the presence of DTBDS under a load of 2200 N*

5.1.6 Scanning electron microscopy (SEM) and the study of extreme-pressure films formed with disulphide additives

The use of scanning electron microscopy in *any* tribological study is now almost mandatory! It has a very large depth of focus compared with optical microscopy. It also has a wide range of *useful* magnifications available at any given part of the specimen surface, making it a very useful technique from low magnifications ($\sim \times 10$) up to very large magnifications ($\sim \times 10\,000$). Let us briefly see how

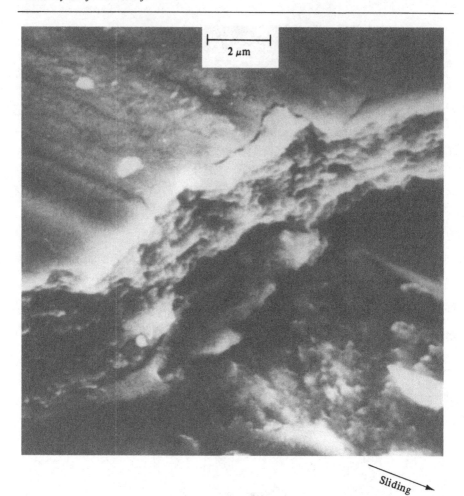

Figure 5.4 *Scanning electron micrograph of an extreme-pressure film formed at 2000 N load using a DTBDS additive blend*

scanning electron microscopy has added to our knowledge of extreme-pressure additive mechanisms (using the paper of Coy and Quinn, 1975, as an illustrative example).

Typically, before initial seizure, all three sulphur additives used by Coy and Quinn (1975) formed smooth, finely-scored scars covered with a *thin* film. After initial seizure, in the extreme-pressure regime of lubrication, the good extreme-pressure additives (DBDS and DTBDS) formed thick reacted layers. Figure 5.4 is typical of the electron micrographs obtained with DTBDS at a load (2000 N) well within the extreme-presure region. It shows a friable structure, some 2 to 4 μm thick. The top surface of this extreme-pressure film is smooth and tracked

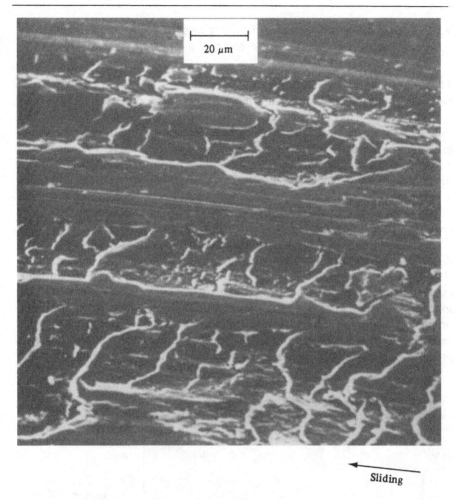

Sliding

Figure 5.5 *Scanning electron micrograph of the surface of a specimen worn at 1000 N using a DPDS additive blend*

showing that it was a load-bearing area. This observation is consistent with the supposition that a thick film of FeS has been formed. When DPDS (a poor extreme-pressure additive) is used in the extreme-pressure condition, the surfaces of the scars are rough and covered with torn areas where localized welding has occurred (see Figure 5.5), indicating that very little protection has been afforded by the DPDS additive when used in the extreme-pressure lubrication region. Thus we see that scanning electron microscopy has served in a confirmatory role to the EPMA and the glancing-angle X-ray diffraction analyses. Let us now see what optical microscopy plus microhardness measurements can tell us about extreme-pressure wear mechanisms.

5.1.7 Optical microscopy plus hardness measurements of specimens worn under extreme-pressure conditions with disulphide additives

The optical microscopy of normal and taper sections through worn surfaces is a very useful way of obtaining information about the substrate. However, it typically does not have the resolution to detect the immediate surface layers, for example, the oxide layers formed in mild-oxidational wear. This probably explains why so many metallurgists seem to ignore the surface crystallography and concentrate mainly upon the morphological structures of the substrates down to about 20 to 30 μm below the surface. Obviously, the substrate morphology has an effect on the load-bearing ability of the surface layers and hence optical microscopy of metallographic sections will always be an important ancillary technique in tribological studies.

Clearly, the hardness of the unworn surfaces must be ascertained before discussing any changes due to wear. Continuing with the work of Coy and Quinn (1975), as a typical example of the multi-disciplinary approach to studying extreme-pressure wear mechanisms, it turns out that the unworn parts of their specimen surfaces were not very different in hardness (850 VPN) from the original hardness of the EN31 steel, both in the anti-wear and extreme-pressure regions. In fact, the hardness of the scars formed in the anti-wear and extreme-pressure regions using DPDS as additive was also around this value, as was the

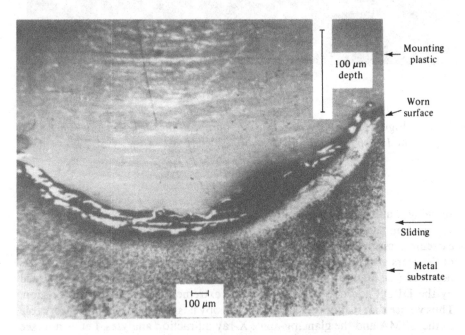

Figure 5.6 *Optical micrograph of a taper section through a scar formed with DBDS at a load of 1800 N*

scar hardness for DBDS and DTBDS in the anti-wear region. However, in the extreme-pressure region, the hardness of the surfaces of the wear scars formed with DBDS and DTBDS decreased with increasing load, the lowest value being 375 VPN on a scar formed at 2000 N with DTBDS. Thus we see that good extreme-pressure additives form a relatively soft surface coating, whilst no detectable change occurs with a bad extreme-pressure additive.

Probably the best way to get the most information about substrate morphology is to use taper sections. Figure 5.6 shows a taper section through a scar worn with DBDS at a load of 1800 N. The light area at the leading edge of the scar has a hardness of 640 VPN falling to 460 VPN in the centre of the scar. In the centre, the hardness increases with depth below the surface until the ordinary martensitic hardness of 850 VPN is reached (at approximately 100 μm below the surface). The hardness of the surface film cannot be obtained even at the smallest indenter loads as the film will crack. See Figure 5.7 for a scanning electron micrograph of a surface film formed with DTBDS at a load of 2000 N, showing extensive cracking due to the hardness indenters.

When a DPDS additive is used, a thick white layer, approximately 30 μm thick is formed on the bulk of the material beneath the scar. This is shown in Figure 5.8.

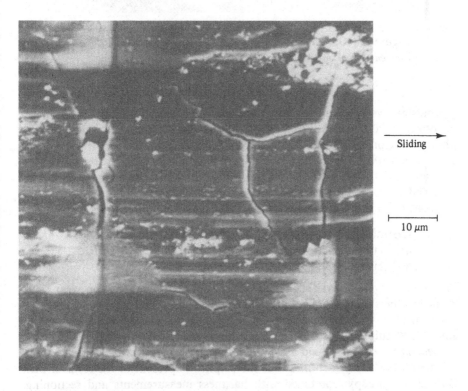

Sliding

$\vdash\!\!\!-\!\!\!-\!\!\!\dashv$
10 μm

Figure 5.7 *Scanning electron micrograph of the surface of a film formed with DTBDS at a load of 2000 N, showing extensive cracking due to the hardness indenters*

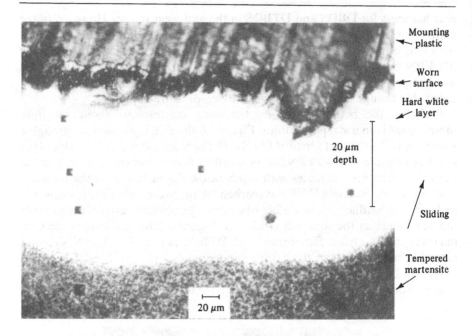

Figure 5.8 *Optical micrograph of a taper section through a wear scar formed with DPDS at a load of 1400 N*

The hardness values of the four indentations on the left of the figures are, from top to bottom, 1070, 910, 830 and 450 VPN. The hardness of the tempered martensite (the dark region below the white layer) increases from 450 to 850 VPN at a depth greater than 100 μm. Thus, it would appear that, when a good extreme-pressure film is formed, the reacted layer sits on a soft substrate; when no appreciable film is formed, a hard, white layer (hardest just below the surface) is produced on top of a soft, tempered martensitic layer.

5.1.8 Concluding remarks regarding extreme-pressure additive protection mechanisms using disulphide additives

The use of several physical analytical techniques to examine the wear scars formed in the sliding four-ball machine can be seen to increase very greatly the information available. For example, while the electron probe shows that major amounts of sulphur exist in the wear scars due to DBDS and DTBDS in the *extreme-pressure region* (and also shows how that sulphur is distributed), X-ray diffraction positively identifies that the sulphur is present in the form of FeS. Optical microscopy, combined with hardness measurements and sectioning, shows that the FeS is a soft, friable layer which not only protects the surfaces from welding but also prevents severe structural damage beneath the surface. The thickness of the film, its friable nature, and the differences in appearance of wear scars produced on the one hand by DBDS and DTBDS, and on the other by DPDS are well brought out by scanning electron microscopy.

In the *anti-wear region*, none of the additives used by Coy and Quinn (1975) formed films of sufficient thickness to be detected by X-ray diffraction, despite the fact that the other techniques all indicate films are indeed formed. The electron probe showed that, with disulphide additives, one is likely to get uniform sulphur distributions, with the amount of sulphur decreasing to a minimum at initial seizure. The scars are all smooth and of similar hardness to that of the unworn metal.

Most investigators (e.g. Allum and Forbes, 1968; Coy and Quinn, 1975) agree that the order of extreme-pressure activity for the disulphide additives is:

$$\text{diphenyl} < \text{di-tert-butyl} \leq \text{dibenzyl}$$

and follows the ease of scission of the carbon–sulphur bond. Sakurai and Sato (1966) found that, in hot-wire experiments, DBDS additives produced FeS while DPDS did not, showing that the reactivity of DPDS is very low. This is an agreement with the X-ray diffraction and electron probe results of Coy and Quinn (1975). The minimum sulphur concentration at the initial seizure load (the boundary between the anti-wear and extreme-pressure regions), indicates there is a change in the mode of action. It seems likely that desorption occurs at initial seizure (due to the more severe conditions of sliding) and chemical reactions take place to form an inorganic sulphide film. Initial seizure could possibly indicate a change from an adsorbed organic film (under anti-wear conditions) to a chemically-reacted inorganic film (under extreme-pressure conditions). The extreme-pressure performances of the disulphides is probably determined by the reactivity and the properties of the film formed. The important required properties of tribochemically-formed extreme-pressure films have been listed by Godfrey (1962). They should have (a) high melting points, (b) low shear strengths and (c) low hardness values. The melting point of FeS is around 1200 °C and its shear strength is approximately 200 N/mm^2 compared with approximately 900 N/mm^2 for steel. Also, the hardness of FeS is approximately 220 VPN compared with 850 VPN for steel. Hence, FeS should be a satisfactory extreme-pressure film material. It is, indeed, found that thick, friable layers of FeS are formed with hardnesses around 400 VPN when good extreme-pressure additives (such as DBDS and DTBDS) are used in the extreme-pressure region. With poor extreme-pressure additives (such as DPDS), no such layers are formed and there is considerable evidence of metal-to-metal contact. Also, the low-load weld indicates that only poor protection is afforded.

It will be recalled (in Section 5.1.4) that small amounts of $\alpha\text{-Fe}_2\text{O}_3$ and Fe_3O_4 were detected by glancing-angle X-ray diffraction. In our discussion we have tended to ignore the role of oxidation of the metal substrates, insofar as it affects the wear under *truly extreme-pressure lubrication conditions*. It is considered most likely that oxidation of the substrate is more relevant to anti-wear mechanisms, and this aspect will be discussed later (Section 5.4). There are several investigators, however, who maintain that the concurrent formation of oxides is an important factor in truly extreme-pressure conditions, as well as the formation of

FeS. For instance, we have already mentioned Godfrey (1962). Other supporters of the role of oxides in extreme-pressure lubrication are Bjerk (1973), Toyoguchi and Takai (1962), Buckley (1974), Tomaru, Hironaka and Sakurai (1977), Wheeler (1978), Debies and Johnston (1980) and Murakami *et al.* (1983). The work of Murakami *et al.* (1983) is particularly notable due to the fact that these investigators used electron probe microanalysis, X-ray photoelectron spectroscopy and electron diffraction techniques. Using these three techniques, they showed how beneficial it could be to run their four ball tests with DBDS, *after* running the same group of balls in a similar test *without* the additive. They claim that the first run provided the specimens with a tribo-chemically-formed wear surface consisting of oxide only. This idea has long been in my mind and, in fact, my colleague (Sullivan, 1986) has produced a paper in which an attempt is made to bring together the oxidational wear theory (see Section 1.4.4) and boundary lubrication mechanisms (see Section 1.3.2).

Let us now change the scene somewhat so that we deal with solid lubricant films that are deliberately rubbed onto the mating surfaces, with the intention of providing a *ready-made* solid lubricant film, rather than one which is *tribo-chemically* or *tribo-physically* formed (such as we have just been discussing with the disulphides forming FeS at the real areas of contact). It turns out that the preferred orientation of the solid lubricant crystallites relative to the sliding surfaces is an important factor in the tribology of solid lubricants. Let us, therefore, discuss this factor in some detail in the next section.

5.2 The analysis of solid lubricant films formed when lamellar solids slide against metals and alloys

5.2.1 Introduction

Lamellar solids consist of polycrystalline composites in which each grain, or crystallite, has a lamellar or layer-like structure. Typical examples are graphite (naturaly- and artificially-formed), molybdenum disulphide, talc, boron nitride, titanium iodide and graphite fluoride. The main function of these lamellar solids is to reduce friction and prevent wear of the surfaces between which these solids are introduced. Sometimes, we require the lamellar solid itself to be also resistant to wear as, for instance, in the case of carbon brushes. In this section of our chapter on the analysis of lubricant films, we will be mainly concerned with *graphite materials*. Some mention, however, will also be made of molybdenum disulphide and graphite fluoride. It will be seen that each lamellar solid must be treated as a separate entity. Although they have similar (i.e. layer-like) structures, each one behaves very differently in response to changes in atomosphere and/or temperature. Let us start by considering the early ideas regarding solid lubrication by lamellar solids.

5.2.2 Early ideas on solid lubrication by lamellar solids

Graphite was the first of the lamellar solids to be used as a solid lubricant, mainly through its use in rotating electrical machinery for collecting or commutating the current. The structure of graphite was first proposed by Bernal (1924) as being

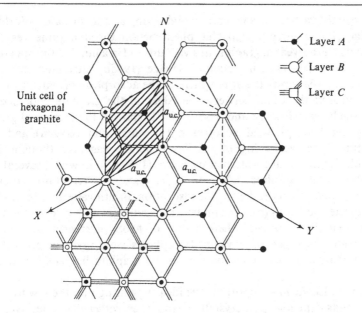

Figure 5.9 *The structure of graphite (plan view)*

hexagonal with a large distance (3.348 Å) between the hexagonal sets of carbon atoms and a relatively small distance (1.42 Å) between the carbon atoms within the hexagonal layers. It turns out that all natural graphites also contain about 20% of a 'rhombohedral modification', first proposed by Lipson and Stokes (1943) nearly twenty years later. Figure 5.9 shows how the two hexagonal layers are disposed with respect to each other in Bernal's hexagonal graphite. Note that layer B is placed 3.34 Å above and below layer A, with each corner of its hexagonal net exactly above and below the centre of the hexagonal net in A. We could describe the Bernal structure as an $ABAB$... layer structure. It can be seen that there is another way of putting a hexagonal net symmetrically with respect to *both A and B*. This would be a net that can be placed with one atom immediately above the points marked by a single open or closed circle in Figure 5.9. One such net is shown in the bottom left-hand corner of Figure 5.9, the corners of the hexagons being denoted by open squares. This structure could be described as an $ABCABC$... layer, and is the structure proposed by Lipson and Stokes (1943). It is, in fact, the rhombohedral modification mentioned above.

Bragg (1928) suggested that the well-known lubricity properties of graphite were due to its structure, with weak bonding *between* the hexagonal planes and strong bonding between the carbon atoms *within* the hexagonal nets. Graphite composites were thought to provide lubricity by virtue of their surface crystallites somehow becoming oriented with their basal planes parallel to the sliding surfaces. The weak bonding between these planes resulted in low friction resistance. What was not explained was why such oriented crystallites should not exhibit really catastrophic wear – after all, what is the criterion for any particular basal plane being the actual shear plane? Catastrophic wear did indeed begin to take place in the carbon brushes used in high-flying aircraft just after World War II. Savage (1948a) investigated this phenomenon through some very good experiments on the wear of graphite in a vacuum and in a humid atmosphere. The wear was extremely high in vacuum. Savage (1948b) attributed this strong dependence of wear upon the atmosphere to the adsorption of water molecules, which decreased the bond between the edges of the graphite and especially between the layers. Bowden and Young (1951) showed the friction of graphite depended on *both* physical adsorption (of water and oxygen) and upon temperature. The weakening of interlayer bonds was also thought to be responsible for the results obtained by Rowe (1960a,b), in which several types of lamellar solids (graphite, BN, MoS_2, TiI_2 and $CrCl_2$) showed frictional behaviour which was dependent upon adsorption. Braithwaite and Rowe (1963), using graphite and MoS_2 (running under water, nitrogen and oxygen), found that graphite wear life *increases*, whereas that of MoS_2 *decreases*, according to whether water, nitrogen or oxygen atmospheres were used. In other words, running under water gave *maximum* wear life for graphite, but *minimum* wear life for MoS_2.

There was also strong support for the idea suggesting that the low friction of lamellar solids depends upon crystallite properties, *rather than* interlayer shear promoted by adsorption. Deacon and Goodman (1958) deposited layers of graphite, BN, MoS_2 and talc upon Pt substrates and showed that the friction of these films (sliding in an air atmosphere) depended upon the temperature of the substrate, this temperature being strongly connected with the oxidation of the film. Deacon and Goodman (1958) suggested that the low friction was due to forces operating *between crystallites*, and not between layers. These ideas were supported by the electron diffraction evidence of Midgley and Teer (1961) and Porgess and Wilman (1960). The latter investigators introduced the concept of rotational slip or surface re-orientation.

The paper by Braithwaite and Rowe (1963) reports on probably the most definitive work carried out before the tremendous increase in the interest in solid lubricants for the 'space race' to the moon. Table 5.5 shows the variation in the wear life of dry buffed-films with particle size and atmosphere (taken from the paper by Braithwaite (1966)).

Each reading of wear life represented the mean of five measurements using cross-cylinders of mild steel loaded at 31 N, running at a linear speed of 0.5 m/s. The mild steel had a hardness of 150 VPN and the surface roughness equal to about 4×10^{-5} mm CLA (i.e. 1.5 μin CLA), before the solid lubricant was applied to each of the cylinders through a buffing procedure.

This table gives some quantitative evidence regarding the decrease in the wear life of graphite lubricant films as the atmosphere is changed from water to oxygen to nitrogen. It also shows the reverse behaviour for MoS_2. The table also indicates that particle size (i.e. surface area) has a marked effect on the wear life of graphite, namely, the wear life increases with particle size (surface area). Its effect on the wear life of MoS_2, on the other hand, is almost non-existent. An interesting result was the adverse effect of intercalating the graphite with fluorine.

Having provided the reader with an overview of the state of the art of solid

Table 5.5. *Wear life of selected solid lubricants*

	Surface area (m^2/g)	Wear life (min)			
		Wet air	Dry oxygen	Dry nitrogen	Dry CO_2
Carbon black	125	0	0	0	–
Graphite	570	95	16	1	90
Graphite	285	47	20	3	–
Graphite	220	21	19	1	–
Graphite	110	4	0	0	36
Graphite	5	0	0	0	–
MoS_2	33	3	13	207	–
MoS_2	6	3	20	209	56
Graphite fluoride	130	5	6	7	–

lubrication as it was at the start of the space age, we will now describe some applications of physical analytical techniques to solid lubrication by lamellar solids and, hopefully, show that this approach is beginning to shed more light on the vexed question regarding the relative importance of crystallite orientation effects compared with adsorption effects. With the current interest in high-temperature tribo-systems, for example the ceramic diesel, some of the controversy may naturally abate, since it is most unlikely that water will be adsorbed at 500 °C or more (the temperature existing in the cylinders of the ceramic diesel). It is under these conditions that the role of oxides, tribo-chemically formed at the real area of contact, may be similar to that of the solid lubricants. Let us now start with an account of how reflection electron diffraction has contributed to our knowledge of graphite lubrication.

5.2.3 Reflection electron diffraction studies of graphite lubrication

Jenkins (1934) carried out some investigations into the surfaces of rubbed natural graphite, using the then new technique of reflection electron diffraction. He obtained patterns similar to the one illustrated in Figure 5.10, from which he deduced that the polycrystalline compact of natural graphite crystallites had taken up a preferred orientation in which the (0001) planes lay parallel to the rubbed surface. The reader is referred to Figure 3.9 (Section 3.1.6) regarding the Miller–Bravias system for describing hexagonal planes with four indices, namely h, k, i and l. As regards the pattern, the reader is also referred to Figure 3.31 (Section 3.3.2) showing the reciprocal lattice of a hexagonal crystal with its thinnest dimension parallel to the [001] direction. (Note we have not used four indices for describing crystal direction, in the interests of simplicity.) If one then realizes that the *preferred orientation* of the graphite crystallites with their (0001)

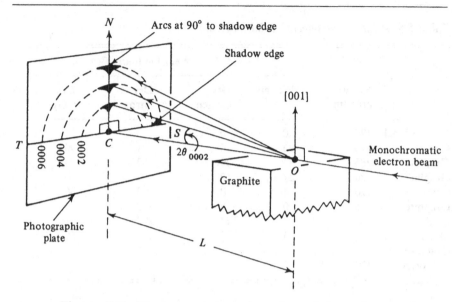

Figure 5.10 *Diagrammatic representation of the reflection electron diffraction pattern arising from a polycrystalline graphite specimen surface, the crystallites all lying with their basal planes (0001) parallel to the rubbed surface. C = centre spot; CN = normal to shadow edge; TS = shadow edge*

planes parallel to the surface is not essentially different from having a *single crystal* with its [001] direction normal to the surface, it becomes obvious that the arcs at 90° to the shadow edge are in fact the 0002, 0004 and 0006 diffraction maxima from the (0001) planes (see Figure 3.32 for the pattern one would expect from a single crystal having the hexagonal structure). Quite often, the preferred orieintation also involves many crystallites having their $(1\bar{1}l)$ and $(\bar{1}1l)$ planes in the position for Bragg reflection, so that one obtains streaks on either side of the 0001 arcs passing through the centre-spot (as shown in Figure 3.32).

For several years after Jenkin's (1934) work was published, it was believed that graphite lubrication was indeed due *entirely* to the tendency of the graphite crystallites to orient themselves so that they presented their planes of easy cleavage (i.e. their (0001) basal planes) *parallel* to the surface. How these crystallites managed to re-orient themselves from the almost random positions in the bulk composite material, and why catastrophic wear did not ensue when the crystallites were so oriented, was never discussed. In fact, Savage (1948b) did show that a single crystal of graphite did actually wear very quickly when the basal planes were parallel to the rubbing surface. The wear rate of this single crystal was considerably reduced by merely tilting the basal (0001) planes about 5° from the horizontal. Although the use of single crystals for studying crystallographic effects on friction and wear is to be avoided (due to the breaking-up of the single crystal surface into many crystallites, thereby producing a randomly-oriented polycrystalline surface), this particular piece of research into the initial stages of graphite single crystal wear does indicate that there must be some crystallites present in worn polycrystalline graphite surfaces that are oriented at some angles other than parallel to the rubbing surface.

It is interesting to see how research into solid lubricants (and in particular, graphite) has blossomed and waned as each new step forward in technological advance has occurred. In the late 1940s and early 1950s, the advent of the jet plane caused interest in carbon brushes that would function properly at high altitudes. In the 1960s, there was an interest in the mechanical properties of the type of graphite used in graphite-moderated nuclear reactors (which we will call 'reactor graphite' for brevity). In the 1960s and 1970s, we also saw an interest in solid lubricants that would function in the extremely high vacuum and very low/high temperatures of space. In the 1980s, there has been an interest in some prototypes of the uncooled diesel, sometimes (incorrectly) called the 'adiabatic diesel'. However, it seems more likely that the best form of 'lubricant' for this tribologically improbable system is one formed naturally on the surface of the tribo-element, for instance, the formation of an oxide on a ceramic element.

Having digressed into the reasons why interest in solid lubricants is still very much alive, let us continue with our account of the research carried out on graphitic systems with the aim of discovering oriented crystallites at angles other than zero with respect to the rubbing surface. Porgess and Wilman (1960) examined the surfaces of reactor graphite specimens after they had been unidirectionally abraded against various grades of abrasive paper. They found, in the reflection electron diffraction patterns from these reactor graphite surfaces, that the 0002, 0004 and 0006 areas tended to make a small angle δ with respect to the normal (drawn through the centre-spot) to the shadow-edge. The angle δ varied from 7° to 22°, according to the average size of the particles of the abrasive paper. They also measured the kinetic coefficient of friction (μ_{kin}) between the reactor graphite and the abrasive paper, and found a relationship between (μ_{kin}) and the angle δ, namely:

$$(\mu_{kin}) = \tan \delta \qquad (5.6)$$

Porgess and Wilman (1960) explain these results as being due to a tendency for the graphite crystallites to lie with their [001] directions along the direction of the resultant of the frictional force and the normal reaction of the abrasive surface upon the graphite brush.

I have also published a paper which indicated the complexity of the situation regarding the orientation of the surface crystallites of rubbed graphite (Quinn, 1963). Using reflection electron diffraction, I showed that the surface crystallites tended to be oriented so that the trailing edges of the (0001) planes made a small angle δ to the direction of motion of the counterface beneath the graphite specimen. In fact, the graphite specimens were made of a common carbon brush material (called 'electrographite') which had been rubbed against a copper disc, in experiments intended to simulate the mechanical (but not the electrical) conditions of a typical slip-ring assembly. Electrographite is an artificially-made graphite in which the crystallites are not all exactly like the ideal Bernal (1924) or Lipson and Stokes (1943) structure indicated in Figure 5.10. In that figure, one can see an exact line-up of the atoms in each layer. In natural graphite, this register is maintained throughout every crystallite. In electrographite, on the other hand, about 80% of the crystallites do not possess this perfect register. Sometimes, these imperfect crystallites are called 'turbo-static carbon' since they have all their basal planes, the (0001) planes, randomly oriented with respect to each other, but still maintaining their parallelism.

There are some very important differences between the tribo-systems investigated by Porgess and Wilman (1960) and those studied by Quinn (1963). The former were essentially examining abraded surfaces of reactor graphite, whereas I was studying the equilibrium contact films formed after repeated sliding on *both* the electrographite brush surfaces *and* on the copper disc surfaces. It was most unlikely that Porgess and Wilman (1960) were examining *equilibrium* films. Nevertheless there are some similarities between the two sets of results. For instance, the 'sense' of the angle of tilt δ in Quinn's (1963) reflection electron diffraction patterns was in accord with the hypothesis of Porgess and Wilman (1960), as shown in Figure 5.11. This figure shows that when the brush (or disc) has a relative motion to the left, then the diffraction arcs (from the (0001) planes) will also tend to be to the left of the normal to the surface. The figure combines the force diagrams relevant to each surface, with the diagrams representing the electron diffraction pattern obtained from each surface. The points C_B and C_D represent the direction of the electron beam passing almost perpendicular to the plane of the paper and almost parallel to the sliding surfaces. The arcs and the shadow edge are how the reflection electron diffraction patterns would appear when viewed along the direction of the electron beam. In fact these features are recorded on a photographic film or plate some 500 mm or so *below* the specimen.

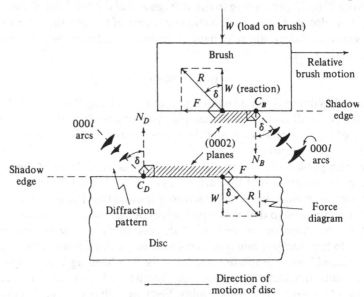

Figure 5.11 *Application of the hypothesis of Porgess and Wilman (1960) to the rubbing of an electrographite brush on a copper disc*

The angle of tilt of the (0001) basal planes, as indicated by the electron diffraction arcs, seems to be a function of the load. Quinn (1963) used various loads (from 0.1 to 60 N) in his electrographite-on-copper experiments and found that:

(a) for loads from 0.1 to 3 N, the angle of tilt was constant at ($20° \leqq 2°$),
(b) for loads between 3 and 10 N, the angle of tilt monotonically reduced to ($6° \pm 2°$),
(c) for loads between 10 and 60 N, the angle approached zero and
(d) for loads greater than 60 N, there was no preferred orieintation at all (i.e. the electron diffraction pattern consisted of the uniform concentric, half-rings normally expected when electrons are passed at grazing incidence across a surface containing randomly-oriented crystallites).

It is interesting to note that, at loads greater than 60 N, the friction coefficient μ was equal to 0.08, which was well *below* the values obtained from experiments in which the surfaces exhibited preferred orieintation, indicating that the orientation of surface crystallites is not always a requirement for lower friction. By plotting the coefficient of friction against the value of the tangent of the angle made by the diffraction arcs with the normal to the shadow-edge, Quinn (1963) showed that, for electrographite at least, the dependence of (μ_{expt}) upon ($\tan \delta$) is given by:

$$(\mu_{expt}) = (0.5)(\tan \delta) + 0.15 \qquad (5.7)$$

The graphs of 'μ versus $\tan \delta$' for both Equations (5.6) and (5.7) are shown in Figure 5.12. It is worth pointing out that there was a definite difference between

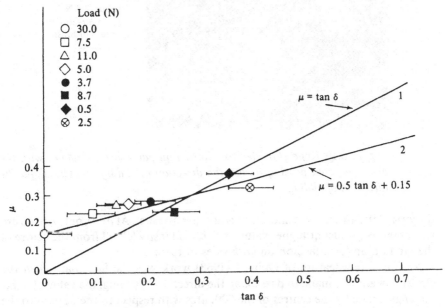

Figure 5.12 *Graphs of coefficient of friction (μ) versus ($\tan \delta$) for electrographite on copper*

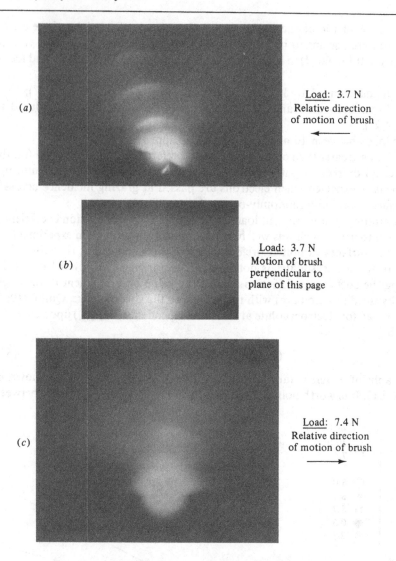

Load: 3.7 N
Relative direction
of motion of brush

←

Load: 3.7 N
Motion of brush
perpendicular to
plane of this page

Load: 7.4 N
Relative direction
of motion of brush

→

Figure 5.13 *Reflection electron diffraction patterns (a) and (c) with the electron beam passing across the direction of sliding and (b) along the direction of sliding*

the *form* of the arcs at $\delta = 0$ and the arcs at other δ-values. At $\delta = 0$, only *half* an arc was obtained, whilst at higher values, the arc intensity fell off from the centre of the arc in a uniform fashion on *both* sides of the arc.

An interesting feature of Quinn's (1963) work was the fact that, when the electron beam was made to pass *along* the direction of sliding, the value of δ (i.e. the angle made by the centres of the $000l$ arcs with respect to the normal to the shadow edge) remained at zero for all loads less than 60 N. These arcs were always symmetrical and were quite different from the $\delta = 0°$ arcs obtained when looking *across* the direction of sliding. Examples of the reflection electron diffraction patterns exhibiting (*a*) whole arcs at non-zero values of δ (looking across the direction of sliding), (*b*) whole arcs at zero values of δ (looking *along* the direction of sliding) and (*c*) half-arcs at $\delta = 0°$ (looking *across* the direction of sliding), are given in Figure 5.13(*a*), (*b*) and (*c*) respectively. These patterns are typical of those obtained from electrographite/copper surfaces. They have been explained on the basis of a mechanical twinning model of friction (as discussed in the next sub-section).

5.2.4 The friction of graphitic tribo-systems and the mechanical twinning model

The 'double orientation' of rubbed electrographite surfaces, as revealed by reflection electron diffraction analysis *along* and *across* the direction of sliding can be explained in terms of *mechanical twinning* of the graphite crystallites in the contact films (formed on both the brush and the copper ring) due to the compressive shear forces occurring during sliding. Similar twinning has been found by Laves and Baskin (1956). They found that, for light pressures combined with sliding, their graphite single crystal specimens twinned at about 22°. The angle remained constant with increasing pressure, up to some (unspecified) large pressure, above which the twin angle reduced to one degree. Frieze and Kelly (1963) showed that similar twinning occurred on the *microscopic* scale. Figure 5.14 is a diagrammatical representation of how the twinned graphite crystallites might tend to lie in the large areas of smooth surface known to exist between ridges running parallel to the direction of sliding in such tribo-systems. The ridges would tend to obscure the *parallel* members of the twinned crystallite when looking *across* the direction of sliding. This also explains why the $\delta = 0°$ arcs have their asymmetrical shape. When the load is so great that the tilted members are forced to lie parallel to the surface, there can only be a relaxation of tilt in a direction away from the surface, that is the arcs cannot extend beyond the normal to the surface. We must realize that none of these crystallites are actually free

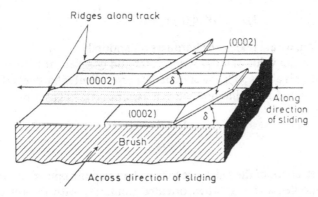

Figure 5.14 *Twinning graphite crystallite hypothesis of the 'double orientation' of electrographite brush surfaces after sliding on copper discs*

entities. Each crystallite, even one which has had its natural twin angle reduced to approximately zero, is held in position by all its neighbouring crystallites in the equilibrium contact film formed on both the brush and disc surfaces.

When the electron beam passes (at grazing incidence) *along* the direction of sliding, the *tilted* members are generally at too large an angle for electron diffraction to occur. Normally, Bragg angles for electron diffraction are of the order of 1°, *not* the values such as 20° (found for the tilt angle at fairly low loads). However, the *parallel* member of each crystallite, *will* be in the right position for Bragg reflection. Hence the arcs found at 90° to the shadow edge (as shown in Figure 5.13(*b*)).

The reflection electron diffraction evidence suggests that the force of friction between electrographite sliding on copper consists of three components. These are (a) the force ($F_{\parallel}$) required to *shear parallel* members in the contact films formed on both surfaces, (b) the force (F_T) required to *move tilted* members up the slope defined by their tilt angle (δ) and (c) the force (F_p) required to *plough randomly-oriented* crystallites on both surfaces through each other. We can, therefore, write an expression for (F_{theory}) as follows:

$$(F_{\text{theory}}) = (F_{\parallel}) + (F_T) + (F_p) \tag{5.8}$$

As we discussed in Section 1.2, the force of friction is always considered to be proportional to the real area of contact. For parallel members, the real area of contact $(A)_{\parallel}$ will be given by:

$$(A)_{\parallel} = (W)/(p_m)_{\perp} \tag{5.9a}$$

where W is the normal applied load and $(p_m)_{\perp}$ is the hardness of graphite measured perpendicularly to the basal plane. Clearly, the real area of contact of the tilted members will be $(A)_T$ given by:

$$(A)_T = (W \cos \delta)/(p_m)_{\perp} \tag{5.9b}$$

For the randomly-oriented crystallites we would expect:

$$(A)_R = (2W)/[(p_m)_{\perp} + (p_m)_{\parallel}] \tag{5.9c}$$

In this last equation, we are taking the hardness of randomly-oriented crystallites to be equal to the mean value of the perpendicular and parallel hardnesses. Now the parallel hardness is about 30% lower than the perpendicular hardness (Datta, 1984). Hence to a good aproximation we can assume $(A)_R$ is given by:

$$(A)_R \approx \frac{1.18W}{(p_m)_{\perp}} \tag{5.9d}$$

If $S_{\parallel}$ is the shear strength of the junctions formed between the opposing surfaces when oriented parallel, and (S_R) when oriented randomly with respect to the surface, then we may write Equation (5.8) in terms of $(A)_R$, $(A)_T$ and $(A)_{\parallel}$ as follows:

$$(F_{\text{theory}}) = (A)_{\parallel}(S_{\parallel}) + (A)_T(S_{\parallel}) + (A)_R(S_R)$$

$$(F_{\text{theory}}) = (\mu_{\text{theory}}\, W) = \frac{W(s_{\parallel})}{(p_m)_{\perp}} + \frac{W(s_{\parallel})\cos\delta}{(p_m)_{\perp}} + \frac{1.18\,W(S_R)}{(p_m)_{\perp}} \tag{5.10}$$

$$(\mu_{\text{theory}}) = \frac{(S_{\parallel})}{(p_m)_{\perp}}(1+\cos\delta) + \frac{1.18(S_R)}{(p_m)_{\perp}}$$

Although values are available for $(p_m)_{\perp}$, the perpendicular hardness, it is not possible to ascribe values to $(S_{\parallel})$ and (S_R) for most electrographites and other artificially-formed graphite materials. The main point of this discussion is to *indicate* the sort of experiments needed to deduce the friction properties of graphite materials from first principles. If we assume $\mu_{\parallel} = S_{\parallel}/(p_m)_{\perp}$ and $\mu_R = (S_R)/(p_m)_{\perp}$, then we may write Equation (5.10) as:

$$(\mu_{\text{theory}}) = (\mu_{\parallel})(1+\cos\delta) + 1.18(\mu_R) \tag{5.11}$$

If we now introduce the concept of the percentage ratio (P_R) of crystallites *randomly* oriented with respect to the sliding surface (see Section 5.2.7), we can write Equation (5.11) as:

$$\mu_{\text{theory}} = (1 - P_R)(\mu_{\parallel})(1+\cos\delta) + P_R(1.18\mu_R) \tag{5.12}$$

Let us examine this equation in the knowledge that $(\mu_{\text{expt}}) = 0.15$ for $\delta = 0°$, that $(\mu_R) = 0.08$ for randomly-oriented crystallites and $(P_R) = 0.09$ for contact films formed by sliding electrographite on copper discs. For $\delta = 0°$ we can write Equation (5.12) as:

$$(\mu_{\text{theory}}) = 0.91(\mu_{\parallel}) + 0.0085 \tag{5.13}$$

Setting $(\mu_{\text{theory}}) = (\mu_{\text{expt}}) = 0.15$ we obtain $(\mu_{\parallel}) = 0.155$. Since $(\mu_{\parallel})$ is unlikely to change from this value for the *non-zero tilt angle experiments*, we get:

$$(\mu_{\text{theory}}) = (1 - P_R)(0.155)(1+\cos\delta) + (P_R)(0.094) \tag{5.14}$$

Again, putting $(P_R) = 0.09$ and $\delta = 22°$, we obtain $(\mu_{\text{theory}}) = 0.28$. This is fairly close to the value of 0.35 *predicted* on the basis of the empirical equation for the coefficient of friction, namely Equation (5.7). It is sufficiently close for one to feel confident that the use of our twinned electrographite crystal model, together with critical experiments involving reflection electron diffraction and the measurement of the various kinds of friction forces (i.e. $(F_{\parallel})$, (F_T) and (F_p)), can lead us towards a true understanding of the frictional behaviour of graphite materials.

5.2.5 The effect of electric current on the orientation of graphite surfaces

One of the most common tribo-systems involving graphite materials is the electric motor or generator. These 'carbon' brushes are expected to carry a wide range of currents. Fisher (1973), in his search for an alternative material to copper for slip rings, carried out an in-depth investigation of the tribological properties of aluminium and aluminium alloys as counterface materials for his electrographite brushes. One aspect of his work involved using reflection electron

Table 5.6 *Values of δ and* μ_{expt} *for electrographite sliding on aluminium at various currents under a load of 0.6 N*

Current (mA)	Angle of tilt (δ)	tan δ	μ_{expt}
0	19° ±0.3°	0.344	0.40±0.02
1	18° ±0.3°	0.325	0.40±0.02
500	12.5°±1°	0.222	0.24±0.02
2000	4.5°±0.5°	0.079	0.12±0.02

diffraction to study the effects of current upon the angle of tilt δ, as revealed by the (0002) diffraction arcs. One set of his results, for 0.6 N load, is shown in Table 5.6. This table also includes the measured coefficient of friction (μ_{expt}). Fisher (1973) also reported that, with the electron beam passing parallel to the direction of sliding, he always found δ=0° for zero current and all currents up to 2 A. This confirmed the results of Quinn (1963) for electrographite on copper, insofar as those results showed zero tilt for all loads when passing the electron beam parallel to the direction of sliding.

Inspection of Table 5.6 shows that (tan δ)≠(μ_{expt}), as might have been expected from the work of previous investigators, for example Porgess and Wilman (1960). Fisher did not, however, find evidence of zero tilt when viewing across, even when he increased the load from 0.6 up to 10.6 N. This is, of course, below the 30 N load at which Quinn (1963) found zero tilt when viewing perpendicularly to the direction of sliding. Hence, it is not possible to apply Equation (5.11) to Fisher's (1973) results, since we do not have values for ($\mu_\parallel$), nor for (μ_R) (the value of the coefficient of friction for randomly oriented electrographite surfaces). It is interesting, however, to see that by plotting 'δ versus current' and '(μ_{expt}) versus current' one obtains similar graphs slightly displaced (along the (μ_{expt}) axis) from each other, as shown in Figure 5.15. This does suggest (μ_{expt})≈0.02+(tan δ), which is similar to *both* Equations (5.6) and (5.7). The important feature of Figure 5.15, however, is the *reduction* of both μ_{expt} and δ with current, for currents above 10 mA. This is equivalent to a current density of about 400 A/m², a value often exceeded in electrical machine slip-ring or commutator assemblies. Perhaps the current helps the crystallites to re-orient themselves to a lower tilt value through its heating effects? What is the cause of the production of aluminium and aluminium oxide particles which Fisher (1973) also detected in his reflection electron patterns from the tilt-oriented surfaces?

The work carried out on aluminium counterface materials, under the influence of different electric currents, raised some interesting questions. It did, however, confirm the existence of tilted crystallites in the wearing surfaces and it confirmed that the angle of tilt is related to the coefficient of friction. The role of metal oxides in the friction and wear of electrographite on metal surfaces is not clear, although it would seem that excessive current densities can cause temperatures that affect both the orieintation of the crystallites in the contact film *and* the oxidation of the

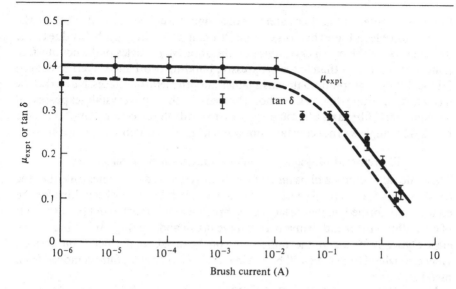

Figure 5.15 *Plot of (μ_{expt}) and (tanδ) versus electric current (electrographite brush on aluminium; load = 0.6 N)*

underlying counterface material. Analysis of graphitic contact films formed in brushes which had been rubbed against a metal counterface always tend to show the presence of substantial amounts of metal oxides in the contact film. For instance, Van Brunt and Savage (1944) found 66% of Cu_2O, 22% C and 12% of a residue containing SiO_2, Al_2O_3, Fe_2O_3 and CaO in the contact films on their copper slip-rings. Once more we see tribo-oxidation as an important factor in the friction and wear process between sliding surfaces, even surfaces that are being lubricated with an (apparently) inert solid lubricant such as graphite. We will be constantly returning to the role of oxidation in the friction and wear of tribosystems. Let us now, however, continue with our analysis of solid lubricant films by considering how much transmission electron diffraction (supported by transmission electron microscopy and X-ray diffraction) has contributed to our current knowledge of the crystallography of contact films formed when graphite materials slide against metal counterfaces.

5.2.6 Transmission electron diffraction of graphite contact films

(i) Introduction

Porgess and Wilman (1960) were the first investigators to show that graphitic contact films contained a *distribution of orientations*. They used reflection electron diffraction to study the orientation as a function of depth below the surface of an abraded reactor graphite brush, as revealed by *burning off* successive layers. From their studies, Porgess and Wilman (1960) proposed that the reactor-graphite crystallites were preferentially-oriented down to a level of approximately 40 μm below the surface. This is many times more than the maximum thickness normally found for contact films (see later). Hence one must assume that most of the orientation effects found by Porgess and Wilman (1960) were probably due to the deformation of the brush crystallites by the abrasive particles of the counterface (emery paper) rather than to the progressive build-up of oriented layers envisaged for the sliding of graphitic materials against smooth metal interfaces. Nevertheless, it is worth reporting that these authors showed that the brush crystallites possessed a simple [001] fibre axis tilted at an angle δ towards the direction of motion of the brush. This angle was not constant, nor was it a function of depth below the surface.

(ii) Removal of graphitic contact films from their substrates

The removal of contact films *intact* from their respective substrates on either the brush or ring is not an easy task. As far as I know, it has never been done for the contact film formed on the surface of the brush (due to transfer and back transfer of crystallites metal and oxides between the brush and the ring). It has been done electrolytically for the contact film formed on the surface of the metal counterface. Hence, in what follows, we will be talking only about contact films removed from metal slip-rings.

Using the same method as Smith (1936) used for removing iron oxide films from iron substrates, Quinn (1964) loosened the contact film by making the ring (or more correctly, a small part of the ring defined by a removable copper plug inserted flush to the ring surface before the experiment) the anode (film uppermost) in an electrolytic bath of dilute (5%) sulphuric acid. A current of about 6×10^{-6} A/mm^2 was passed for about 30 s. Easy access of the electrolyte to the copper was provided by scoring the track into small squares before placing in the bath. When the squares floated to the surface, they were quickly transferred to a dish of distilled water, caught on electron microscope grids, and allowed to dry slowly. It should be noted that subsidiary experiments using this electrolytic method on freshly-etched copper specimens revealed that this method probably produces *some* cuprous oxide as a by-product of the electrolytic action. This underlines the importance of always being aware of the possibility of artefacts appearing in one's electron diffraction patterns arising from the electrolytic removal technique, rather than from tribological interactions.

(iii) Measurement of contact film thickness

The thickness of contact films is best determined when the specimens are on the specimen stage of the electron microscope, *prior* to being analysed by selected area electron diffraction. This is done by measuring the *contrast* of the *transmission electron micrographs* obtained from these specimens (as described in some detail in a later sub-section, namely Section 5.3.3). It has been shown (Quinn, 1971) that the thickness of typical electrographite contact films are about $2\frac{1}{2}$ times the surface roughness of the copper substrate for *rough* surface finishes, but about a factor of *sixteen* times the roughness of the *smooth* copper substrate. Having determined the thickness with the instrument in the electron microscope operating mode, one can readily change over to the selected area electron diffraction mode of operation in order to obtain an analysis of the crystallography of the contact films.

Table 5.7. *Graphitic structure parameters for reactor graphite and electrographite*

Material	$2\theta_{0002}$	d_{0002} (Å)	p	L_a(Å)	L_c (Å)	L_a/L_c
Reactor graphite	30.72°	3.387	0.54	318	168	1.89
Electrographite	30.22°	3.431	0.90	66	84	0.786

(iv) X-ray diffraction analysis of the brush material

The main aim of this sub-section is to describe how the preferred orientation of the contact films can be analysed. We will use the early work of Quinn (1964) and the more recent work of Datta, Sykes and Quinn (1990) to illustrate the type of methods best suited to this task. As a first step in this process, however, one should characterize the basic crystallography of the graphite material used to form the contact films. Datta *et al.* (1990) used the X-ray powder diffraction method in order to obtain this characterization for their electrographite and reactor graphite specimens. Using CoK$_\alpha$ radiation, of wavelength equal to 1.7902 Å, they obtain values for the diffraction angle ($2\theta_{0002}$), the interplanar spacing (d_{0002}), the crystallite sizes in the x-direction (L_a) and in the z-direction (L_c), together with the ratio (L_a/L_c) as shown in Table 5.7.

The crystallite sizes in the x- and z-directions can best be estimated from the following equations:

$$L_a = \frac{1.84\lambda}{w_c(\cos\theta_{hk})}; \quad L_c = \frac{0.9\lambda}{w_c(\cos\theta_{0002})} \qquad (5.15)$$

where (w_c) is the diffraction line broadening due to crystallite size, (θ_{hk}) is the Bragg angle relating to the (hk) bands (arising from the lack of three-dimensional ordering of the (hkl) planes), and λ is the X-ray wavelength. The symbol p (in Table 5.7) relates to the probability that mismatch occurs between any two adjacent basal planes and is related to (d_{0002}) through

$$d_{0002} = 3.44 - 0.086(1 - p) \qquad (5.16)$$

There is no general agreement on how one should interpret the probability p. It *could* mean that *each crystallite is imperfect* in the sense that there is a mismatch between a proportion p of the basal planes in that crystallite. I *favour* the interpretation that a proportion p of the total crystallites irradiated by the X-ray beam comprises crystallites with completely random orientation between adjacent planes, together with a proportion $(1 - p)$ of perfect graphite crystallites.

The width (w_c) of the (0002) maximum is related to the measured width [$(w_c)_m$] by the relation:

$$(w_c)^2 = [(w_c)_m]^2 - [(w_c)_s]^2 \qquad (5.17)$$

where [$(w_c)_s$] is the width of the (sharp) lines of the standard material (i.e. a material whose crystallites have diameters greater than 1000 Å.

(v) The tribological history of the formation of contact films

In order to make any electron diffraction analysis relevant to the tribology of the system producing the contact films, one should know the tribological history of their formation. For example, one should know the wear rate relevant to the experiment during which they were formed and the friction between the contact films in both opposing surfaces. Datta *et al.* (1990) could only measure the wear rates in their experiments with reactor graphite and electrographite. Quinn (1964) only measured the friction. Although there may be practical reasons for omitting one of these two tribological parameters, it is strongly recommended that one measures *both* parameters. Researchers must avoid putting too much emphasis on their physical analysis, unless it is well supported by good tribological measurements. There is a temptation to present the results of one's physical analysis of one or two selected specimens (a) as if it were typical of a whole range of experimental conditions and (b) without any attempt to relate the analysis in a *quantitative* manner to the tribological behaviour. Tribology already has suffered from too many *qualitative*, ad hoc, explanations in which each tribo-system is treated separately, with no attempt to bring any cohesion to the subject. For instance, some researchers see no need to relate their experimental results with similar experiments, carried out under identical conditions but with slightly different specimen materials. Surely, when one is wearing any given pair of steels against each other, one should always try to relate one's work to the work on ferrous materials as a whole. Also, if one uses a lubricant, the results should be compared with results without any lubricant. Tribologists are moving towards a more quantative and analytical approach to their subject, as indicated by the current interest in the mechanistic modelling of wearing tribo-systems (Nichols, 1988). It is hoped that the publication of this book will also help towards making the complex subject of tribology more well-defined and understood.

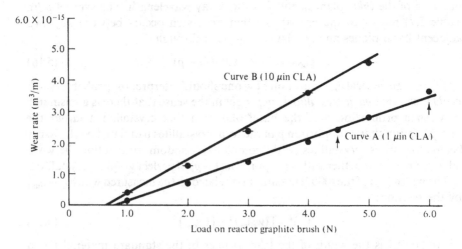

Figure 5.16 *Wear rate of reactor graphite versus load (copper counterface)*

Having made such a strong point about supportive tribological evidence, we must clearly report on the wear results of Datta *et al.* (1990) before discussing their transmission electron diffraction analysis. Using copper discs that had been lapped (with randomly-oriented scratch marks) down to 1 μin (0.025 μm) CLA (smooth) and 10 μin (0.25 μm) CLA (rough), they obtained equilibrium wear rates that varied linearly with load, as shown in Figure 5.16. Note that the rough copper disc gave rise to a larger specific wear rate, that is $(1.08 \pm 0.06) \times 10^{-15}$ m^3/(m–N) than the smooth disc, namely $(0.66 \pm 0.05) \times 10^{-15}$ m^3/(m–N). This is not entirely unexpected.

(vi) The representation of the preferred orientations of a contact film using a pole figure

Most text books deal only with the construction of pole figures using X-ray diffraction. The extremely small thickness and low atomic number of the contact films formed by graphitic materials on metal counterfaces makes them virtually transparent to X-rays, that is X-rays are not significantly scattered by the contact films. Hence, for completeness, we will describe how one can analyse a contact film by electron diffraction so that it reveals its preferred orientation. We will also discuss how we present that information in the form of a pole figure.

Films are placed in various orientations about the axis of the specimen stage of the electron microscope. It turns out that the electron diffraction patterns obtained from these films in the selected area electron diffraction mode of operation are affected in characteristic ways by the rotation (ϕ_R) of the contact film about an axis perpendicular to the beam, depending on the orientation of the direction of sliding relative to that axis. The sign convention of (ϕ_R) (and of the direction of the arcs relative to an axis perpendicular to both the beam direction and the axis of specimen rotation, namely θ_A) is illustrated in Figure 5.17.

It should be noted that diffraction arcs were found at all angles of tilt (ϕ_R) of the specimen by Datta *et al.* (1990), using reactor graphite running on copper. Quinn (1964), however, has found that arcs were not detected until (ϕ_R) was more than about 25° for his experiments with electrographite. To give the reader some idea of the actual numbers (and signs) involved, Table 5.8 is reproduced from Quinn's (1964) paper.

Quite clearly, the electron diffraction patterns described in Table 5.8 could only have arisen from a complex *distribution* of orientations of crystallites within the film. Crystallographers call this distribution a 'sheet texture'. The first step in elucidating such a texture is to measure the integrated intensity through the centre of gravity of an arc (i.e. through AA' in Figure 5.17) relative to the integrated intensity through the ring (along the line YY') for rotation (a) perpendicular and (b) parallel to the track direction. The relative intensity at Y and Y' arises from crystallites with their basal planes lying approximately normal to the axis of specimen rotation. Rotation of the specimen about the manipulator axis has very little effect on this intensity, since the number of crystallites giving rise to diffraction at Y and Y' is approximately the same for all rotations. The relative intensities at Y and Y' will have suffered the same amount of attentuation due to the increase in the effective film thickness at large values of (ϕ_R) and so it is

Table 5.8. *The effect of specimen rotation* (ϕ_R) *on arc angle* (θ_A) *for a contact film formed by an electrographite brush on a copper disc with 2 μin (0.05 μm) CLA*

Angle of tilt (ϕ_R) (deg)	Tracks anti-parallel to axis of tilt		Tracks perpendicular to axis of tilt	
	Positive rotation (deg)	Negative rotation (deg)	Positive rotation (deg)	Negative rotation (deg)
0	Uniform rings	Uniform rings	Uniform rings	Uniform rings
10	Uniform rings	Uniform rings	Uniform rings	Uniform rings
30	−28	+24	−2	Uniform rings
40	−23	+22	−1	Uniform rings
60	−22	+22	−4	Uniform rings
75	−20	–	−4	Uniform rings

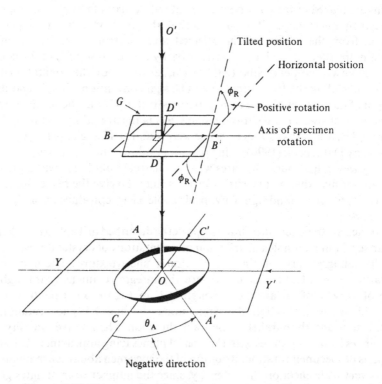

Figure 5.17 *Diagram showing sign convention for* (θ_R) *(specimen tilt angle) and* (θ_A) *(arc angle) for transmission electron diffraction patterns from the contact films*

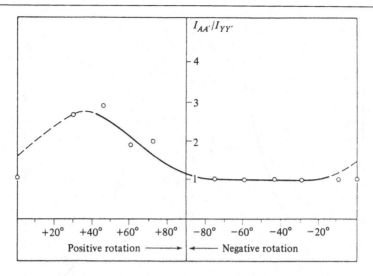

Figure 5.18 *Graph of $(I_{AA'}/I_{YY'})$ versus a single of rotation (θ_R). (Axis of tilt perpendicular to tracks)*

valid to take the mean of (I_Y) and $(I_{Y'})$ (denoted by the symbol $I_{YY'}$). The relative values of (I_A) and $(I_{A'})$, however, will have been unequally affected by a change in specimen tilt (ϕ_R). Hence, it is usual to take the most intense arc as being more representative of $I_{AA'}$. The appropriate relative intensity at A, or A', (denoted by the symbol $(I_{AA'})$ can be compared wtih $I_{YY'}$ and the ratio plotted as a function of the angle of rotation (ϕ_R). Figure 5.18 is a graph of this ratio for rotation of the contact film with the tracks *perpendicular to* the axis of specimen rotation, whereas Figure 5.19 is a graph of this ratio for rotation within the tracks *parallel to* the axis of specimen rotation.

In both figures, the broken lines are speculative estimates made for the sake of internal consistency between the two figures at $(\phi_R) = \pm 90°$. The reason for plotting the abscissa in the way illustrated will become clearer in the next few paragraphs. Similar graphs have been obtained by Datta *et al.* (1990) using reactor graphite, the main difference being the smaller values for $(I_{AA'})/(I_{YY'})$, thereby indicating there were fewer crystallites in the reactor graphite contact film lying in preferred orientations compared with the electrographite contact film. Both films were formed on approximately the same roughness counterfaces (between 1 and 2 μin CLA, that is between 0.025 and 0.050 μm CLA). The differences *could* have been due to the differences in the p-values between reactor graphite and electrographite.

The best way to represent in two dimensions the three-dimensional distribution of orientations of the crystallites in a contact film is to construct the [001] pole figure. In this, the plane of the specimen surface is taken to be the diametral plane (i.e. the stereogram) of the stereographic projection (see Section 3.1.4) of the [001] poles of all the crystallites present in the contact film. Now the number of crystallites with their [001] poles in a given orientation determines the intensity

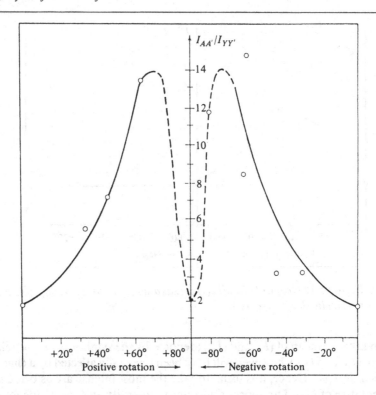

Figure 5.19 *Graph of* $(I_{AA'}/I_{YY'})$ *versus angle of rotation* (θ_R) *(Axis of tilt parallel to tracks)*

of the electron beam diffracted by the (001) planes. Hence, with the contact film perpendicular to the beam (i.e. making an angle $(\phi_R) = 0°$), the integrated relative intensity of the 002 diffraction ring at all positions around its circumference, is a measure of the density of [001] poles lying in the plane of the stereogram. In the contact film studied by Quinn (1964), the integrated intensity around the circumference of the 002 ring was uniform. Hence we may plot this intensity around the circumference of the stereogram. From Table 5.8, we see that, when the film was tilted through $\pm 10°$, there was no appreciable difference from the $(\phi_R) = 0°$ electron diffraction pattern, regardless of the direction of the tracks. This means we can plot a region of uniform pole density for all orientations (about the NS axis) of [001] poles making an angle of $90 \pm 10°$ with that axis. This is shown in Figure 5.20.

In order to plot pole densities from the electron diffraction patterns obtained at angles of tilt greater than $+ 10°$ and less than $- 10°$, one again assumes that electrons can only be diffracted by the (001) planes when their poles, that is [001], are approximately in the diametral plane. The intensity of the 001 diffraction ring at various azimuths around the perimeter is plotted around the perimeter of the diametral plane. In order to relate this to the position of the pole density in the stereogram, we must now re-plot these intensities in the appropriate position

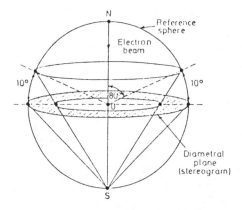

Figure 5.20 *Construction of [001] pole figure for all tilts between 0° and ±10°*

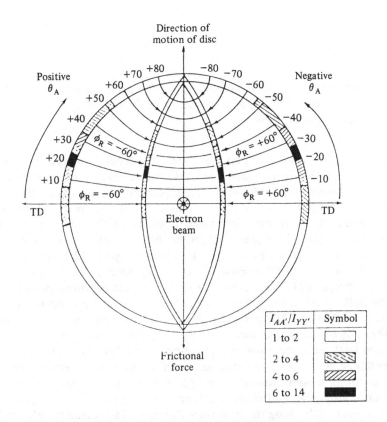

Figure 5.21 *Construction of [001] pole density distribution for direction of action of disc anti-parallel to the axis of rotation, when tilted through ±60°*

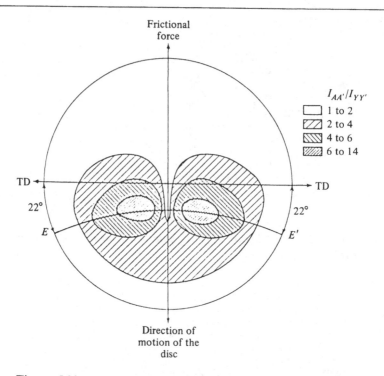

Figure 5.22 *Approximate pole figure showing the distribution of* [001] *poles of an electrographite contact film formed on an initially smooth surface (Quinn, 1964)*

prior to tilting, that is we must plot these intensities around the great circle which lies at $(\phi_R)°$ to the diametral plane. This is shown for $(\phi_R) = \pm 60°$ in Figure 5.21, the axis of rotation being anti-parallel to the direction of motion of the ring. In this way, it is possible to construct a pole density figure direct from the measurements of the relative integrated intensity of the 002 ring about its perimeter for various tilts (ϕ_R) and various orientations of the direction of motion of the ring with respect to the axis of tilt, as shown in Figure 5.22. From this figure, one can readily observe the existence of an *asymmetry* with respect to the transverse direction (TD) and *symmetry* with respect to the direction of motion. For the bulk of the electrographite contact film, there is an accumulation of crystallites with their [001] poles clustering about the points approximately 22° on either side of direction of motion of the ring. These points also lie about 22° measured back from the transverse direction, in the direction of motion of the copper disc. Figure 5.18 represents a section of the orientation distribution *along* the direction of motion, whereas Figure 5.19 represents a section of this same distribution along the small circle (*EE′*) connecting all points in the region of 22° from the great circle along the transverse direction. The reason for plotting the abscissa in the way illustrated in Figures 5.18 and 5.19 should now be self-evident.

This was the first time that it was shown that a 'sliding texture' could be evolved

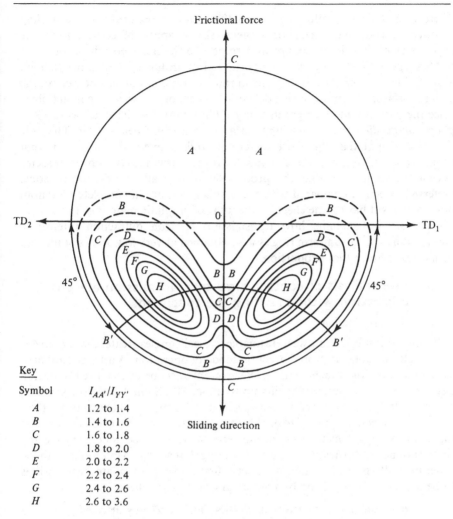

Figure 5.23 *Approximate pole figure showing the distribution of* [001] *poles of the reactor graphite crystallites within a contact film from an initially smooth copper surface (Datta* et al., *1990)*

in a wear track. More recently, Datta *et al.* (1990) have shown that reactor graphite also takes up a similar texture, as shown in Figure 5.23. Apart from the lower density of crystallites in special orientations with respect to the surface of the reactor graphite contact film compared with that of the electrographite contact film, one can see that the reactor graphite crystallites also tend to cluster with their poles at a certain angle on either side of the direction of motion of the ring and at the same angle (measured back from the transverse direction) to the direction of motion of the copper disc. The discrepancy between the 45° and 22° positions of maximum [001] pole densities in Figures 5.22 and 5.23 is insignificant compared with the similarities between the two 'sliding textures'.

There is a striking similarity between these sliding textures and the rolling textures of cold-rolled hexagonal metal sheets apart, of course, from the asymmetry of the sliding texture with respect to the transverse direction.

These pole figures are most important. For instance, they show that an appreciable number of crystallites lie in random orientation with respect to the sliding surface. This must have a deleterious effect on the wear of contact films, since the presence of some graphite crystallites with their hard, abrasive (001) planes *perpendicular* to the sliding surfaces must cause heavy wear. The pole figures also hint that the processes of cold-rolling metal sheets is somewhat similar to the sliding of electrographite (or reactor graphite) on copper. It seems reasonable to assume that the proportion of crystallites oriented in some preferred direction compared with the number that are in completely random orientation will have some effect on the wear of graphitic materials. In the next sub-section, we examine how transmission electron diffraction intensities, measured across the uniform ring system obtained at very low angles of tilt, can be used to deduce that proportion.

5.2.7 The analysis of contact films by transmission electron diffraction without tilting the specimen

(i) Introduction

In the previous sub-section, we were unable to use tilt angles greater than about 45° in our selected area electron diffraction mode of operation. With a general area electron diffraction attachment, the electron microscope can be used to obtain electron diffraction patterns at tilts up to about 80°. Some laboratories will be without tilt stage or general area electron diffraction attachments to their microscope. Hence, it is useful to be able to analyse the preferred orientations from the uniform ring intensities obtained at zero tilt. This can be done by proposing simple orientation distributions and comparing the calculated intensities with the experimentally-measured intensities. Let us first, however, calculate the intensities to be expected from randomly-oriented graphitic crystallites.

(ii) Calculated relative intensities for randomly-oriented graphitic crystallites

The intensity of the *hkl* maximum is given by the equation:

$$(I_{hkl})_{ED} \propto |(F_e)_{hkl}|^2 (m)(d_{hkl}) \tag{5.18}$$

assuming that the kinematical theory of electron diffraction holds for high energy electron diffraction by thin graphitic films. In Equation (5.18), $|(F_e)_{hkl}|^2$ is the square of the structure factor for electron scattering, m is the multiplicity factor and (d_{hkl}) is the interplanar spacing. $[(F_e)_{hkl}]$ is related to the differential cross-section for elastic scattering of electrons, $[E(\theta)]$ by the expression:

$$|(F_e)_{hkl}|^2 = \left\{ \sum_{i=1}^{N} E(\theta)[\cos 2\pi(hx_i + ky_i + lz_i)] \right\}^2$$

$$+ \left\{ \sum_{i=1}^{N} E(\theta)[\sin 2\pi(hx_i + ky_i + lz_i)] \right\}^2 \tag{5.19}$$

Table 5.9. *Summary of parameters affecting electron diffraction intensities from randomly-oriented Bernal graphite*

hkl	d (Å)	$\sin\theta/\lambda$ (Å^{-1})	$E(\theta)^2$ (arbitrary units)	G^2	m
002	3.40	0.147	1.00×10^4	16	2
100	2.12	0.236	3.60×10^3	1	6
101	2.02	0.248	3.10×10^3	3	12
102	1.81	0.276	2.28×10^3	1	12
004	1.69	0.296	1.82×10^3	16	2
103	1.56	0.321	1.41×10^3	3	12
110	1.23	0.407	6.20×10^2	16	6
112	1.16	0.451	5.00×10^2	16	12
006	1.12	0.447	4.40×10^2	16	2

where x_i, y_i, z_i are the coordinates of the N atoms in the unit cell. For 'Bernal graphite', these coordinates are 000, $00\frac{1}{2}$, $\frac{1}{3}\frac{2}{3}0$ and $\frac{2}{3}\frac{1}{3}\frac{1}{2}$.

Since random orientation is assumed, then m is merely the number of equivalent planes having the same interplanar spacing. $E(\theta)$, which is related but not identical to the $f_e(\theta)$ in Chapter 3, can be deduced from the table of $E(\theta)$ versus $(\sin\theta)/\lambda$ values given in Table VII of Thomson and Cochrane (1939) for most of the common elements. Equation (5.19) can be written as:

$$|F_e|^2 = [E(\theta)]^2 G^2 \tag{5.20}$$

where

$$G^2 = \left[1 + \cos\pi l + \cos\frac{2\pi}{3}(h+2k) + \cos\frac{\pi}{3}(4h+2k+3l) \right]^2$$

$$+ \left[\sin\pi l + \sin\frac{2\pi}{3}(h+2k) + \sin\frac{\pi}{3}(4h+2k+3l) \right]^2 \tag{5.21}$$

Using the values of $[E(\theta)]^2$, G^2 and m given in Table 5.9 for Bernal graphite, together with Equations (5.18) to (5.21), it is possible to deduce the intensities expected from randomly-oriented Bernal graphite crystallites. The calculations are tabulated in Table 5.10, from which we see that the calculated electron intensities follow the same trend as the experimental X-ray diffraction intensities.

It is interesting to compare the relative electron diffraction intensities *expected* from randomly-oriented natural graphite crystallites with the typical 'intensity versus distance from centre-spot' plot for electrographite, as shown in Figure 5.24. By taking similar plots at several electron accelerating voltages between 61 kV and 160 kV, it was possible to obtain a reliable table of both peak intensities (I_{pe}) and integrated intensities (I_{we}), as shown in Table 5.11. This table also includes the results of analysing a contact film formed by rubbing electrographite

Table 5.10. *Calculated integrated intensities from randomly-oriented Bernal graphite*

hkl	I_{hkl} (arbitrary units $\times 10^4$)	I_{hkl} (relative to I_{002})	X-ray values
002	109×10^4	100	100
100	4.6×10^4	4.2	2
101	22.5×10^4	20.6	3
102	5.0×10^4	4.5	1
004	9.8×10^4	9.0	8
103	7.9×10^4	7.2	2
110	7.3×10^4	6.7	6
112	11.1×10^4	10.2	6
006	1.6×10^4	1.5	2

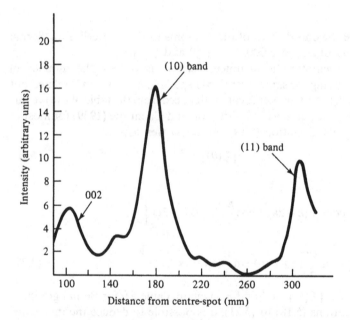

Figure 5.24 *Plot of intensity versus distance from centre-spot for an electron diffraction pattern from a contact film formed by an electrographite brush on a 2 μin CLA copper disc (Quinn, 1964)*

on a rough (30 μin CLA surface finish) copper disc. Significant differences exist (a) between the relative intensities and those expected from randomly-oriented crystallites and (b) between the relative intensities from the *thick* film (formed on the 30 μin CLA disc) and the *thin* film (formed on the 2 μin CLA disc). It will be noted that the 100, 101 and 102 diffraction maxima combine to give the 10 band, and similarly the 110 and 112 combine to give the 11 band.

Table 5.11. *Experimental values of (I_{pe}) and (I_{we}) for contact films formed on* 2 µin *and* 30 µin *CLA copper discs by electrographic brushes*

Specimen	2 µin CLA film			30 µin CLA film		
hkl	002	10	11	002	10	11
(I_{pe})	100	292 ± 17	163 ± 13	100	11 ± 17	53 ± 5
(I_{we})	100	304 ± 45	193 ± 33	100	129 ± 15	94 ± 20

Quite clearly, Figure 5.24 and Table 5.11 show that these contact films consist of many small crystallites with very strong preferred orientations. Quinn (1984) has analysed these preferred orientations in terms of two model distributions, namely one in which there is a tendency for the (001) planes to be at 22° (as indicated by the sliding texture for electrographite illustrated in Figure 5.22), and the other where these (001) planes tend to lie parallel to the surface. He then compares the relative intensities expected for these model distributions with the measured experimental values given in Table 5.11 in order to deduce the proportion P of crystallites in random orientation compared with those in a preferred orientation. Let us discuss the 22° distribution first.

(iii) Electron diffraction intensities expected from a contact film with Bernal graphite crystallites lying with their (001) planes at 22° to the film surface

Let us assume that the distribution of [001] poles, that is the distribution of normals to the (001) planes, is as given in Figure 5.25. This is a model in which one assumes a simple distribution with the [001] poles of equal numbers of crystallites in all orientations (σ) with respect to the electron beam, except for a range of values (ϕ) centred around 22°. In fact, $n(\sigma) \, d\sigma$ is the number of [001] poles lying between angles (σ) and ($\sigma + d\sigma$) with respect to the electron beam. The ratio of (B/A) and ψ can be adjusted, within certain limits, so that they are consistent with the experimental electron diffraction intensities from the film, ψ is related to the length of the diffraction arcs obtained from tilted specimens. Hence we can estimate that ψ will lie between 10° and 30°. From (B/A) values we can deduce the proportion of preferred-to-randomly-oriented crystallites.

Not all arbitrary values of A and B will be suitable. ψ is restricted, by practical considerations, to values between 10° and 30°. Obviously, for completely randomly-oriented crystallites, $B = 0$ and $\psi = \pi$. For an almost ideal preferred orientation of the [001] poles at 22° to the specimen normal, then $(A/B) \rightarrow 1$ and $\psi \rightarrow 0$. Now any given crystallite will only give rise to an (hkl) reflection provided that:

$$\sigma \geq \frac{\pi}{2} - (\rho_{hkl} + \theta_{hkl} + \Delta\sigma) \tag{5.22}$$

251

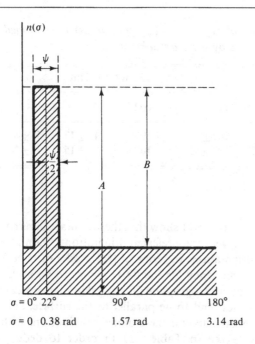

Figure 5.25 *Simple distribution of n(σ) dσ for a contact film with graphitic crystallites tending to lie with their [001] poles at σ with respect to electron beam*

and

$$\sigma \le \frac{\pi}{2} + (\rho_{hkl} + \theta_{hkl} + \Delta\sigma) \tag{5.23}$$

where (ρ_{hkl}) is the angle made by the normal to the (hkl) plane with the normal to the (001) plane in any graphite crystallite, θ_{hkl} is the Bragg angle for the (hkl) plane and $\Delta\sigma$ is the spread of angles (σ) for which Bragg 'reflection' can occur. This last parameter is directly related to the 'half-width' of the (hkl) diffraction ring. Equations (5.22) and (5.23) have been derived from Figure 5.26 for cases in which $0 < (\rho_{hkl}) < \pi/2$. In this figure, we have neglected the values of θ_{hkl} and $\Delta\sigma$ compared with the (generally) much larger values of (ρ_{hkl}). In the case of diffraction from the (001) planes, however, the angle (ρ_{hkl}) is zero, so we then have to take these parameters into account. Equations (5.22) and (5.23) provide us with upper and lower limits for σ respectively.

When $\pi/2 < (\rho_{hkl}) < \pi$, then Figure 5.27 is relevant and the limits become:

$$(\sigma_{max}) = \frac{3\pi}{2} - \rho_{hkl} \tag{5.24}$$

and

$$(\sigma_{min}) = (\rho_{hkl}) - \frac{\pi}{2} \tag{5.25}$$

Assume $0 < \rho_{hkl} < \pi/2$

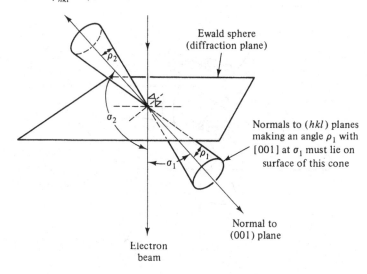

When $\sigma < \pi/2$ (e.g. σ_1 and $\rho_{hkl} = \rho_1$)

$(\sigma_1 + \rho_1)$ must be $\geq \pi/2$ for diffraction to occur from (hkl) plane

that is $\sigma_{\min} = \dfrac{\pi}{2} - \rho_{hkl} \ldots (5.22)$

When $\sigma > \pi/2$ (e.g. σ_2 and $\rho_{hkl} = \rho_2$)

$(\pi - \sigma_2 + \rho_2)$ must be $\geq \pi/2$ for diffraction to occur

that is $\sigma_{\max} = \dfrac{\pi}{2} + \rho_{hkl} \ldots (5.23)$

Figure 5.26 *Diagrammatic derivation of Equations (5.22) and (5.23) for* $0 < \rho_{hkl} < \pi/2$ *(neglecting θ_{hkl} and $\Delta\sigma$)*

In these equations, the angles θ_{hkl} and $\Delta\sigma$ have again been ignored. If it is necessary to allow for these small angles, one must write $(\rho_{hkl} + \theta_{hkl} + \Delta\sigma)$ in place of (ρ_{hkl}) in these two equations.

It is clear that the total number N of crystallites that will give rise to an (hkl) diffraction ring when the beam is at normal incidence to the contact film is given by:

$$N = \int_{\sigma = \pi/2 - \rho_{hkl}}^{\sigma = \pi/2 + \rho_{hkl}} n(\sigma)\, d\sigma \qquad (5.26)$$

for $(\rho_{hkl}) < \pi/2$. In order to deduce the effect of any preferred orientation upon the relative intensities given by Equation (5.18), we must multiply that expression by N, taking care to multiply N itself by a weighting constant (C_{hkl}) to ensure that

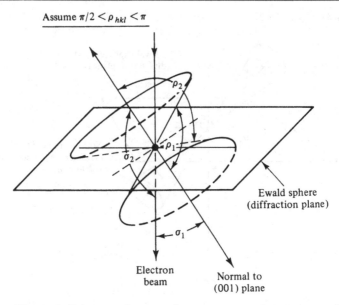

Assume $\pi/2 < \rho_{hkl} < \pi$

Ewald sphere
(diffraction plane)

Electron beam

Normal to (001) plane

When $\sigma < \pi/2$ (e.g. σ_1 and $\rho_{hkl} = \rho_1$)

$(\sigma_1 + \pi - \rho_1)$ must be $\geqslant \pi/2$ for diffraction to occur from (hkl) plane

that is $\sigma_{min} = \rho_{hkl} - \pi/2 \ldots (5.25)$

When $\sigma > \pi/2$ (e.g. σ_2 and $\rho_{hkl} = \rho_2$)

$2\pi - (\sigma_2 + \rho_2)$ must be $\geqslant \pi/2$ for diffraction to occur

that is $\sigma_{max} = 3\pi/2 - \rho_{hkl} \ldots (5.24)$

Figure 5.27 *Diagrammatic derivation of Equations (5.24) and (5.25) for* $\pi/2 < \rho_{hkl} < \pi$

the expression for (I_{hkl}) returns to its 'random' value for $B=0$ and $\psi=\pi$ in the proposed distribution (Figure 5.25). This weighting constant will be different for each hkl ring. Let us now calculate the appropriate intensities on the basis of Figure 2.25.

1 Calculation of the 002 diffraction intensity

Let us write I_{002} as follows, remembering that $(\rho_{hkl})=0$ for this diffraction ring:

$$I_{002} = (I_{002})_{random}(C_{002}) \int_{\sigma = \pi/2 - (\theta_{002}+\Delta\sigma)}^{\sigma = \pi/2 + (\theta_{002}+\Delta\sigma)} n(\sigma)\,d\sigma$$

For random orientation we have $B=0$, $\psi=0$ and we require:

$$(C_{002}) \int_{\sigma = \pi/2 - (\theta_{002} + \Delta\sigma)}^{\sigma = \pi/2 + (\theta_{002} + \Delta\sigma)} n(\sigma) \, d\sigma = 1$$

Now we can see from Figure 5.25 that, if $B = 0$ then the integral equals the area $2A(\theta_{002} + \Delta\sigma)$.

Therefore

$$(C_{002}) = \frac{1}{2A(\theta_{002} + \Delta\sigma)}$$

Hence we may write our original equation for I_{002} as:

$$I_{002} = \frac{(I_{002})_{\text{random}}}{2A(\theta_{002} + \Delta\sigma)} \int_{\sigma = \pi/2 - (\theta_{002} + \Delta\sigma)}^{\sigma = \pi/2 + (\theta_{002} + \Delta\sigma)} n(\sigma) \, d\sigma$$

provided $(\psi/2) < \pi/2 - 0.38$, that is $(\psi/2) < 1.19$ radians, then we can write

$$I_{002} = \frac{(I_{002})_{\text{random}}}{2A(\theta_{002} + \Delta\sigma)} [2(A - B)(\theta_{002} + \Delta\sigma)] \tag{5.27}$$

that is

$$(I_{002}) = (I_{002})_{\text{random}} \left[\frac{(A - B)}{A} \right]$$

Inserting $(I_{002})_{\text{random}}$ from Table 5.10, the intensity of the (002) ring is given by:

$$(I_{002}) = 1.09 \times 10^6 (1 - B/A) \tag{5.28}$$

Clearly this gives the random value for $B = 0$ (as required). If $B/A = 1$, then the preferred orientation is very strong, and $I_{002} = 0$. This is because there would not be *any* crystallites in the right position for diffraction, when the beam was passed normal to the surface of the specimen. For $(B/A) = 0.90$, we get $I_{002} = 1.09 \times 10^5$; for $(B/A) = 0.99$, then $I_{002} = 1.09 \times 10^4$. Obviously, I_{002} can be made as small as we choose by making (B/A) approach unity. It is also obvious that ψ has *no effect* on I_{002}.

2 Calculation of the 101 diffraction intensity

In this case, the value of (ρ_{hkl}) is $72° 22'$ (i.e. 1.26 rad). Hence we may neglect θ_{101} and $\Delta\sigma$ in the limits of the integration, and write I_{101} as follows:

$$I_{101} = (I_{101})_{\text{random}} (C_{101}) \int_{\sigma = \pi/2 - 1.26}^{\sigma = \pi/2 + 1.26} n(\sigma) \, d\sigma$$

For $I_{101} = (I_{101})_{\text{random}}$ we require $B = 0$ so that the integral becomes $(2.52A)$. Hence $(C_{101}) = 1/2.52A$. From Table 5.10 and the above we get:

$$I_{101} = 2.25 \times 10^5 \left[(1 - B/A) + \frac{\psi}{2.52} (B/A) \right] \tag{5.29a}$$

for $\psi/2 < 0.07$ radians, and

$$I_{101} = 2.25 \times 10^5 \left[(1 - B/A) + \left(\frac{\psi - 0.07}{2.52} \right) (B/A) \right] \qquad (5.29b)$$

for $\psi/2 > 0.07$ radians.

3 Calculation of I_{110} diffraction intensity

In this case, the value of $\rho_{hkl} = \pi/2$, and so $\Delta\sigma$ and θ_{110} may again be neglected with very little error. Hence we may write (I_{110}) as follows:

$$(I_{110}) = (I_{110})_{\text{random}}(C_{110}) \int_{\sigma=0}^{\sigma=\pi} n(\sigma) \, d\sigma$$

It can readily be shown that $(C_{110}) = 1/\pi A$, so that:

$$I_{110} = 7.3 \times 10^4 \left[(1 - B/A) + \frac{\psi}{\pi} (B/A) \right] \qquad (5.30)$$

This equation holds for all ψ.

4 Calculation of I_{100} diffraction intensity

It is interesting to note that the 100 diffraction maximum also has $(\rho_{hkl}) = \pi/2$ and hence $[C_{100}] = 1/(\pi A)$. Hence, we may readily write the expression for I_{100}, namely:

$$I_{100} = 4.6 \times 10^4 \left[(1 - B/A) + \frac{\psi}{\pi} (B/A) \right] \qquad (5.31)$$

In order to compare the above theoretical expressions with experimental values, we must make assumptions regarding the contributions from the (100), (101) and (102) planes to the 10 band, the contributions from the (110) and (112) planes to the 11 band, and the best value to take for ψ. Probably, the easiest way of obtaining a quick order of magnitude estimate is to assume that ψ must be strongly related to the lengths of the diffraction arcs that occur when the contact film is titled. These arcs vary from approximately $10°$ to approximately $30°$. If we assume $\psi \approx 20°$, that is $\psi \approx 0.35$ radians, very little error should arise. Also, inspection of Table 5.10 indicates that the 101 maximum makes the strongest contribution to the 10 band. Hence, we will assume that:

$$(I_{10})_{\text{calc}} = (I_{100})_{\text{calc}} + (I_{101})_{\text{calc}} \qquad (5.32a)$$

It is obvious that 11 band is a mixture of 110, 112 and 006. Due to the ambiguities arising from identifying the strongest contribution, it is not worth using the data for I_{11}. Let us briefly discuss using Equation (5.32a) as a means of estimating the proportion P of randomly-oriented crystallites.

Since $\psi/2$ equals 0.175 radians, which is greater than 0.07 radians, we use Equation (5.29b) for $(I_{101})_{\text{calc}}$. $(I_{100})_{\text{calc}}$ is given by Equation (5.31) and $(I_{002})_{\text{calc}}$

by Equation (5.28). From these three equations, we can deduce a theoretical value for $(I_{10}/I_{002})_{calc}$, namely:

$$(I_{10}/I_{002})_{calc} = [2.71 \times 10^5] + [0.3 \times 10^5/(A/B - 1)] \qquad (5.32b)$$

Equation (5.32b) has been obtained using $\psi = 0.35$ rad. If we now set $(I_{10}/I_{002})_{calc}$, equal to 3.04 (the ratio of $(I_{10}/I_{002})_{expt}$) for the 2 μin CLA film, it can be readily shown that $(B/A) = 0.990$. Putting $(I_{10}/I_{002})_{calc}$ equal to 1.29, the experimental ratio for the 30 μin CLA film, one readily obtains $(B/A) = 0.974$.

These values of (B/A) can be used to give an approximate indication of the percentage of randomly-oriented crystallites present in each of the contact films. If one assumes that the number of randomly-oriented crystallites is proportional to $\pi(A - B)$ approximately, and the number of crystallites in preferred orientations is proportional to (ψA), then the *percentage ratio of randomly-oriented to preferentially-oriented crystallites* (P_R) is given by:

$$(P_R) = \frac{\pi(1 - B/A) \times 100}{\psi} \% \qquad (5.33)$$

Taking $\psi = 0.35$ radians (as before) for both contact films, the following values of (P_R) emerge:

$$2 \text{ μin CLA film: } P_R = 9\%$$
$$30 \text{ μin CLA film: } P_R = 23\%$$

Clearly the thicker film has a higher proportion of crystallites *not* in a special position relative to the sliding direction. These are probably those crystallites originally removed from the brush, which tend to take up randomly-oriented positions in the valleys of the disc surface until the valleys become filled, after which preferred orientation ensues.

(iv) Electron diffraction intensities from a contact film with Bernal graphite crystallites lying with their (001) planes parallel to the film surface

Just for completeness, we will discuss the orientation most previous investigators have assumed to be taken up by graphitic crystallites when rubbed against a metal substrate, namely the orientation indicated in Figure 5.28. Obviously, the intensity of the 002 diffraction ring will be given by the same equation as for the idealized distribution in Figure 5.25, that is by Equation (5.28). For the sake of brevity, we will only consider the 101 intensity, which is given by:

$$I_{101} = 2.25 \times 10^5 (1 - B/A) \qquad (5.34)$$

for $\psi/2 < 0.31$, and

$$I_{101} = 2.25 \times 10^5 \left[(1 - B/A) + \frac{\psi - 0.62}{2.52}(B/A) \right] \qquad (5.35)$$

for $\psi/2 > 0.31$. For our assumed value of ψ equal to 0.35, it is clear that Equation (5.34) is relevant. Hence, from Equations (5.28) and (5.34) we get:

$$(I_{101}/I_{002}) = [2.25 \times 10^5 (1 - B/A)]/[1.09 \times 10^6 (1 - B/A)]$$

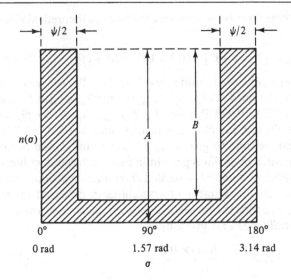

Figure 5.28 *Simplified distribution of* [001] *poles of a specimen in which the (001) planes are parallel to the surface of the specimen*

that is

$$I_{101}/I_{002} = 0.212 \tag{5.36}$$

This value is *much less* than the experimental values (1.93 and 0.94) obtained for (I_{11}/I_{002}) from both contact films (see Table 5.11). It therefore seems most unlikely that the contact films take up the orieintation distribution shown in Figure 5.28.

5.2.8 Summarizing remarks

Quite clearly, we have seen that the analysis of solid lubricant films presents a formidable task to the tribologist. In this sub-section, we have concentrated mainly on the *graphitic* type of lamellar solids. What has been said about the application of physical techniques to the analysis of graphitic contact films also applies to *other lamellar solids*. Certainly, I have studied molybdenum disulphide with electron diffraction and electron microscopy, supported also by X-ray diffraction of the bulk materials, and found these techniques most informative. There is a resurgence of interest in solid lubrication at high temperatures. This is where graphite has the edge over molybdenum disulphide. I have always maintained that preferred orientation is the *cause* of good lubrication properties rather than the *result*. Provided one can build up contact films (on both the lubricating brush and the metal counterface) which are oriented so that their crystallites are not *all* parallel to the surface, then one has the conditions for *both low friction* and *low wear*. Contact films are the essential ingredient of a solid lubricant system and the analysis of such films should lead to an understanding of how they provide their protection. Obviously, the *thickness* of these films must be related to the quality of that protection. Hence the next sub-section on measuring lubricant film thicknesses.

5.3 The application of physical methods of analysis to the measurement of lubricant film thickness

5.3.1 Introduction

In Chapter 7 we will discuss the problem of contact fatigue of liquid-lubrication tribo-systems. We will often use the concept of the *D*-ratio, which is the combined surface roughness of both tribo-elements divided by the lubricant film thickness, *as calculated* by *elasto-hydrodynamic lubrication theory*. This is a theory which has been developed to explain the anomalously thick films obtained in concentrated contact conditions, for example point and line contact, when *classical hydrodynamics* predicts that the lubricant film would be of a thickness very much *less* than the combined surface roughness of the tribo-elements. Obviously the *D*-ratio, which depends on a theoretical estimate of the lubricant film thickness, should not be used if there is any doubt about those thicknesses. This is where optical interferometry can be used, provided one of the tribo-elements is transparent to light! In this sub-section, we will explain the elastohydrodynamic theory of a rolling-point contact system and then show how two-beam and multiple-beam interferometry have been used to measure oil film thicknesses in a typical ball-on-flat tribo-system.

Another important form of lubrication is solid lubrication by graphitic materials (as we have just seen in Section 5.2). The thickness of the contact film formed on the metal counterface should be related to the life of the system, as well as the tribological (and electrical) properties of the system as, for example, in the graphite brush/copper slip ring system. We have seen that the contact film can be removed for examination by selected area, transmission electron diffraction. In this sub-section, we will describe how the thickness of this contact film can be deduced from the *contrast* of the electron micrograph of the area selected for electron diffraction. Since such a micrograph is generally taken before switching the electron microscope from its microscopy mode to its diffraction mode of operation, this method of measuring film thickness can be effected with very little extra effort on the part of the investigator. We will briefly give the theory of the contrast of electron micrographs, since this method could, in principle, be applied to a wide range of solid lubricant films.

5.3.2 Elasto-hydrodynamic film thicknesses

(i) Introduction

It is not the purpose of this section to duplicate much of the very good treatments of elasto-hydrodynamic lubrication (EHL) already available (see, for example, the work of Dowson and Higginson (1966), Crook (1961a, b), Snidle and Archard (1972) and Archard, Gair and Hirst (1961) relating to the pioneering stages of research into this subject). Rather, we wish to concentrate on the optical analysis of EHL films formed between moving surfaces. In this introduction, we will very briefly describe the essential elements of liquid lubrication. Let us start with 'fluid film lubrication', which is the normal conforming contact situation occurring in journal bearings, where the 'wedge action' of the shaft in the journal enables the oil to support the load (provided the two surfaces are in relative motion). The thickness of the oil film is many times that of the combined surface roughnesses, so that intermetallic contact is impossible. Clearly, wear cannot take place under fully hydrodynamic lubrication conditions. However, when the load is concentrated into a non-conforming contact, such as in gears, cams and tappets, and rolling-contact bearings, the application of the basic Reynold's equation (which relates *only* to hydrodynamic lubrication) leads to a predicted film thickness much less than the sum of the roughnesses of the opposing surfaces. Why then do non-conforming contacts not wear out? It is because of a new mechanism of lubrication (known as elasto-hydrodynamic lubrication) in which the local deformation of the surfaces, together with the large increase in lubricant viscosity with pressure, combine to provide the conditions for fluid film lubrication, where previously this was thought impossible.

To give some idea of numbers, using a lubricant of conventional viscosity (say a fraction of 1 poise under laboratory conditions) and steel specimens (discs of 1 in radius), it has been shown (Archard and Kirk 1961) that EHL persists down to surface speeds less than 0.01 m/s, where the film thickness will be about a few hundred ångström units. Normally, however, the film would be about 10^{-6} m, which is just a little more than the combined surface roughness of a typical gear system, for example 10 µin CLA or 0.25 µm. It seems likely, therefore, that there will be occasional metal–metal contacts through the lubricant film, which will eventually cause wear. Also, in places where metal–metal contacts do not occur, it is possible that some form of fatigue mechanism could promote failure. We shall see that contact fatigue of this sort is thought to be responsible for the pitting of gears as well as cams and tappets (see Chapter 7).

This introduction has been deliberately kept clear of mathematics, in the hope that the reader will understand the basic ideas of EHL theory without getting lost in specifics. However, in order to discuss the application of optical interferometry to the measurement of EHL oil film thickness, we must introduce the basic equations as they relate to the particular case of the nominal point contact of a ball on flat.

(ii) Elasto-hydrodynamic lubrication of rolling point contact
Reynold's equation is:

$$\frac{\partial}{\partial x}\left(\frac{(h_{\text{lub}})^3}{\eta}\frac{\partial P}{\partial x}\right) + \frac{\partial}{\partial y}\left(\frac{(h_{\text{lub}})^3}{\eta}\frac{\partial P}{\partial y}\right) = 12\bar{U}\left(\frac{\partial h_{\text{lub}}}{\partial x}\right) \tag{5.37}$$

where (h_{lub}) is the oil film thickness, P is the pressure in the film, η is the local viscosity of the lubricant and x and y are coordinates in the plane of the contact. $\bar{U}$ is the combined surface velocity which is the mean of the velocities (U_1 and U_2) of the bearing surfaces. The viscosity is assumed to vary with pressure according to:

$$\eta = \eta_0 \exp[\alpha_P)P] \tag{5.38}$$

where η_0 is the ambient viscosity of the lubricant and (α_P) is the pressure viscosity coefficient. By introducing the term 'reduced pressure' q defined as:

$$\frac{1}{(\alpha_P)}\{1 - \exp[(\alpha_P)P]\} \tag{5.39}$$

we can rewrite Reynold's equation as:

$$\frac{\partial}{\partial x}\left((h_{\text{lub}})^3 \frac{\partial q}{\partial x}\right) + \frac{\partial}{\partial y}\left((h_{\text{lub}})^3 \frac{\partial q}{\partial y}\right) = 12\bar{U}\eta_0 \frac{\partial h_{\text{lub}}}{\partial x} \tag{5.40}$$

Grubin and Vinogradova (1949) applied this equation to line contact (e.g. two cylindrical discs rolling against each other) by assuming (a) the shape outside the contact was Hertzian, and (b) that $q \approx [1/(\alpha_P)]$, everywhere inside the contact. Of course, for line contact, $\partial q/\partial y = 0$. Grubin and Vinogradova's analysis has been extended to point contacts by Cameron and Gohar (1966), Archard and Cowking (1965), Cheng (1970), and Wedeven, Evans and Cameron (1971). We will summarize the treatment by Wedeven *et al.* (1971) in what follows. The film thickness (h_{lub}) at any point is the sum of the central film thickness $(h_{\text{min}})_0$ and the Hertzian deformation for dry contact. Thus:

$$(h_{\text{lub}}) = (h_{\text{min}})_0 + \frac{3W}{2\pi a E''}\left[\left(\frac{r^2}{a^2} - 1\right)^{1/2} - \left(2 - \frac{r^2}{a^2}\right)\cos^{-1}\left(\frac{a}{r}\right)\right] \tag{5.41}$$

where r is the radius from the centre of the Hertzian region, a is the Hertzian radius and E'' is the reduced elastic modulus given by Equation (1.24). The contact geometry is shown in Figure 5.29.

Following Wedeven *et al.* (1971), write $r^* = (r/a)$, $h^* = (h_{\text{lub}})/(h_{\text{min}})_0$, $q^* = q(h_{\text{min}})_0^2/12\eta_0\bar{U}a$. Since the pressure falls rapidly with distance from the centre of the Hertzian region, let us choose a new distance $m^* = \ln(r^*)$ and the new angle ϕ (defined by $\phi\pi = 2\theta$, where θ is the angular displacement from the line of fluid motion on the axis of symmetry). Applying these transformations to Reynold's equation, one gets the following form:

$$\frac{\partial^2 q^*}{\partial m^{*2}} + \frac{r}{\pi^2}\left(\frac{\partial^2 q^*}{\partial \phi^2}\right) + \left(\frac{3}{h^*}\right)\left(\frac{\partial h^*}{\partial m^*}\right)\left(\frac{\partial q^*}{\partial m^*}\right)$$

$$= \exp(m^*)\cos\left[\frac{\pi\phi}{2}\left(\frac{1}{h^*}\right)^3\left(\frac{\partial h^*}{\partial m^*}\right)\right] \tag{5.42}$$

The solution of the above equation cannot be done analytically. Making several assumptions about the boundary conditions, mainly so that the equation can be solved, Wedeven *et al.* (1971) arrive at the following formula for the central oil film thickness $(h_{\text{min}})_0$:

$$\frac{(h_{\text{min}})_0}{R_r} = 1.73\left(\frac{\alpha_P\eta_0\bar{U}}{R_r}\right)^{5/7}\left(\frac{W}{E''(R_r)^2}\right)^{-1/21} \tag{5.43}$$

where R_r is the reduced radius of contact, defined by:

$$\frac{1}{R_r} = \frac{1}{R_1'} + \frac{1}{R_2'} \tag{5.44}$$

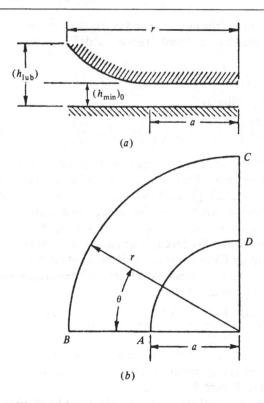

(a)

(b)

Figure 5.29 *Contact geometry for EHL contact between a ball and a flat*

and where R'_1, R'_2 are the radii of the two contacting bodies (in the instance of a ball on a flat, of course, $R'_2 = \infty$ and $R_r = R'_1$, the radius of the ball).

Having obtained an expression which should be relevant for rolling point contact, let us see how far Equation (5.43) can be validated by experiment, that is by direct observation, in the next sub-section.

(iii) Optical elasto-hydrodynamics of rolling point contact

Wedeven *et al.* (1971) used an optical arrangement, specially designed to measure the oil film thickness between a ball and an optical crown glass flat. This arrangement was first described in the paper by Foord *et al.* (1969). Let us examine the optical design of that arrangement in more detail.

The ball used by Foord *et al.* (1969) was a conventional steel ball-bearing ball, with a very highly finished surface (better than 1 μin CLA) to give the required reflectivities of 75% in air, reducing to 60% in oil. The transparent plate was optical crown, free of striations and flat to one fringe per 200 cm². The nominal Hertzian stress which this glass can carry under lubricated conditions is about 700 MPa. For higher stresses, a sapphire plate can be used. In fact, sapphire is better from the thermal conductivity point of view, since it is only 20% lower than that of steel. In addition, thermal diffusivity of sapphire and steel are equal at 37 °C.

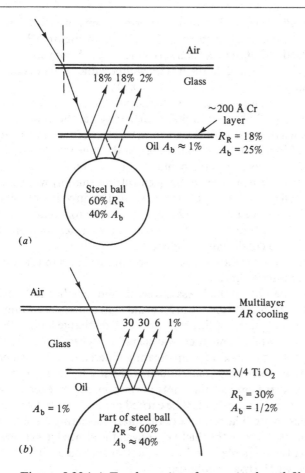

Figure 5.30*(a) Two-beam interference in the oil film between the glass plate and the steel ball of the optical arrangement used by Foord et al. (1969); (b) Multiple beam interference in the oil film in (a) above*

The optical arrangement is a modification of Newton's Rings by reflection as described in Section 2.1.4 and illustrated in Figure 2.9(a). The steel ball replaces the plano-convex lens and the light source and viewing system must now be such that the light is incident on the side of the optical flat remote from the convex surface. This is shown in Figure 5.30(a). In order to increase the reflectivity of the glass, a layer of chromium is frequently deposited on to the glass when hot ($\sim$ 140 °C) to give a very abrasion-resistant surface and provide the oil film with a metal interface between it and the surface, which will be not very different (tribologically speaking) from the oil/steel interface on the other side of the film. The optimum reflectivity (R_R) of the chromium layer is approximately 20%. This layer also has a 20% absorption coefficient (A_b). We did not discuss absorption in Section 2.1.4. Essentially, however, we require ($T_T + R_R + A_b$) to be equal to 100%, so that whatever is not reflected nor transmitted, must have been absorbed! The steel reflects 60%. The two interfering beams therefore have the same intensity (18% of the incident intensity) as shown in Figure 5.30(*a*). The second and third reflections (at the steel ball surface) produce beams so very weak that only two beams interfere and so the fringes will be rather broad (similar to the fringe pattern shown in Figure 2.11 for $R_R = 0.30$).

If a dielectric, such as TiO_2, is used as the semi-reflecting layer, then the absorption is very low ($\sim 0.5\%$) and a 30% reflectivity can be achieved. The reflected rays are shown in Figure 5.30(*b*), where several reflections are achieved, giving multiple beam fringes, which are much sharper than the two-beam fringes, as shown in Figure 5.31. This shows the contact region (defined by the circular area inside the ninth fringe from the zeroth (central) maximum) within which some very interesting oil film thickness contours occur. The direction of rolling of the ball was probably from the bottom of Figure 5.31 to the top, that is the inlet region is probably at the bottom. The authors do not make it clear in their paper just what were the precise conditions of sliding, their objective being mainly to show how much sharper are the dark fringes under multiple beam interferometry than under two-beam interferometry.

In order to deduce the optical thickness of the oil film, one has to know the order n of the fringes. For this reason, the most usual light sources are white light incandescent lamps of the quartz–iodine type or xenon discharge lamps. With a continuous spectrum of white light, each wavelength independently interferes with itself. The fringes thus formed overlap creating different colour sequences, thus enabling the fringe *order* to be determined without counting fringes. The colours are impure and the optical thickness corresponding to each must be found by calibration. The calibration must be carried out *in air* with the steel ball loaded against the glass plate. The radius x of the interference ring outside of the central contact region depends on the thickness (d_a) of the air wedge between the chromium surface on the glass plate and the spherical surface of the steel ball, through the well-known Hertzian equation:

$$d_a = \left[\frac{(1+v_1^2)}{2E_1} + \frac{(1-v_2^2)}{2E_2}\right][ap_{max}]\left[\left(\frac{x^2}{a^2} - 2\right)\cos^{-1}\left(\frac{a}{x}\right) + \left(\frac{x^2}{a^2} - 1\right)^{1/2}\right] \quad (5.45)$$

where E_1, E_2, v_1, v_2 are the Young's moduli and Poisson's ratios of the glass and steel; p_{max} is the maximum Hertzian pressure in the region of the circular zone of contact of radius a. For the *ideal* condition of zero load, then the thickness (d_a) corresponding to each fringe of radius x is given by Equation (2.15), namely:

$$d_a = x^2/2(R_L) \quad (2.15)$$

where (R_L) is now the radius of the steel ball (instead of the optical lens). In order to deduce the thickness (d_a) for zero load, Foord *et al.* (1969) measured the diameters $2x$ of the fringes for loads decreasing from 4.90 N down towards zero. The fringe radius x was then plotted against the calculated Hertzian (i.e. elastic) radius a as shown in Figure 5.32, for the first seven dark fringes of a monochromatic green ($\lambda = 5461$ Å) system. As the load approaches zero, the curve of a versus x should approach the asymptote $x = $ constant. We see that the

Figure 5.31 *Comparison of two-beam and multiple-beam interferographs:*
(a) using a TiO₂ semi-reflecting layer and a helium–neon laser
$(\lambda = 6328\,\text{Å})$; *(b) using a chromium semi-reflecting layer and a mercury-*
vapour lamp $(\lambda = 5461\,\text{Å})$ $[Load = 11\,N;\ Lubricant:\ 5\text{-}phenyl\text{-}4\text{-}ether]$
(from Foord et al., 1969)

experimental points deviate from this condition for loads below about 1 N
($a = 20\,\mu\text{m}$), probably due to the surfaces being held apart by micro-asperities
which, at high loads, were squeezed flat. The curves of a versus x can be derived
from the equation for (d_a) (i.e. Equation (5.45)), using the measured value of x at
the load (2 N) giving rise to a Hertzian radius of 68 μm as a reference. These
curves are also shown in Figure 5.32.

Two other factors must be determined before one can use this interferometer to
measure oil film thicknesses. These are (a) the phase change occurring at the
chromium semi-reflecting layer and (b) the change in refractive index of the
lubricant with pressure from that of the atmosphere up to about 700 MPa. For
more details, the reader is referred to the literature on this subject.

Let us now return to our discussion of the results obtained by Wedeven *et al.*

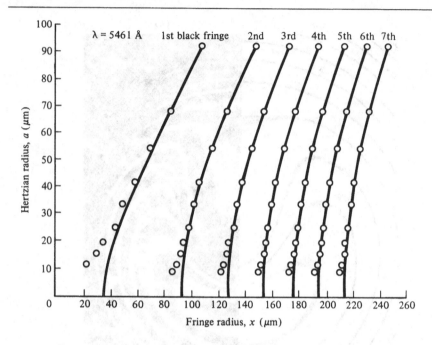

Figure 5.32 *Graph of the Hertzian radius a plotted against the radius x of the first seven dark fringes for the interferometric system shown in Figure 5.30 (a) without motion and with air replacing the lubricant*

(1971) in their experiments to deduce oil film thicknesses. Figure 5.33 shows the measured film shapes for a flooded ball-on-flat geometry. This figure was obtained using a high-speed photomicrographic technique, under conditions which were considered to simulate a ball-bearing. The film shape in the direction of motion showed a constriction in the exit region. A similar constriction is also found in line-contact geometries. In the *transverse* direction, however, Wedeven *et al.* (1971) have shown that the minimum lubricant film thickness $(h_{\min})$ is equal to about 0.27 μm, which is significantly less than the central film thickness $(h_{\min})_0$ of about 0.5 μm. By plotting $(h_{\min})_0$ against (h_{min}) Wedeven *et al.* obtained a linear relationship, as follows:

$$(h_{\min}) \approx [(h_{\min})_0 - 0.114]\,\mu m/1.40 \tag{5.46}$$

This is an important practical relationship, since $(h_{\min})$ must surely be relevant to the local breakdown of the lubricant and hence lead, eventually, to contact fatigue and hence pitting.

5.3.3 The thickness of solid lubricant contact films formed on metal substrates

(i) Introduction

In Section 5.2.6, it was shown that the contact films formed by sliding electrographite brushes on copper slings could be removed electrolytically and

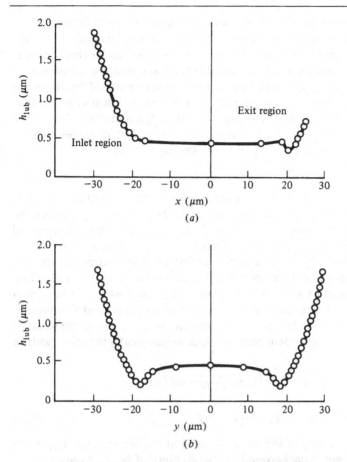

Figure 5.33 *Measured oil film shape (a) along the centre line in the direction of motion and (b) in the transverse direction, at the location of minimum film thickness Wedeven et al., 1971)*

then mounted on an electron microscope grid for examination by electron diffraction. It was mentioned, without proof, that the thicknesses of the contact films formed on the rough copper surfaces were about twice the surface roughness of the copper substrate. This statement was based on measurements of the contrast of the transmission electron micrographs obtained from contact films electrolytically removed from the copper. In this sub-section, we will briefly discuss the theory behind the contrast of electron micrographs and then show how it has been used to deduce the thicknesses of graphitic contact films formed on copper.

(ii) The contrast of transmission electron micrographs of amorphous specimens

The term 'amorphous' is intended to cover a wide range of specimens from truly amorphous thin films (such as the evaporated carbon films used for replicas) to pseudo-crystalline thin films (such as vacuum-evaporated metal films in which the crystallite size is often less than about 100 Å). Certainly, the graphitic contact films formed on copper could be considered 'amorphous' under this criterion, since electron diffraction experiments reveal crystallite sizes approximately 40 Å (Quinn, 1963, 1964). Any specimen in which a large percentage of the atoms are effectively in *random* positions with respect to each other, will scatter an electron beam *isotropically*, that is, without any appreciable diffraction effects. We will be concerned with the contrast of electron micrographs of just such specimens.

It is usual to define contrast (C) be the following relation:

$$C = \ln(I_0/I_c) \tag{5.47}$$

where I_0 is the intensity of the electron beam incident on the specimen and I_c is the intensity of the electron beam emerging from the specimen. I_c includes the undeviated transmitted beam and parts of the incident beam that are *scattered through small angles*, that is angles less than the objective lens aperture.

If we define the visibility (V_f) of a particular feature of an electron microscope image as the difference in contrast between that part of the image corresponding to the feature (C_f) and its adjacent background (C_B), then we would have zero visibility if the contrast of the feature equalled that of its background. Conversely, the visibility would be infinitely great if the feature could be seen as the result of *complete* scattering of the incident beam through angles greater than the aperture angle. Thus:

$$V_f = C_f - C_B = \ln[I_0/(I_c)_f] - \ln[I_0/(I_c)_B]$$

i.e.

$$V_f = \ln[(I_c)_B/(I_c)_f] \tag{5.48}$$

where $(I_c)_f$ is the intensity of the electron beam at the feature and $(I_c)_B$ is the intensity at the adjacent background. If the visibility of features separated by distances well within the resolution limit is *poor*, then it will *still* be difficult to achieve high resolution, due to the practical difficulties involved in focusing. Hence the importance of contrast, and the factors affecting contrast, in the use of the electron microscope at high resolution. Let us now discuss the contrast of electron micrographs, based on the paper by Halliday and Quinn (1960).

Consider a beam of electrons of intensity (I_0) passing through a thin film of thickness d_t. Let K_S be the fraction of the electrons singly scattered into *all angles*, *by all scattering processes*, per unit specimen thickness. Consider an elementary layer of thickness (dx), at a distance x from the original face of the specimen, and at which the electron beam intensity is $[I(x)]$. Now the change in intensity (dI_x) in passing through the lamina is given by:

$$dI(x) = -K_s[I(x)]\,dx \tag{5.49}$$

If we do not allow for *any* scattered electrons returning to the beam, we get an expression for the intensity $(I_c)_0$ transmitted throughout the whole layer (of thickness d_t), namely:

$$(I_c)_0 = I_0 \exp[-(K_s)(d_t)] \tag{5.50}$$

K_s is given by:

$$K_s = \frac{1}{\lambda_T} = \frac{1}{\lambda_e} + \frac{1}{\lambda_i} = \frac{1}{\lambda_e}\left(1 + \frac{\lambda_e}{\lambda_i}\right)$$

that is

$$K_s = \frac{1}{\lambda_e}\left(1 + \frac{\sigma_i}{\sigma_e}\right) \tag{5.51}$$

where λ_e, λ_i are the electron mean free paths for elastic and inelastic scattering, σ_e and σ_i are the elastic and inelastic electron scattering cross-sections, and λ_T is the mean free path for total electron scattering. λ_e can be found from the transparency thickness $(\rho\lambda_e)$ versus accelerating voltage curve of Lenz (1954) as shown in Figure 5.34(a). The ratio (σ_i/σ_e) can be found from the (σ_i/σ_e) versus atomic number (Z) curve, also of Lenz (1954), as shown in Figure 5.34(b). ρ is the density of the specimen.

Applying Equation (5.50) to an electron beam passing through a collection of atoms in an amorphous specimen of thickness d_t will not give us the intensity (I_e) emerging from the exit side of the film, because that equation does not take into account the electrons scattered through small angles, but which still are able to pass through the objective aperture and reach the final image.

In order to allow for small-angle scattering, let us define (p_s) as the fraction of the electrons singly-scattered into small angles per unit specimen thickness, that is into angles less than the objective aperture angle. The intensity reaching the lamina at x (namely $I(x)$), in which all the previous scatters have resulted in scattering through angles greater than the aperture angle, is given by Equation (5.50), so that:

$$I(x) = I_0 \exp\left[-(K_s)x\right] \tag{5.52}$$

Hence the intensity scattered through small angles on passing through the lamina of thickness dx at x is $(dI'_K)_x$ given by:

$$(dI'_K)_x = (p_s)(K_s)[I(x)](dx) \tag{5.53}$$

In general, only a proportion $(dI_K)_x$ of these electrons will pass through the objective aperture into the final image, because some of them will be scattered into large angles between the lamina and the exit side of the specimen. It can be shown that all electrons which are singly scattered into small angles stand the same chance of going through the aperture when re-scattered, as those being scattered for the first time. Hence we must substitute $[K_s - (p_s)(K_s)]$, that is $K_s(1-p_s)$ for K_s when we apply the experimental decay formula [Equation (5.52)] for evaluating $(dI_K)_x$, the *intensity scattered by the lamina into small angles* and which eventually ends up emerging from the exit side of the specimen

$$(dI_K)_x = (dI'_K)_x\{\exp - [K_s(1-p_s)(d_t - x)]\} \tag{5.54}$$

Integrating Equation (5.54) for all such laminar, we get

$$I_K = I_0 \exp\left[-(K_s)(d_t)(1-p_s)\right]\{1 - \exp\left[-(K_s)(p_s)\right]\} \tag{5.55}$$

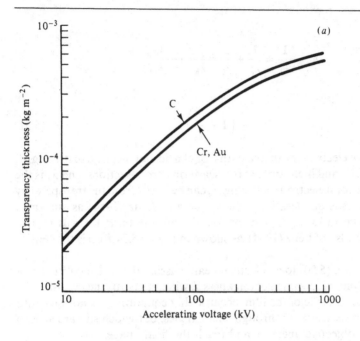

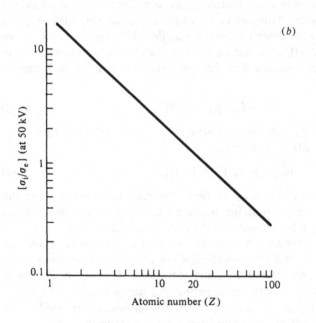

Figure 5.34 *(a) Transparency thickness ($\rho\lambda_e$) versus accelerating voltage (Lenz, 1954); (b) (σ_i/σ_e) versus atomic number (Z) at 50 kV (Lenz, 1954)*

Table 5.12. *Experimental* (p_s)-*values* $[(p_s)_{expt}]$
for common electron microscope materials (Halliday et al., 1960)

Material	Z	$(p_s)_{expt}$
Ge	32	0.56
Cr	24	0.57
Cu	29	0.67
Pd	36	0.70
U	92	0.76
Pt	78	0.79
Ag	47	0.83
Au	79	0.85
SiO	11 (Mean)	0.86
Al	13	0.89
Be	4	0.90
C	6	0.95

Effective aperture angle approximately 10^{-3} rad.

where I_K is the total number of electrons scattered through small angles on passing through the film of thickness d_t. Now we know that:

$$I_c = I_K + (I_c)_0 \qquad (5.56)$$

Hence, from Equations (5.50), (5.55) and (5.56) we obtain:

$$I_c = I_0 \exp - \{(K_s)(d_t)[1 - p_s]\} \qquad (5.57)$$

This equation forms the basis of *any* evaluation of thickness or density from the contrast C, since we know that:

$$C = \ln(I_0/I_c) = (K_s)(d_t)(1 - p_s) \qquad (5.58)$$

If we have a specimen of constant density ρ, then we can deduce the thickness *directly* from the contrast, provided we known K_s and p_s. K_s can be evaluated from Equation (5.51), together with Figures 5.34(a) and (b). On the other hand, p_s is not so readily ascertained from basic scattering theories. In fact, a reliable theoretical value of the proportion of electrons scattered through small angles still awaits a thorough revision of current electron scattering theory. Until a valid expression for p_s is forthcoming, the best way to use Equation (5.58) is to obtain appropriate p-values from experiments with known materials of known thicknesses. Table 5.12 gives the experimental p_s-values $[(p_s)_{expt}]$ of some of the more widely used materials in electron microscopy. It is clear that there is no obvious connection between atomic number Z and $[(p_s)_{expt}]$.

Let us now discuss how Quinn (1971) used the above described theory of electron micrograph contrast, together with the experimental value of $(p_s)_{expt}$ equal to 0.95 for carbon, to determine the thicknesses of the contact films forming the subject of his papers on the electrographite/copper tribo-system (Quinn, 1963 and 1964).

The analysis of lubricant films

(iii) The application of the theory of contrast of electron micrographs to the measurement of contact film thicknesses

In order to obtain representative images of the contact films, it was necessary to use only the condenser lens of the electron microscope, with the objective and projector lenses turned off and their respective apertures removed (see Figure 2.13), to obtain a focused point-source a few millimetres in front of the contact film (in its usual position for electron microscopy). This is the best way to obtain a very low magnification (about × 10) using an electron optical system. Such a low magnification image will tend to 'average out' the variations in specimen thickness. The effective aperture of the instrument in this mode of operation is about 10^{-3} rad. The accelerating voltage was varied from 160 down to 61 kV in four stages. At each accelerating voltage, micrographs were taken of each contact film, on the same photographic plate, together with other exposures taken for the purposes of deducing the characteristic curve between the density of blackening D and the exposure Σ of the plate at that voltage. Σ is defined as the product of the intensity of radiation causing the blackening of the plate times the time t of exposure. The characteristic curves for the four photographic plates are shown in Figure 5.35. It is clear that considerable error would be involved if one assumes that D is proportioned to Σ, since none of the curves is parallel to the 45° line drawn in Figure 5.35 as a guide to interpretation. It is possible to have an instrument which gives the electron intensity in a more direct manner than through the intermediary of a photographic film or plate. Nevertheless, this method is so readily adapted to measuring thickness at the same time as the electron micrograph is obtained, that it is probably worth the effort involved in constructing a characteristic curve for each photographic film or plate. Provided developing and fixing strengths and times are standardized, and the same batch of films or plates are used, there is no reason why one characteristic curve should not cover the whole batch.

The method for measuring thickness (d_t) from the contrast (C) is simply to plot the experimental contrast versus (K_s) for the various voltages between 61 and 160 kV. According to Equation (5.58), this graph should give a straight line with a slope equal to $[(d_t)(1 - p_s)]$. Assuming $p_s = (p_s)_{expt} = 0.95$ (from Halliday and Quinn, 1960), this means the slope of the line will give $(0.05d_t)$, from which d_t (the film thickness) is readily ascertained. In order to deduce C_{expt} from the experimental values of the exposure Σ, the density of blackening was measured for a region of the electron microscope grid in which no specimen intervened between the electron beam and the photographic plate. This enabled a value of Σ_0 to be obtained (from the appropriate characteristic curve) which is proportional to $(I_0 t)$, the product of incident intensity and the time of the exposure (assuming that the reciprocity law between intensity, time and exposure still holds). The density of blackening was then measured across several electron microscope grid squares containing uniform pieces of film and an average value of Σ_f thus determined. It was sometimes necessary to subtract a general background (Σ_b) from this (Σ_f) to give the required (Σ_c) values. (Σ_c) is proportioned to (I_c), where the factor of proportionality is the same as for (I_0). Table 5.13 gives a summary of

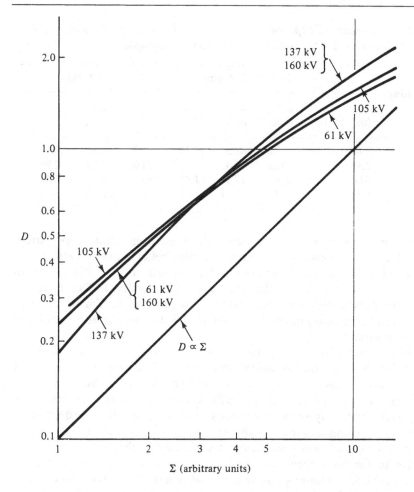

Figure 5.35 *Characteristic curves for the photographic plates used in the contrast experiments*

the (Σ_c) and (Σ_0) values obtained from the contact films removed from smooth (0.05 μm CLA) and rough (0.76 μm CLA) copper surfaces, together with the appropriate K_s-values. The units of (Σ_c) and (Σ_0) are arbitrary, but are consistent for any particular voltage. From this table, together with Equation (5.58), it can be deduced that the thickness of the film formed on the 0.05 μm CLA copper surface was (0.81 ± 0.18) μm, whereas the thickness of the contact film on the 0.76 μm CLA copper surface was (1.92 ± 0.72) μm. It is tempting to speculate that, for the *rough* surface, the electrographite crystallites removed from the carbon brush tend to first fill the 'empty' volume beneath the peaks of the copper surface, that is the contact film thickness is just a little below the peak-to-valley roughness (R_t) – see Section 2.3.2. Although not actually measured, R_t will obviously be about $2(2)^{1/2}$ (CLA), which is 2.15 μm. For the smooth surface, R_t was about 0.14 μm, which is only about one-sixth of the contact film thickness. In view of the

Table 5.13. *Summary of experiments to measure the effect of changing electron accelerating voltage on the contrast of electron micrographs*

Accelerating volts (kV)	K_s $(m^{-1}) \times 10^7$	0.05 µm CLA film			0.76 µm CLA film		
		$\Sigma_o/$ time	$\Sigma_c/$ time	C	$\Sigma_o/$ time	$\Sigma_c/$ time	C
160	2.16	12.8	4.05	1.15	14.2	0.64	3.1
137	2.46	31.0	10.0	1.12	23.0	1.34	2.96
105	5.60	108	17.6	1.81	86	2.8	3.42
61	8.58	11.6	0.89	2.57	6.2	0.04	5.09

extremely long running times to which these electrographite/copper tribo-systems had been subjected, this result indicates better bonding between the electrographite layers within the contact film formed on the initially smooth copper surface compared with that from the rough surface. In fact, if we can believe the values of thickness calculated from the contrast, the film on the rough copper surface is probably not an equilibrium film in the same sense as the film on the smooth surface.

The remark made in the last paragraph about the credibility of the thickness calculated from the contrast measurements obtained in the electron microscope needs some further discussion. We see that, for the more usual surface finish found in copper slip rings (~ 0.05 µm CLA), the thickness is calculable to an accuracy of $\pm 22\%$. This is relatively low accuracy compared with (say) the more destructive method of ion etching the surface in an Auger electron spectrometer. It is however its less *destructive* and more *convenient* nature that makes the contrast method so attractive to the researcher who has access only to a transmission electron microscope. Time and time again one finds that, in a research situation, it is often better to be satisfied with a very rough estimate of a parameter which one knows can never be estimated better than about $\pm 5\%$, even with the most sophisticated apparatus available. The rough estimate often enables the researcher to proceed with his/her research apace, leaving the 'loose ends' to be tidied up later! Researchers cannot afford the luxury of being perfectionists, and this is especially true of researchers into tribology.

5.3.4 Concluding remarks

In this sub-section, we have concentrated on (a) optical interferometry and its use for providing experimental support for the elastohydrodynamic lubrication theory, and (b) contrast measurements of electron micrographs of contact films formed in tribo-systems subjected to solid lubrication. Lubrication is all about the protection of sliding surfaces by the formation of surface films, whether they be liquid or solid lubricants deliberately added to the tribo-system, or solid lubricious films formed by interaction between the tribo-elements and the ambient atmosphere (i.e. oxide films and extreme-pressure and anti-wear films).

If we can deduce the thickness of these protective films, we can estimate how long their protection will last. Chapter 8 is devoted entirely to oxidational wear, a form of wear in which the metal of the specimen and the oxygen in the air interact with others at the real areas of contact to form oxides, some of which are more protective than others. Knowledge of oxide thickness, measured in situ by a scanning electron microscope with tilt facilities, has enabled the oxidational theory of mild wear to be developed to such a stage of sophistication that it is now accepted as one of the more important mechanisms of wear (see Lim and Ashby, 1987).

Other methods for measuring oil film thickness have been used. Crook (1961a,b) was one of the first people to use the electrical capacitance (between loaded rollers and flat steel pads riding on the oil films carried away from the conjunction of two rolling discs) as a means of measuring the oil film thickness in elastohydrodynamic lubrication. Kannel, Bell and Allen (1965) were among the first researchers to obtain direct measurements of pressure, using a manganin transducer instead of the electrode used by Crook (1961a). They could not detect the pressure spike mentioned earlier in this chapter. Furey (1966) used electrical resistance to determine the proportion of metallic contact between a fixed metal ball and a rotating cylinder. He did *not* deduce the thickness of the lubricant from the existence of the oil film (which is a difficult task anyway!) Sibley and Orcutt (1961) used X-ray absorption to measure lubricant film thickness while the rollers were actually moving. The method can only give average oil film thicknesses. Kannel *et al.* (1965), who also used the X-ray technique for measuring oil film thickness, concluded that the technique is not very sensitive to small changes in oil thickness along the contact.

5.4 The analysis of surface films formed in the presence of anti-wear additives

5.4.1 Introduction

In Section 5.1.1 it was suggested that an anti-wear additive is one which *increases* the initial seizure load in the conventional sliding four-ball test (involving one-minute runs with standard ball-bearing steels) compared with either a straight mineral oil or a mineral oil containing an extreme-pressure additive. It may be recalled that it is *often* found that a good extreme-pressure additive (which is characterized by high weld loads) has an initial seizure load not very different from that obtained with a non-additive oil. Most of the so-called anti-wear additives used in service, for example the zinc dialkyldithiophosphates (ZDDPs), will perform in the sliding four-ball test in the manner described at the beginning of this paragraph. Since the ZDDPs have been the dominant anti-wear additives used in automobile oils over the past 30 years (Rounds, 1985), and because more physical analysis has been carried out on these additives than any other, they are obvious candidates for discussion in this part of the chapter.

There is another term which is used to describe an additive's ability to reduce wear when introduced into a tribo-system, and that is 'lubricity'. It is normally used in connection with certain additives that are used in engine fuels, in particular, aviation fuels. Clearly, a lubricity additive is also an anti-wear additive. The main difference between these terms probably arises from the fact that fuel lubricity is generally measured in a dwell-tester (which is a friction testing device) whereas anti-wear behaviour in a lubricant blend is normally measured in a sliding four-ball tester (which is a wear testing device). In view of the numerous physical analyses of the protective mechanisms of lubricity additives for aviation fuel, this will be the subject of the final sub-section of this chapter.

5.4.2 Analysis of surface films formed during the anti-wear activity of zinc dialkyldithiophosphate (ZDDP)

(i) Introduction

The use of zinc dialkyldithiophosphate (ZDDP), as an additive to lubricating oils to reduce engine wear and the oxidation of the oil, is almost universal. Its popularity as an anti-wear/anti-oxidant lubricant additive arose as a result of empirical tests in service, and in the laboratories of the oil companies. The mechanism of the anti-wear activity of ZDDP has been related to an initial decomposition, followed by a reaction of the decomposition products with the surface to form a surface coating on those parts of the tribo-elements which actually come into contact, namely the real areas of contact. *Elemental analysis* of these surface coatings has been the subject of many investigations, using a variety of techniques, to show that the atomic ratios for zinc, phosphorus and sulphur found on the surface are significantly different from those for the original additive. Typical values for the worn surface are 1Zn:1.5P:1S, compared with 1Zn:2P:4S for the original additive. For a complete understanding of the mechanism of anti-wear protection, however, it is necessary to know the *compounds*, rather than the elements that are present in the surface coatings. The difficulties in identification are due

(a) to the extremely thin nature of the surface layers,
(b) to the fact that these coatings are often amorphous (so that X-ray or electron diffraction cannot give the required information) and
(c) to the problems involved in actually applying many techniques of analysis to worn surfaces. Indeed, many proposed reaction schemes for anti-wear additives are based on static immersion tests in which layers are formed on metal plates, wires or powders by heating them in additive solutions.

Although these static immersion tests produce thick films on more suitably-shaped specimens, it has been shown (Bird and Galvin, 1976) that very different compounds are formed from the same additive solution in rubbing end immersion tests.

Some of the earliest work on the analysis of anti-wear films formed (on steels) in the presence of ZDDP was carried out using radio-active additives in engine tests by Larson (1958). He detected surface films in engines in which the zinc content was enhanced in comparison with the phosphorus and sulphur contents, and the phosphorus content enhanced with respect to the sulphur. Loeser, Wiquist and Twiss (1959) carried out both immersion and engine tests. Their engine tests showed a much enhanced phosphorus content, namely

Zn :P :S = 1 :10 :0.7 (instead of the original Zn :P :S = 1 :2 :4). Furey (1959), who also carried out engine tests, concluded that it was the phosphorus (and not the zinc or sulphur) that was the key to the effectiveness of such additives. He suggested that phosphorus was present in several forms, namely as adsorbed additive, thermal decomposition products, lacquers or varnishes and metal--phosphorus reaction products. Forbes, Allum and Silver (1968) using electron-probe micro-analysis estimated the relative abundance of Zn, P and S to be 1 :1.9 :0.31 in what they define as the anti-wear region. Their definition of an anti-wear additive is one that performs well for 30 min in the four-ball machine at 147.1 N (the load of 15 kg mass). This is a very arbitrary definition, but it does seem to have some relevance to the service performance of so-called anti-wear additives, such as ZDDP. Coy and Quinn (1975), using the definition of the anti-wear region given in Section 5.1.1, found ratios of Zn, P and S of 1 :4.9 :1.4. They found that, in the very thin films formed in the anti-wear region, the elements were unevenly distributed, the Zn and P tending to occur together and the S concentrating in other areas.

One can see from the previous paragraph that, although one can make intelligent guesses at what form the composition of the surface layers *could* take, very little actual analytical evidence has been produced regarding the compounds actually formed on the surfaces that have been provided with anti-wear properties, when run in the presence of a ZDDP additive blend.

The best method for deducing the chemical composition of anti-wear surface films is through analysis by X-ray photoelectron spectroscopy (XPS), since this method can, under favourable circumstances, not only allow positive identification of the presence of an element, but also of its state of combination, for example sulphur can be shown to be present as a sulphide or a sulphate. By way of illustrating the power of XPS for studying anti-wear additive protection mechanisms, we will discuss in some detail, the pioneering work of Bird and Galvin (1976) who used XPS to study the surfaces formed in the presence of ZDDP. Although other investigators have also used XPS to analyse anti-wear surface films (e.g. Baldwin, 1975, Georges *et al.*, 1979), the work of Bird and Galvin (1976) deserves to be highlighted in a book devoted to the application of physical techniques to tribology. For completeness, we will also give a summary of the recent position as regards ZDDP as an anti-wear additive given in the paper by Jahanmir (1987). This paper was mainly concerned with surface analysis by Auger microprobe.

(ii) X-ray photoelectron spectroscopy

It has already been mentioned that Bird and Galvin (1976) showed how unreliable immersion tests can be when dealing with a complex anti-wear additive, such as ZDDP. Hence, this summary of their work will tend to concentrate on their rubbing tests. These tests were very unconventional, since they involved manually rocking the flat end of an EN31 steel roller backwards and forwards 12 times/min against the periphery of a 5 cm diameter × 1 cm thick disc of EN31 steel rotating at 200 rev/min (i.e. 0.52 m/s) for a test period of 5 min. The load was varied between 196 and 981 N. The reason for this somewhat unusual

Table 5.14. *Spectral data on reference materials relevant to the anti-wear behaviour of ZDDP (from Bird and Galvin, 1976)*

Material	Binding energies (in eV)				
	S 2p	Zn 3s	P 2p	O 1s	Fe 3p
Commercial ZDDP	162.6	140.6	133.6	532.9	
Pure ZDDP	162.8	140.4	133.7		
Ferric phosphide			129.9		54.8
Zinc orthophosphate		141.1	134.2	532.1	
Zinc phosphide		140.1	128.9		
Ferrous sulphate	169.6			532.4	55.9
Ferric sulphate	169.7				58.0
Ferrous thiosulphate	168.7 164.1				56.4
Zinc sulphate	169.8	141.1		532.7	
Zinc sulphite	167.9	140.8		532.1	
Ferrous sulphide	162.8				55.8
Zinc sulphide	162.9	140.1			
Ferric oxide				532.5 530.4	55.9
Zinc oxide		139.6			
Cleaned unused roller					55.5
					53.5
Freshly-abraded steel				531.6	55.2,
					52.4
Metallic zinc		139.6			
Ferric orthophosphate			134.0	531.7	57.4
Ferric pyrophosphate			133.7	531.6	55.0

test method was to provide a sufficiently large area (1 mm × 10 mm) of film to fill the area being examined by XPS. Modern instruments have a somewhat improved ability to locate on small areas, but nowhere near the resolution of the Auger microprobe. Hence, the complementary nature of XPS and Auger electron spectroscopy (AES), especially if the latter has a scanning facility.

The spectrometer used by Bird and Galvin (1976) was a 300 W instrument with AlK$_\alpha$ exciting radiation. The binding energies were calibrated with reference to the 1s line of carbon, taken to be equal to 285 eV. These authors were very meticulous regarding the effect of the small amount of sulphur in the EN31 steel (approximately 0.01 wt% sulphur) or even sulphur from atmospheric contamination. Both could be important, compared with the amounts of sulphur put down by the additives. They prepared blank experiments with rollers prepared in the absence of additives and found evidence of the presence of sulphur in the form of sulphate and sulphide, namely two 2p peaks between 168.2 and 168.8 eV binding energies. Bird and Galvin (1976) produce a very useful table of spectral data on reference materials which is reproduced in Table 5.14. To give the reader some idea of the complexities involved in analysing anti-wear films formed in

Table 5.15. *Spectral observations from the surfaces of EN31 steel run at various loads in additive blend of commercial ZDDP and white oil (from Bird and Galvin, 1976)*

				Binding energies (eV)		
Load (N)	S 2p		Zn 3S	P 2p	Fe 3p	O 1s
196	–	162.3	140.1	133.6	–	–	–
245	168.7	162.3	140.1	133.4	55.8	–	–
245	168.8	162.2	140.1	133.5	55.6	531.5	530
490	169.1	162.2	140.1	133.5	55.6	531.5	–
490	–	162.6	140.6	133.9	55.7	531.6	–
736	168.8	162.8	140.3	133.8	55.5	531.6	529.9
736	168.6	162.4	140.3	133.7	55.4	–	–
981	168.9	162.6	140.4	133.8	56.0	531.6	–
981	169.0	162.8	140.3	133.6	55.4	–	–

rubbing experiments, Table 5.15 has been compiled from Table 3 from Bird and Galvin (1976). It gives the spectral observations (in terms of binding energies alone) from the surfaces of EN31 steels worn at various loads in an additive blend of commercial ZDDP and white oil.

Let us first consider the possibilities for zinc compounds. Zinc orthophosphate, with its Zn3s peak at 141.4 eV, its P2p at 134.2 eV and its O 1s peak at 532.1 eV, is clearly not present in any of the surface films, whilst zinc phosphide with its P 2p peak at 128.9 eV is also unlikely to be present. It will be noticed that the Zn 3s peaks are not very dependent upon the chemical state from which they originated, so are of little analytical value. However, the $L_3 M_{4,5} M_{4,5}$ zinc Auger spectra were found to be more informative. By comparing spectral traces from zinc, commercial ZDDP, zinc sulphate, zinc oxide and zinc sulphide, Bird and Galvin (1976) concluded that the state of combination of zinc in one of the surface films was similar to that of the original additive. The shapes of the spectra were very similar, with maxima at 987.9 eV kinetic energy for the zinc in the original additive and 987.6 eV in the surface of the rubbed specimen. This is one of the few examples discussed in this book where Auger peaks in the XPS spectrum are used to augment the main analysis. Additional evidence can be obtained from the Fe 3p peak. Table 5.15 shows peaks around (55.6 ± 0.2) eV, which rules out ferric orthophosphate (with its Fe 3p peak at 57.4 eV) and makes ferric pyrophosphate very unlikely (with its Fe 3p peak at 55.0 eV). The P 2p and Fe 3p peaks of ferric phosphide do not relate very closely to the (133.6 ± 0.1) eV and the (55.6 ± 0.2) eV binding energies of the peaks obtained from the worn surfaces. We are left, therefore, with the conclusion that the electron binding energies shown in Table 5.15 are compatible with iron in the form of oxide, sulphate and sulphide.

From an analysis of the spectra similar to that given above, Bird and Galvin

(1976) conclude that *polymeric* films are formed when ZDDP additive blends are used in the anti-wear region of lubrication. The films are variable in composition, and contain the zinc, phosphorus and part of the sulphur of the original additive. There is no evidence, from XPS examination of the worn surfaces, for the presence of zinc phosphide, zinc phosphate, zinc sulphide, zinc sulphate, zinc oxide or iron phosphate. Let us now see what Auger electron spectroscopy (combined with scanning electron microscopy) has to tell us about the anti-wear activity of ZDDP.

(iii) Auger electron spectroscopy

In this part of the sub-section on wear reduction and surface layer formation by a ZDDP additive, we concentrate on the recent work of Jahanmir (1987). He used a steel block pressed against a rotating ring at loads ranging from 147 N up to 2943 N and at angular speeds from 10 to 500 rev/min. The diameter of the wear track is not given but, from the measuremens given in the paper, an upper limit can be determined for the linear speed, namely from linear speeds in a range from slightly below 0.74 m/s down to slightly below 0.015 m/s. With these conditions, Jahanmir (1987) was able to simulate different lubrication conditions from boundary to elasto-hydrodynamic lubrication. He used a constant oil temperature of 100 °C and ran each test for 2 h.

Jahanmir (1987) used a scanning Auger microprobe to analyse the chemical composition of the wear tracks on the test blocks. Large areas, with a diameter of 20–30 μm were analysed, both before and after sputter cleaning of the surface with an argon ion beam. Two typical Auger energy spectra of wear tracks with the base oil and the base oil plus ZDDP are shown in Figures 5.36(a) and (b), respectively. These spectra were obtained after approximately 2 min of sputtering, during which about 10–30 Å of surface material was removed. The author claims that

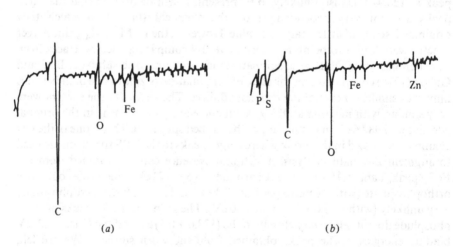

(a)   (b)

Figure 5.36 *Typical Auger energy spectra on wear tracks formed (a) in base oil only and (b) in base oil plus ZDDP additive (from Janhanmir, 1987)*

this removes the surface layers formed by 'contamination' with the atmosphere and the hydrocarbon of the oil. It would have been better to examine the original surface before doing this sputtering, just in case the top few Ångströms are relevant to the mechanism of anti-wear activity. Not surprisingly, the major elements in the wear track with the base oil were Fe, O and C. Further spluttering of the wear track showed the presence of an oxide layer on the surface of about 200 Å thickness, which was twice the oxide thickness off the track.

Figure 5.36(*b*) shows P, S and Zn, in addition to the elements found on the track formed with base oil only, on these AISI 52100 steel test blocks. High resolution spectra of P and S from the wear track indicate these elements were present as phosphates and sulphides. Off-track analysis showed only the steel composition and a thin (100 Å) layer. In order to determine the thickness of the surface layer and the dependency of chemical composition upon depth, Jahanmir (1987) sputtered off a further 400 Å. The depth profiles of O, Zn and P are plotted against depth in Figure 5.37(*a*), whilst the profiles of Fe, S and C are plotted in Figure 5.37(*b*), the object being to show the possible coexistence of some elements. It is suggested that, since O, Zn and P have the same depth distribution, this indicates that zinc phosphate is being formed (see Figure 5.37(*a*)). It is also suggested that there is probably iron sulphide present (as indicated by the S and Fe lines following each other in Figure 5.36(*b*) down to about 150 Å below the surface). From this evidence, Jahanmir concludes that ZDDP additives will react with steel surfaces to form a combination of different products down to several hundred ångströms below the surface. The reacted layer may be composed of phosphates, sulphides and oxides of zinc or iron.

In a confirmatory experiment carried out on wear particles collected from an experiment at 294 N, 100 rev/min and 100 °C for 2 h, the author showed the presence (using electron probe micro-analysis in the scanning electron microscope, i.e. EDX) of Zn, P, S and Fe. Since the typical EDX attachment cannot readily detect oxygen and carbon, it is possible that these elements were also present, thereby confirming that the wear particles do originate from the reacted surface layer.

(iv) Concluding remarks

The use of XPS and AES has produced evidence regarding the reaction of ZDDP with the surfaces of steel specimens in wear experiments designed to reproduce the conditions of extreme-pressure lubrication in the anti-wear regime. Although this evidence is not completely consistent, one can probably set most reliance upon the XPS analyses of chemical composition. Where AES scores is in the very fine beam which can be used to deduce composition in respect of the worn surface topography. Clearly, there is more work required before the anti-wear activity of ZDDP is completely explained. I have often said that there should be a theoretical explanation of how surfaces wear in the presence of an additive, which is similar to the oxidational wear model, but with P or S, (or whatever element is in the additive), combining with the oxygen to form an iron–phosphorus–oxygen compound or an iron–sulphur–oxygen compound. Such an explanation is still some time away!

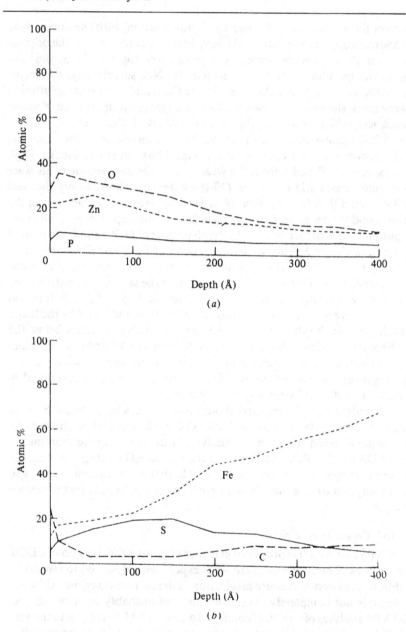

Figure 5.37 *Dependency of chemical composition of surface layer on depth in the presence of a ZDDP additive lubricant blend: (a) O, Zn and P profiles; (b) Fe, S and C profiles*

5.4.3 Analysis of surface films formed in the presence of lubricity additives in aviation fuel

(i) Introduction

The problem of low lubricity in aviation fuels began to show itself in the form of aircraft fuel system malfunctions in the mid-1960s. These systems either involved gear pumps and valves (in the USA) or piston pumps (in Europe). A consensus emerged with respect to the cause of the problem, namely, the need for extraordinary refinery processing of poor quality crudes to meet the stringent fuel specification requirements. In particular, due to the increased availability of high-sulphur crudes, it was necessary to subject them to severe hydrotreating so that the sulphur content could be reduced to an acceptable level. Unfortunately, hydrotreating will also *reduce* (or *remove completely*) the proportion of polar, surface-active constituents of the fuel which are believed to provide improved lubricating characteristics. The malfunctions mentioned above took the form of seizure between the piston and the bore, which sometimes caused slipper removal and piston fracture in the European fuel pumps with their cadmium-plated aluminium–bronze rotors. In the USA, the failures included sticking valves in some USAF aircraft and gear tooth scuffing in gear pumps operating in Jet B fuel.

A possible solution to the fuel pump problems was the addition to the fuel of approved corrosion inhibitors, which had been found to alleviate those lubricity problems associated with sticking valves. Santolene C (more recently re-named as Hitec E515), added in small quantities appeared to allow the piston and gear pumps to operate satisfactorily under unfavourable conditions. This solution was resisted on technical grounds, because of the detrimental effect on thermal stability and its effect on specification tests, and on economic grounds because of cost and problems with joint fuel supplies to aircraft operators.

An alternative solution would be to re-design the fuel pump to meet any difficulties arising from poor lubrication. This was done for the European piston pump, by fitting carbon liners to new units, and retrofitting old units where necessary. This solved the problem as far as the piston pump manufacturers were concerned. Nevertheless, corrosion inhibitors similar to Hitec E515 are still added to aviation kerosene, presumably to provide lubricity to other parts of the fuel system and, of course, to the actual fuel pumps (in the case of the gear pumps normally fitted to US aircraft).

Although the effects of the additive have been known for some time, the mechanisms responsible for the reduction in wear rates of the materials used in fuel pump tribo-elements are not understood. Aird and Forgham (1971) postulated that a boundary lubricating film is formed by adsorption of fuel constituents on the metal surfaces. In order to test a fuel for differences in its lubricity (i.e. its resistance to breakdown of the boundary lubricating film), they proposed a test that could be easily correlated with data from actual pumps. This test was *not* a wear test. It was, in fact, what has become known as the 'Lucas dwell test', after the manufacturer of the piston fuel pumps. It measures the durability of mono-molecular boundary lubricants. It uses a stationary pin loaded on to a flat rotating disc of surface finish similar to that of pump pistons. The surface durability of a film of fuel on the disc was measured by the number of disc rotations that occurred before the coefficient of friction reached a pre-determined level. The number of rotations was called the 'dwell number'.

The 'Lucas dwell test' has been the subject of much controversy over its relevance to actual pump performances and failures. Factors affecting this relevance are evaporation, adsorption by the metal surface and the choice of fluid for zero calibration. For instance, the dwell number of a pure hydrocarbon liquid is a measure of the time taken for evaporation to cause the thickness to drop to a level that allows significant asperity contact between the rubbing specimens. In the same way as for pitting contact fatigue in lubricating oils (see Chapter 7), the important factor is the D-ratio, that is, the ratio of the combined surface roughness to film thickness. Even if the pin and disc are changed for each fuel to be tested, there could well occur a significant change in surface roughness and an accumulation of wear debris, either during a test or as a result of the zero lubricity calibration test, thereby producing unpredictable contact conditions. Also, some kerosene fuels contain high molecular weight compounds, which are deposited on the disc surface during evaporation, producing a solid film, thereby giving a high dwell number.

The adsorption of pro-lubricity polar species, either naturally present in the fuel or in the form of additives such as Hitec E515, is inhibited by the short time available during spin-off. Under conditions where the surfaces are significantly roughened during the test, the response to polar compounds would be small. In the choice of fluid for zero lubricity one should make an allowance for the change of lubricity with time. The general consensus seems to be that the 'Lucas dwell test' suffers from lack of repeatability and reproductibility and is not suitable for a refinery control test.

Although the general consensus was not in favour of the dwell test as a lubricity test, it was also not really sure what was the best way to measure this quantity for such fluids as aviation fuel. Nevertheless, by producing a Stribeck curve of coefficient of friction (μ) versus Bearing Number ($\eta U/P$, (see Section 1.3.3) it should be possible to select those conditions under which boundary lubrication occurs for a conventional pin-on-disc wear machine, and thereby produce a model to explain the protection mechanisms of lubricity additives such as Hitec E515. Obviously the choice of metallurgy should reflect the tribo-elements used in the actual fuel pumps. This was the method chosen by Poole and Sullivan (1979, 1980). Since they were also among the first investigators to use Auger electron spectroscopy to analyse their surfaces, their work will form the main part of our next sub-section.

(ii) Physical analysis of surface films formed on surfaces of aluminium–bronze and steel sliding in the presence of aviation fuel

Poole and Sullivan (1979, 1980) used an additive consisting of 45% of a dimeric acid of the type:

$$CH_3(CH_2)_5-CH-CH-CH-CH(CH_2)_7-COOH$$

$$CH_3(CH_2)_5-CH \qquad CH-CH(CH_2)_7-COOH$$

$$CH-CH$$

plus 5% of a phosphate ester in kerosene or fuel oil. Although the aircraft fuel pump to be simulated consisted of KE961 steel running in cadmium-plated aluminium–bronze bores, it had been suggested that the cadmium coat is removed by the wear-in process. Hence, the authors used alumium–bronze pins running on a KE180 steel disc (this is a steel containing 13% chromium and is equivalent in most ways to the actual steel used in the pump). They measured wear rate and friction as functions of load and speed, that is they deduced the wear patterns for this combination, care being taken to ensure the conditions were in the boundary lubrication region of the Stribeck curve. It cannot be emphasized too strongly that the establishing of both the wear patterns and the Stribeck curve should be the fundamental and invariable way of carrying out *any* investigation into lubricated wear. The present unsatisfactory state of much of our knowledge of lubricated wear must surely arise from the lack of adherence to these fundamental rules! A striking feature of the experiments carried out by Poole and Sullivan (1979, 1980) was the use of several physical methods of analysis, namely Auger electron spectroscopy, ellipsometry, electron probe micro-analysis, scanning electron microscopy, X-ray diffraction and replica electron microscopy. Figure 5.38 shows the Stribeck curves for the fuel with additive for two different surface roughnesses, and for iso-octane. The curve for iso-octane was for a surface roughness of 0.2 μm CLA (see curve (*C*)). The coefficient of friction (μ) was always around the dry value of approximately 0.7. Compare this with curve (*a*), which was for the same surface roughness, but in the presence of fuel with the lubricity additive mentioned earlier. Hence we see the three distinct regions of hydrodynamic lubrication, mixed lubrication and boundary lubrication. From these curves, the authors were able to choose the loads and speeds that gave friction coefficients greater than 0.07, thereby ensuring the wear tests were carried out under boundary lubrication conditions.

Ellipsometry was used to estimate the thickness of any boundary film formed on the steel disc. This technique has not been used very much in tribological studies. Hence it was not discussed in Section 2.1, our introductory section on optical microscopy and interferometry. Ellipsometry has been used in corrosion studies by Kruger and Hayfield (1971), so it is probably relevant to measuring thin oxide films on metals. Poole and Sullivan (1979) used ellipsometry in a rather indirect way to estimate the thickness of the optically transparent surface film formed on their steel and aluminium–bronze surfaces. They took several additive mixtures from 100, 10, 1, 0.1 and 0.01% concentration, and formed films on silver-plated glass slides (by immersion and gentle heating, to drive off the non-polar components). The ellipsometer is a means of measuring the changes in amplitude

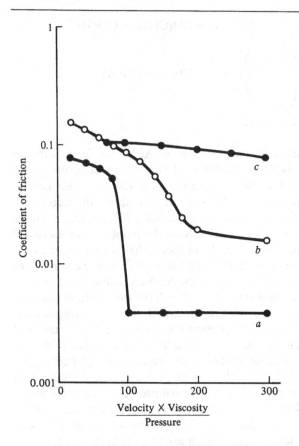

Figure 5.38 *Stribeck curves (a) for fuel with additive and a disc roughness of 0.2 μm CLA; (b) for fuel with additive and a disc roughness of 1120.6 μm CLA; and (c) for trimethylpentane (iso-octane) and a disc roughness of 0.2 μm CLA (from Poole and Sullivan, 1979)*

and phase of a light wave when it is reflected at an optically transparent or absorbing medium, such as a surface film. By measuring the changes in amplitude and phase between a clear silver–glass plate and the same plate with an additive film, Poole and Sullivan (1979) estimated that the surface film on their metal surface would be about 130 Å thick and that it would desorb at surface temperatures (T_S) of about 80 °C. This estimate must be treated with caution, however, since the same blend is likely to react differently with chemically-active elements such as the iron of the steel and the aluminium and copper of the bronze, compared with the silver on the glass plate.

Figure 5.39 is a typical wear rate versus load curve for aluminium–bronze pins sliding at a speed of 0.6 m/s (in the presence of fuel with additive) against steel discs of 0.6 μm CLA roughness (Figure 5.39(a)) and 0.2 μm CLA roughness (Figure 5.39(b)). Both curves show an initial increase in wear rate, up to about 80 N for the high surface roughness, and 125 N for the low surface roughness.

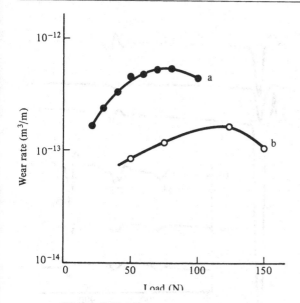

Figure 5.39 *The wear pattern (wear rate versus load curve), for 0.6 m/s and fuel with additive, of aluminium–bronze pins sliding on steel discs of (a) 0.6 μm CLA roughness and (b) 0.2 μm CLA roughness (from Poole and Sullivan, 1979)*

Using rough estimates of the 'hot-spot' temperatures (estimated from the frictional force and speed), it was found that the turnover point in the wear rate versus load curve occurred at the same hot-spot temperature, indicating that the reduction in wear rate was due to surface heating and its effect on either oxidation of the surface or an increased activity of the additive. It should be emphasized that the system was boundary lubricated over the entire load and speed range.

Figure 5.40 shows Auger Spectra for the worn aluminium–bronze surfaces. The important spectral trace is the one shown in Figure 5.40(c), the pin worn against the rougher disc at a load of 100 N. Here, we see peaks due to phosphorus and chlorine. Xenon sputtering, with 400 eV xenon ions at normal incidence and a pressure of 10^{-4} torr, of the pin surfaces revealed that copper, iron, carbon and sulphur were present in detectable quantities down to about 700 Å, but no phosphorus nor chlorine could be detected at any time during the sputtering process. What was most surprising, however, was the non-appearance of aluminium, since the aluminium–bronze pin contained at least 10% aluminium in its original composition. This apparent anomaly caused some discussion when their paper was presented at the ASME–ASLE Lubrication Conference in Kansas City, Missouri in October 1977. Because of this, and to complete their investigations, Poole and Sullivan (1980) carried out experiments with hydro-fined fuel only (instead of iso-octane) in order to deduce more about the action of the additive on the wearing of the tribo-elements of fuel pumps. They also subjected the worn surfaces and the wear debris to a thorough examination by physical analytical techniques.

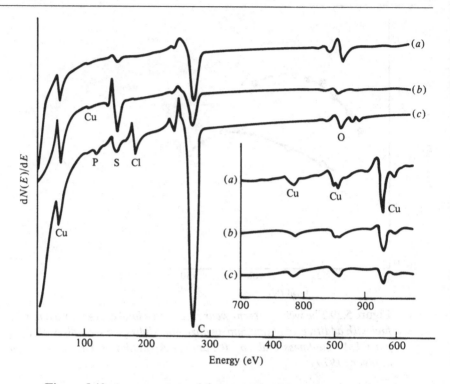

Figure 5.40 *Auger spectra of the worn aluminium–bronze pin surface run against steel in (a) iso-octane, (b) fuel with additive and 0.2 μm CLA roughness and (c) fuel with additive and 0.6 μm CLA roughness (from Poole and Sullivan, 1979)*

Using X-ray diffraction, these authors showed that the wear debris from an experiment with just the hydrofined fuel, gave interplanar spacings that probably were due to the presence of bronze particles only. The wear debris from the experiments with fuel plus the additive, however, were much more interesting. At 60 N, the X-ray diffraction analysis indicated the presence of aluminium ortho-phosphate and copper phosphate in the wear debris. At 100 and 125 N, the debris tended to change to aluminium phosphide. Cuprous oxide was present at all loads. Aluminium was present only at the lower loads whilst copper was present only at the higher loads. The bases of these interpretations are given in Table 5.16.

(iii) Concluding remarks

The problem of low lubricity of hydrofined fuels, has, to some extent, been solved by (a) the use of corrosion inhibitor as a lubricity additive and (b) by including carbon liners in the bore of the piston-type fuel pumps (where the problem first manifested itself).

The way in which the corrosion inhibitor acts as a lubricity additive has been partially revealed by the application of several physical methods of analysis to surfaces worn under boundary lubrication conditions in the presence of (a)

Table 5.16. *The interplanar spacings and possible identification of chemical constituents of wear debris formed in the wear of aluminium-bronze on steel in the presence of aviation fuel plus additive (at various loads)*

Interplanar spacings (Å)			
60 N	100 N	125 N	Possible identification
		4.94	?
4.05			AOP
3.37			AP
3.14	3.19	3.11	AOP, APD
3.04	3.04		?
2.92			CPA
2.83			CPA
		2.63	?
2.57			CPA
2.48	2.49	2.48	Cu_2O, AOP, CPD
2.33	2.29		Al
		2.17	Cu_2O
2.13	2.10	2.09	Cu_2O, Cu
2.03	2.02		Al, Fe, Ni
	1.93	1.93	CPD
	1.81	1.81	AOP, Cu
1.75			Ni, Fe
1.60		1.65	APD
1.54	1.52		Cu_2O
	1.45		Fe, Al
1.35	1.31	1.31	?
1.28	1.28	1.28	Cu, Cu_2O, Fe, Ni

Abbreviations: AOP, Aluminium ortho-phosphate; CPA, Copper (II) phosphate; CPD, Copper phosphide; APD, Aluminium phosphide.

aviation kerosene plus the additive and (b) aviation kerosene alone. In a recent paper, reviewing the work carried out on aluminium–bronze/steel interfaces with and without dimeric acid as the lubricity additive, Sullivan and Wong (1986) come to the following conclusions regarding this particular tribo-system:

(a) Oxidational wear is the dominant mode of wear for aluminium–bronze on tool steel under boundary lubrication conditions in the presence of aviation kerosene alone.

(b) With the dimeric acid present in the kerosene, there is a competing action between acid and oxygen molecules from the fluid and the surface, leading to a thinner or more patchy oxide film than that developed in additive-free kerosene. Hence, there is a slight pro-wear effect of the additive under relatively mild conditions.

(c) Under more severe conditions, when oxide film destruction and metal-to-metal contact would normally occur, the adsorbed fatty acid surface film affords sufficient protection to allow oxide films to be maintained, and hence prevent seizure.

From this we see that the major function of the additive is to preserve and maintain a protective oxide film. The time is obviously ripe for a theoretical approach to predicting oxidational wear under boundary lubrication conditions involving lubricity additives (in much the same way as for the more conventional anti-wear additives discussed in an earlier section of this chapter).

This ends our discussion of the analysis of lubricant films. We will now turn our attention to another important factor in the behaviour of tribo-systems, namely, the analysis of surface temperatures.

6 The analysis of surface temperatures in tribo-systems

6.1 Thermal aspects of sliding

6.1.1 Introduction

There have been many attempts to measure the temperature of the interfaces of sliding surfaces, both by direct and indirect measurements. Early direct measurements tended to involve the dynamic thermocouple, for example Shore (1925), Herbert (1926) and Furey (1964). Such measurements tend to be restrictive as regards the choice of sliding material combinations. They are also insensitive to the very rapid transients that occur during sliding. More recent direct measurements involve the use of optical photography through the transparent member of a sliding pair (Quinn and Winer, 1985) or the use of an infra-red detector (Meinders, Wilcock and Winer, 1984). We will describe both types of measurements in more detail in later sub-sections. The indirect measurements of surface temperature have tended to be a mixture of theory and experiment, for example see Ling and Pu (1964) and Archard (1959). Occasionally, temperature-dependent transitions in tribo-element phases have been used to give an indirect indication of the temperatures of sliding, for example see Quinn (1968), a paper in which the proportions of the various iron oxides in the wear debris were used to estimate the most probable temperature of formation of those oxides. We will describe the X-ray analysis of wear debris in a later sub-section (Section 6.2).

There seems to have been only a few attempts to deduce surface temperatures from measurements of heat flow. The first of these attempts appears in the paper by Grosberg, McNamara and Molgaard (1965). This was a theoretical paper, dealing with the rather special case of an unlubricated 'ring-traveller' moving at very high speed (30 m/s). In a later paper, Grosberg and Molgaard (1966–67), applied that theory to some high speed, light load experiments with a pin-and-disc wear machine, designed to simulate the conditions that occur in the ring-traveller tribo-system. In terms of Lim and Ashby's (1987) wear mechanism map for steels (see Figure 1.5) this velocity puts the work of Grosberg and Molgaard (1966–67) into either 'melt wear' or 'severe-oxidational wear'. It seems probable that, in fact, 'melt wear' was occurring but, since our main interest in this chapter lies in temperature determination, we will not differentiate between these two mechanisms of wear.

In the experiments carried out by Grosberg and Molgaard (1966–67), the division of heat (δ_{theory}) at the rubbing surfaces of the pin and the disc was measured by strategically-placed thermocouples along the length of the pin. This measured value was then compared with the theoretical value expected from a *surface model* consisting of N circular contacts of radius a, each with a thin layer of oxide ξ on the pin surface only. These authors showed that the best correlation (between experimental and theoretical values of the division of heat) is obtained if one assumes a value of N between 1 and 10, and a value of ξ around $0.5\,\mu\text{m}$.

Grosberg and Molgaard (1966–67) did not relate their values of N and ξ (obtained from the division of heat experiments) to the actual wear rates of their pins. This was because there was no suitable theoretical wear equation available for their high-speed sliding conditions. The oxidational theory of mild wear had only just been published (Quinn, 1962) and it was not until 1978 that an attempt was made to apply the heat flow analysis method of Grosberg and Molgaard (1966–67) to some oxidational wear experiments using low-alloy, medium-carbon steels sliding against themselves (without lubrication) at various loads between 5.8 and 29.4 N and a speed of 5.11 m/s (Quinn, 1978). Comparison with Lim and Ashby's (1987) wear mechanism map for steels, shows us that these conditions were on the border between mild wear and severe oxidational wear mechanisms.

In the following sub-sections we will therefore discuss (a) the analysis of heat flow between sliding tribo-elements in a tribo-system exhibiting oxidational wear, and (b) the application of heat flow analysis and the mild-oxidational wear theory to the sliding of that tribo-system. Finally, we will discuss (c) the potential of wear mechanisms and temperature maps for classifying and predicting sliding wear and flash temperatures.

6.1.2 Analysis of heat flow between sliding tribo-elements in a tribo-system exhibiting oxidational wear

We will restrict ourselves to the wear of steels, since this is where much of the definitive work in oxidational wear has been carried out. There seems to be no reason why the following should not apply to any metal that readily oxidizes during sliding.

Archard (1959) considered the heat flow between a moving and a stationary heat source, *without taking into account the effect of oxides on both the pin and disc of his tribo-system*. He had a very simple wear machine, with no facilities for actually measuring heat flow into the pin (or the disc). He also assumed that the general temperature of the surface of the pin $(T_S)_p$ was not different from that of the disc surface $(T_S)_d$, so that he could apply the conventional thermal resistance analogue for obtaining the temperature excess (θ_m) at the real areas of contact, over the general surface temperature (T_S) of each member of a sliding pair, so that:

$$\theta_m = \frac{t_p t_d}{t_d + t_p} \tag{6.1}$$

In Equation (6.1), $\theta_m = (T_c - T_S)$, t_d is the (fictitious) excess temperature of the disc contact areas, assuming (for the sake of discussion) that *all* the heat generated at the interface passes into the disc, and t_p is the (fictitious) excess temperature of a pin asperity, assuming (again for the sake of discussion) that all the frictionally-evolved heating passes into the pin. Neither assumption is correct, of course. These assumptions are only made with the object of deducing the excess temperature (θ_m) without any actual temperatures at all!

Archard (1959) deduced expressions for (t_d) and (t_p) which did not allow for the presence of oxide films on either the pin and disc asperities. Allen, Quinn and Sullivan (1986) modified Archard's (1959) expressions so as to include (ξ_d) and (ξ_p), the thicknesses of the oxide on the disc and pin, respectively, as follows:

$$t_d = \frac{(\alpha_{Pe})(p_m)(UF)}{W}\left(\frac{a}{K_{sd}} + \frac{\xi_d}{K_o}\right) \tag{6.2}$$

$$t_p = \frac{(p_m)(UF)}{W}\left(\frac{a}{K_{sp}} + \frac{\xi_p}{K_o}\right) \tag{6.3}$$

The symbol (α_{Pe}) is related to the Peclet number and takes into account the fact that the disc sees the pin as a moving source of heat. It has the form (Archard, 1959):

$$\alpha_{Pe} = 0.86 - 0.10\left(\frac{Ua}{X_o}\right) \tag{6.4}$$

where U is the speed of sliding at the pin, a is the radius of contact at a typical asperity and X_o is the diffusivity of the oxide (equal to $K_o/(\rho_o c_o)$, where K_o is thermal conductivity, ρ_o is the density and c_o is the specific heat capacity of the oxide film). p_m is the hardness of the metal substrate, F is the frictional force, the subscript (sd) relates to the property of the steel of the disc, (sp) to the property of the steel of the pin, (o) to the property of the oxide, and W is the normal load. Archard (1959) assumed that, for ($Ua/2X_o$) between 0.1 and 8.0, α_{Pe} varied linearly with that quantity, which is the basis for Equation (6.4) (see Rowson and Quinn, 1980). It should be emphasized that we must not use this equation for values of ($Ua/2X_o$) outside the range from 0.1 to 0.8.

If one wishes to deduce the excess temperature (θ_m) for the surfaces of tribo-elements in a tribo-system not instrumented to measure temperatures nor the division of heat (δ_{expt}) into at least one of the tribo-elements, then Equations (6.1), (6.2), (6.3) and (6.4) can be used, provided we know values to be assigned to the individual contact radius a and the oxide film thicknesses (ξ_d and ξ_p). These parameters are not readily accessible to the typical design engineer who wishes to know what sort of temperatures to expect at his (or her) sliding interfaces. There is an obvious need for a method for deducing θ_m in terms of measured or well-defined quantities. We will discuss the approach of Lim and Ashby (1987) to this problem in the final sub-section of this chapter. For the present, let us return to the approach involving the comparisons of theoretical and experimental division of heat and wear rate for tribo-elements experiencing mild-oxidational wear.

Allen, Quinn and Sullivan (1986) made no attempt to use Equations (6.2) and (6.3) to deduce the excess surface temperature. Instead, they decided to replace (UF) in Equation (6.3) with H_1, where H_1 is the rate of heat flow along the pin at the interface between the pin and the disc. Then, instead of the fictitious temperature (t_p), they used $[T_c - (T_S)_p]$ which is the excess temperature of the contact over the general temperature of the surface of the pin, and re-wrote Equation (6.3) as:

$$[T_c - (T_S)_p] = \frac{(p_m)(H_1)}{W}\left(\frac{a}{K_{sp}} + \frac{\zeta_p}{K_o}\right) \qquad (6.5)$$

Similarly, they modified Equation (6.2) to allow for the heat transferred to the disc $(UF - H_1)$ and hence obtained:

$$[T_c - (T_S)_d] = \frac{(\alpha_{Pe})(p_m)(UF - H_1)}{W}\left[\frac{a}{K_{sd}} + \frac{\zeta_d}{K_o}\right] \qquad (6.6)$$

These equations do not suffer from the difficulties incurred in the derivation of (θ_m) in Equation (6.1) which assumes (for its validity) that the general surface temperatures of both the pin and the disc are the same.

The measurement of H_1 is not an easy task. It can be done, however, if the pin and pin-holder are instrumented as shown in Figure 6.1. Rowson and Quinn (1980) have used this specimen geometry to deduce (H_1) and hence (δ_{expt}) from the relation:

$$(\delta_{expt}) = \frac{H_1}{UF} \qquad (6.7)$$

Essentially, these authors were dealing with steady state heat flow conditions, when the friction has settled down to its equilibrium value. They divide their analysis into two parts related to the insulated and exposed portions of the pin. In the insulated portion, it is assumed that the axial heat flow rate entering a cylindrical element must equal the axial heat flow leaving that element, together with the radial heat flow rate through the sides of the element. This gives H_3, as shown in Figure 6.1. In the exposed portion of the pin, it was assumed that the radial heat flow rate transferred from an element to the surroundings was given by an empirical equation involving the convective heat transfer coefficient (h_c) and the temperature difference $(T(x) - T_E)$, where $T(x)$ was the temperature at

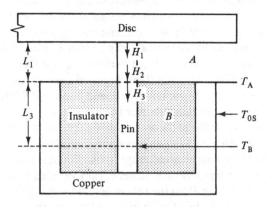

Figure 6.1 *Heat flow diagram for a pin loaded against a rotating disc*

distance (x) from the heat source (H_1) and (T_E) was the temperature of the air flowing past the cylindrical pin. To a good approximation, one could assume that (T_E) was not very different from (T_{OS}), the temperature of the copper calorimeter holding the insulated portion of the pin. Rowson and Quinn (1980) show that (H_1) is given by:

$$H_1 = \left\{ \frac{\pi(R_t)(K_{sp})}{(Z_h)} \ (T_A - T_{OS}) \sinh \left[\frac{L_1}{(R_t)(Z_h)} \right] \right\} + \left\{ H_2 \cosh \left[\frac{L_1}{(R_t)(Z_h)} \right] \right\}$$

(6.8)

where:

$$Z_h = \left[\frac{K_{sp}}{2(R_t)h_e} \right]^{1/2}$$

(6.9)

In these last two equations, H_2 stands for the heat flow rate entering the insulated portion of the pin, and is equal to H_3 plus the heat conducted away by the thermocouple measuring T_A. Obviously, H_3 depends on T_A, T_B and T_{OS}, (R_t) is the radius of the pin and (L_1) is the length indicated in Figure 6.1.

Rowson and Quinn (1980) also calculate the general surface temperature, $(T_S)_p$, of the pin as follows:

$$(T_S)_p = \left\{ (T_A - T_{OS}) \cosh \left[\frac{L_1}{Z_h(R_t)} \right] \right\} + \left\{ \left[\frac{Z_h H_2}{(K_{sp})\pi(R_t)} \right] \sinh \left[\frac{L_1}{Z_h(R_t)} \right] \right\} + T_{OS}$$

(6.10)

The general surface of the disc $(T_S)_d$ is not so readily calculable in terms of heat flow into the disc. Sufficiently accurate estimates of $(T_S)_d$ can be obtained by embedding a thermocouple in the disc at a radius just beyond the wear track (Allen *et al.* 1986).

For continuity purposes, we must assume that the temperature of the contact (T_c) is the same regardless of whether one uses an expression involving pin terms only (i.e. Equation (6.5)) or disc terms only (i.e. Equation (6.6)). Hence we can equate these two equations and thereby obtain a quadratic in a, the contact radius, as follows:

$$Aa^2 + Ba + C = 0$$

(6.11a)

where

$$A = \frac{0.1U(UF - H_1)(p_m)}{2(X_o)W(K_{sd})}$$

(6.11b)

and

$$B = -\frac{(p_m)}{W} \left\{ \frac{0.86(UF - H_1)}{K_{sd}} - \frac{H_1}{K_{sp}} - \frac{0.1U(UF - H_1)(\xi_d)}{2(X_o)(K_o)} \right\}$$

(6.11c)

and

$$C = -\frac{(p_m)}{W} \left\{ \frac{0.86(UF - H_1)\xi_d}{(K_o)} - \frac{H_1\xi_p}{(K_o)} \right\} + (T_S)_p - (T_S)_d$$

(6.11d)

Using H_1, from Equation (6.8), $(T_S)_p$ from Equation (6.10), $(T_S)_d$ from the thermocouple measurement near to the wear track on the disc, and assuming $\xi_d = \xi_p$, it is possible to solve Equation (6.11) for a, for a particular value of ξ, with K_o and X_o set equal to the values of these properties at the general surface temperature $(T_S)_p$ of the pin. One can then make the (reasonable) assumption that there are N equal asperity contact areas (of radius a) comprising the real area of contact, so that:

$$a = \left[\frac{W}{N\pi(p_m)} \right]^{1/2} \tag{6.12}$$

This gives a value of N. We can also use the solution of Equation (6.11) to obtain an estimate of T_c, from Equation (6.5). These are not necessarily the *true* values of ξ, N or T_c. Hence we calculate new values of K_o and X_o at this value for T_c, and repeat the whole procedure for new values of a, N and T_c, until two consecutive values of (T_c) are within 1% of each other. At this stage, a value of (δ_{theory}) can be calculated from:

$$\delta_{theory} = \frac{t_d}{t_d + t_p} \tag{6.13}$$

together with Equations (6.2) and (6.3) (remembering that we have assumed $\xi_d = \xi_p = \xi$). This can be compared with (δ_{expt}) (given by Equation (6.7)) and the whole interaction repeated until a reasonable comparison is obtained between (δ_{expt}) and (δ_{theory}).

Sullivan and Athwal (1983) have carried out the above-described iterative process on their data from the oxidational mild wear of AISI 52100 steel pins sliding with lubrication against discs of the same material at 2 m/s. The 'best' values which they obtained from their calculations are shown in Table 6.1.

We shall see (in later sub-sections) how these values compare (a) with those expected from analysis of wear debris (for T_c), (b) with those values expected from optical analysis of hot-spots (for N) and (c) with electron microscopy of worn surfaces (for ξ). For the present, we will merely point out that by heat flow analysis together with suitable thermocouple measurements, we been able to deduce values of these quantities that correlate reasonably well with the physical analysis of laboratory tribo-systems involving the dry wear of steels. Let us see what the theory of mild-oxidational wear, combined with heat flow analysis, can tell us about the microcosm known as the real area of contact.

6.1.3 The application of heat flow analysis and theory of mild-oxidational wear to the sliding of steel tribo-systems

Let us re-write the mild-oxidational wear equation (Equation (1.35b)) so that T_o is in units of degrees Celsius:

$$(w_{theory}) = \left[\frac{d(A_p)(A_{real})}{\xi^2 \rho_o^2 f_o^2 V} \right] \exp \left[-\frac{Q_p}{R(T_o + 273)} \right] \tag{6.14}$$

This expression was based on the assumption that no appreciable oxidation

Table 6.1. *Values of N, a, T_c, ξ and δ obtained (after iteration) from data generated at 2 m/s (using AISI 52100 steel specimens) (Sullivan and Athwal, 1983)*

Load (N)	N	a (μm)	T_c (C)	ξ (μm)	δ_{theory} (%)	δ_{expt} (%)
9.8	30	6.5	481	3.0	8.8	8.5
19.6	61	6.5	438	2.0	13.7	11.4
24.5	78	6.4	498	4.5	7.7	6.8
29.4	134	5.4	506	2.0	16.1	13.9
34.3	142	5.6	564	4.0	9.4	8.5
39.2	161	5.2	576	4.5	8.2	7.3
58.9	281	5.2	637	6.0	7.3	6.7
78.5	457	4.7	703	6.5	6.9	6.1
98.1	576	4.7	675	5.0	10.4	8.4

occurs when the wearing areas are 'out-of-contact' with the opposing surface. If oxidation of the 'out-of-contact' areas becomes significant, then the tribo-system is wearing by a 'severe-oxidational wear' mechanism, using the terminology of Lim and Ashby (1987).

It is reasonable to assume that the real area of contact during oxidational wear is determined by the hardness of the steel substrate (at the general surface temperature, T_S) through Equation (1.7), namely $A_{real} = W/(p_m)$. We also note that the oxide film thickness ξ appears explicitly in Equation (1.35b). T_o is generally assumed to be not very different from T_c (the contact temperature between the real areas of contact on both sliding surfaces). The radius a of each of the N circular contacts making up the real area of contact is implicitly included in the term for the distance d of a sliding contact, namely:

$$d = 2a = 2 \left(\frac{W}{\pi N p_m} \right)^{1/2} \tag{6.15}$$

Equation (6.15) arises from Archard's (1961) suggestion that the distance from complete conjunction between two circular areas of contact, to the point where the contact is just over, should be considered as a good estimate of the distance of a sliding asperity contact. Thus, the mild-oxidational wear equation is a rather complex function of N, ξ, a, T_c, and the operating parameters. We should not expect, therefore, that it would be an easy task to find the values of N, ξ, a and T_c that are consistent with (w_{theory}) and (δ_{theory}). As we shall see (in later subsections), our physical analytical techniques help us in confirming theoretically derived values of N, ξ, a and T_c, thereby giving credibility to one of the few *really quantitative* theories of wear.

One aspect of the mild-oxidational wear equation which has not yet been

Table 6.2. *Oxidation constants* (*derived by the static oxidation of iron* (*Caplan and Cohen, 1966*))

Temperature (T_o) range (°C)	<450	450 to 600	>600
Arrhenius constant (A_p) (kg^2/m^4–s)	1.5×10^6	3.2×10^{-2}	1.1×10^5
Activation energy (Q_p) (kJ/mol)	208	96	210

discussed is to decide what values should we use for the Arrhenius constant (A_p) and the activation energy for parabolic oxidation (Q_p) in that equation. If we examine the *static oxidation* literature for iron (not very different from low-alloy steels as far as oxidational behaviour is concerned), we find that Caplan and Cohen (1966) carried out experiments in which it is possible to deduce *three* ranges of parabolic oxidation, namely the α-Fe$_2$O$_3$ (rhombohedral) range for $T_o < 450$ °C, the Fe$_3$O$_4$ (spinel) range for $450° < T_o < 600$ °C, and the FeO (Wüstite) range for $T_o > 600$ °C. Table 6.2 details the oxidation constants as derived from Caplan and Cohen (1966).

Although Q_p only changes by a factor of two (and back again) when the temperature increases from below 450 to beyond 600 °C, the Arrhenius factor changes by eight orders of magnitude (and seven orders of magnitude back) for the same temperature range and the same oxidizing conditions. Sullivan, Quinn and Rowson (1980) suggest that the oxidational activation energies (Q_p) will remain the same in both static and tribo-oxidation, and hence statically-derived values may be applied to tribological situations, if the correct temperatures are chosen. The authors give possible physical reasons why this should be so. They also use the same reasoning to suggest that the Arrhenius constants will be very different in the tribological situation.

In order to deduce the tribological Arrhenius constants, Sullivan *et al.* (1980) chose a particular (starting) value for ξ (the oxide film thickness) and then evaluated a value for the theoretical division of heat (δ_{theory}) and the theoretical value of T_c (from Equations (6.1), (6.2), (6.3) and (1.36)), for a wide range of N values. If the computed value of (δ_{theory}) was within 1% of the value of (δ_{expt}) measured by those authors in their pin-on-disc experiments with a low-alloy steel (EN31), then the computer used those values of ξ, T_c and N to derive a set of (A_p) values which would be consistent with Equation (1.35b), written with d given by Equations (6.15), (A_{real}) given by Equation (1.7), T_c written for T_o, Q_p given by Table 6.2 (according to the particular value of T_o produced by the computer search) and, of course, theoretical wear rate (w_{theory}) replaced by the value for the experimentally-measured wear rate (w_{expt}).

The above approach was intuitive insofar as certain criteria were used for choosing a particular set of ξ, T_c and N. These criteria were:
(a) $0.5 \, \mu m < \xi < 15 \, \mu m$
(b) $1 < N < 2000$
(c) $T_S + 150 °C < T_c < 1000 °C$

Table 6.3. *Oxidation constants relevant to the mild-oxidational wear of EN31 steel*

Temperature (T_c) range (°C)	<450	450 to 600	>600
Tribological Arrhenius constant (kg^2/m^4–s)	10^{16}	10^3	10^8
Activation energies (Q_p) (kJ/mol)	208	96	210

The first criterion was based on electron microscope evidence (see Section 7.1) regarding the height (about $3\,\mu m$) of the plateaux of contact between sliding surfaces. The limits of ξ are arbitrarily centred around this value. It is also intuitively considered that d will be of the same order of magnitude as ξ, that is about 10^{-6} m. From Equation (6.15), this means that N will be less than 2000 for most of the situations involved. The third criterion arises from the X-ray diffraction evidence (to be discussed in Section 6.3) which indicated that, for similar conditions to those used by Sullivan et al. (1980), the oxidation temperature was about 200 °C above T_S for $T_o < 450$ °C. The criterion that T_c ($= T_o$) be less than 1000 °C was arbitrary. Recent work by Quinn and Winer (1985), however, indicates that perhaps the upper limit should have been 1200 or 1300 °C!

Apart from intuitively knowing that the tribological values of (A_p) should be *greater* than the static values given in Table 6.2, there was no guide as to the correct order of magnitude of A_p produced by the above-mentioned computer search technique. Fortunately, A_p seems to be rather insensitive to slight increases in the wear rate with increasing load (for a given oxide range), so that fairly good average values for the *tribological* Arrhenius constants for the three temperature ranges over which the three different iron oxides predominante, were able to be obtained. These (A_p) values (together with the appropriate Q_p values) are given in Table 6.3.

It will be noted that the Arrhenius constant, for the formation of FeO during mild oxidational wear of steel, is three orders of magnitude greater than the static oxidation value, whilst for Fe_3O_4 there are five orders of magnitude increases between static and tribological values. The dramatic increase of ten orders of magnitude in the Arrhenius constant for the formation of α-Fe_2O_3 during wear compared with the static value is somewhat unexpected. It should be noted that α-Fe_2O_3, however, is the predominant oxide at temperatures less than about 450 °C. The contact temperature (T_c) is nearly always greater than this value for steels sliding at normal speeds and normal loads in the mild-oxidational wear regime. An example of low values of T_c is provided by the work (with 9% Cr steel sliding in a carbon dioxide atmosphere at an ambient temperature of around 300 °C) carried out by Sullivan and Granville, 1984. These authors used a reciprocating slider with a domed end (of radius 6 mm) at average speeds between 25 and 160 mm/s and a normal load of 22 N. The wear debris was predominantly the rhombohedral oxide (α-Fe$_2$O$_3$, Cr$_2$O$_3$). Clearly, the speed of sliding was so slow that no appreciable frictional heating occurred, so $T_c \approx T_s \approx 300\,°C$. The surprising feature of this work was the thick ($\sim 6\,\mu m$) oxide plateaux formed during the mild wear of these steels, even though the contact temperatures were not sufficiently high to produce any appreciable oxidation.

Having obtained tribological values for the Arrhenius constants relevant to the mild-oxidational wear of low-alloy (EN31) steel, the path was now opened for applying these values to other steels. Quinn, Rowson and Sullivan (1980) used these new tribological values for the oxidation constants to deduce values of ξ, N, a and T_c for another low-alloy steel, namely EN8. They did this through the derivation of an expression for the mild-oxidational wear rate in terms of measurable quantities and only one of the above surface model parameters, namely the contact radius a. The derivation is not difficult – it is just too long and complicated for the purposes of this sub-section. Hence, we will merely quote the expression for (w_{theory}), as follows:

$$w_{\text{theory}} = \left[\frac{J}{C^2 a - 2CEa^2 + E^2 a^3}\right]\exp\left(-\frac{Q_p}{Ma - Sa^2 + V}\right) \quad (6.16)$$

where $J = 2WA_p/[U(p_m)f^2\rho_o^2 B^2]$ (where U = velocity of sliding); $M = RG$; $B = \pi K_o/[4(\delta_{\text{expt}})(K_s)]$; $S = RI$; $C = 0.86 \ (1 - \delta_{\text{expt}}) - (\delta_{\text{expt}})$; $V = R(T_S + 273)$; $E = 0.10U(1 - \delta_{\text{expt}})/(2\chi_s)$.

In addition

$$G = \frac{(\delta_{\text{expt}})(H_{\text{total}})(p_m)}{W}\left[\frac{\pi}{4K_s} + \frac{BC}{K_o}\right]$$

and

$$I = \frac{(\delta_{\text{expt}})(H_{\text{total}})(p_m)BE}{K_o W}$$

The symbols M, G, E, S, V, J, B, C and I relate only to Equation (6.16) and are not to be confused with their meanings when used elsewhere in this book. In order to use Equation (6.16), we set $w_{\text{theory}} = w_{\text{expt}}$ and re-write Equation (6.16) in a more readily calculable form, namely:

$$-\left(\frac{Q_p}{Ma - Sa^2 + V}\right) = \ln[C^2 w_{\text{expt}}a) - (2CEw_{\text{expt}}a^2) + (E^2 w_{\text{expt}}a^3)] - \ln J \quad (6.17)$$

The derivation of Equation (6.17) can be found in the original reference (Quinn, Rowson and Sullivan, 1980), together with the computer programme (in FORTRAN) which they used for the iterative process for deducing consistent values of contact radius a, number of contacts beneath the pin N, oxide thickness on the real areas of contact on the pin ξ and the contact/oxidation temperature (T_c). Since this programme can be modified to suit other tribo-systems experiencing mild-oxidational wear, it was considered worth reproducing in this book (see Figure 6.2).

This programme will produce values of the required surface model parameters for any tribo-system where facilities are provided:

(a) for deducing δ_{expt} from thermocouples placed as indicated in Figure 6.1,

(b) for measuring the friction and hence (H_{total}) and

(c) for measuring the wear rate (w_{expt}).

Another pre-requisite is that one should know what values should be assigned to the tribological oxidation constants for one's system. As indicated in the previous paragraphs, these values can only be obtained by reference to actual wear and heat flow measurements from *similar* tribo-systems. For low-alloy steels we can use the values given in Table 6.3. Inspection of Figure 6.2 near to lines 66, 100 and 110 reveals that we have indeed included the tribological values for A_p and Q_p (conditional upon the type of wear debris identified, namely α-Fe_2O_3, Fe_3O_4 or FeO).

It should be pointed out that Quinn et al. (1980) did not take into account the effect of the oxide thickness (ξ_d) on the real areas of contact on the disc. Hence the terms B, C and E, all of which arise from an evaluation of (δ_{theory}) from Equation (6.13), are slightly in error. The correction will not alter the trends reported by Quinn et al. (1980), some of which are indicated in Table 6.4. From this table, it will be noted that the iteration gives values of T_c between 371 and 448 °C for those experiments in which Fe_3O_4 was the predominant oxide present in the wear debris. This is slightly inconsistent with the temperature ranges given in Tables 6.2 and 6.3. This may be due to the slight errors in B, C and E mentioned above. However, it is worth pointing out that the transition between the α-Fe_2O_3 and Fe_3O_4 has been given as low as 300 °C by some investigators of the static oxidation of iron. The work-hardening due to sliding may also affect this transition (see Caplan and Cohen, 1966).

We can perceive the *trends* indicated by Table 6.4 in the following graphs of wear rate (w_{expt}) versus load (Figure 6.3(a)), surface temperature (T_S) versus load (Figure 6.3(b)), number of contacts (N) versus load (Figure 6.3(c)) and the contact (or oxidation) temperature (T_c) versus load (Figure 6.3(d)) for EN8 steel at 2 m/s. If we examine these figures we see that wear rate is indeed proportional to the load until the oxidation or contact temperature (T_c) reaches approximately 400 °C (at a load of about 15 N), at which stage the type of oxide changes from α-Fe_2O_3 to Fe_3O_4 and the wear rate suddenly drops to 20% of the wear rate at 9.85 N. As the load is increased above 15 N, however, wear rate is again proportional to the load, the factor of proportionality now being about 16% of the factor related to the experiments at the lower loads. When the load is sufficient to produce a contact temperature (T_c) around 580 °C, we again get a sudden increase in wear rate coinciding with the appearance of FeO in the wear debris, and a return to a dependence upon load that could be interpreted as being a return to the factor of proportionality (the K-factor) pertaining to low temperature sliding.

As regards the surface temperature (T_S), we see there is some evidence for a transition at 35 N. The transition at 15 N (if it exists) is not clearly defined. Nevertheless, it is gratifying to see that wear rate and surface temperatures exhibit the same trends, as also does the graph of T_c versus load (i.e. Figure 6.3(d)). The

```
      MASTER OXYWEAR
      REAL KI, KS, L1, L3, M, L, KO, N, N1, N2, NM
      DIMENSION U(20), W(20), FF(20), TA(20), TB(20), TC(20), HTOT(20),
    2 DELTAE(20), TS(20), DELTA(20), TF(20), XN(20), WR(20), H1(20)
      DATA PI, XKS/3.14159, 9.656E - 06/
      DATA KO, KS, P/ 2.1, 37.20, 3.26E 09/
      DATA AP, RHO, F, R/ 1.0E 08, 7.0E 03, 0.3, 8.314/
      DATA KI, RA, RT, C, H/ 0.1045, 0.00795, 0.003175, 0.00001174, 114.1/
      DATA L1, L3/ 0.0067, 0.02/
      M = SQRT(2*KI/((KS*(RT**2))*ALOG(RA/RT)))
      READ(1, 10) NEXP
   10 FORMAT (I0)
      DO 20 I = 1, NEXP
      READ (1,30) U(I), W(I), FF(I), TA(I), TB(I), TC(I), WR(I)
   30 FORMAT (6F0.0, 1PE9.3)
   20 CONTINUE
      WRITE (2,40)
   40 FORMAT (1H1,6HEXP NO,4X,5HSPEED,5X,4HLOAD,4X,8HFRICTION,3X,
    2 2HTA,4X,2HTB,4X,2HTC,4X,2HH1,4X,2HH2,4X,2HH3,3X,4HHTOT,
    3 5X,4HDEXP,5X,2HTS)
      I = 0
      DO 50 I=1,NEXP
      RRE=(2.0*RT*U(I)*1.165)/(1.9015E - 05)
      IF(RRE.LE.4.0) GO TO 41
      IF(RRE.LE.40.0) GO TO 42
      IF(RRE.LE.4000.0) GO TO 43
      IF(RRE.LE.40000.0) GO TO 44
      IF(RRE.LE.250000.0) GO TO 45
   41 RNU = 0.891*(RRE**0.33)
      GO TO 46
   42 RNU = 0.821*(RRE**0.385)
      GO TO 46
   43 RNU = 0.615*RRE**0.466)
      GO TO 46
   44 RNU = 0.174*(RRE**0.618)
      GO TO 46
   45 RNU = 0.0239*(RRE**0.805)
   46 H = RNU/0.2337
      Z=SQRT(KS/(2.0*RT*H))
      A=L1/(Z*RT)
      B = TA(I) - TB(I)
      D = TB(I) - TC(I)
      E = TA(I) - TC(I)
      H3= KS*PI*(RT**2)*M*((E*COSH(M*L3)) - D)/SINH(M*L3)
      H2 = ((C*E)/(RA - RT))+H3
      H1(I)=(PI*RT*(KS/Z)*(E*SINH(A)))+(H2*COSH(A))
      HTOT(I) = FF(I)*U(I)
      DELTAE(I) = H1(I)/ HTOT(I)
      TS(I) = (E*COSH(A))+((Z*H2/(KS*PI*RT))*SINH(A)+TC(I)
      WRITE (2,60) I,U(I),W(I),FF(I),TA(I),TB(I),TC(I),H1(I),H2,H3,
    2 HTOT(I),DELTAE(I),TS(I)
   60 FORMAT(1H, 1X,I2,6X,F6.2,3X,F6.2,4X,F6.2,2X,F6.1,1X,F5.1,1X,
    2 F5.1,1X,F5.2,1X,F5.2,1X,F5.2,1X,F6.2,2X,F5.3,3X,F6.1)

      DO 2 K=1,NEXP
      WRITE(2,65) K
```

```
 65   FORMAT(1H0,'EXP. NO = ',I3)
      WRITE(2,66)
 66   FORMAT(5X,'TH',11X,'RAD',13X,'N',12X,'TO',11X,'AP',12X,'QP')
      RAD = 2.0E − 06
      AP = 1.0E 16
      QP = 208000.0
      F = 0.3006
      RHO = 5.24E 03
      I = −1
 75   B=(PI*KO)/(4.0*DELTAE(K)*KS)
      C=(0.8605*(1.0 − DELTAE(K)) − DELTAE(K))
      E=0.1021*U(K)*(1.0 − DELTAE(K))/(2.0*XKS)
      G=DELTAE(K)*HTOT(K)*P*((PI/(4.0*W(K)*KS))+((B*C)/(W(K)*KO)))
      RI=(DELTAE(K)*HTOT(K)*P*B*E)/(KO*W(K))
      RJ=(2.0*W(K)*AP)/(U(K)*P*F*F*RHO*RHO*B*B)
      RM=R*G
      S=R*RI
      V=R*(TS(K)+273.0)
 70   Z1=(C*C*WR(K)*RAD) − (2.0*C*E*WR(K)*RAD*RAD)
    2   +(E*E*WR(K)*RAD*RAD*RAD)
      IF(RJ) 72,72,74
 74   IF(Z1) 72,72,76
 72   WRITE(2,78)
 78   FORMAT(1H , 'NO SOLUTION FOUND')
      GO TO 92
 76   Z = ALOG(Z1) − ALOG(RJ)
      RAD1=−(QP/(Z*RM))−(V/RM)+(S*RAD*RAD/RM)
      IF(ABS(RAD1 − RAD).LE.ABS(RAD/1000.0)) GO TO 80
      RAD = RAD1
      GO TO 70
 80   N = W(K)/(PI*P*RAD1*RAD1)
      TH = B*RAD1*(C−(E*RAD1))
      TO = (G*RAD1)−(RI*RAD1*RAD1)+TS(K)
      WRITE(2,90) TH, RAD1, N, TO, AP, QP
 90   FORMAT(1H, 1PE10.3,4X,1PE10.3,4X,0PF10.1,4X,F10.1,4X,1PE10.1,
    2   4X,0PF10.1)
 92   IF(I) 100,110,2
100   I = I + 1
      AP = 1.0E 03
      QP = 96000.0
      F = 0.2885
      RHO = 5.21E 03
      RAD = 2.0E − 06
      GO TO 75
110   I = I + 1
      AP = 1.0E 08
      QP = 210000.0
      F = 0.2277
      RHO = 5.7E 03
      RAD = 2.0E − 06
      GO TO 75
    2 CONTINUE
      STOP
      END
```

Figure 6.2 *Computer programme relating to the mild-oxidational wear of low-alloy steels*

Table 6.4. *Experimentally and theoretically derived data for EN8 steels sliding at 2 m/s*

Load (N)	H_1 (W)	H_{total} (W)	δ_{expt}(%)	T_S(°C)	$(w_{expt}) \times 10^{13}$ (m³/m)	a (μm)	N	ξ (μm)	T_c (°C)
3.94	0.53	9.19	5.8	39.9	2.06	1.46	180	0.82	226
6.89	0.32	9.81	3.3	39.0	1.94	2.53	105	2.67	239
9.85	0.97	17.89	5.4	52.0	4.03	1.76	310	1.07	227
9.85	0.75	16.68	4.5	39.0	3.04	2.05	222	1.54	230
					Mainly α-Fe$_2$O$_3$ in the wear debris				
14.78	1.54	19.62	7.8	65.0	0.68	4.47	72	1.70	371
14.78	1.69	24.52	6.9	65.8	1.49	3.94	93	1.75	409
19.70	1.78	28.02	6.3	76.4	3.00	4.94	79	2.40	443
24.63	2.19	33.35	6.6	86.3	4.41	5.16	90	2.40	449
29.551	1.93	27.57	7.0	83.9	4.42	7.84	47	3.24	448
					Mainly Fe$_3$O$_4$ in the wear debris				
34.47	6.00	35.32	17.0	162.7	10.45	9.43	38	1.14	583
39.40	4.95	34.82	14.2	174.6	17.89	11.37	30	1.75	615
					Mainly FeO in the wear debris				

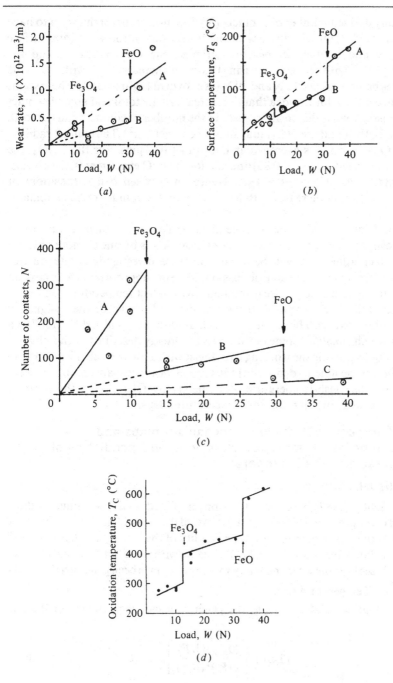

Figure 6.3 *(a)(w$_{expt}$) versus load;(b) T$_S$ versus load;(c) N versus load;(d) T$_c$ versus load for EN8 steel pin sliding at 2 m/s con EN8 steel disc)*

theoretically derived number of contacts (N) does not, at first sight, seem to have much of a trend. However, in view of the transitions in the wear rate, surface temperature and contact temperature graphs, one can indeed line up the graphical points with three (decreasing) rates of increase of N with load, the boundaries being at about 15 N and 30 N (as shown in Figure 6.3(*c*)). There seems to be some evidence for the idea that the increase in the total real area of contact with increasing load is due to an increase in the number of contacts and *not* to an increase in both N and the individual areas of contact (πa^2). The contact radii for the α-Fe$_2$O$_3$ experiments averaged at 1.95 µm (± 0.34 µm). For the Fe$_3$O$_4$, the values of a averaged at (5.27 ± 0.83) µm, and for the FeO experiments, the average contact radius was (10.40 ± 0.97) µm. Hence, in between oxide transitions, it would appear that N increases with load, whereas the contact radius a remains constant.

All the indications and suggestions detailed in this sub-section owe their origin to the investigation of worn surfaces and/or wear debris by one or more physical analytical techniques. We will be discussing these investigations later in this chapter. Before starting on that discussion, it seems appropriate to discuss an approach to the prediction of wear mechanisms, contact temperatures and wear rates which will have an appeal to the practical tribologist or the designer of practical tribo-systems. This is the approach used by Lim and Ashby (1987), an approach which *simplifies* some of the analysis already described in this chapter in an attempt to provide the working tribologist with a 'wear mechanism map', so that he/she will be able to deduce (a) the type of wear mechanism, (b) the wear rate and (c) the 'flash-temperature' (T_c) merely by reference to the position ($\tilde{U}, \tilde{W}$) in the wear mechanism and temperature maps.

6.1.4 The potential of wear mechanism maps and temperature maps for classifying and predicting sliding wear and flash temperatures

(i) Introduction

Lim and Ashby (1987) base their maps on simplified analyses similar to those described in the previous sub-section. It is beyond the scope of this book to examine the effect of this simplification upon the validity of their analysis. Hence, we will merely give the basic equations, so that the reader with more fundamental interests can roughly see how the wear and temperature maps have been generated for steels.

(ii) Temperature maps

The first equation relates to $(T_S)_p$, the general surface temperature of the pin, namely:

$$(T_S)_p = (T_{OS}) + \left[\frac{(\delta_{expt})(UF)}{\pi(R_t)^2(K_{sp})} \right](L_1 + L_3) \qquad (6.18)$$

All the symbols in Equation (6.18) have been chosen so that they approximately are the same as in the previous sub-section. Since Lim and Ashby postulate a pin held in a normal chuck (and not in a calorimeter such as we have considered in Figure 6.1), then $(L_1 + L_3)$ represent the length of the pin from the sliding surface to where it is held by the chuck (and where the temperature is T_{OS}). This equation is very much simpler than the equivalent Equation (6.10) given in Section 6.1.2. Lim and Ashby then proposed that $(T_S)_d$ is given by:

$$(T_S)_d = T_{OS} + \left\{ \frac{2(2)^{1/2}(1-\delta_{expt})F(UX_S)^{1/2}}{[(K_S)_d(\pi R_t)^{3/2}]} \right\} \tag{6.19}$$

No attempt was made in Section 6.1.2 to estimate $(T_S)_d$, so we cannot comment on this equation, apart from pointing out that the temperature of the heat sink attached to the disc is unlikely to be the same as T_{OS} (the heat sink to which the pin is attached). Where Lim and Ashby (1987) significantly depart from the analysis of the previous sub-section is when they equate $(T_S)_p$ to $(T_S)_d$, to obtain their equation for the bulk surface temperature (let us denote it by the symbol T_S) as follows:

$$T_S = (T_{OS}) + \left\{ \frac{\mu(T^*)\beta}{[2+\beta(\pi\tilde{U}/8)^{1/2}]} \right\} (\tilde{W}\tilde{U}) \tag{6.20}$$

where $\tilde{W}$ and $\tilde{U}$ are the normalized pressure and velocity given by Equations (1.34b) and (1.34a) respectively. β is a dimensionless number given by:

$$\beta = (L_1 + L_3)/(R_t) \tag{6.21}$$

T^* is a convenient way of denoting the collection of terms $((X_S)(p_m)_o/(K_S))$, where (K_S) is written for $(K_S)_d$ and $(K_S)_p$, since we are assuming that the pin and disc are made of the same material. In the previous sub-section, we assumed that $(T_S)_d$ was *different* from $(T_S)_p$, an assumption based on physical experience with asymmetrical pin-and-disc configurations running without any external heating. It is suggested that this is an important aspect of oxidational wear occurring at normal loads and speeds. However, when dealing with orders of magnitude, Lim and Ashby's assumption may be essentially correct.

The coefficient of friction μ has long been known to be dependent upon speed of sliding. By taking the friction data from several investigations into the wear of steels, Lim and Ashby (1987) shown that μ is a linear function of the normalized velocity $\tilde{U}$, hence:

$$\mu = 0.78 - 0.13[\log_{10}(\tilde{U})] \tag{6.22}$$

This equation can be used in Equation (6.20) to calculate the expected bulk temperature (T_S).

The 'flash temperature' discussed by Lim and Ashby (1987) is, of course, the contact temperature (T_c) discussed in the previous sub-section. It will depend on the number of contacts per unit area n_c, which is related to N through the relation

$$n_c = N/[\pi(R_t)^2] \tag{6.23}$$

The authors show that, assuming it is the *number* of contacts of radius a that increases with the load and not their *size*, one can express (n_c) as follows:

$$n_c = \left[\left(\frac{R_t}{a}\right)^2 (\tilde{W})(1-\tilde{W}) \right] + 1 \tag{6.24}$$

Finally, Lim and Ashby (1987) show that, by assuming (δ_{expt}) equals 0.5 and ignoring the oxide film on the disc, it is possible to obtain a very much simplified expression for (T_c), namely:

$$T_c = T_S + \left[\frac{\mu(T_c^*)\beta}{2(n_c)^{1/2}} \right] (\tilde{W})^{1/2} (\tilde{U}) \qquad (6.25)$$

where (T_c^*) is given by:

$$(T_c^*) = (\chi_S)(p_m)_o/(K_e) \qquad (6.26)$$

In Equation (6.26), (T_c^*) is the equivalent of (T^*), and (K_e) is the equivalent thermal conductivity, which allows for the effect of an oxide film upon the transfer of heat.

In order to calibrate Equation (6.25), the authors plotted flash temperature contours on a graph with $(\tilde{W})$ as the ordinate and $(\tilde{U})$ as the abscissa. They chose three temperatures, namely 750, 900 and 1527 °C (the melting point of pure iron). The first two temperatures represent the limits between which the peak temperatures must lie in order to give martensite. Considering those wear experiments in which the martensitic transformation was detected (either by X-ray diffraction of the wear debris or metallographic examination of the surface layers on the pin or disc), Lim and Ashby found that the data points fell close to, or between, the contours for 750 and 900 °C. These authors also quote the results of Quinn and Winer (1985), who slid tool steel pins against sapphire at 1.7 and 2.6 m/s and values of $(\tilde{W})$ between 1.2×10^{-4} and 6.6×10^{-4}. It can be seen, from Figure 6.4(a), that these values of U and $\tilde{W}$ lie almost exactly on top of the plotted 900 °C contour. We will be discussing Quinn and Winer's (1985) experiments in a later sub-section. For the present, however, it should be said that, although they found values of T_c between 950 and 1150 °C indicated by the approximate colour-temperature calibration, they were very *cautious* about whether those hot-spots were the *only ones present* in the contact zone.

The temperature map shown in Figure 6.4(b) is clearly of great utility for the designer of a tribo-system, especially as regards the bulk temperatures (T_S). *Simplification*, however, seems to have provided values of T_c (contact temperatures) *much higher* than those obtained in the previous sub-section from *simultaneous* application of *both heat flow/transfer analysis* and the *oxidational wear equation*. Does this show that the oxidation occurs at some temperature between (T_S) and the flash temperatures indicated by Figure 6.4(b)? We shall see that the X-ray diffraction evidence supports the lower temperatures (~ 300 to 600 °C) as being the most relevant ones for the formation of oxides during oxidational wear. Let us briefly review what Lim and Ashby (1987) have to say about wear mechanisms.

(iii) Wear mechanism maps

Lim and Ashby (1987) consider four broad classes of mechanisms, namely:
(a) Seizure
(b) Melt wear
(c) Oxidation-dominated wear, and
(d) Plasticity-dominated wear

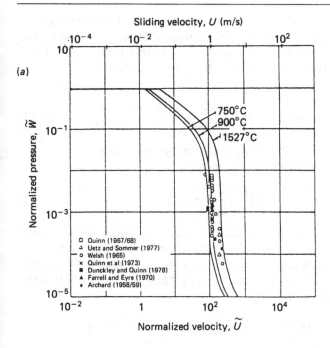

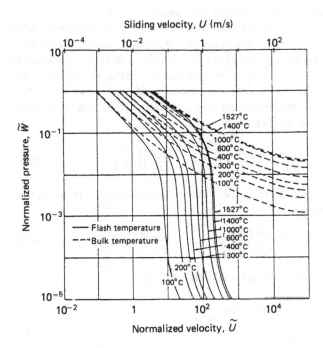

Figure 6.4 *(a) The martensitic calibration of flash temperature for steel;
(b) the temperature maps (from Lim and Ashby, 1987)*

Seizure occurs when the real area of contact (A_{real}) equals the nominal area of contact (A_n), which equals $\pi(R_t)^2$ for a cylindrical pin of radius (R_t). This type of wear mechanism is really a *failure mechanism* which, possibly, is the limit to which the other three mechanisms approach. We will merely quote the *seizure condition for steel*, namely:

$$\tilde{W} = \left(\frac{1}{[1+(\alpha_l)(\mu)^2]^{1/2}}\right)\left\{1 - \frac{(T_S - T_{OS})}{20(T_m)}\left[\ln\left(\frac{10^6}{\beta\tilde{U}}\right)\right]\right\} \qquad (6.27)$$

where T_m is the melting point of the steel, T_S is given by Equation (6.20), and (α_l) is given by Tabor's (1959) junction growth equation:

$$\left(\frac{W}{A_{real}}\right)^2 + (\alpha_l)\left(\frac{\mu W}{A_{real}}\right)^2 = (p_m)_{local}^2 \qquad (6.28)$$

where $(p_m)_{local}$ is the local hardness of the material, (which may be very hot!).

Melt wear is what happens when solids slide on ice or snow, when friction-induced melting generates a lubricating layer of water. Lim and Ashby use this problem as an analogue for the melt wear of steel. These authors show that the normalized wear rate ($\tilde{w}$) is given by:

$$\tilde{w} = \left(\frac{T_m - T_{OS}}{T^*}\right)\frac{(p_m)_o}{(L_F)}\left(\frac{1}{\beta\tilde{U}}\right)\left[(\delta_{expt})\mu\tilde{W}\tilde{U}\left(\frac{T^*\beta}{T_m - T_{OS}}\right) - 1\right] \qquad (6.29)$$

where (L_F) is the latent heat of fusion per unit volume of steel.

In the case of oxidation-dominated wear, Lim and Ashby distinguish between *mild*-oxidational wear (where flash temperatures are enough to cause oxidation but the oxide is, for most of the time, cold and brittle) and *severe*-oxidational wear (where the oxide film is thicker, more continuous, hotter and more plastic than with mild-oxidational wear). For the case of *mild-oxidational wear*, they produce an expression for the normalized wear rate ($\tilde{w}$), without having recourse to the Archard Wear Law, as given by:

$$\tilde{w} = \left[\frac{(C_w)^2(A_p)(R_t)}{\xi X_s}\right]\exp\left(-\frac{Q_p}{RT_c}\right)\left(\frac{\tilde{W}}{\tilde{U}}\right) \qquad (6.30)$$

To compare this with Equation (6.14), we have to write Equation (6.30) in non-normalized terms, namely:

$$w = \frac{(C_w)^2 W(A_p)}{\xi(p_m)U}\exp\left[-\frac{Q_p}{RT_c}\right] \qquad (6.31)$$

Thus, Equation (6.31) and (6.14) are identical, provided:

$$(C_w)^2 = \frac{d}{\xi\rho_o^2 f_o^2}$$

that is if

$$(C_w)^2(\rho_o)^2 = \left(\frac{d}{\xi}\right)\frac{1}{(f_o)^2} \qquad (6.32)$$

Lim and Ashby (1987) define (C_w) on the assumption that the oxide formed in mild-oxidational wear is always Fe_3O_4 which, as we saw in the last sub-section, is probably correct for most normal conditions of sliding. This means that (C_w), which appears in the parabolic equation:

$$\xi^2 = (C_w)^2 (k_p)(t_o) \tag{6.33}$$

where t_o is the time required to build up to a critical oxide film thickness (ξ) at the real areas of contact, will be given by:

$$(C_w) = \frac{3(M_{Fe})}{2(M_o)(\rho_{Fe})} \tag{6.34}$$

where M_{Fe} and M_o are the molecular weights of iron and oxygen respectively. From Equations (6.34) and 6.32), we obtain:

$$\left[\frac{3}{2} \left(\frac{M_{Fe}}{M_O} \right) \left(\frac{\rho_o}{\rho_{Fe}} \right) \right]^2 = \left(\frac{d}{\xi} \right) \left(\frac{1}{f_o^2} \right) \tag{6.35}$$

Since f_o is the mass fraction of oxide which is oxygen, then Equation (6.35) tells us that Lim and Ashby's Equation (6.30) is identical to Equation (6.14), *provided* $d \approx \xi$. If one is discussing wear *in broad terms*, then it is quite reasonable to make this assumption.

Lim and Ashby (1987) address the problem of the oxidation constants for Equation (6.30). They took the static oxidation activation energy for iron from an older reference than Caplan and Cohen's (1966) reference, namely Kubaschewski and Hopkins (1962). This value was 138 kJ/mol, which is somewhere near to the mean of the 208–210 kJ/mol value for FeO and the 96 kJ/mol value for Fe_3O_4 given in Table 6.2. Following the same line of thinking as discussed in the last sub-section, these authors took the value for the tribological Arrhenius constant (A_p) to be $10^6 \, kg^2/m^4 \, s^{-1}$, which is close to the geometric mean of $10^3 \, kg^2/m^4 \, s^{-1}$ for Fe_3O_4 and $10^8 \, kg^2/m^4 \, s^{-1}$ for FeO (see Table 6.3).

Assuming $\xi = 10 \, \mu m$, taking $(C_w) = 3.4 \times 10^{-4} \, m^3/kg$ (for Fe_3O_4), Lim and Ashby (1987) drew contours of constant normalized wear rate $(\tilde{w})$, calculated from Equation (6.30), using Equation (6.25) for the contact temperature, and showed that much of the data obtained by Quinn and his co-workers, and that of Archard and Hirst (1956), could be superimposed on these contours. As Lim and Ashby (1987) point out, behaviour in this regime is complicated by the martensite formation. They try to overcome this by assigning two values of wear rate for each contour. The first are those given by Equation (6.30) with $(p_m) = 1 \, GPa$; the second (in parenthesis) are 100 times smaller, reflecting the increase in (p_m) caused by the martensite transformation. This is shown in Figure 6.5. Although I would not agree in *detail* with the actual values of $(\tilde{w})$ and (T_c) predicted by Equations (6.30) and (6.25) respectively, there is no doubt that mild-oxidational wear can be predicted as likely to occur if one's values of $(\tilde{U})$ and $(\tilde{W})$ lie somewhere within the area of Figure 6.5 covered by these contours. The actual wear rate might well be an order of magnitude different from the predicted value, but this is sufficiently close for many design purposes.

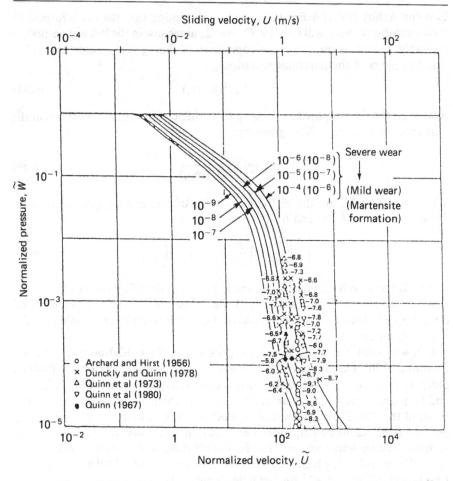

Figure 6.5 *Plot of contours of constant normalized wear rates ($\tilde{w}$) based on Equation (6.30) [the numbers given against the points are $\log_{10}(\tilde{w})$] (from Lim and Ashby, 1987)*

Lim and Ashby's models for *severe-oxidational wear* and *plasticity-dominated wear* are not so well-developed as the models of mild-oxidational wear, melt wear and seizure, and hence need not concern us at the time of writing. Nevertheless, it is clear that wear mechanism maps and temperature maps can be useful aids to designers of tribo-systems. They give order-of-magnitude estimates for wear rates and surface temperatures. What is most important is that they help the designer to predict what *type* of wear is likely to occur under his/her conditions of ($\tilde{W}$) and ($\tilde{U}$). The *complete* wear mechanism map for steels has still to be developed. In the meantime, the map given in Figure 1.5 is sufficiently accurate for most of our purposes. Certainly, all future wear experiments carried out with steels, will now have to be related in some way to this map. Perhaps, the concept of wear mechanism maps will bring about the cohesion that has been lacking in this field for so long. Let us now continue this chapter by describing how physical analytical techniques have been used to determine the hot-spot temperatures through the crystallographic changes they cause at the real areas of contact.

6.2 The crystallographic changes caused by hot-spot temperatures during sliding wear

6.2.1 Introduction

It has already been mentioned that different oxides are formed at different hot-spot temperatures in the mild-oxidational wear of steels (see Section 6.1.3). We also saw that there were changes in the substrate metal from austenite to martensite. These changes in the surface film and substrate crystallography were detected by X-ray diffraction analysis of the wear debris. Now there is an implicit assumption that these crystallographic changes occur during the time that the wear debris was actually *in situ* on the wearing surface. By analysing the crystallography of the surfaces, it is possible to show that the oxide and martensite transformations detected in the wear debris must have occurred before the debris particle left its position as part of the real area of contact. There is still the uncertainty as to whether the transformation occurs *while* the real areas of contact are *in* conjunction or *immediately after* conjunction. In the previous section, we assumed that the transformations occur at T_c *during* conjunction. This assumption has *not* yet been proved to be incorrect in all the experiments so far carried out with the objective of studying the crystallograhic changes brought about by hot-spot temperatures during wear.

In this sub-section, we will discuss

(a) the use of X-ray and electron diffraction for *identifying transitions* in both the worn surface and the wear debris structures [and hence the changes in hot-spot temperatures (T_c)] with changing load (or with changing speed);

(b) the use of X-ray diffraction techniques for analysing *proportions* of the various constituents in wear debris and, hence, the most likely value of the hot-spot temperatures before, between or after the transitional temperatures;

(c) the use of glancing-angle X-ray diffraction techniques for studying the crystallography of *worn surfaces*;

(d) the use of Auger electron spectroscopy, plus depth profiling, for studying the elemental distribution as a function of depth *below the worn surface*; and

(e) the limitations of using electron diffraction for studying the crystallography of worn surfaces or wear debris.

6.2.2 Transitions in wear debris structures as revealed by X-ray and electron diffraction

Possibly the most convincing evidence of the strong correlation between transitions in wear debris structures and transitions in the hot-spot temperatures (or, more correctly, in the temperature indicated from the bulk static oxidation of the materials being worn) can be found in the work of Quinn *et al.* (1973). These authors used steels containing up to 12 wt % of chromium in unlubricated pin-and-disc sliding experiments using loads between 5 and 140 N at a speed of 1 m/s.

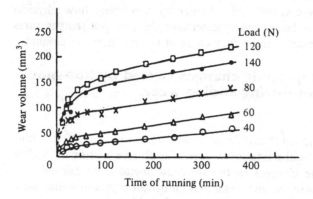

Figure 6.6 *Wear volume versus time of sliding for 2 wt% chromium steel at various loads above 30 N*

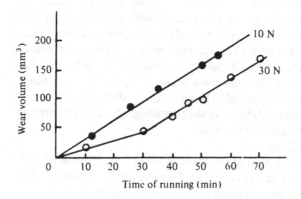

Figure 6.7 *Wear volume versus time of sliding for 2 wt% chromium steel at loads of 30 N and 10 N*

For loads greater than 30 N, Quinn *et al.* (1973) found 'wear-volume-versus-time-of-running' curves characteristic of mild-oxidational wear. A typical set of mild wear curves is shown in Figure 6.6 for 2 wt% chromium steel sliding against itself in a conventional pin-and-disc machine. Note the initial severe wear for the first hour, followed by equilibrium mild wear, the slope of the mild wear line increasing with increasing load. This means that the wear rate (the volume removed per unit sliding distance) for these 1 m/s speeds increased with increasing load. For loads of 30 and 10 N, Quinn *et al.* (1973) found that the initial severe wear was never ameliorated, the whole of the wedge portion of the pin being worn away at the end of the first hour (see Figure 6.7).

Plotting the wear rates versus applied load for steels with zero, 0.9, 2.0 and 3.0 wt% chromium, it was found that there was a transition load of about 30 N at which severe wear changed to mild wear as the load increased. This transition had

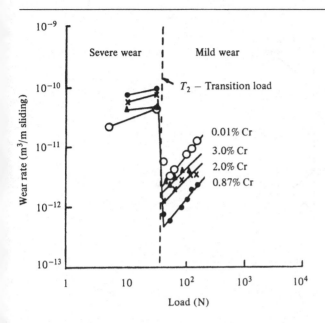

Figure 6.8 *Volume wear rate versus applied load for 0.01, 0.87, 2.0 and 3.0 wt% chromium steels*

already been noted by Welsh (1965). It will be seen, from Figure 6.8, that the slopes of all four mild wear lines are about 45° showing that, for a log–log plot, *the mild wear rate was proportional to the load*, for loads greater than the 30 N transitions loads (as might be expected from the first wear law (see Section 1.4.2)). Below the transition load (which we will call the T_2 transition, following Welsh, 1965), the slope is almost zero, indicating that (possibly) severe wear is independent of the load. This is *not* what would be expected from the first wear law!

The results with 8 and 12 wt% chromium steels were even more unexpected. We have already mentioned the work of Hong *et al.* (1987) (in Section 1.4.4), in which AISI 316 stainless steel was worn against itself and was found to exhibit oxidational wear that was more consistent with a linear dependence of oxidation upon time than the normal parabolic dependence. Clearly, large amounts of chromium affect the wear behaviour in rather unexpected ways. For the 8 and 12 wt% Cr steels of Quinn *et al.* (1973), the wear rates showed transitions at about 10 and 50 N, as shown in Figure 6.9. The most striking result was that for the 12 wt% Cr steel where, after a T_2 transition from severe to mild wear at 10 N, the wear rate proceeded approximately proportional to the load until about 50 N. At loads from 50 up to around 100 N, the wear rate decreased with increasing load, in complete contrast to normal wear behaviour of low chromium and other low-alloy steels.

The friction versus load curves from the chromium steels containing less than 8 wt% chromium showed a decrease with increasing load at a slope of

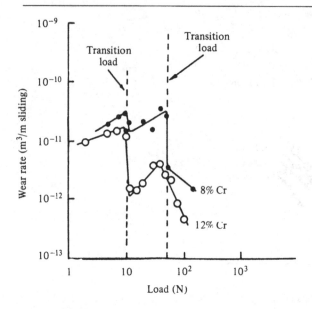

Figure 6.9 *Volume wear rate versus applied load for 8 and 12 wt%
chromium steels*

$(-6.7 \times 10^{-3} \text{N}^{-1})$ for loads up to about 80 N, after which the coefficient of
friction μ tended towards a constant value of about 0.35 ± 0.05. There was no
marked effect of chromium content. However, for the 8 wt% and 12 wt%
chromium steels, there was a definite change in the coefficient of friction. The
12 wt% chromium steel result is shown in Figure 6.10, with its very low values at
loads less than 10 N, changing to very high values (> 1) for loads greater than
10 N. Note the return to more normal values for loads greater than 30 N.

Having provided the reader with an adequate data base from which we will be
able to make sensible judgements, let us now see what electron and X-ray
diffraction examination of the worn surfaces and the wear debris told Quinn *et al.*
(1973) about the crystallographic changes occurring at the very marked changes
(transitions) in the wear rate curves.

The structure of the top 10 µm of the worn surfaces was revealed by the
glancing-angle, X-ray diffraction film technique described in Section 3.2.4, while
the top 20 Å or so was revealed by reflection electron diffraction (see Section
3.3.3). The wear debris was first analysed by placing it in the capillary tube
situated along the axis of a cylindrical X-ray diffraction powder camera (see
Section 3.2.4) (for detecting the structures within the larger (i.e. about 1 µm)
particles of debris). It was also examined by using the general area electron
diffraction attachment described in Section 3.3.5. The smaller particles of wear
debris were dispersed on to a carbon film on an electron microscope grid and
examined by transmission general area electron diffraction at accelerating
voltages around 100 kV. The electron diffraction patterns from such debris show

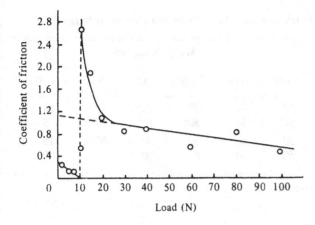

Figure 6.10 *Friction coefficient versus applied load for 12 wt% Cr steel*

up minority constituents much more readily than the X-ray diffraction patterns from large particle agglomerates of the same debris.

Typically, it was found that a worn surface revealed the presence of higher temperature oxides, if it had been worn at higher loads. There seemed to be a definite transition load at which the higher temperature oxide was first noted. For example, with the 2 wt% Cr steel, and at loads of 10 and 30 N, the authors obtained X-ray and electron diffraction patterns of a mixture of α-Fe and the room-temperature rhombohedral oxide. The authors made no attempt to determine the precise nature of the solid solution of α-Fe$_2$O$_3$ and Cr$_2$O$_3$ which they found. They were only interested in comparing the interplanar spacings of the diffraction lines they obtained from their worn surfaces (or from their wear debris) with those spacings obtained in the bulk static oxidation experiments carried out on these same chromium steel alloys by Baig (1968). (As an aside, it is interesting to note that oxidation and/or some crystallographic experts will spend endless hours over such esoteric differences, when all the tribologist need know is that the wear debris or worn surface contains rhombohedral oxide. The tribologist must avoid such experts when seeking help with the analysis of his/her specimens!) To return to the theme of this paragraph, Quinn *et al.* found that, at 40 N, the next higher temperature oxide, the spinel oxide, was also noted in the diffraction patterns from the worn surfaces and the wear debris obtained with 2 wt% Cr steel specimens. The spinel oxide was some solid solution between Fe$_3$O$_4$ and FeCr$_2$O$_4$. At 50 N, the highest-temperature oxide, wüstite was detected, as well as α-iron, and the rhombohedral and spinel oxides. The wüstite was some form of non-stoichiometric FeO. Although the first appearance of the higher oxide was sometimes detected by electron diffraction rather than by X-ray diffraction, it is difficult to draw any conclusions about this, except to point out that electron diffraction is more sensitive to minority constituents than X-ray diffraction. The results of the various experiments are shown in Table 6.5.

From this table, it can be seen that increasing the chromium content up to

Table 6.5. *Transition loads* (*in newtons*) *and chromium content* (*see text*)

Type of transition	Type of specimen	Wt% chromium					
		0.0	0.9	2.0	3.0	8.0	12.0
Spinal (W_S)	Pin	40	40	40	30*	10	15
	Debris	30[a]	30[a]	30[a]	30[a]	10[a]	5
Wüstite (W_W)	Pin	50	50	50	40	50	15
	Debris	50	50[a]	50	40	–	100

[a]From electron diffraction evidence

3 wt% has negligible effect as far as the spinel transition (W_S) or the wüstite transition (W_W) are concerned. The spinel transition at 30 N was also the load at which these steels change from severe to mild wear. The low load (somewhere between 5 and 15 N) at which the spinel oxide was formed for the 12 wt% Cr steel coincides with the first transition in the wear rate versus load curve at about 10 N (see Figure 6.9).

The appearance of the spinel oxide at certain loads indicates a possible critical temperature occurring at the real areas of contact. From bulk (static) oxidation tests carried out on these chromium steels [Baig (1968)], it is possible to compile a table of the temperature at which the spinel oxide and wüstite phases first appear (see Table 6.6). Whether it be detected by electron or X-ray diffraction is of no consequence for our purposes.

From these experiments, one can see that electron and X-ray diffraction analysis has revealed invaluable evidence (not readily available from other methods of analysis):

(a) that the presence of the spinel oxide is indeed an ameliorative feature as regards the wear of low-chromium steels;

(b) that the transition from rhombohedral oxide to spinel oxide occurs at about 300 °C, thereby telling us that, at the transition load of 30 N, the oxidation temperature during sliding was about 300 °C and;

(c) that for chromium contents greater than 3 wt%, one can expect anomalous wear behaviour from chromium steels. The role of wüstite in the wear of all of these chromium steels is also unclear. It is worth pointing out, however, that there is no doubt that the appearance of wüstite in the worn surfaces or wear debris from these steels must indicate temperatures of the order of 700 °C or more.

It will be apparent that, although the detection of high-temperature phases indicate certain minimum temperatures of formation, mere identification cannot give any idea of the *actual* temperature at which the phase was formed. To do this, we need to measure the proportions of the various phases present, and this is discussed in the next sub-section.

Table 6.6. *Temperature (in °C) at which the spinel oxide and wüstite phases first appear in bulk (static) oxidation tests (Baig, 1968)*

Type of oxide	Percentage chromium					
	0.0	0.9	2.0	3.0	8.0	12.0
Spinel oxide	300	300	300	350	250	300
Wüstite	700	700	700	700	800	800

6.2.3 The use of X-ray diffraction techniques for analysing the proportions of the various constituents in wear debris

(i) Introduction

In this sub-section, we will be mainly concerned with how proportional analysis of wear debris can lead to estimates of the temperature at which that debris was formed (when it was part of the real areas of contact). We will also give an example, however, of how proportional analysis can actually be used to estimate the relative wear rates of pins and discs comprised of different base materials. Proportional analysis is concerned with relative integrated intensities of various X-ray diffraction lines. We will first show how the technique devised by Averbach and Cohen (1948) for determining the amount of retained austenite in hardened tool-steel, can be modified so that it can be applied to the proportional analysis of wear debris. We will then discuss some examples, namely the proportional analysis of wear debris produced in the sliding wear of electrographite on aluminium and copper and in the sliding wear of steels upon themselves. The tribological implications of these examples will be discussed in some detail, especially as regards the use of this technique for estimating wear rates and contact temperatures.

The reader may ask why no mention has been made of using *electron diffraction* for proportional analysis of wear debris or worn surface structures. The reason is two-fold. Firstly, the accuracy of electron diffraction methods, especially reflection electron diffraction, does not compare with X-ray diffraction. Secondly, with X-ray diffraction one has a wide basis of general theory regarding the intensities of scattered X-ray beams, whereas with electron diffraction we sometimes do not know whether one should use the Dynamical Theory or the Kinematic Theory (as discussed in Section 3.3.8). Electron Diffraction has its place as a worthwhile physical method of analysis for tribological purposes, but *not* for proportional analysis. Let us now start with a discussion of Averbach and Cohen's (1948) method.

(ii) Determination of the amount of austenite retained in hardened tool-steel

Essentially, Averbach and Cohen (1948) irradiated a flat polished specimen of austenitized plain carbon steel at a glancing angle of 60°. At this angle, the line broadening due to the obliqueness of the irradiated surface was not excessive. If the authors had used the edge-irradiated modification of the glancing-angle technique (see Section 3.2.4), they would have obtained information about the surface layers. They were mainly interested in the crystallography of the bulk of their specimen, so this large glancing angle would have given them the information they required (as well as slightly sharpening the diffraction lines). Their glancing-angle pattern contained lines from both austenite and martensite, the martensite lines being much stronger in intensity than the austenite lines.

Averbach and Cohen (1948) used Equation (3.28) to deduce the relative integrated intensity (I_n) from the component (n) of a multi-component X-ray diffraction pattern,

$$I_n = \frac{[(F_X)_{hkl}]_n^2 (m_n)(LP)_n \left\{ \exp\left[-\frac{2(B_{DW})\sin^2\theta}{\lambda^2} \right] \right\}_n}{(v_{uc})_n^2} [V_n A(\theta)] \quad (6.36)$$

where all the symbols relate to those parameters defined in Section 3.2.6, the subscript (n) merely referring to value of these parameters for the nth component. Let us re-write Equation (6.36) as:

$$I_n = (R_n)(V_n)A(\theta) \quad (6.37)$$

where

$$R_n = \frac{[(F_X)_{hkl}]_n^2 (m_n)(LP)_n \left\{ \exp\left[-\frac{2(B_{DW})\sin^2\theta}{\lambda^2} \right] \right\}_n}{(v_{uc})_n^2}$$

$$(6.38)$$

The main problem in using these two equations relates to calculating (R_n), since we need to know all about the structure of the components we expect to find. This means knowing the positions of the atoms in the unit cell. Sometimes those positions will not be available, in which case we will not be able to use this method for deducing proportional amounts! I_n can be measured fairly easily via (a) a microdensitometer trace of the diffraction pattern recorded on a photographic film or (b) the scintillation counter or proportional counter used for directly recording the output of the X-ray diffractometer. V_n is our required quantity and is obviously a constant for a given diffraction pattern. Hence, if one plots (I_n/R_n) versus θ (the Bragg angle), one is effectively plotting the dependence of the absorption factor $[A(\theta)]$ upon θ. It seems reasonable to assume that, for a given multi-component specimen, the absorption factor has the same dependence upon θ for each component, irrespective of which component actually causes the diffraction.

If another component (m) is present in equal volume (i.e. $V_n = V_m$), then the plot of (I_m/R_m) will lie on the same curve $(A(\theta))$. If, however, V_m does not equal V_n, then one will obtain a curve of the same form as (I_n/R_n) versus θ, but displaced from it along the (I/R) axis equal to the ratio (V_n/V_m). In other words, the ratio of the ordinates of the two curves (at any given value of θ) is the ratio of their volumes. If one plots $\log (I/R)$ for each component, then the ordinates of the curves will always be equidistant from each other. Furthermore, if only two or three diffraction lines are available for a given component, it is possible to draw the best curve through those points by ensuring it is equidistant from the good curve drawn through the component with the most diffraction lines (normally taken to be the internal standard).

In Averbach and Cohen's (1948) work, the predominant martensite was taken to be the standard. By comparing the ordinates of (I/R) for austenite at two given Bragg angles, they were able to obtain estimates of the volume percentages of the amount of retained austenite which compared very favourably with determinations by alternative methods. For further details of their technique, the reader is referred to the original reference. By way of illustration, let us discuss the application of Averbach and Cohen's (1948) technique to two fairly simple tribo-systems, namely the sliding of electrograhite on copper and electrographite on aluminium. The reader will be able to see the method without getting too heavily involved in any complex crystallography. This will be a fairly easy introduction to the more complex tribo-systems, namely steel sliding upon steel, where there are many possible constituents all of which will change with changing contact temperature (T_c).

(iii) Proportional analysis of wear debris in an aluminium-electrographite tribo-system

It may be recalled that, in Section 5.2.5, we discussed Fisher's (1973) work arising from his search for an alternative material to copper for slip-rings. As part of that work, Fisher was interested in finding the proportionate amounts of metal and electrographite in the severe wear regime, that is at loads greater than about 3 N, as shown in Figure 6.11. This is an interesting graph. First of all, if we compare it with the graph of the wear of reactor graphite versus load (see Figure 5.16), we see that for a load of (say) 5 N, reactor graphite has a wear rate of about $4 \times 10^{-15}\,\text{m}^3/\text{m}$. This is characteristic of mild wear for electrographite on aluminium. We shall see (in Figure 6.15) that this is also the wear rate (at 5 N) for electrograhite on copper. Clearly, the counterface material seems more important in the wear process than slight changes in the crystallinity of the artificially-manufactured graphites. The oxide that forms on aluminium is thin and prohibits further oxidation. It is also a well-known abrasive, namely $\alpha\text{-}Al_2O_3$. The oxides of copper, however, are not prohibitive of further oxidation and have structures that are much more amenable to easy glide than $\alpha\text{-}Al_2O_3$. Once again we see oxidation is an important factor in wear, even when one of the sliding pair (graphite) is relatively inactive.

Fisher (1973) collected the wear debris as it was thrown clear of the aluminium disc. This could mean that any proportional analysis will not be representative of the system, since the heavier (metallic) particles will be thrown more energetically than the non-metallic particles (i.e. the electrograhite particles). This means that the calculated proportions will represent minimum amounts for the lightest constituent and, hence, maximum proportions of the heaviest component, in this case, the aluminium. The debris was contained within a capillary tube and mounted in the standard X-ray diffraction powder camera. The sample was radiated with CuK_α radiation, the accelerating voltage being 40 kV and the beam

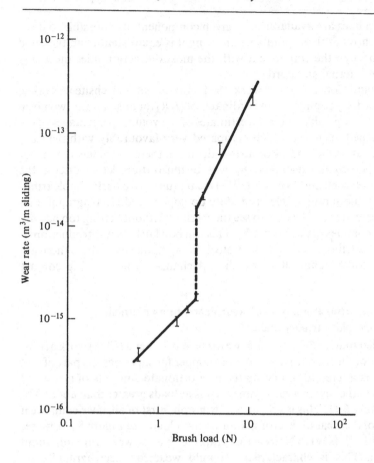

Figure 6.11 *Variation of wear rate of electrographite with applied normal load when sliding on an aluminium disc*

current 20 mA respectively. The radiation from the CuK_α line, obtained using a Ni filter, has a wavelength λ of 1.542 Å.

In order to deduce (I_n) from Equation (6.37), we need to deduce (R_n) from Equation (6.38) for graphite, aluminium (and copper, see later), using Equation (3.24) to obtain the square of the structure factor for X-rays, $[F_x)_{hkl}]^2$, Figure 6.12 for the atomic scattering factor, Table 3.3 for the multiplicity factor m, Equation [3.28(a)] for the Lorentz-Polarization Factor (LP) and Table 6.7 for the Debye–Waller temperature factor. The volumes of the unit cells of aluminium, copper and carbon are 66.4 Å^3, 47.24 Å^3 and 35.2 Å^3, respectively. (We have included data for copper, since we will need it in the next sub-section.) Both Al and Cu have face-centred cubic unit cells, so tht h, k, l must be all odd or all even for the 'allowed' diffraction lines. Graphite is hexagonal and has already been discussed in Section 5.2.7 dealing with electron diffraction. Table 6.8 gives the summary of the calculations for C and Al.

Table 6.7. *Values of* B_0 *and* B_T *for the Debye–Waller temperature factor*

Atom	B_0	B_T (273 K)	$B_{DW} = B_0 + B_T$
Al	0.26	0.55	0.81
C	0.13	0.38	0.51
Cu	0.14	0.39	0.53

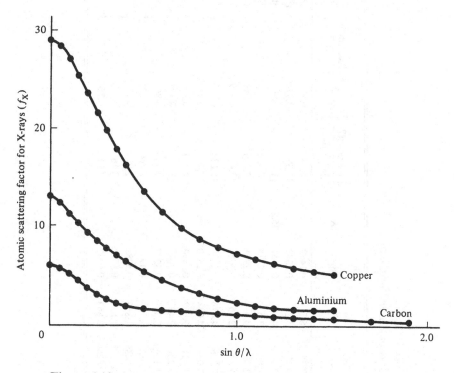

Figure 6.12 *Atomic scattering factors of aluminium carbon and copper*

The next step is to deduce the relative integrated intensity. We have selected the experiment carried out at 10.6 N, since this is well into the severe wear regime. The speed of sliding was about 9 m/s at the brush. Figure 6.13 is a micro-densitometer trace of the X-ray diffraction pattern from the selected debris. Analysis of the 2θ values for this pattern showed that the major constituents were aluminium and carbon, with traces of Al_2O_3 (corundum). The 'noise' of the trace has been smoothed out and the background level filled in. The integrated intensities of the maxima from the aluminium were measured by the product 'peak height' × width of the peak at half its height. Since the exposure time was restricted to a value within the linear range of the characteristic curve of 'exposure versus density of blackening', it can be assumed that the density of blackening is strictly proportional to the intensity of X-rays causing that blackening.

Table 6.8. *Calculation of R values for aluminium and graphite*

(hkl)	(m)	θ (Bragg angle)	2θ	d_{hkl} (Å)	$2d_{hkl}$ (Å)	$(\sin\theta)/\lambda$	v_{uc} (Å)³	$\left[\dfrac{1+\cos^2(2\theta)}{(\sin^2\theta)\cos\theta}\right]$	$\exp\left[-\dfrac{(B_{DW})\sin^2\theta}{\lambda^2}\right]$	$[(F_X)_{hkl}]^2$	$(f_X)_{hkl}$	R
C 002	2	13.3	26.6	3.35	6.7	0.1493		34.94	0.989	$16f^2$	4.40	17.09
C 100	6	21.22	42.44	2.13	4.26	0.2347		12.68	0.974	f^2	3.17	0.586
C 101	12	22.21	44.42	2.04	4.08	0.2451	35.2	11.43	0.969	$3f^2$	3.00	2.086
C 102	12	25.36	50.72	1.80	3.6	0.2778		8.420	0.962	f^2	2.75	0.5708
C 004	2	27.4	54.8	1.70	3.4	0.2941		7.086	0.959	$16f^2$	2.55	1.090
C 110	6	38.73	77.46	1.23	2.46	0.4065		3.43	0.919	$16f^2$	1.93	0.836
Al 111	8	19.3	38.5	2.338	4.676	0.2139		15.62	0.965		9.0	34.0
Al 200	6	22.4	44.78	2.024	4.048	0.2470		11.20	0.951		8.5	15.8
Al 220	12	32.6	65.19	1.431	2.862	0.3494		4.809	0.907		7.3	9.20
Al 311	24	39.2	78.3	1.221	2.442	0.4095	66.4	3.361	0.874	$16f^2$	6.6	9.70
Al 222	8	41.3	82.52	1.169	2.338	0.4277		3.106	0.863		6.4	1.74
Al 400	6	44.4	88.86	1.012	2.024	0.4938		2.860	0.825		5.7	1.39
Al 331	24	56.1	112.18	0.9289	1.858	0.5382		2.974	0.792		5.2	4.40
Al 420	24	58.4	116.7	0.9055	1.811	0.5522		3.166	0.785		5.1	4.40
Al 422	24	68.8	137.7	0.8266	1.653	0.6042		4.916	0.750		4.6	5.10

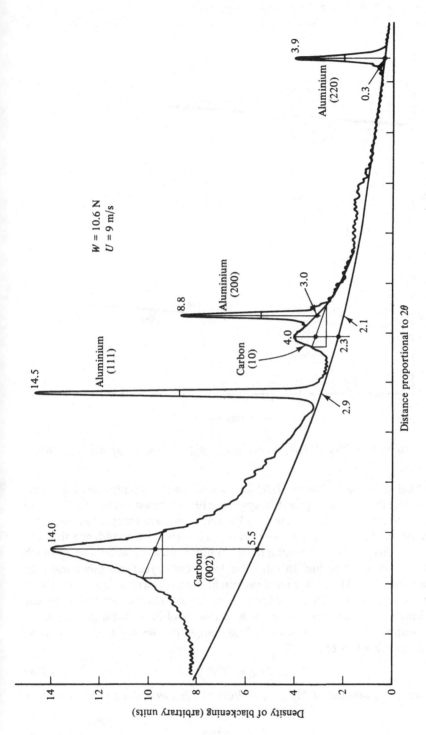

Figure 6.13 Micro-densitometer trace of density of blackening of X-ray film versus (2θ) for wear debris formed when electrographite slides against aluminium at 10.6 N and 9 m/s

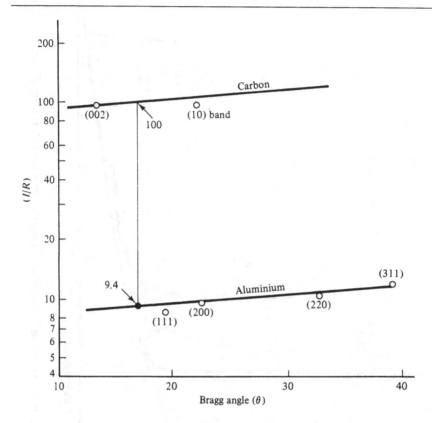

Figure 6.14 *Plot of (I/R) versus Bragg angle (θ) (derived from Figure 6.13)*

The final step is to calculate (I/R) for each line and then plot (on log–linear graph paper) those values against Bragg angle (θ), as shown in Figure 6.14. Note we have used the aluminium as our standard, since we have sharper lines from the aluminium in the debris (and more lines!). This gives us confidence that the straight line through the aluminium points in Figure 6.14 is a 'good' line, so much so that we draw the 'best line' through the (less accurate) carbon lines *parallel* to the aluminium line. The (10) band of carbon has been drawn in with some 'poetic licence'. Nevertheless, the shape of the band is similar to the carbon (002) line and we do know that three lines merge to give this band in artificial graphites.

Choosing $\theta = 17°$ as our common Bragg angle (θ), we see that the volume percentages are given by:

$$V_C/V_{Al} = 100/9.4 \tag{6.39a}$$

If we assume the amount of Al_2O_3 is negligible in this wear debris, then we know that

$$V_C + V_{Al} = 100 \tag{6.39b}$$

where V_C and V_{Al} are now the percentage volumes. From Equation (6.39a), we readily compute $V_{Al} = 9\%$ and $V_C = 91\%$. From Figure 6.11, it can be seen that the wear rate at a load of 10.6 N is $2.8 \times 10^{-13}\,m^3/m$. This is the wear rate of the electrographite. From our proportional analysis of the wear debris, we would suggest that the wear rate of the aluminium disc was about $1.9 \times 10^{-14}\,m^3/m$. This is very low and not at all what was expected when aluminium was chosen as a model of just the *wrong sort* of material to select for an alternative counterface material. It turns out, in fact, that aluminium alloys *do* make good counterface materials, as shown by Fisher (1973). Let us briefly now consider the more conventional copper–electrographite tribo-system.

(iv) Proportional analysis of wear debris in the copper–electrographite tribo-system

For the sake of comparison with the aluminium–electrographite tribo-system, the wear rates of electrograhite on copper are shown in Figure 6.15, together with the graphs of wear-rate versus load for hardened steel and electrographite discs. This graph is interesting for many reasons. Firstly, we see that all the points lie on

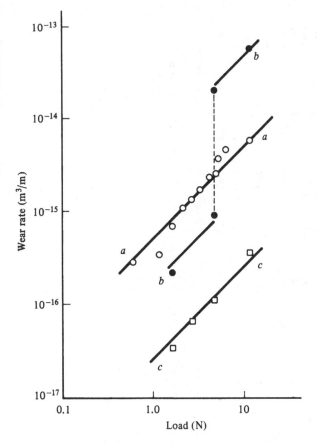

Figure 6.15 *Variation in the wear rate of an electrographite brush with load when sliding on (a) copper, (b) hardened steel and (c) electrographite discs*

Table 6.9. *Calculation of R-values for copper and cuprous oxide*

(hkl)	(m)	θ_{hkl} (Bragg angle)	$2\theta_{hkl}$	d_{hkl} (Å)	$2d_{hkl}$ (Å)	$(\sin\theta)/\lambda$	v_{uc} (Å)^3	$\left[\dfrac{1+\cos^2(2\theta)}{(\sin^2\theta)(\cos\theta)}\right]$	$\exp\left[-\dfrac{(B_{DW})\sin^2\theta}{\lambda^2}\right]$	$[(F_X)_{hkl}]^2$	$(f_X)_{hkl}$	R
Cu 111	8	21.67	43.34	2.088	4.176	0.239		12.03	0.972		22.0	396.6
Cu 200	6	25.24	50.48	1.808	3.616	0.277		8.573	0.962		20.4	142.0
Cu 220	12	37.10	74.20	1.278	2.556	0.391		3.701	0.927		16.7	76.32
Cu 311	24	45.01	90.02	1.090	2.180	0.458	47.24	2.828	0.900	16f^2	14.6	84.02
Cu 222	8	47.62	95.24	1.044	2.088	0.479		2.742	0.891		14.1	31.25
Cu 400	6	58.53	117.06	0.9038	1.8076	0.554		3.175	0.860		12.4	16.19
Cu 331	24	68.37	136.74	0.8293	1.6586	0.603		4.812	0.835		11.4	75.02
Cu 420	24	72.50	145.00	0.8083	1.6166	0.618		6.109	0.825		11.2	89.76
Cu$_2$O 111	8	18.22	36.45	2.465	4.930	0.202	77.86	17.78	0.980	16f$^2_{cu}$	23.6	201

Intensity of (220) reflection of Cu$_2$O determined by comparison with ASTM index card.

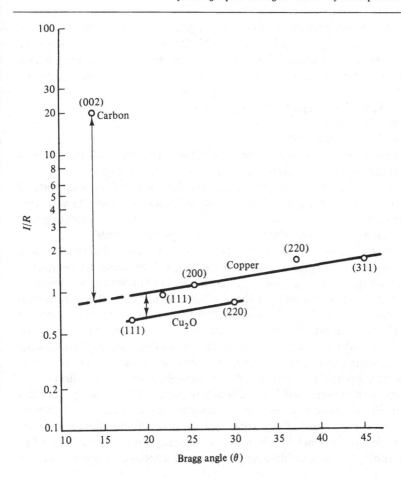

Figure 6.16 *Graph of (I/R) against Bragg angle for wear debris produced during the sliding of electrographite on copper at a load of 10.6 N*

straight lines at 45° to the axes, showing that wear rate is indeed proportional to the load. Secondly, it can be seen that the hardened steel showed a transition at about 4.5 N, from 10^{-15} m^3/m to 2×10^{-14} m^3/m, which is almost the same increase in wear rate as we saw for electrographite on aluminium. Thirdly, there is no transition (within the load range 1.0 to 10 N) in the wear rates of the electrographite brush on either the copper or electrographite discs. Clearly, if one could arrange for electrographite to slide against itself with no contaminants from the substrate metal of the disc, we would have a very satisfactory wear situation.

The proportional analysis of debris from the electrographite on copper systems involves knowing the R values for graphite, copper and cuprous oxide (Cu_2O). We already know the R values for graphite (see Table 6.8). Without going into further details, let us accept that the R values given in Table 6.9, are sufficiently accurate for our purposes. Figure 6.16 shows the result of plotting (I/R) versus θ

(on log–linear graph paper) for wear debris produced during the sliding of electrographite on copper at a load of 10.6 N. Obviously, we must choose the copper line as our standard. From Figure 6.16, therefore, we can deduce $V_C = 23.25$ (V_{Cu}) and $V_{Cu_2O} = 0.66$ (V_{Cu}). Assuming $V_C + V_{Cu_2O} + V_{Cu} = 100$, we obtain:

$V_{Cu} = 100/(23.25 + 0.66 + 1) = 4.01\%$

$V_{Cu_2O} = 0.66 \times 4.01 = 2.65\%$

and $V_C = 23.25 \times 4.01 = 93.23\%$

This calculation indicates that the wear rate of the copper disc, whether it be by oxidation or by mechanical removal, is $[(100 - 93.23)/100](5.5 \times 10^{-15})$ m^3/m that is 3.7×10^{-16} m^3/m. This value is remarkably close to the wear rate of electrographite against itself at 10.6 N. It is also very much less than the wear rate of the aluminium disc at 10.6 N (namely 1.9×10^{14} m^3/m), thereby indicating why copper is such a good counterface material for slip-ring assemblies.

In this sub-section, we have not explicitly shown that X-ray diffraction can help us determine hot-spot temperatures with electrographite–metal tribo-systems. Clearly, oxidational wear has occurred in both systems, but only on a small scale compared with mechanical wear of the electrographite brush. We have shown that proportional analysis of the wear debris from these systems can lead to estimates of the wear rate of the counterface material. Furthermore, although Fisher's (1973) results were internally consistent, he made no attempt to check his results by preparing samples of *known proportions* of aluminium and elec-trographite or copper, copper oxide and electrographite, to see what effects these samples had on their combined X-ray diffraction patterns. In the last part of this sub-section related to the crystallographic changes caused by hot-spot tempera-tures, we will describe some of the work done on the oxidational wear of steels over the past few years; work that is not only internally consistent but which has also been subjected to considerable amounts of checking against standard mixtures.

(v) The proportional analysis of wear debris produced in the oxidational wear of steels

Possibly the first published account of X-ray diffraction being used to analyse wear debris into its proportionate parts was based on some oxidational wear experiments carried out on EN26 steels at various speeds (Quinn, 1967). Since that time, the technique has been refined and the R-values of the constituents of the wear debris re-calculated. The new values of R for α-Fe are considerably different from those used in the original publication, whereas the α-Fe$_2$O$_3$, Fe$_3$O$_4$ and FeO values only differ in detail from the previously published values. Since α-Fe is often only a minor constituent in oxidational wear, this revision will not alter the trends already reported. However, we will briefly put the record straight, so that newcomers to the subject can readily appreciate the potential of this approach to their problems in tribology. Let us start with Table 6.10(a), the revised R-values for α-Fe, FeO and Fe$_3$O$_4$ and Table 6.10(b), the revised R-values for α-Fe$_2$O$_3$. I gratefully acknowledge the debt I owe to my former colleague at Aston University, Dr David M. Rowson, for all the hard work he put into these

Table 6.10(a). *The revised R-values for α-Fe, Fe_3O_4 and FeO (for cylindrical X-ray diffraction powder cameras)*

	α-Fe			FeO			Fe_3O_4	
hkl	d (Å)	R	hkl	d (Å)	R	hkl	d (Å)	R
110	2.0268	196.20	111	2.486	46.52	111	4.85	5.84
200	1.4332	28.48	200	2.153	68.28	220	2.967	26.62
211	1.1702	68.13	220	1.523	37.41	311	2.532	58.35
220	1.0134	35.77	311	1.299	17.70	400	2.099	12.66
310	0.9064	192.96	222	1.243	12.94	422	1.715	6.47
Structure BCC			400	1.077	8.24	511+333	1.616	19.56
$a_o = 2.8664$ Å			331	0.988	18.18	440	1.485	26.94
ASTM card 6-0696			420	0.963	42.79	620	1.328	1.31
			Structure NaCl cubic			533	1.287	5.95
			$a_o = 4.307$ Å			444	1.212	2.15
			ASTM card 6-0615			642	1.122	2.20
						731+553	1.093	18.96
						800	1.050	6.00
						822+660	0.9896	3.31
						662	0.9632	3.91
						840	0.9386	7.55
						Structure spinel (Al_2MgO_4)		
						$a_o = 8.396$ Å		
						ASTM card 19-0629		

Table 6.10(b). *The revised R-values for α-Fe_2O_3 (for cylindrical X-ray diffraction powder cameras)*

hkl	d (Å)	R	hkl	d (Å)	R
110	3.686	13.48	23$\bar{1}$	1.1416	4.47
211	2.703	36.18	024	1.042	4.70
$\bar{1}$10	2.519	28.92	22$\bar{2}$	1.077	0.71
210	2.208	6.68	235	1.0571	6.01
220	1.843	15.10	400	1.0398	1.73
321	1.6966	17.71	512+31$\bar{2}$	0.9900	3.38
332+12$\bar{1}$	1.6013	4.19	14$\bar{1}$+554	0.9600	13.27
310	1.4873	12.76	511+12$\bar{3}$	0.9521	8.80
$\bar{2}$11	1.4543	12.76	$\bar{2}$23+104	0.9315	0.48
224	1.3514	1.10	154	0.9090	20.60
334	1.3133	5.01			
2$\bar{2}$0	1.2595	3.29			
303+114	1.2285	1.48	Structure rhombohedral		
31$\bar{1}$	1.2145	0.33	$a_r = 5.4367$ Å		
134+10$\bar{3}$	1.1908	2.61	$\alpha = 55.2043°$		
244	1.1645	2.89	ASTM card 24-0072		

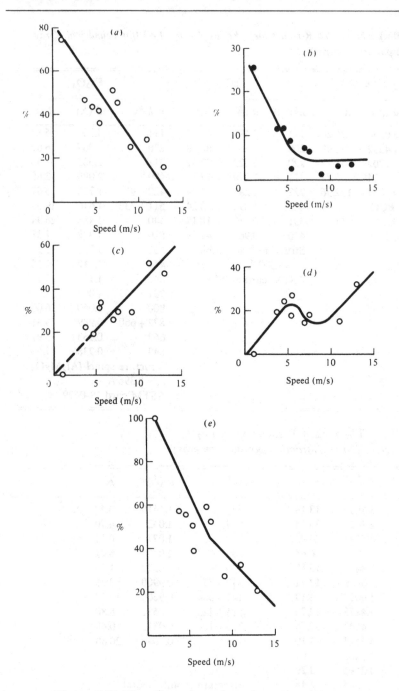

Figure 6.17 *The effect of speed on the proportionate amounts of (a) α-Fe$_2$O$_3$, (b) α-Fe, (c) Fe$_3$O$_4$, (d) FeO and (e) α-Fe + α-Fe$_2$O$_3$ in wear debris from EN26 steels sliding at 5 N load in a pin-and-disc wear machine (no lubrication)*

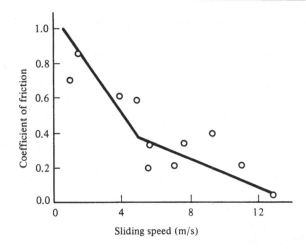

Figure 6.18 *Variation of coefficient of friction with sliding speed for EN 26 steels (normal load, 5 N)*

revised values. Dr Rowson was also very much involved in the experiments undertaken to check the method against standard mixtures of these oxides.

Using these revised R-values it is possible to use Table I of Quinn's (1967) paper to show that the percentage volumes of α-Fe and α-Fe$_2$O$_3$ in the debris tended to *decrease* with *increasing speed* (from 1 to 13 m/s) as shown in Figures 6.17(a) and (b), while the percentage of Fe$_3$O$_4$ tended to *increase* (see Figure 6.17(c)). The percentage of FeO seemed more unstable, but they also might be thought of as increasing with a transition occurring at a speed somewhere around 5 m/s. The curved line drawn in Figure 6.17(d) is only tentative. If we put α-Fe$_2$O$_3$ and α-Fe together we obtain Figure 6.17(e). If we now look at the graphs of friction coefficient (μ) versus speed (Figure 6.18) and wear rate (w) versus speed (Figure 6.19), we can see a strong similarity between these graphs of mechanical properties and the graph of volume percentages of α-Fe$_2$O$_3$ and α-Fe in the wear debris. There is also some evidence for a transition in the wear debris composition at around 5 m/s (especially with the α-Fe graph) that seems to coincide with the transitions in Figure 6.18 and 6.19. We also can see that the volume percentages of Fe$_3$O$_4$ (the spinel oxide) is zero at 1 m/s. It does not appear in significant amounts until the speed reaches a value around 5 m/s, the transition speed for the α-Fe graph.

It is clear from these graphs that Fe$_3$O$_4$ is to be associated with low wear and low friction, whereas α-Fe$_2$O$_3$ (and α-Fe) is associated with high wear and high friction. It is also clear that this is due to the increased speed of sliding. One might guess that the hot-spot temperature at 5 m/s was just sufficient (at this light load) for the spinel oxide (Fe$_3$O$_4$) to be formed at the real areas of contact. One might also assume that the actual temperatures of contact at speeds above 5 m/s are proportional to the amount of Fe$_3$O$_4$, or inversely proportional to the amount of

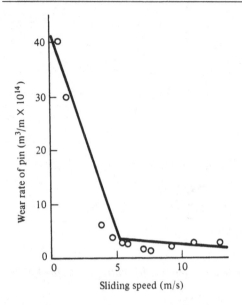

Figure 6.19 *Variation of wear rate with sliding speed for EN 26 steels (normal load, 5 N)*

(α-Fe_2O_3 + α-Fe). It was this idea that prompted the work carried out with a low-alloy, medium carbon steel (AISI 4340) at a constant, fairly high speed of 6.25 m/s in a specially-designed tribometer (Quinn, 1968).

In this apparatus, the bulk temperatures of the pin and the disc were measured continuously, as also were the frictional forces and the decreases in pin height (i.e. the volume wear of the pin), during experiments carried out at loads between 10 and 37 N. Similar experiments were then carried out, at the slower speed of 0.05 m/s. It was assumed that frictional heating would be negligible compared with that due to a speed of 6.25 m/s (two orders of magnitude more). As far as was practically possible, the bulk temperatures were maintained (by external heating) at the same values as for the initial, frictionally heated experiments (at each given load). The wear debris was collected and examined by means of an X-ray diffractometer technique. This has an effect on the R-values, through the Lorentz-Polarization Factor. In fact the Lorentz-Polarization Factor for the diffractometer (where the powder is packed into a flat 1 mm depression, with sides equal to 1 cm × 2 cm, and the beam is at glancing incidence) is given by:

$$(LP)_{\text{diffractometer}} = \frac{1 + \cos^2 2\theta}{\sin \theta \cos \theta} \qquad (6.40)$$

Comparing this equation with Equation (3.28a), we see that the diffractometer value of (LP) is $\sin \theta$ times the powder camera value. This means that the diffractometer R-values are given by Tables 6.10(a) and 6.10(b) by multiplying each powder camera R-value by the appropriate value of $\sin \theta$. For example, for α-Fe, the (R_{110}) for the powder camera is 196.20. Since $\theta = 26.21°$ for CoK_α

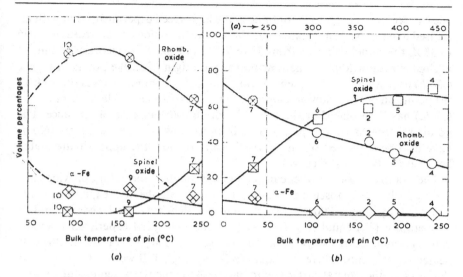

Figure 6.20 *Volume percentages of wear debris components versus measured bulk temperature $[(T_B)_m]$ of the pin for (a) externally-induced ambient temperatures, (b) frictionally-induced ambient temperatures*

radiation ($\lambda = 1.7902$ Å), this means that (R_{110}) for the diffractometer is 86.65. For the (R_{200}) of α-Fe, we get 17.79 instead of 28.48. Now (R_{110}/R_{200}) equals 6.89 for the powder camera and 4.87 for the diffractometer. There is therefore a 70% error if one uses the incorrect R-value, so intending users of this technique must be alert to this often overlooked difference between the diffractometer and the powder camera. Let us return to our main theme.

The analyses of the debris are best presented in the form of two graphs of volume percentages versus measured bulk temperature of the pin for the externally-induced ambient temperature experiments (Figure 6.20(a)) and the frictionally-induced ambient temperature experiments (Figure 6.20(b)). These graphs are based on the same R-values used by the author in his earlier experiments (Quinn, 1967). The revised R-values given in Tables 6.10(a) and (b) will, of course, change the actual volume percentages, especially those of the α-Fe components. Fortunately, the diffraction lines from α-Fe were always very faint in the patterns from the wear debris obtained when oxidational wear was occurring, so that the effect upon these graphs will be marginal. Since the surface temperatures of the pin and disc are the same for Figure 6.20(a) and (b), and since the volume percentages of α-Fe, rhombohedral oxide (α-Fe$_2$O$_3$) and spinel oxide (Fe$_3$O$_4$) are clearly different, the differences must be due to the increased speed at which the experiments were carried out, that is due to contact temperatures well above ambient. One can see (from Figure 6.20(a) that the relative volume percentages of the rhombohedral oxide and α-Fe are decreasing with increasing temperature above about 150 °C, whereas the volume percentage of spinel oxide begins to increase. Figure 6.20(b) has been placed alongside Figure 6.20(a) so that one can see that these trends continue in the high speed experiments, but with increased magnitude. For instance, the rhombohedral oxide line decreases from 46 to 35%, the spinel increases from 53 to 71%, and the α-Fe decreases to almost negligible proportions. If one now moves the origin of all the points in Figure 6.20(b) back about 200 °C, then it is possible to draw smooth curves through all the points on both the low and high speed set of experiments. The scale of Figure 6.20(a) has been placed at the top of Figure 6.20(b) for ease of reference. The points relating to volume percentages from the low-speed experiments (No. 7) have also been plotted on this revised scale to show the approximate match between the two sets of curves.

The fact that smooth curves can be drawn through *all* the points, provided one assumes that the abscissae relating to frictionally-heated experiments are increased by about 200 °C, means that the debris is produced at contact temperature (T_c) equal to $[(T_B)_m + \alpha]$ for the low speed experiments, and at contact temperatures (T_c) equal to $[(T_B)_m + \alpha + 200]$ °C for the frictionally heated experiments, where α is equal to $[(T_S)_p - (T_B)_m]$. If we assume that $\alpha = 0$, that is $(T_S)_p$, the general surface temperature of the pin is not significantly above the temperature $(T_B)_m$ measured by the thermocouple embedded a little way behind the sliding surface of the pin, then we get, for the frictionally-heated, (high speed) runs the following:

(6.43)

$$T_c \approx (T_B)_m + 200 °C \qquad (6.41)$$

Since $(T_B)_m$ was shown to be linearly dependent on load (w) through the relation:

$$(T_B)_m = 4.7W + 60 \qquad (6.42)$$

where W is the load (in newtons) and $(T_B)_m$ is in °C, it follows that, *for this set of experiments only*, T_c is given by:

$$T_c \approx 4.7W + 260$$

Thus we have contact temperatures equal to 305, 365, 393 and 448 °C for experiments 6, 2, 5 and 4 (of Figure 6.20) respectively. These temperatures are consistent with the temperatures at which the spinel oxide is expected to be formed – see for instance the 0% Cr (i.e. pure iron) temperature of 300 °C for the first appearance of the spinel oxide in Baig's (1968) bulk static oxidation experiments (Table 6.6).

As a final postscript to this rather lengthy (but nonetheless, important) subsection on the proportional analysis of wear debris by X-ray diffraction, some mention must be made of the work carried out by Athwal (1982) and Rowson (1982) on the study of standard mixtures of α-Fe, α-Fe$_2$O$_3$, Fe$_3$O$_4$ and FeO. There are many difficulties involved in mixing powdered oxides, even after one has ensured that one truly does have the oxide described on the commercial package. These authors had several difficulties, especially over 50/50 percentage samples of the various combinations of two oxides. Taking into account anomalous X-ray scattering only seemed to affect both curves of log (I/R) versus θ by the same amount, thereby leaving the ratio of the two ordinates the same. Most of the trouble seems to have resided in certain diffraction lines which were often prominent in the diffraction pattern from the mixture but which were never in line with the other points on the log (I/R) versus θ curves. This is best seen in Figure 6.21, which shows plots of (I/R) versus Bragg angle (plotted on log–linear graph paper) for 100% standards of α-Fe, α-Fe$_2$O$_3$, Fe$_3$O$_4$ and FeO. Note that the 220 point of Fe$_3$O$_4$, the 210 and 2$\bar{2}$0 of α-Fe$_2$O$_3$ are very much out of line.

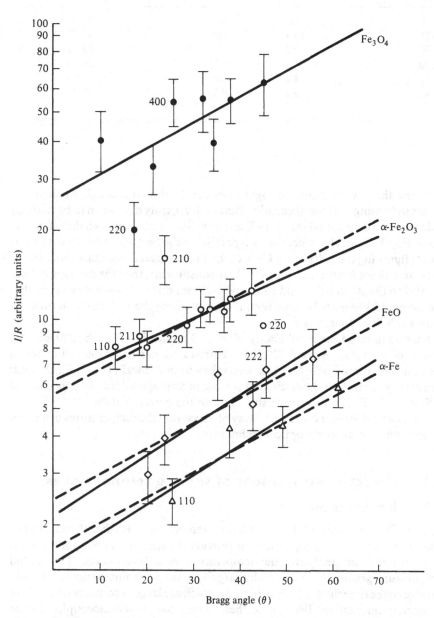

Figure 6.21 *(I/) versus Bragg angle (θ) for 100% standards of α-Fe, α-Fe$_2$O$_3$, Fe$_3$O$_4$ and FeO*

Table 6.11. *Proportional analysis of* (α-Fe_2O_3/Fe_3O_4) *mixtures* (*Athwal, 1982*)

Fe_2O_3/Fe_3O_4 standard (% volume)	(I/R)		Calculated volume (Fe_2O_3/Fe_3O_4)%± 8%
	α-Fe_2O_3	Fe_3O_4	
80/20	16.0	3.1	83/17
75/25	10.5	3.9	73/27
60/40	9.2	4.2	68/21
50/50	7.8	6.1	56/44
25/75	4.2	8.0	34/66
20/80	2.2	9.8	18/82

Ignoring these 'wild' points straight lines can be drawn through all the other points indicating that the absorption factor ($A(\theta)$), truly does seem to be virtually independent of the actual material. The dotted lines drawn through the α-Fe, FeO and α-Fe_2O_3 points have been drawn parallel to the best line through the Fe_3O_4 points (ignoring the 220 point). If Fe_3O_4 had been taken as standard, and if these lines were those from a mixture of all four constituents, then the drawing of lines parallel to the standard would entail insignificant error. (It should be noted that the points of Figure 6.21 have been displaced along the (I/R) axis, in order to show each line separately.)

Bearing in mind the unreliability of the 220 point of Fe_3O_4, the 210 and $2\overline{2}0$ points of α-Fe_2O_3, Athwal (1982) has produced fairly reliable estimates of standard mixtures. Table 6.11 has been taken from Athwal's (1982) Ph.D. thesis and clearly shows that the calculated volume percentages of standard mixtures of α-Fe_2O_3 and Fe_3O_4 are reasonably close to the correct values.

Let us now discuss a more *direct* way of measuring the temperatures of a tribo-system, namely by using optical techniques.

6.3　Direct measurement of surface temperatures

6.3.1　Introduction

The traditional method for measuring temperature is the thermocouple. Unfortunately, thermocouples do not provide the spatial and temporal resolution required for analysing the temperature of tribo-contacts. The spatial temperature gradients are generally large and the time fluctuations are comparable to the time fluctuations of friction, which are large in contacts where there is asperity interaction. Because of these requirements, thermocouples can be expected to give little more than a local bulk temperature estimate, even when very small thermocouples are placed near to the surface. Even the dynamic thermocouple, where the contact is between two dissimilar metals, does not have the resolution required. For the sake of completeness, however, we will discuss thermocouple methods in the next sub-section.

Early work at the Cambridge Physics and Chemistry of Solids (PCS) Laboratories in which hot-spots were viewed directly through a transparent member of a sliding pair was, surprisingly, not followed up until quite recently. This was because of the tremendous interest in lubricated contacts, already covered in this book in Section 5.3.2 under the heading 'Optical Elasto-hydrodynamics'. Even the early work involving the use of an infra-red detector was mainly confined to lubricated contact. In this sub-section, therefore, we will be discussing optical analysis of hot-spots formed at the surface between sliding tribo-elements without lubrication. This will then be followed by a discussion of the use of infra-red techniques in the study of surface temperatures during sliding and/or rolling, with and without liquid lubrication.

6.3.2 Thermocouple measurements of surface temperatures

Most thermocouple measurements involve inserting a small diameter stainless steel sheath into a hole drilled so that the end of the hole is close to the surface to be investigated. In the pin-on-disc machine, this involves drilling a hole along the axis of the cylindrical pin so that, even after the expected amount of wear has occurred, the end of the thermocouples (in good thermal contact with the end of the stainless steel sheath) are not part of the actual contact. Provided the inclusion of particles of stainless steel is not considered harmful to one's tribo-system, it does not matter if the wear is sufficient to reach the embedded thermocouple. From personal experience, it can be said that the temperatures recorded show no significant change when the wearing surface passes through the end of the stainless steel sheath! To attach a thermocouple to the disc requires some form of rotating contact to the disc. Alternatively, a light-loaded trailing thermocouple can be fixed so that it detects a temperature fairly close to the wear track. From the above discussion, it can be seen that more sophisticated measuring techniques are required if we are ever to get a reasonably accurate estimate of the temperature of the real areas of contact.

The dynamic thermocouple might seem to hold more promise, since we then have our wearing pin surface in contact with a heat source, which in turn is in contact with the different material of the disc. Bowden and Ridler (1935, 1936) were probably the first people to use the dynamic thermocouple in a pin-on-disc configuration. These authors chose lead sliding on mild steel and constantan sliding on mild steel. Typical results are shown in Figures 6.22 and 6.23, in which the observed rise in temperature is plotted against the speed of sliding. The choice of lead in Figure 6.22 is deliberate. The authors wanted to use a material that was extremely soft, did not work-harden, and also had a low melting point. Typically one would expect severe wear of the lead, with the real area of contact approaching the nominal area of contact. Hence T_c would be very close to $(T_s)Pb$, that is the contact temperature would be not much different from the general surface temperature of the lead.

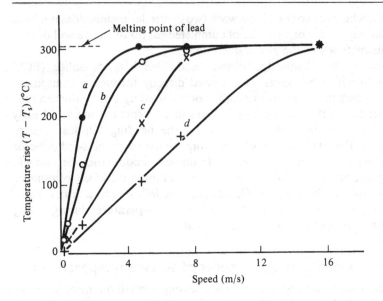

Figure 6.22 *Dynamic thermocouple measurements for lead sliding on mild steel at loads of (a) 1.04 N (b) 0.81 N (c) 0.58 N and (d) 0.35 N (initial temperature, 17 °C*

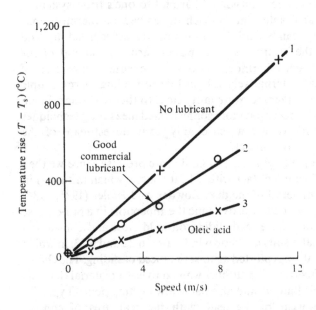

Figure 6.23 *Dynamic thermocouple measurements for constantan sliding on mild steel, with and without lubricant (load = 1.04 N; Initial temperature, 17 °C)*

Using metals of high melting point, Bowden and Ridler (1935, 1936) obtained measurements well over 1000 °C. A typical example is shown in line 1 of Figure 6.23, where at a speed of about 11 m/s (a very high speed and a light load, making this a very *unstable* tribo-system) they obtained a reading in excess of 1000 °C. Even with lubricants, they obtained values of up to 450 °C, at the highest speeds, (as shown in lines 2 and 3 of Figure 6.23). It was this early work of Bowden and Ridler (1935, 1936) that influenced the calculations of hot-spot temperatures for many years after. The authors point out that all the points on the surface may not be rubbing at the same temperature. The electromotive force measured will be the integrated value of a number of thermocouples in parallel, so that many of the points of contact may be at a higher temperature than the recorded value.

More recent examples of the use of dynamic thermocouples can be found in the work of Furey (1964). He used a fixed constantan ball loaded against a rotating steel cylinder under unlubricated conditions of sliding, and with a fairly conventional electronic circuit was able to measure temperatures ranging from 80 to 290 °C, for loads between 0.6 N and 2.5 N and speeds between 0.14 to 2.24 m/s. These values are, of course, considerably lower than those obtained by Bowden and Ridler (1935, 1936), and well below those predicted theoretically by Archard (1959). Remember, however, that Archard did not take into account the effect of oxide films. Furthermore, Dayson (1967) has shown that, even for severe wear (where presumably the real areas of contact are not oxide–oxide contact) the effect of junction growth will considerably lower the estimate of hot-spot temperatures. Also, if one extrapolates Furey's (1964) Figure 11 to predict the expected value for a speed of 6.25 m/s, a load of 2.5 N, and a constantan-steel sliding system, one obtains $(T_B)_m$ about 125 °C and T_c (as indicated by occasional excursions to high readings) would be about 325 °C. The difference between these readings is approximately 200 °C which is surprisingly close to the difference predicted from the X-ray diffraction analysis of the wear debris in the last sub-section. The actual values predicted by Equations (6.42) and (6.43) are 72 and 272 °C. The 53 °C difference between the absolute values given by the thermo-couple extrapolation and those deduced from the wear debris composition, could be due to the different thermal parameters. We will now consider how optical techniques have been used for *directly* studying hot-spots.

6.3.3 Optical analysis of the hot-spots formed at the interface between tribo-elements during sliding without lubrication

In this sub-section we take the term 'optical analysis' to mean the use of electro-magnetic radiation in the visible wavelength range. This means that most of the pioneering work at the Physics and Chemistry of Surfaces Laboratory at Cambridge (see Bowden and Tabor, 1954), insofar as it related to hot-spots, is best left over until the next sub-section. As far as I am aware, very little research has been carried out by direct viewing (by visible light) of the contact region, the aim of the research being to determine the number and tem-perature of the various contact spots by conventional camera photography, combined with colour comparison against heated coupons of the same material as that being rubbed against the transparent medium. Most of the research carried out under the title 'Optical Elasto-hydrodynamics' was invariably concerned only with using interference methods for deducing the thickness of the lubricant film between sliding contact, and *not* with the analysis of any hot-spots that might be present at certain places in the film. In view of the apparent lack of other papers in this very restricted field, the author is forced to describe the work that he has been associated with, namely Quinn and Winer (1987). As Dr Carpenter said, in his contribution to the discussion, 'these authors had carried out pioneering work which, like any good pioneering work, raises more questions than it answers and is subject to hindsight criticism'. Perhaps Quinn and Winer (1987) had found a gap in the knowledge, a surprising gap in view of the simplicity of their approach.

Quinn and Winer (1987) used a sapphire disc which was a single crystal of industrial quality rotated at 1330 rev/min, giving a sliding speed of 1.68 m/s at the inner edge and 2.58 m/s at the outer edge of the wear track. Loads were arbitrarily chosen so as to give hot-spots that could be observed with the naked eye through the sapphire disc, namely 18 N for the experiments run with unhardened tool steel and 26 N for the hardened tool steel. The application of the load was an interesting feature. The pin was held vertically in a chuck that was fixed to the frame of the machine. The sapphire disc was attached to a vertical spindle, to give the usual pin-on-disc geometry. The spindle was free to move in the vertical direction, thereby allowing the disc to be loaded (by dead weights) against the pin. This particular geometry left the back of the sapphire disc, that is the side furthest from the wear track, to be viewed by microscope or camera attachments (see Figure 6.33 for a diagram of this particular pin-on-disc apparatus).

The use of single crystal sapphire discs is not as impractical as one might, at first, suppose. Apart from its transparency to optical and infra-red radiation, sapphire also has similar thermal and elastic properties to those of steel. The simulation is, therefore, reasonably close to the practical steel-on-steel tribo-system. Furthermore, Al_2O_3 (the chemical composition of sapphire) is a ceramic. Hence the results obtained with the steel/sapphire tribo-system could be relevant to other ceramic/metal combinations thought to be suitable for the ceramic diesel (see Chapter 7).

Quinn and Winer (1987) show that mild-oxidational wear can indeed be obtained with tool steel sliding on single crystal sapphire. They also show that the hot-spots can be seen and photographed for size analysis. Three examples are given in Figures 6.24, 6.25 and 6.26. The authors point out that, except for the shortest exposure times, the disc travels through several pin diameters during the photograph. Even during the shortest exposures, the disc travels approximately one pin diameter. If the hot-spots were loose debris, or attached to the moving sapphire disc, then one would have obtained 'streaking' (so characteristic of moving car headlights in a timed exposure of the photograph of a city at night!). No such streaking can be seen in Figure 6.26, showing that these hot-spots are on the pin.

The actual number of hot-spots seen in any of the three photographs of Figures 6.24, 6.25, 6.26 depended on the exposure time. By plotting the average number of

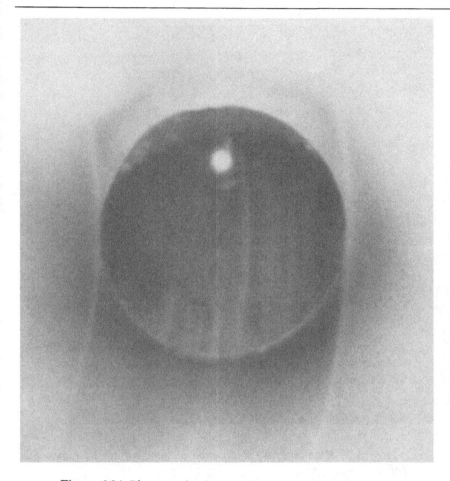

Figure 6.24 *Photograph of pin surface taken through sapphire discs 3 s after start of run. Diameter of pin: 6 mm (pin sliding at 2 m/s from top to bottom of photograph)*

spots per photograph (as determined by measurements taken from a high magnification projected image of the pin surface upon a screen) as a function of exposure time, Quinn and Winer (1987) found that the average number of spots tended towards about seven. They also carried out an analysis of spot size distribution and found that, regardless of the exposure time, most of the spot diameters were between 50 and 100 μm. Making the usual assumption about the dependence of the real area of contact upon the hardness (p_m) and the normal load (W) (see Equation (1.7)), one would expect A_{real}/A_n to have the value of 1.1×10^{-4} (for a load of 26 N and a hardness of 8 GPa). Quinn and Winer (1987) point out that, for the very short exposure photographs, this ratio is 1.1×10^{-3}. This is an order of magnitude *greater* than the expected value. This result is at

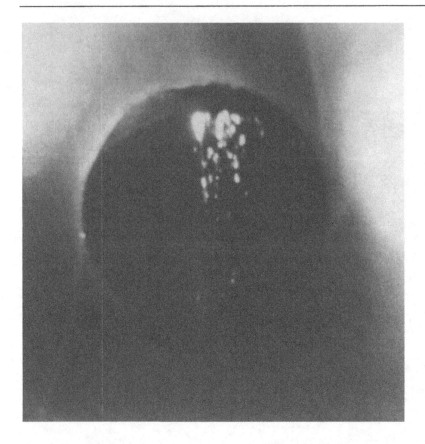

Figure 6.25 *High contrast photograph of same surface as Figure 6.24 (after 10 min)*

variance with the view (which seems intuitively correct) that the hot-spots represent only the most intense of the contact temperatures and that there are significant areas of contact not emitting light. The authors predict several reasons for the unexpectedly high values of the hot-spot area compared with that expected from the plastic deformation theory of asperity contact, including the suggestion that Equation (1.7) may not apply to moving contact conditions. They emphasize, however, that their experiments using direct observation of hot-spots were only of a preliminary nature and that much more evidence would be required before one could entertain the possibility that this cornerstone of tribology might be limited to stationary contact only!

Quinn and Winer (1987) estimated the temperatures of their hot-spots by comparing their photographs with those of heated coupons of tool steel, due recognition being accorded to the effects of exposure times and the developing and printing procedures. In Figure 6.26, for example, the two large bright areas near to the leading edge had the same colour as the standard tool steel

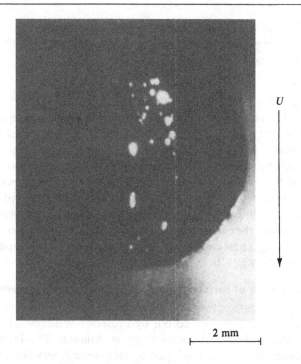

U

2 mm

Figure 6.26 *Very high contrast photograph of same surface as Figure 6.24 (after 25 min)*

photographed at 1200 °C, whereas the remaining spots, dull red on the original photographs, were between 950 and 1000 °C. All these photographs gave much the same information.

From the limited amount of research carried out on the optical analysis of hot-spots formed during unlubricated sliding, it is clear that *direct* observations (as Dr Carpenter said in his contribution to the discussion of the paper) raise more questions than they answer. For example, perusal of either Table 6.1 (for AISI 52100 steel) or Table 6.4 (for EN8 steel) shows that, at 2 m/s and approximately 30 N, one would expect (from heat flow analysis and/or the oxidational wear theory) the number of contacts to be somewhere between 50 and 130 and the contact temperatures between 450 and 500 °C. These numbers are consistent with the direct observation of hot-spots if one assumes that the hot-spots represent only a fraction of the actual contacts giving rise to oxidational wear. This is where infra-red analysis, with its ability to measure temperatures less than the temperature at which the metal begins to emit visible radiation, can possibly resolve the situation (see the next sub-section). Let us leave any further discussion of this aspect until we have given some details of this powerful surface analytical method.

6.3.4 Infra-red analysis of the interface between tribo-elements during sliding and/or rolling with and without liquid lubrication

(i) Introduction

As mentioned in Section 6.3.3, Bowden and Tabor (1954) were perhaps the first researchers to use an infra-red cell to measure the radiation through a transparent moving surface in both dry and lubricated sliding contact. They measured temperatures around 600°C in the dry experiments. In some instances with their pin-on-disc experiments, the surfaces were covered with a mixture of glycerin and water, which undoubtedly resulted in what we would now call a mixed mode of elastohydrodynamic lubrication (see Section 1.3.2). Although the extent of high temperature flashes in such lubricated conditions is greatly reduced, the presence of the liquid film did *not* prevent the occurrence of extremely high local temperatures due to local frictional heating. These early experiments used glass as the transparent moving medium – not a very practical material!

(ii) Infra-red anaysis of surface temperatures in lubricated contacts

Some of the more useful contributions to the study of surface temperatures during *lubricated sliding* have been carried out by Professor Winer and his co-workers at the Georgia Institute of Technology in Atlanta. The infra-red temperature measurement technique developed by these researchers (Turchina, Sanborn and Winer, 1974) and further improved by Nagaraj (1976) consisted of a 31.8 mm diameter chrome steel (AISI 52100) ball rotating and loaded against a sapphire flat 1.6 mm thick as shown in Figure 6.27. The infra-red radiation emitted at this contact was measured with an infra-red radiometric detector having a spot-size resolution of 36 μm with a × 15 objective and a response time of 8 μs in the AC mode of operation. The oil bath temperature was monitored with a thermocouple and held constant by a constant-temperature bath and an oil circulation system. This apparatus enables measurements to be made which do not interfere with the tribo-contact, and can be used at more realistic and severe conditions. The loading of the flat sapphire disc against a steel counterface is a good simulation of the normal metal–metal contact of rotating machinery. The mechanical and thermal properties of sapphire are very similar to steel, as shown in Table 6.12. An important property of the sapphire disc is its transparency to infra-red radiation in the wavelength range from 0.25 to about 5.5 μm. It is this transparency that makes it possible to measure contact temperatures by means of infra-red radiometry.

Nagaraj, Sanborn and Winer (1977) measured the temperatures in elasto-hydrodynamic (EHD) contacts using the apparatus shown in Figure 6.27. Since the spot size of the infra-red microscope was about 38 μm, this gave a resolution of about 13 points on the centre line of the EHD contact diameter of 0.5 mm. The distinct emission spectrum of the fluid made it possible to separate the radiation contributions of the steel ball and the lubricant film by using interference filters. The ball was assumed to be a diffuse, grey body. The emissivity of the lubricant is a function of the film thickness and the fluid temperature, requiring extensive

Table 6.12. *Mechanical and thermal properties of steel and sapphire*

Property	AISI 52100 steel	Sapphire
Hardness (GPa)	14.5	19.6
Thermal conductivity at 23 °C ($\mathrm{W\,m^{-1}\,K^{-1}}$)	37.0	41.9
at 100 °C ($\mathrm{W\,m^{-1}\,K^{-1}}$)	34.7	25.1
Thermal diffusivity at 100 °C ($\mathrm{m^2/s}$)	9.56×10^{-6}	13.9×10^{-6}
Modulus of elasticity (GPa)	207	365
Poisson's ratio	0.3	0.25

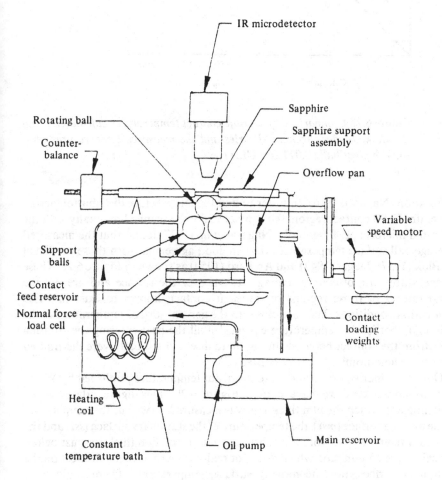

Figure 6.27 *Sliding elasto-hydrodynamic test apparatus*

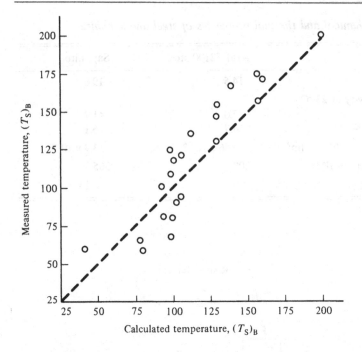

Figure 6.28 *Comparison of average contact temperatures calculated using the Blok–Jaeger–Archard theories and the measured temperatures on a smooth steel ball (0.011 μm CLA)*

calibration (Nagaraj, 1976). In order to find the temperature distribution across the entire EHD contact region, the contact had to be scanned manually with the infra-red microscope. Essentially, Nagaraj *et al.* (1977) found that the measured average ball surface temperatures were in good agreement with those predicted by Blok (1937), Jaeger (1942) and Archard (1959) as shown in Figure 6.28. These temperatures are, of course, averages taken over the surface and are *not* the temperatures of those regions of transient contact, known to exist in EHD lubrication conditions. They are closer to the general surface temperature of the ball, (T_S), that is, the temperature expected if all the heat entering the rotating ball (from the contact between the sapphire disc and the ball) were distributed over the whole nominal area of contact.

Out of the many graphs related to lubricant temperature published by Winer and his co-workers, we have chosen Figure 6.29. This figure relates to an experiment in which the film thickness was measured, as well as the temperature of the moving surface (*mst*), the temperature of the stationary surface (*sst*) and the lubricant temperature. Note the constriction in the oil film thickness just before the exit of the conjunction which does not really seem to have much effect on the stationary surface, nor the moving surface, temperatures. There could be a connection, however, between the lubricant temperature and this constriction

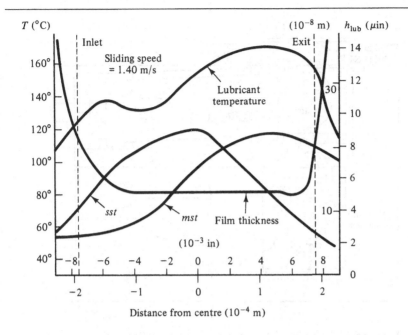

Figure 6.29 *Lubricant and ball surface temperatures plotted against distance from the centre of the contact region (Load = 67 N ; maximum Hertzian pressure = 1.02 GPA ; oil bath temperature = 40°C)*

thickness. The main conclusion that we can draw from their infra-red analysis of the temperatures of sliding lubricated surfaces, is that the method does not give us more than the steady state average value arising from the dissipation of the frictional and/or viscous energies. The more recent work to emerge from the Tribology Group at the Georgia Institute of Technology seems to indicate, however, that with more sophisticated instrumentation, for example an AGA Thermovision 750 infra-red camera associated with improved data acquisition and storing facilities, it is becoming more feasible to analyse the transients we know must exist in EHD contacts. The above-mentioned camera is a scanning instrument. The video signal from this camera can be sampled with a digital oscilloscope and then transferred to a computer for storage and subsequent analysis of the data. Computer programs have been developed for (a) converting video voltages into temperatures and (b) three-dimensional representations of the temperature distributions and the corresponding isotherms. For details of such programs the reader is referred to the work of Griffioen (1985).

An example of the power of this technique for analysing surface temperatures in test rigs more closely simulating the practical situation of lubricated contact is the paper by Bair, Griffioen and Winer (1986) related to the tribological behaviour of an automotive cam and flat lifter system. In the automobile engine, valve actuation is achieved with a cam and follower system. The follower is either

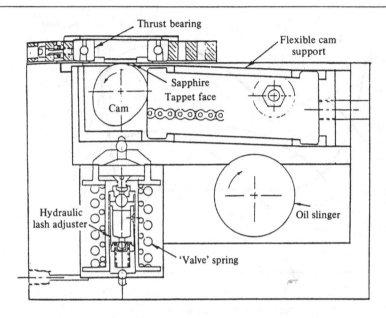

Figure 6.30 *Diagram of the Georgia Institute of Technology cam-lifter test rig*

nominally flat or of the roller type. The cam/follower contact is a significant source of loss of energy in an automobile (e.g. 510 W or 0.7 hp on a 4-cylinder engine). Bair *et al.* (1986) constructed a laboratory test rig to simulate the cam/tappet contact, one in which the lifter is stationary, except for rotation. Instead of the lifter moving, it is arranged for the cam to move through the lift curve in such a way that the contact remains essentially in the horizontal plane. The contact may be viewed through a transparent sapphire lifter foot by an optical or infra-red microscope. Figure 6.30 is a diagram of the Georgia Institute of Technology cam–lifter test rig. Three components of the contact force on the lifter are measured. If required, the sapphire flat can be replaced by a roller tappet. The automotive valve spring rate was simulated by a double compression spring of 97.2 kN/m rate pushing against the bottom end of the flexible frame which holds the cam.

The complete temperature distribution of a typical infra-red scan field is shown in Figure 6.31. The corresponding isotherms are presented in Figure 6.32. The 0° line in Figure 6.32 indicates the contact line between the cam and the lifter at the moment of peak lift.

Other investigators have also used infra-red analysis in a practical, lubricated situation. For instance, Wymer and MacPherson (1975), applied an infra-red technique to a spiral bevel gear pair in a helicopter tail rotor gear box. This was a splash-lubricated system which had been the subject of a large amount of previous research. The surfaces to be observed were the flanks of the 16 pinion teeth, a quarter of a revolution after contact. For details, the reader is referred to

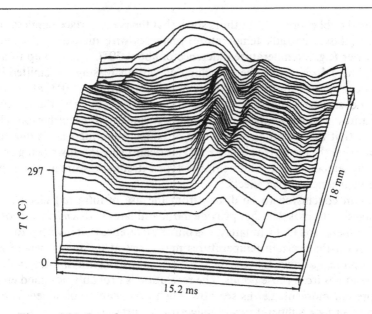

Figure 6.31 *Complete temperature distribution at cam–lifter contact (cam speed = 1500 rev/min; oil temperature = 100%C; lubricant: Amoco LF 6707)*

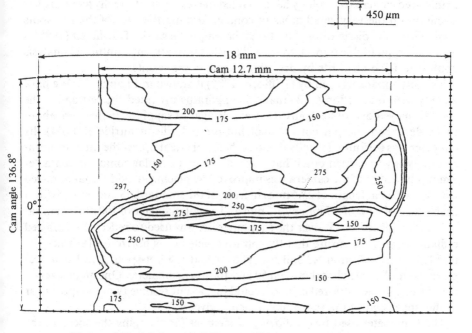

Figure 6.32 *Isotherms of Figure 6.31*

the original publication. The authors show that the mean surface temperature of the teeth (above the bulk temperature) increases with increasing transmitted horse power (e.g. a temperature difference of about 30 °C from 20 up to about 170 hp). There does seem to be a levelling off with increasing transmitted horse power over about 170 hp. Since the oil temperature was about 80 °C at this point, clearly the surface temperatures were about 110 °C. In fact, the maximum temperatures would be above this value. Wymer and MacPherson (1975) suggested that the reduction in the rate of temperature increase is due to the temperature at the local areas where EHD lubrication breaks down being around 150 °C, the temperature at which sulphur-type extreme-pressure additives begin to have their ameliorative effects.

The main conclusion to be drawn from the use of infra-red techniques in lubricated situations is that they provide the investigator with a *minimum* possible surface temperature. Unfortunately, it is only by inference that investigators have tried to show that certain temperatures must exist at the points where EHD lubrication breaks down. Obviously, the lubricant will have a 'smothering' effect on any readings from those points. The resolution is probably not good enough at the present moment. Let us see how much more can be obtained from the analysis of surface temperatures of *unlubricated* surfaces.

(iii) Infra-red analysis of surface temperature in dry sliding contacts

Although infra-red techniques have been used on an occasional ad hoc basis to look at surfaces immediately after they have been in dry sliding contact (for instance, see the paper by Santini and Kennedy, 1975), the more recent advances in infra-red camera technology have enabled the researcher to really focus his/her attention on the very small areas of contact forming the sites of the hot-spots arising from the dissipation of frictional heating. The work of Griffioen (1985) is very important in this respect, so we will finish our treatment of infra-red analysis applied to tribo-elements by describing some of his results.

The test apparatus used by Griffioen (1985) is shown in Figure 6.33. We have already seen (in Section 6.3.3) that this apparatus provided the setting for the optical analysis of tool steel pins also wearing against sapphire. In the work about to be described the pin consisted of hot-pressed silicon nitride (HPSN). Its diameter was 6.35 mm. Initially it had an hemispherical tip. At the time the infra-red analysis was carried out, it had worn down to a circular contact of diameter 2 mm. The optics of the camera was improved by placing an additional infra-red lens (with a focal length of 25.4 mm) in front of the standard 33 mm lens, giving a usable field of view of 2.7 mm by 1.5 mm and a spot diameter of 100 μm. An infra-red mirror, placed at 45° over the sapphire disc, was used to reflect the infra-red radiation (this is to ensure that the infra-red detector is horizontal!). It may be recalled that, in Section 6.3.3, it was shown that the hot-spots seen by optical means were all somewhere between 50 and 100 μm in diameter. One might expect, therefore, that the infra-red camera would just about resolve two hot-spots near to the upper end of the range mentioned above.

The tribometer used by Griffioen had facilities for changing the speed of the disc. This was measured with a tachometer. The bulk temperature of the pin was

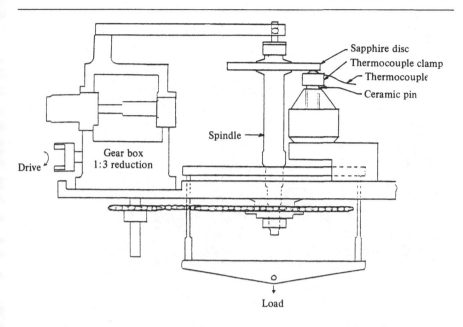

Figure 6.33 *The pin-and-disc apparatus used by Griffioen (1985) for infra-red analysis of a ceramic pin, as seen through the sapphire disc against which it is wearing*

measured by a thermocouple placed with its tip about 1 mm below the plane of contact. This thermocouple played an important role in determining when the running-in of the ceramic pin had occurred. When the heat flow has reached a steady state, it is most likely that the wear mechanism, whether it be mild or severe wear, will also have reached equilibrium. Griffioen (1985) does not identify the wear mechanism operating in his experiments. Having reached an equilibrium state, the infra-red camera was first used in the 'area scan' mode, so as to obtain the temperature distribution across the field of view. The speed with which the temperatures change is illustrated vividly in the two consecutive area scans of Figures 6.34(*a*) and (*b*). These scans are only separated by 1.2 s. At 1.53 m/s, the disc has made 14 revolutions during the time that has elapsed between each complete area scan. The hot-spot around 1800 °C in Figure 6.34(*a*) had reached around 2400 °C within those 1.2 s (as we can see from Figure 6.34(*b*)). New hot-spots have appeared to the left of the centre and the hot-spots to the right have already gone! Isotherm charts for both of these figures indicate that most of the surface is at temperatures less than 200 °C. They also indicate that, apart from the extremely high values for the main peak, most of the other peaks lie between about 300 and 600 °C, a value not inconsistent with the value of $T_c = 481$ °C given in Table 6.1 for 52100 steel specimen sliding at 2 m/s and 9.8 N load. This should not surprise us since silicon nitride has very similar thermal and mechanical properties to sapphire, which in Table 6.12, we saw had very similar properties to steel. For example, thermal conductivity of silicon nitride is 24.0 W/mK

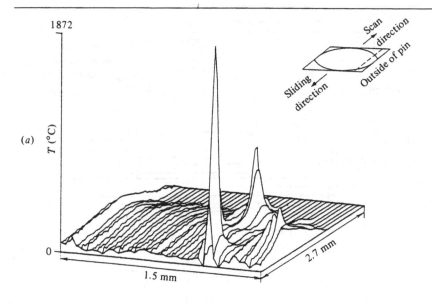

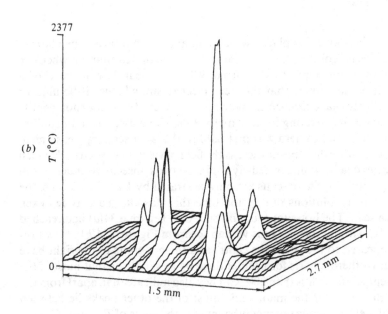

Figure 6.34 *(a) Temperature distribution of an area of ceramic pin surface (speed of sliding = 1.53 m/s load = 8.90 N); (b) temperature distribution of same surface as Figure 6.34(a) taken 1.2 s later*

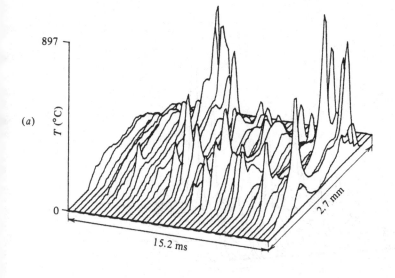

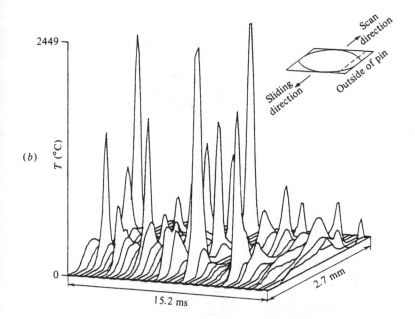

Figure 6.35 *Temperature distribution along a line near to edge of field of view repeated 40 times in 15.2 ms that is repeated every 400 μs (approx.); (b) temperature distribution along same line as Figure 6.35(a) (1 s later);*

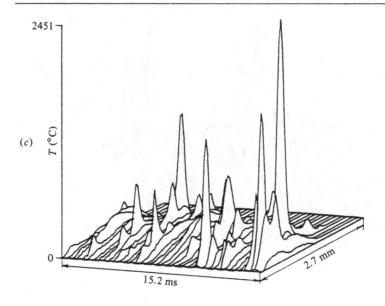

(c)

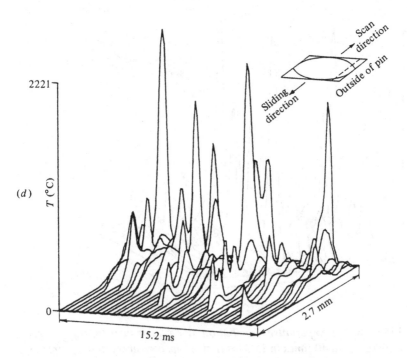

(d)

(c) temperature distribution along same line as Figure 6.35(a),(3 s later);
(d) temperature distribution along same line as Figure 6.35(a),(4 s later);

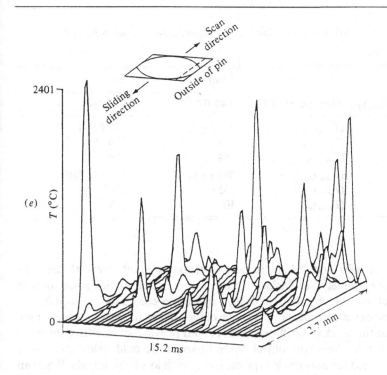

2401

(e) T (°C)

0

15.2 ms

2.7 mm

(e) temperature distribution, line scan [6 s later]

(compared with 27.0 W mK for sapphire), the modulus of elasticity is 310 GPa for silicon nitride (compared with 365 GPa for sapphire) and the hardness of silicon nitride is 17.0 GPa (compared with 19.6 GPa for sapphire).

It is worth emphasizing that the 400 μs time step between data points on *adjacent scan lines* is much larger than (the 5 μs) between consecutive data points on the *same scan line*. A temperature plot, and consequently the corresponding isotherms plot, does not represent a truly *instantaneous* temperature distribution. The temperature distribution is subject to distortions in time due to the scanning method for collecting data points. Figures 6.35(*a*) to (*e*) present a sequence of temporal temperature distributions established from a *fixed* line scan near the outer edge of the ceramic pin. The approximate location of the scan line is indicated by the broken line in the small inserted diagram. It can be seen that the rapid temperature fluctuations have transient times of less than the 400 μs it takes to complete a scan line.

Griffioen (1985) computed the mean of the maximum temperatures for experiments at 0.305 and 1.53 m/s and loads of 4.45 and 8.90 N. He also computed the lowest and highest maximum temperature and his results are shown in Table 6.13. The maximum values of 2353 and 2703 °C are remarkably high and clearly do not contribute to the main wear process. As far as oxidational wear is concerned, if it does happen with silicon nitride sliding on sapphire, it would seem more probable that the mean values of T_{max} (i.e. 1080 and 396 °C for the 1.53 m/s

Table 6.13 *Statistical analysis of hot-spot temperatures (from infra-red measurements)*

Sliding speed (m/s)	Temperature (°C)	Load	
		4.45 (N)	8.90 (N)
0.305	$(T_{max})_{mean}$	105 (± 3)	117 (± 6)
	$(T_{max})_{max}$	148	200
	$(T_{max})_{min}$	94	98
1.53	$(T_{max})_{mean}$	396 (± 87)	1080 (± 180)
	$(T_{max})_{max}$	2353	2703
	$(T_{max})_{min}$	102	116

experiments at 8.90 and 4.45 N respectively) are relevant. Nevertheless, the occurrence of extremely hot flashes of temperature could provide some form of triggering action whereby the *tribological* oxidation constants are so different from the *static* oxidation constants. In view of the interest in the wear of ceramics, it is clear that the work of Griffioen (1985) could provide an incentive for more work along similar lines, the object being to see if the mild oxidational wear theory can be used for non-oxide type ceramics, such as silicon nitride. If we can get independent (reliable) measurement of the number of contact points (N) and their temperatures (T_c), we can readily 'calibrate' theory in terms of the oxidation content relevant to tribological oxidation, *without* the somewhat complicated procedures involving heat flow analysis given earlier in this chapter.

6.4 Concluding remarks on the relevance of physical analytical techniques for analysing surface temperatures in tribo-systems

The essential fact about surface temperatures is that they are very difficult, even impossible, to measure *directly*. In this chapter, we have seen how even the latest developments in infra-red microscopy still leave us with many questions unanswered. The time of the interaction between contacting high spots is so small that even 5 µs is too long for us to follow individual hot-spots. The temperature of the hot-spot is also remarkably high in some instances. Direct measurement is full of problems related to the high-temperature gradients and the short duration of the hot-spots. These problems are even more pronounced when trying to use thermocouples, whether they be embedded, trailing or dynamic.

Until the *direct* methods have been improved still further, there seems to be no alternative for us other than the crystallographic changes caused by changes in the temperatures of the surfaces of tribo-elements during sliding and/or rolling. The proportional analysis of wear debris by X-ray diffraction seems to hold some promise for analysing surface temperatures of materials in dry sliding contact, provided the results obtained with steels can also be extended to other systems, in particular to ceramic/ceramic or ceramic/steel tribo-systems. Glancing-angle X-ray diffraction and reflection high-energy electron diffraction (RHEED) could also be helpful in identifying surface constituents that could only be present provided some transition temperature has been exceeded.

The thermal aspects of sliding cannot be analysed without analysing the *heat flow* between sliding tribo-elements and relating that via some *wear theory* (e.g. the oxidational wear theory) to the temperatures of sliding. The wear mechanism maps of Lim and Ashby (1987) and, in particular, their temperature maps, could be helpful in getting some idea of the type of wear and, hence, the range of contact temperatures between the surfaces that are wearing, provided they can be modified to suit other systems than steels sliding upon steels. There is also a need for the oxidational wear theory to be extended to other materials than steels and to elevated temperatures. A similar extension to the heat flow analysis could also be undertaken. Underlying all these proposals there must be a firm foundation of physical methods of analysis. The present good position as regards the wear of steels could not have been achieved without the extensive use of physical methods of analysis.

7 The analysis of pitting failures in tribo-systems

7.1 Failure in rolling contact bearings in nominal point contact

7.1.1 Introduction

Czichos (1980) defines a tribo-system as one in which motion, energy and materials are transmitted, in various relative amounts, according to the required function of the system, from clearly prescribed inputs to desired outputs. Invariably, motion is a characteristic of any tribo-system. Sometimes the purpose of the system may be to change the rate of motion or to eliminate it altogether (e.g. in brakes). Such changes involve undesired outputs, such as frictional heating and the undesired removal of material from the surface through which the motion is transmitted. Gears are intended to transfer motion and power in rotating machinery, but sometimes unwanted transfer of material may also occur. The wheels of a railway engine are intended to transmit force to the rail and hence produce motion, but again we find that the unwanted transfer of material, and the loss of energy due to slip, will make the actual output somewhat different from that which was desired. Cams and tappets, valves and valve seats, piston–cylinder systems, hot and cold rolling mills, sheet-forming and wire-drawing dies, metal cutting tools, dry bearings and current collectors are further examples of tribo-systems which involve some degree of transfer of motion, energy and materials. All such tribo-systems are said to have failed when the actual output deviates significantly from the intended output.

In this section of the chapter on one of the most insidious forms of tribological failure, namely pitting, we will discuss the pitting failure of ball-bearing systems. We will see that the life expectancy of any given ball-bearing system is not known with any certainty. We will deal with the statistical description used by bearing manufacturers to characterize their products. We will also describe the test machines used for predicting bearing lives. We will leave over until a later section (i.e. Section 7.5), the discussion of how physical analysis has been applied to the pitting failures of rolling-contact bearings in nominal point contact.

7.1.2 The statistical nature of ball-bearing lives

Ball-bearing systems always fail by contact fatigue which manifests itself by small pits, causing noisy and uneven running and, eventually, complete failure. Unfortunately, this type of failure is unpredictable for any given bearing. For a batch of bearings, however, it is possible to talk about a statistical life. Manufacturers often talk of the 'B_{10} life', which is the life which 90% of the bearings will reach or exceed without failure. They also rate bearings in terms of the basic dynamic capacity (C_{dyn}), which is the load corresponding to a B_{10} life of

10^6 revolutions. For any other load (W), the B_{10} life can be obtained from the empirical equation:

$$B_{10} = [(C_{dyn})/W]^n \times 10^6 \text{ revolutions} \qquad (7.1)$$

where $n = 3$ for ball-bearings and 10/3 for roller-bearings (Lester, 1973).

7.1.3 The rolling four-ball machine

Rolling-contact fatigue of ball bearings has been extensively studied with this machine over the past 20 years. It is a modification of the sliding four-ball lubricant tester (mentioned in Section 5.1.1) such that the lower three balls (see Figure 7.1) are allowed to rotate in a race driven by the upper test ball, held in a chuck, to provide a combination of rolling and sliding, similar to that experienced in angular contact bearings. The upper ball simulates the inner race of a conventional bearing and is easily and cheaply replaced. The lower race can be used for a considerable number of tests. Normally, the machine runs at 1500 rev/min, but special machines are available to provide speeds over a range of

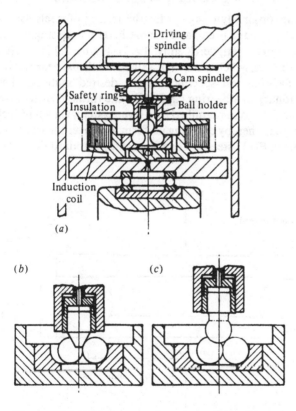

Figure 7.1 (*a*) *Diagrammatic arrangement of the rolling four-ball machine;* (*b*) *cone/three-ball arrangement;* (*c*) *ball-ended specimen/ three-ball arrangement*

3 000 to 60 000 rev/min. Different materials, not available in ball form, can be compared using a standard lubricant, if the upper test ball is replaced by a simple conical specimen, easily prepared from a small quantity of material (see Figure 7.1(b)). A ball-ended specimen may be used in place of a conical specimen (see Figure 7.1(c)) and thus the kinematic system remains unaltered from the rolling four-ball arrangement, enabling any material to be compared with commercially-available balls.

Typically, the loads used in sliding four-ball tests are always in excess of those used in practice. Scott and Blackwell (1971) claim that the results of such tests generally conform to the inverse cube law relating B_{10} life with load (W) given in Equation (7.1) and hence provide support for their extrapolation down to normal service loads. The criterion for failure is the time taken from the start of the test to the appearance of the first failure pit on the surface of the upper ball. This is detected by a change in the noise characteristics of the machine or by a vibration detector (an accelerometer), which allows automatic cut-out on failure and the recording of the test duration or number of revolutions.

7.1.4 The Unisteel rolling contact fatigue machine

The Unisteel Machine, originally designed for the testing of steels for rolling-contact bearings, has been found useful as a test method for assessing the effect of lubricants on rolling-contact fatigue (Kenny, 1977). Figure 7.2 is a schematic diagram of the test assembly, which consists of a standard thrust bearing with one race replaced by a flat-ring test specimen of the desired material. The test specimen is held stationary in a bearing cap, whilst the thrust bearing assembly rotates at 1500 rev/min. The load is applied by a lever system, via a half-inch ball bearing on the bearing cap thereby ensuring uniform load over the test assembly. Only nine of the eighteen EN31 steel ball bearings of the standard thrust bearing

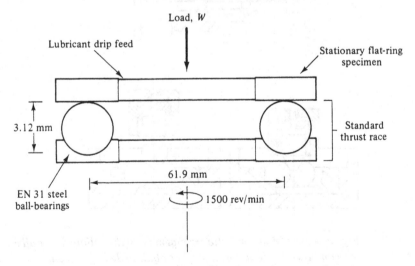

Figure 7.2 *A schematic diagram of the Unisteel test configuration*

are used in order to generate a slight increase in the maximum Hertzian compressive stress (p_{max}). This quantity, which is generated at the specimen–ball interface, can be calculated from the standard equation for point contact conditions (Timoshenko and Goodier, 1951), namely:

$$p_{max} = \frac{1}{\pi}\left\{\frac{3}{2}(W)\left(\frac{E''}{R_1}\right)^2\right\}^{1/3} \tag{7.2}$$

where W is the load, R_1 is the radius of the ball, and E'' is the reduced elastic modulus already defined in Chapter 1 (see Equation 1.24). When the test piece fails through pitting fatigue, vibrations within the stationary specimen housing are picked up by accelerometers. The resulting signals are processed by a control unit that activates a 'cut out' in the motor supply. The threshold value for cutting out can be set manually. A typical run on a Unisteel machine is about 60 h. Hence the need to run several machines concurrently so as to generate enough data to enable a statistical analysis (known as the Wiebull Analysis) to be carried out and values obtained for both B_{10} and B_{50} lives.

7.2 Basic research into the contact fatigue of ball bearings

There has been very little research carried out on the *fundamental* reasons behind contact fatigue and pitting of ball bearings in nominal point-contact conditions. *Most attention* has been given to the *factors that will give an increased B_{10}* life, such as hardness, heat treatment, fibre orientation, the steel-making process, the material combination, the lubricant and the environment. Scott (1968) examined these seven factors in some detail and concluded that there were several distinctly different modes of rolling-contact fatigue, each one of which can cause cracks to nucleate and propagate independently at various rates. The nature of the lubricant and the environment can have a dominant effect on failure. Scott (1968) recommends that significant increases can be obtained in the contact fatigue life of ball bearings provided (a) due care is taken in the steel-making process, (b) that the metal is heat treated to 'optimum' hardness with the 'correct' microstructure, and (c) the final ball has the 'proper' surface finish.

One is left with a distinct feeling that, despite the wealth of data obtained with the rolling four-ball machine, the understanding of contact fatigue and pitting is *unlikely* ever to emerge from such a complex geometry. The lower three-balls, upon which the upper test specimen is made to rotate, cannot possibly be considered to be in a state even approximating to *pure* rolling. The amount of slip between all the contacting surfaces is significant and yet it is unknown and uncontrollable. Some of the more recent work on rolling-contact fatigue of point contacts has involved attempts to get away from the conventional four-ball geometry. For instance, Diaconescu, Kerrison and MacPherson (1975) have developed a 'Twin-Head Five-Ball Machine', whilst Dalal, Chiu and Rabinowicz (1975) used a unidirectional thrust bearing for their rolling-contact fatigue tests.

There has been a small amount of research carried out on the Unisteel Rolling Contact Fatigue Machine (as opposed to mere tests). Some particularly

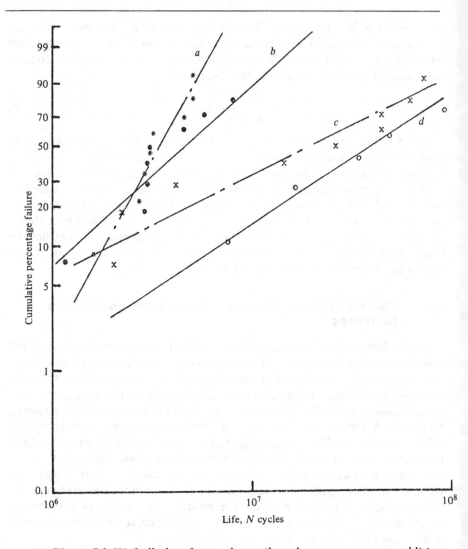

Figure 7.3 *Weibull plots for two base oils and extreme-pressure additive blends tested in the Unisteel machine (Phillips, 1979), (a) *, Risella 17 + 5 wt% extreme-pressure additive; (b)* ●, *Risella 17; (c)×, Risella 32 + 5 wt% extreme-pressure additive; (d)* ○, *Risella 32*

interesting work on the Unisteel Machine has been reported by Phillips (1979). He was interested in the rather unexpected behaviour of extreme-pressure additives (which we have seen in Chapter 5 are included in lubricants to help prevent scuffing, that is premature failure of gears, cams and tappets and other non-conforming tribo-elements). These additives were thought to cause *pitting* at an earlier stage than if the tribo-system were run without them. Obviously, no manufacturer of expensive gear systems would be willing to prove this idea in an actual service test. Hence the use of laboratory apparatus. Using two base oils

(Risella 17 and Risella 32) and 5 wt% extreme-pressure additive blends with these oils, Phillips (1979) fitted the results from his Unisteel tests to the Weibull cumulative distribution function, namely:

$$F(t_B) = 1 - \exp\left\{ -\left(\frac{t_B}{t_c}^{b_w}\right)\right\} \tag{7.3}$$

where $F(t_B)$ is the probability of failure of a percentage of the population in a lifetime (t_B), t_B is the individual bearing life, t_c is a characteristic life (the life at which 63.2% of the population will have failed) and b_w is the Weibull slope. These Weibull plots are given in Figure 7.3. The base oils were highly-refined medicinal white oils, containing mainly saturated cyclic hydrocarbons with side chains and less than 20 ppm of sulphur. The extreme-pressure additive was a commercial sulphur and phosphorus type, representative of gear lubricant additives in common use.

The arithmetic mean lives for Risella 17 and Risella 17 + 5 wt% extreme-pressure additive were 5.65×10^6 and 3.55×10^6 cycles respectively. The confidence number that the base oil gave a longer mean life than that obtained with the extreme-pressure additive blend was greater than 95%. The mean lives for Risella 32 and Risella 32 + 5 wt% extreme-pressure additive were 5.4×10^7 and 3.1×10^7 cycles respectively, with the confidence number related to the longer life with the base oil equal to 88%. In both instances, therefore, the blending of the extreme-pressure additive to the base oil has significantly reduced the mean fatigue life. We will be returning to this significant finding of Phillips (1979) after we have discussed the strongly connected topic of contact fatigue in gears and other tribo-systems in nominal *line* contact.

7.3 Failures in gears and other tribo-systems in nominal line contact

7.3.1 The scuffing and pitting of gears

Theory of elasto-hydrodynamic lubrication (EHL), when applied to gears or other line contact tribo-systems, predicts oil film thicknesses of the order of the combined roughness of the surfaces being lubricated. The measure of *possible* contact between surface asperities on opposing surfaces is given by the D-ratio, which we have already defined in Equation (2.36) (see Section 2.5.3). It may be recalled that this was the ratio of the total initial surface roughness divided by the (theoretical) EHL film thickness. Lubrication experts tend to favour the 'λ-ratio', which is merely the inverse of Equation (2.36). There is no reason to prefer either definition, except that the lubrication expert might say that the larger the λ-value, the better the lubricating conditions. However, we will use the D-ratio, since this has traditionally been favoured by researchers into the pitting resistance of gears and other line-contact tribo-systems, and much of what we will be discussing is related to the pitting failures of such systems.

It is clear that when $D > 1$, then the combined initial surface roughness of both surfaces is greater than the film thickness predicted on the basis of EHL theory. Hence, *marginal contact must occur* between the asperities (i.e. high spots) on opposing surfaces. In view of all we have already said about mild oxidational wear, we realise that a certain amount of protection would be provided by the tribologically-formed oxide film at the contacting asperities. If the rate of removal of these films exceeds the rate of formation, however, we have the conditions for severe metallic wear, or 'scuffing' as the gear technologists call it. In fact, scuffing is defined as the 'gross damage of the working surfaces caused by the formation and shearing of local welds', a definition that is equally applicable to the severe wear of unlubricated surfaces.

Scuffing of gears is a complex problem still not fully understood despite many investigations into its *unexpected* occurrence in actual tribo-systems. It occurs abruptly, accompanied by frictional heat and, sometimes, noise. The surfaces become roughened and, in severe cases, the whole of the load-carrying surface of the gear is destroyed. Although advances in gear technology have made scuffing a rare occurrence nowadays (Saunders-Davies, Richards and Galvin, 1980), the reasons for its onset in any particular tribo-system remain somewhat obscure.

Temperature has long been used as a criterion related to the load-carrying capacity of the lubricant used between two gears transmitting load (Blok, 1937, 1970). Essentially, the criterion states that oil breakdown will occur when the maximum temperature in the real areas of contact reaches an initial value, which will be constant for a given combination of materials and lubricant. This maximum temperature is the sum of the general surface temperature (T_S) of the hotter of the two bodies and the temperature 'flash' experienced by points on their surfaces as they pass through the contact zone. This idea seems to be relevant to tribo-systems operating in non-additive oils. For oils containing extreme-pressure additives, there has been less agreement about the criterion. This may be due to the fact that the chemical reaction of the additive (or its derivative) with the load-carrying surface is much more important, from the wear point of view, than the *loss* of hydrodynamic support of the oil at its breakdown temperature (see Chapter 5).

In much the same way as with ball bearings, gear teeth can exhibit signs of pitting due to contact fatigue. The roughness, hardness, thickness and composition of the surface layers and the presence of tensile or compressive stresses in these layers, and of defects or inclusions in or near the surface, all affect the resistance to pitting of gear teeth (Saunders-Davies *et al.* 1980). Pitting of gear teeth due to contact fatigue takes many forms, but all forms belong to one or another of two general categories, namely *incipient pitting* or *progressive pitting*. Most incipient, or initial, pitting is attributable to the presence of high spots on the flank of the gear tooth. The pits are small, shallow and randomly distributed. Once the gears are run-in, this type of pitting generally stops. However, due to causes not really well understood, it sometimes happens that the pitting does not stop, but progresses until the lost area is too great for any remedial action to be taken.

7.3.2 The scuffing and pitting of cams and tappets

Cam–follower systems are extensively employed in engineering, particularly in the automotive valve train. The contact conditions are generally nominal line contact, although some systems may involve nominal point contact. We have already discussed the cam–lifter contact temperatures (in Section 6.3.4), where we saw peaks around 290 °C in a commercial lubricant maintained at a general temperature of 100 °C. Although no comment was made, at that stage, on the significance of these temperature peaks, 290 °C is well above the maximum contact temperature of 125 °C recommended by Summers-Smith in Section E10 of the *Tribology Handbook* (Neale, 1973) for rolling bearings. One might expect more metal–metal contacts would occur under these conditions, or possibly, any extreme-pressure additive present would become more reactive. Whatever happens at cam–follower contacts, we know that the behaviour of this complex tribo-system is a function of oil film thickness. If one ignores the effect of temperature, basic elasto-hydrodynamic lubrication theory can be used to show that a cam with a smaller 'nose' radius (8 mm) is more effective in producing a hydrodynamic film in the region of the nose contact compared with a cam with a 15.28 mm nose radius (Naylor, 1967). His results are shown in Figure 7.4.

It should be emphasized, however, that *either thickness* is comparable with the best surface finish that can be produced by normal engineering methods, showing that the ratio of roughness to oil film thickness (the *D*-ratio) is indeed an important factor in these tribo-systems. Friction is only important in that it produces unwanted heat. It is interesting to note that Naylor's (1967) predicted

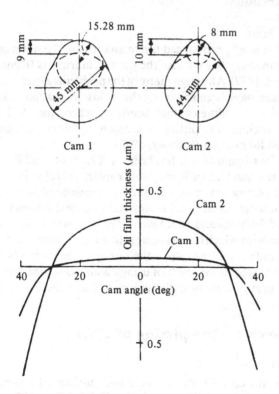

Figure 7.4 *Oil film thickness versus cam angle (Naylor, 1967)*

superiority of performance of the small-nosed cam was found to be consistent (Muller, 1966) with actual engine performance. In an engine using a steel camshaft and chill cast-iron tappets, the large-nosed profile suffered severe scuffing followed by pitting after a very short time in operation, whilst the small-nosed profile camshaft was found to be satisfactory.

Scuffing and pitting do seem to occur sequentially in many tribo-systems involving sliding and rolling in the presence of a lubricant. We shall see, in a later section, that *some* sliding wear is always involved in these systems. Perhaps it is when the sliding wear is no longer of the mild, reactive, kind (e.g. mild-oxidative wear), that the surfaces become scuffed and give all the appearances of severe mechanical wear? Although we will not be able to answer this question as regards cams and tappets, we will see that current knowledge of line-contact tribo-systems in general supports this hypothesis.

7.3.3 The pitting and cracking of rolling/sliding contact in the presence of an aqueous lubricant

In certain industries, the use of mineral-based lubricating oils is not possible due to the risk of ignition. Hence, fire-resistant fluids have been developed to replace them. The four major classes of fire-resistant fluids are:
(a) oil-in-water emulsions,
(b) water-in-oil invert emulsions,
(c) synthetic fluids and
(d) water-glycol-based fluids.
The fluids that exhibit the most pronounced fire resistance are the water-glycol-based solutions. Unfortunately, they are also the poorest lubricants (Kenny and Yardley, 1972 and Knight, 1977). Although many of the problems associated with sliding wear and corrosion have been solved by the inclusion of additives, these fluids still perform badly in rolling-contact bearings, where the steel tribo-elements suffer fatigue cracking and pitting at a much earlier stage than with conventional mineral oil lubrication (Cantley, 1977).

Although the rolling four-ball machine has been used by Scott and Blackwell (1971) and the Unisteel test machine by Kenny and Yardley (1972) to investigate the factors affecting the pitting failure of steels in the presence of water-glycol fluids, full instrumentation cannot be easily achieved. In the next sub-section, we will discuss the work of Sullivan and Middleton (1985) in which they used a rolling two-disc machine, with which it was possible to monitor wear and friction continuously and to identify cracks and study their progress to form pits. These authors were mainly interested in the effect of using an aqueous lubricant. This will, therefore, be only part of a general discussion on basic research into the pitting of discs.

7.4   Basic research on the pitting of discs

7.4.1   Introduction

In this sub-section, we will concentrate on the basic mechanical engineering research into the pitting of discs in (a) normal lubricants and (b) aqueous lubricants. Although some contention *still* exists regarding the equivalence between *actual* tribo-systems (such as gears, cams and tappets) and the rolling two-disc machine, most practicing tribologists would now accept results obtained on that machine as being a good indication of the likely behaviour of those tribo-elements in actual service.

We have already discussed (in Section 2.5.3) the pioneering work of Way (1935) and the more definitive work of Dawson (1962). It may be recalled that we obtained the following equation from Dawson's (1962) paper:

$$\log_{10}(N_R) \stackrel{\sim}{=} -0.67[\log_{10}(D)] + 5.67 \tag{2.37}$$

Clearly, N_R is inversely dependent upon the D-ratio, such that high D-ratios give short pitting lives (N_R), and vice versa. High D-ratios occur when theoretical (EHL) film thickness is much *less* than the combined initial roughness of the discs. Dawson (1962) attributes the short pitting life to the increased intermetallic contact between the discs.

7.4.2 Comparison of pitting lives of gears and discs in conventional lubricants

Before discussing the more recent work on the pitting of discs in base oils, and base oils plus extreme-pressure additives, we should be aware of the work of Onions and Archard (1975) who carried out pitting tests with *both discs and gears* using base oils only. They used the same materials for the discs and gears as Dawson (1962) and then plotted pitting lives versus D-ratio, the definition of which they took as being given by *any* one of the following equations:

$$D_1 = (Ra_1 + Ra_2)/(h_{min})_0 \tag{7.4a}$$
$$D_2 = (2Ra_2)/(h_{min})_0 \tag{7.4b}$$
$$D_{F_3} = (Ra_3 + Ra_4)/(h_{min})_0 \tag{7.5a}$$
$$D_{F_4} = (2Ra_4)/(h_{min})_0 \tag{7.5b}$$

where Ra_1 and Ra_2 are the respective *initial* surface roughnesses of the EN25 and EN32 steel gears and Ra_3 and Ra_4 are the respective *final* surface roughnesses, whilst $(h_{min})_0$ is the *minimum* oil thickness calculated from elasto-hydrodynamics. Their results are shown in Figure 7.5, which is actually derived from Onion's PhD thesis (1973). The variation in D-ratio was obtained by using two base oils of widely different viscosities, but produced from the same base stock. The various definitions shown in Equations (7.4) and (7.5) are merely the various ways in which different groups of researchers have chosen to define the combined surface roughnesses. The main difference between gears and discs is the *significantly shorter lives of gears compared with discs*. For instance, Figure 7.5 shows that a forecast based on disc machine tests at $D = 1$ (say) would overestimate the pitting life by a factor of a hundred! Notice that the difference gets progressively smaller as D increases. Another feature of this figure is that for the disc geometries, there does not seem to be much of an effect of whether one uses initial or final surface roughness, nor whether one uses EN25 or EN34 steel. However, using the gear machine, there is clearly an effect of using D_1 (relating to the initial surface roughnesses of EN25 steel gears) compared with using the other D-ratios.

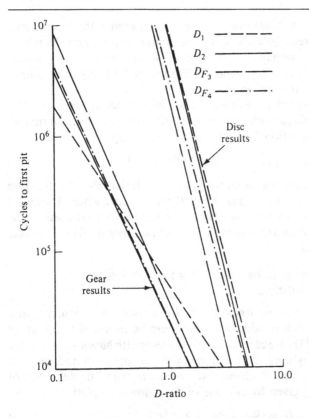

Figure 7.5 *Comparison of gear results with disc results (Onions, 1973):
plots of cycles to first pit versus D-ratio (base oils only)*

However, the significance of this is somewhat marred by the alignment of D_{F_3} (for EN34) with the other definitions for the D-ratio.

Onions (1973) was aware that his results would once again resurrect the old arguments about the merits of the simulation of gear behaviour by disc machines or other laboratory equipment. Hence he discusses and eliminates the following factors as possible causes of the discrepancy so vividly indicated by Figure 7.5, namely (a) relative radius of curvature, (b) character of surface topography, that is the effect of circumferential or axial grinding, (c) slightly different drive ratios (1/1 for discs and 29/30 for the gear rig which he used), (d) differences in the nominal (apparent) areas of contact between gear teeth and discs, and (e) statistical averaging. He suggests that the divergence arises from more fundamental differences between the disc and gear tests, and these differences are probably associated with the film thickness or with dynamic loads. We are still not very sure about the applicability of EHL theory for deducing the film thickness between meshing gear teeth, and techniques for experimentally measuring it have not always been widely accepted as valid. Also, because pitting is critically dependent upon stress, it is conceivable that the differences shown in Figure 7.5 arise from dynamic loading, which probably increases with speed.

The above mentioned problem of possible discrepancies between behaviour in service and in laboratory simulations will constantly recur in tribological investigations. We must not be too hasty, however, in rejecting the results from a particular simulatory system simply because it falls down in one respect. It is interesting to note that Bell and Kannel (1970) consider that it might be valid to use coned *discs* to simulate the geometry and lubrication of any particular chosen *ball* bearing. In my opinion the controlled nature of the two-disc machine makes up for most of the disadvantages it may have as regards absolute values. Provided it reproduces the same trends and enables us to understand the mechanisms of pitting, then it is surely worth putting up with the sort of discrepancy which is apparent in Figure 7.5.

7.4.3 The effect of extreme-pressure additives on the pitting lives of rolling/sliding discs

In Section 7.1.4 we saw how Phillips (1979) used the Unisteel machine, with its *point-contact* geometry, to show that the addition of a commercial sulphur and phosphorus type of extreme-pressure lubricant reduced the arithmetic mean lives by about 40% from that of the plain mineral oils. In fact, Phillips (1979) used the Archard and Cowking (1965–66) equation for the minimum oil film thickness (h_{min}) for *point-contact*, namely:

$$(h_{min}) = 1.40[(\alpha_P)(\eta_0)\bar{U}]^{0.740}(R_r)^{0.407}(E''/W)^{0.074} \qquad (7.6)$$

where (α_P) is the pressure coefficient of viscosity, η_0 is the ambient viscosity of the lubricant, $\bar{U}$ is the combined surface velocity (which is the mean of the velocities, U_1 and U_2 of the bearing surfaces). R_r is the reduced radius of contact given by Equation (5.44). E'' is the reduced elastic modulus (given by Equation (1.24) and W is the load.

In their work with the rolling disc machine, Phillips and Quinn (1978) used the same base oils and base oils plus additives as Phillips (1979) in his research with the Unisteel (point-contact) machine (see Section 7.1.4). These authors used EN26 steel discs, the harder (560 VPN) disc being the faster, driving disc, whilst the softer (260 VPN) disc was the slower, pitting disc. There was 10% slip between the discs, which were loaded at 2668 N, giving a maximum Hertzian compressive stress of 1.2 GPa (similar to that used by Dawson, 1962). Their results are summarized in the regression lines shown in Figure 7.6. Linear regression of $\log_{10}(N_R)$ on $\log_{10}(D)$ gives the following best fit lines:

$$\log_{10}(N_R)_{BO} = -0.93[\log_{10}(D)] + 6.16$$
$$\log_{10}(N_R)_{EP} = -0.44[\log_{10}(D)] + 6.07 \qquad (7.7)$$

where $(N_R)_{BO}$ relates to the lives using the base oils only and $(N_R)_{EP}$ relates to the lives obtained using the extreme-pressure (EP) blends. D is defined by Equation (7.4a). We can re-write Equations (7.7) in the power form, namely:

$$(N_R)_{BO} = (1.44 \times 10^6)D^{-0.93} \qquad (7.8)$$

$$(N_R)_{EP} = (1.18 \times 10^6)D^{-0.44} \qquad (7.9)$$

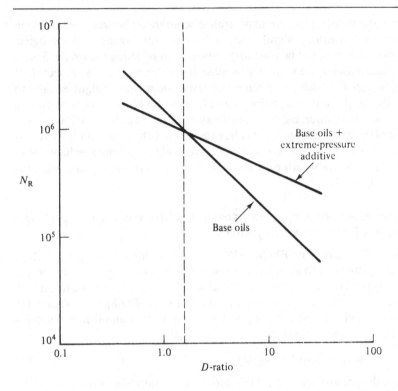

Figure 7.6 *The life to first pit (N_R cycles) as a function of initial D-ratio of two discs rolling with 10% slip using both the base oils and the extreme-pressure additive blends, (the regression lines only are drawn)*

In order to obtain a sufficiently wide range of D-ratios, Phillips and Quinn (1978) had to vary both surface roughnesses and viscosities. The theoretical (EHL) oil film thickness $[h_{min}]_0$ for *line-contact* has been derived by Dowson and Higginson (1966), based on the assumptions that (a) pure rolling occurs (no frictional heating due to sliding), (b) the lubricant is an incompressible Newtonian fluid and (c) the elements are ideal rolling elements. Hence:

$$(h_{min})_0 = 1.6(\eta_0 \bar{U})^{0.70}(\alpha_P)^{0.60}(E'')^{0.03}(R_r)^{0.43}(W)^{-0.13} \tag{7.10}$$

where all the terms have already been defined.

Phillips (1979) also plotted these results against a D-ratio involving *final* surface roughnesses (actually it was D_{F_3} as given in Equation (7.5a), and the resulting regression lines are shown in Figure 7.7. Regression analysis on the final D-ratio (D_{F_3}) gives the following equations:

$$\log_{10}(N_R)_{BO} = -1.04[\log_{10}(D_{F_3})] + 6.08$$
$$\log_{10}(N_R)_{EP} = -0.12[\log_{10}(D_{F_3})] + 6.03 \tag{7.11}$$

Note that all four regression lines (Equations (7.9) and (7.11)) appear to pivot

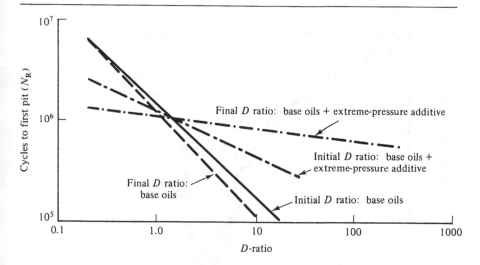

Figure 7.7 *Comparison of regression lines obtained using initial and final D-ratios for the base oil tests compared with those obtained with the base oil plus 5 wt% extreme-pressure additive (Phillips, 1979)*

around a D-ratio of approximately 1.5. For the sake of comparison, let us re-write Equations (7.11) in the power form, namely:

$$(N_R)_{BO} = (1.20 \times 10^6)(D_{F_3})^{-1.04}$$

$$(N_R)_{EP} = (1.07 \times 10^6)(D_{F_3})^{-0.12}$$

(7.12)

It is clear that, under conditions of elasto-hydrodynamic lubrication (EHL), where $D < 1$, the use of the definition of D involving *final* surface roughness (D_{F_3}) means a more *pessimistic* estimate of pitting life when a lubricant containing an extreme-pressure additive is used. For mixed and boundary lubrication ($D > 1$), the *final* surface roughness predicts a more *optimistic* estimate of pitting life for the extreme-pressure lubricant. Using just the base oil, we see very little difference using either *initial* or *final* roughness. Since most of the pitting failures will occur when $D > 1$, this led Phillips (1979) to conclude that the *initial surface roughnesses* were the most important factors in determining pitting lives. It is now generally accepted that the D-ratio given in Equation (7.4a), namely $D_1 = (Ra_1 + Ra_2)/(h_{min})_0$, should be used in any research involving pitting.

7.4.4 The effects of aqueous lubricants on the pitting and wear of steel discs

Since aqueous lubricants are so very different from conventional lubricants, one should always establish in which lubrication regime one is working, using a Stribeck curve, that is, a graph of coefficient of friction versus bearing number (see Section 1.3.3). A typical curve for water-glycol is shown in Figure 7.8. It may be recalled that the bearing number is the dimensionless parameter ($\eta \bar{U}/P$), where η is the viscosity, $\bar{U}$ is the average speed of the lubricant passing through the

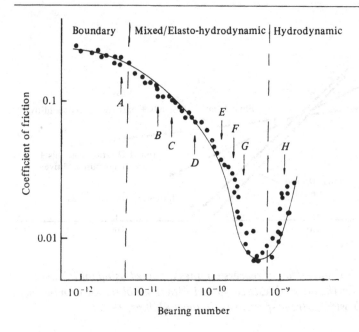

Figure 7.8 *The Stribeck curve for a typical water–glycol fluid (Sullivan and Middleton, 1985)*

conjunction of the two discs and P is the contact pressure (i.e. the elastic stress at the nominal area of contact). This is the classical shape clearly indicating boundary, mixed-elasto-hydrodynamic, and hydrodynamic lubrication. The letters A to H indicate the eight regions of lubrication selected by Sullivan and Middleton (1985) for further investigation. These authors were probably the first researchers to show that wear rates of the softer, slower (pitting) disc (as measured by a linear voltage/distance transducer located on the top disc housing) were inversely related to the bearing number, as shown in Figure 7.9, which is a graph of wear rate versus bearing number for rolling/sliding discs of SAE 52100 steel lubricated with water–glycol. The wear rate (w) in units of (m^3/m) is given by the relation:

$$w = 3.24 \times 10^{-20} (\eta_0 U/P)^{-0.76} \tag{7.13}$$

The letter and number against each point relate to the lubrication regime of Figure 7.8, and the average number of detectable cracks at the appearance of the first pit, respectively. It is interesting to note that the wear rate was found to vary as $(\eta_0 U)^{-0.76}$ (since the contact pressure, P, was kept constant at $0.8\,\mathrm{GPa}$ throughout the tests A to G). Comparing with Equation (7.10), with the load constant at 900 N for these tests and E'', R_r also being constant, we see that, *for line contact*, the minimum film thickness $(h_{\min})_0$ is proportional to $[(\eta_0 U)^{0.70}(\alpha_P)^{0.6}]$. If we examine Equation (7.6), *for point contact*, we see that $(h_{\min})_0$ is proportional to $[\eta_0 U(\alpha_P)]^{0.74}$. Sullivan and Middleton (1985) suggest that the closeness of these indices to the (0.76) in Equation (7.13) could indicate

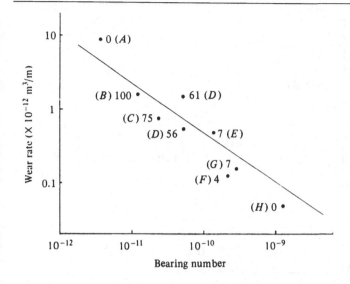

Figure 7.9 *Graph of wear rate versus bearing number for water–glycol fluid lubricated, rolling/sliding discs of SAE 52100 steel (10% slip)*

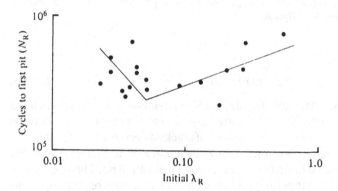

Figure 7.10 *Graph of cycles to first pit versus initial lambda factor (Sullivan and Middleton, 1985)*

that the wear rate is actually *inversely* proportional to the film thickness. This seems intuitively correct.

Sullivan and Middleton (1985), in contradistinction to most other pitting researchers, plotted cycles to first pit against the lambda factor (λ_R), which is, of course, the reciprocal of the D-factor. Their graph of initial lambda factor [i.e. $(1/D_1)$ as given by Equation (7.4a) and cycles to first pit is reproduced in Figure 7.10. The expected increase of (N_R) with λ_R did not occur until $\lambda_R \geq 0.05$, when the 'best-fit' line gave a relationship of:

$$\log_{10}(N_p) = 0.42 \log_{10} \lambda_R + 5.91 \tag{7.14a}$$

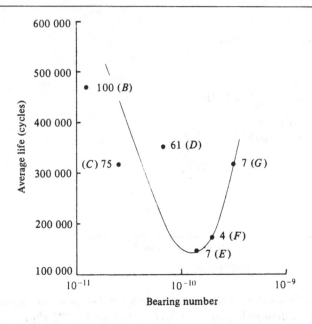

Figure 7.11 *Graph of average life to first pit against bearing number (Sullivan and Middleton, 1985)*

or

$$N_R = (0.81 \times 10^6)D^{-0.42} \qquad (7.14\,b)$$

The fact that N_R for similar steels running under extreme-pressure additives has a similar dependence upon D as the water glycol (see Equation (7.9)) may be fortuitous. What is interesting is the number of cracks detectable (by the magnetic crack detector discussed in Section 2.5.3) in the surface of the pitting disc at the time the run is terminated upon the creation of the first actual pit. This is shown in Figure 7.11, where we see that for runs in the mixed lubrication regime (B, C and D) the numbers of detectable cracks are respectively 100, 75 and 61, whereas in the elasto-hydrodynamic regime, (i.e. E, F and G) the numbers of cracks are 7, 4 and 7 respectively. Obviously, there is an effect of bearing upon the onset of cracking in the mixed lubrication regime. This *could* be the first stages of a form of mild oxidational wear, but clearly, more research is needed.

7.5 Physical analysis of contact fatigue pitting in tribo-systems

7.5.1 Introduction

In this sub-section, we will describe some of the ways in which physical analytical techniques have helped to further our understanding of contact fatigue pitting. We will nearly always be concerned with *steel* tribo-elements. We will divide the discussion into three main parts, namely (a) physical analysis and point-contact tribo-systems, (b) physical analysis and line-contact tribo-systems and (c) a summary of the present position in contact fatigue pitting.

7.5.2 Physical analysis and point-contact tribo-systems

As already stated at the start of Section 7.1.2, there has been very little research carried out on the fundamental reasons for the contact fatigue and pitting of tribo-elements in *nominal point contact*. This also includes the application of physical analytical techniques. We will not go into any detail about any particular technique, since none of the physical analysis has been of a definitive nature. In what follows, we will give a broad review of the application of surface analytical techniques to the study of failed ball bearings, thereby providing the reader with sufficient information and references for him/her to assess the potential of these techniques for his/her particular point-contact problem.

Surface analysis of failed ball bearings has tended to be confined to transmission electron microscopy of replicas (Scott and Scott, 1960), transmission electron microscopy of extraction replicas and tin foils (Swahn, Becker and Vingsbo, 1976), and scanning electron microscopy (Syniuta and Carrow, 1970). These references are merely quoted as being typical of their genre. It is, in fact, very difficult to be definitive if one's research involves microscopic analysis of any sort. How can one extract any numbers from optical or electron micrographs which can be plotted on curves, evaluated and tested critically? How can one be sure the regions visible in the electron micrographs are truly representative of the surface as a whole? Rabinowicz (1972) claims that there is a technique that involves the detection of exoelectrons, which does not suffer from the disadvantages implied in the two questions mentioned earlier. Using specimens which had run to failure in the rolling four-ball tester, he was able to reconstruct the history of a bearing-ball failure by observing the rate of emission of exoelectrons and their location at intervals during a fatigue test, as shown in Figure 7.12. The initial measurements were made by placing the unworn ball in an evacuated chamber and scanning with a focused ultra-violet light beam in order to stimulate exoelectron emission (see Figure 7.12(a)). Normally, the damage and failure occurred on a relatively narrow wear track on the upper ball which is removed every 10 min during a test, tested for exoelectron emission and returned to the tester for further stressing. Figure 7.12(b) shows the emission as a function of rotation (or distance along wear track) after 20 min with three large peaks at 4, 10 and 16 mm around the wear track. Figure 7.12(c) shows that one of these peaks, the one at about 10 mm, coincides with what eventually proved to be a fatigue pit site.

Rabinowicz (1972) claims that, besides enabling him to locate, in advance, the position of an eventual surface-fatigue failure, exoelectrons also throw light on the mechanism of the failure process. For instance, the technique (he claims) can answer the question of whether the crack grows in the direction of rolling or away from it. Rabinowicz also claims that the studies of his group at the Massachusetts Institute of Technology have shown that the leading edge of the eventual pit (or spall, as he calls it) shows a prominent peak early in the test but, later on, that

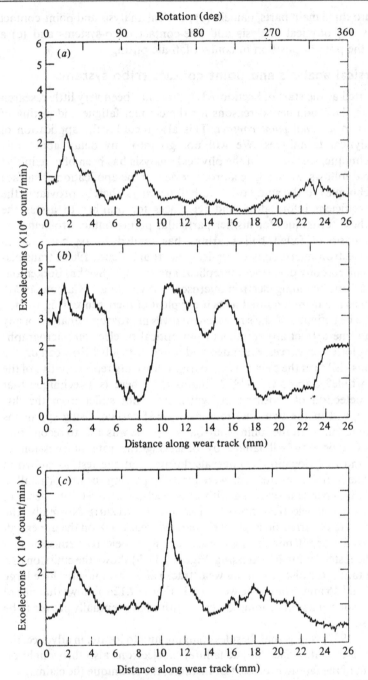

Figure 7.12 *Exoelectron emission from the upper test ball in the rolling four-ball machine (Rabinowicz, 1972): (a) initial emission from unworn ball, (b) emission along wear track after 20 min (c) emission along wear track after failure (at 34 min)*

peak decays, presumably as the crack extends under the surface and parallel to it. In contrast, the trailing edge shows hardly any exoelectron emission above the background rate, until actual pitting takes place.

Apart from some work done by Goldblatt and his colleagues at Exxon Research Laboratories in New Jersey, however, the great future predicted by Rabinowicz for this technique has not materialized. He foresaw that, by the early 1980s, it should be possible to detect exoelectrons in air and thus be able to make routine inspections of entire wing structures of airplanes, searching for growing fatigue cracks. Similarly, he saw the technique being useful for detecting incipient flaws in wheels and rails. He even foresaw it being applied in earthquake prediction by indicating which cracks in the surface of the earth are actively growing! The reason for the lack of progress is probably due to the need for the specimen to be placed in a vacuum for the exoelectrons to be detected. Although this technique (and several other surface analytical techniques) have proved to be rather inconclusive when applied to ball-bearing systems, this should not prevent us from persisting in our efforts to find ways in which we can detect the likely onset of surface contact fatigue in these very practical geometries. The crack detector mentioned in Section 2.5.3 would also be of wider utility if it could be applied to continuous strip, for example to detect incipient cracks in rails or wheels. Much of the trouble arises from the geometry of ball-bearing systems, since it is often difficult to get a probe into the wear scar and efficiently collect the excited radiation, whether it be exoelectrons, X-rays, secondary electrons, Auger electrons, etc. This is especially crucial for the Magnetic Crack Detector with its floating sensing head. Perhaps, like Swahn *et al.* (1976), we should concentrate more on analysing the bearing ring of actual ball-bearing systems or put more effort into using the Unisteel fatigue testing geometry than the rolling four-ball machine? Certainly, there seems more chance for these techniques to be successful at both detecting and analysing contact fatigue and pitting, if we have an in situ instrument probe to monitor the wear track on the bearing ring.

In Section 7.4.4, we discussed the measurement of wear in conjunction with pitting lives and the number of cracks detectable in a typical *line-contact* system subjected to rolling plus sliding in the presence of a lubricant. As far as *point-contact* is concerned, the work of Tallian *et al.* (1965) was probably the first to show that wear occurred in the rolling contacts of hard steel balls. In fact, these authors showed that a linear wear rate applied to these contacts, operating with a fixed degree of spin, in the presence of partial EHL. Tallian *et al.* (1965) did not, however, relate this wear to the pitting life of the spindle ball in the conventional four-ball rolling machine. It is interesting to note that they used a radioactive trace method to deduce the wear rate. They assumed that all the debris worn away came from the spindle ball. The debris particles, mainly platelets of constant thickness for a given load and lubricant, were collected from the lubricant, from the washings of the test configuration and by etching the inactive balls, so that both transferred and loose wear particles were retrieved. The total mass of radio-activated steel worn away in terms of micrograms per thousand spindle ball revolutions was thereby deduced. It is not clear how long their typical experiment lasted but, judging from the large number of tests and variables examined by them, it seems unlikely that the average test duration was long enough to give pitting by contact fatigue, even in an accelerated test.

Tallian *et al.* (1965), in their definitive paper on lubricant film thickness and wear in rolling point contact, produced an equation for the wear rate of the spindle ball in a rolling four-ball machine, namely:

$$(w_m) = (K_r)(W)^{4/3}[F(\lambda_R)] \tag{7.15}$$

where (w_m) is the mass removed per unit rolling distance, W is the total load to produce Hertzian (i.e. elastic) contact, $[F(\lambda_R)]$ is the fraction of the total elastic area, across which asperity contacts occur and K_r is a constant depending on ball diameter, the elastic modulus of the steel, the hardnesses, roughnesses and cleanliness of the surfaces. In other words, K_r is their 'adjustable constant'! (All theories need such constants in order to be applicable to particular systems, whether they are tribo-systems or not.) Tallian *et al.* (1965) claim that this equation relates to the mild wear that exists after a few hours 'running-in' of their rolling four-ball machine, after which they assume that elastic asperity deformations predominate.

$F(\lambda_R)$, where λ_r is the lambda ratio, is related to the average time fraction (t/t_0) during which no contact occurs. Tallian *et al.* (1964) have shown, by measurements of *electrical conductivity* across one ball-to-ball contact, that surface asperities penetrate the EHL film and contact each other at relatively slow speeds. The proportion of such contacts is progressively reduced for any given load and lubricant, if the speed is increased, until a full EHL film forms, completely separating the surfaces. The measurements yield an average time fraction (t/t_0) during which no contact occurs. Tallian *et al.* (1964) used the conductivity data, in conjunction with a statistical analysis of surface geometry data, to yield the relationship between (t/t_0) and λ_R (as shown in Figure 7.13). From this graph, $F(\lambda_R)$ can be deduced from the relation:

$$1 - (t/t_0) = F(\lambda_R) \tag{7.16}$$

Using Equations (7.16) and (7.15), together with Figure 7.13, Tallian *et al.* (1965), were able to deduce values for the rolling wear constant (K_r), by plotting wear rates (obtained from their radioactive tracer experiments) against the lambda ratio (λ_R). A typical graph is shown in Figure 7.14 for a four-ball test at 812 N spindle load, the lubricant being an ester type of fluid. If we take, for instance, a lambda value of 3.0, we see from Figure 7.13 that (t/t_0) is approximately 0.9. From Equation (7.16), this means $F(\lambda_R) = 1.0$. From Figure 7.14, the mass wear rate is about 60×10^{-13} kg/m. Substituting for $F(\lambda_R)$ and (w_m) into Equation (7.15) we can deduce a value for K_r, namely $K_r = 1.66 \times 10^{-14}$ $(\text{kg}^{-13})(\text{m}^{-1})$.

It should be noted that Equation (7.15) differs from the *sliding wear law* of Archard and Hirst (1956) in that it predicts a *four-thirds* dependency of wear rate (instead of a *linear* dependency) upon the applied normal load (W). This difference is probably due to the simplistic model used by Tallian *et al.* (1965). Although a more sophisticated model could now be used, it is unlikely to change the ranking of the effectiveness of a lubricant in terms of its rolling wear constant

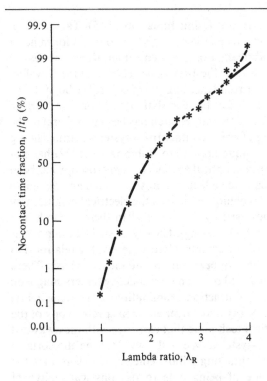

Figure 7.13 *The 'no-contact' time fraction (t/t_0) as a function of lambda ratio (λ_R) for 812 N spindle load (Tallian et al., 1964)*

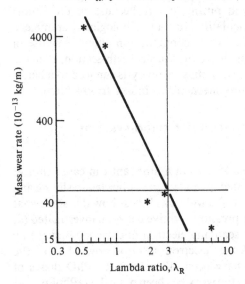

Figure 7.14 *Mass wear rate (w_m) as a function of lambda ratio (λ_R) for an ester lubricant in a rolling four-ball test at 812 N spindle load (Tallian et al., 1965)*

(K_r) in any given point-contact system. Using Equation (7.15), Tallian *et al.* (1965) show that, at $\lambda_R = 3.0$ (for instance), the rolling wear constant for mineral oil is 12.6 times that for the ester lubricant. For polythenylether, K_r is 5.2 times the ester value and for a diester, it is 1.1 times the ester value. The larger the K_r value, the less effective as a lubricant in a rolling/sliding point-control tribo-system.

From the foregoing paragraphs, it can be seen that, apart from the work of Tallian *et al.* (1965), most of the fundamental research has been peripheral to the solution of the problem of pitting of point-contact tribo-systems during sliding with rolling. The exoelectron technique applied to Rabinowicz (1972) has not lived up to its early promise. The use of optical and electron microscopy has been of a conventional nature. There has been a lack of really thorough and definitive research using any other physical techniques, such as the electrical conductivity and radio-active tracer techniques used by Tallian *et al.* in their research into wear during sliding and rolling of ball bearings. Clearly, wear is an important factor in contact fatigue of such tribo-elements. Unfortunately, the relationship between that wear and pitting life has not been put on a quantitative basis. There is a real need for more work to be carried out using radio-active tracers, magnetic crack detectors, X-ray and electron diffraction examination of the wear debris collected from the lubricating oil, X-ray and Auger electron spectroscopy of the surfaces, and any physical analytical technique that can cope with the awkward geometries of point-contact tribo-systems. We will now describe the sorts of analysis that *can* be used in line-contacting tribo-elements. It is here that the reader will begin to see the value of being able to use physical analytical techniques to *either confirm* previously-held views on possible pitting mechanisms, *or suggest* new mechanisms. I have always used line-contact geometries in my research into contact fatigue and pitting, partly because of the simple geometrical considerations when calculating various tribological parameters, but mainly because this particular geometric configuration allows easy use of physical analytical techniques. We shall see, in the next sub-section, that the combination of line-contact geometry with physical analysis can lead to a better understanding of contact fatigue pitting mechanisms in *any* tribo-system.

7.5.3 Physical analysis and line-contact tribo-systems

(i) Introduction

We have seen how extreme-pressure additives in a lubricant can cause shorter pitting lives in both point and line-contact tribo-systems subjected to sliding and rolling in the presence of that lubricant. We will now discuss how this somewhat unexpected effect of using an extreme-pressure additive has been investigated (*a*) by applying magnetic detection and scanning electron microscopy and (*b*) by examining the pitting surfaces by Auger electron spectroscopy. *Most* of the results described in this sub-section have been taken from the PhD theses of Phillips (1979) and Wu (1980). *Some* of the work has been published (Phillips and Quinn, 1978, Phillips and Chapman, 1978, Phillips *et al.*, 1976). Consequently, there will be a mixture of both new and published results. There should be more cohesion in this sub-section than is normally possible when one summarizes an extensive body of work, due to the fact that it is an account of the pitting research carried out in the Tribology Laboratories at Aston University over several years. It is hoped that the reader also finds this (possibly) parochial approach more illuminating than an account which attempts to do justice to *all* the other investigators into this subject. However, I am unaware of *any* other tribology group with such a strong bias towards the application of physical analytical techniques to line-contact fatigue and pitting.

(ii) The application of magnetic detection and scanning electron microscopy to the study of line-contact fatigue pitting

The magnetic crack detector described in Section 2.5.3 can be used to deduce a quantity strongly related to the crack propagation rate, called by Phillips (1979) the 'crack propagation time' (CPT). This is an important parameter in pitting failures. Clearly, if the CPT can be *increased* (i.e. the crack propagation rate *decreased*), then the *pitting life should therefore be increased.* Phillips (1979) used this technique to deduce the time taken for a detected contact fatigue crack to propagate to a pitting failure in his experiments on the effects of extreme-pressure additives upon the pitting lives of discs. He was expecting that the crack propagation time (CPT) would provide a quantitative measurement for comparing pitting failures for different *D*-ratios and different lubricants. Tables 7.1 and 7.2 catalogue these measurements for the base oils and extreme-pressure additive blends respectively, together with the *total* life to pitting (LTP) failure, and the CPT expressed as a percentage of the total life. These tables show that, in general, the major portion of the fatigue life (LTP) of a pitting tribo-system is consumed in *initiating* a fatigue crack. Once initiated, the crack propagates to failure in a relatively low number of cycles. This means that one can use the magnetic crack detector to measure the 'life to first crack' and be fairly certain the the 'life to first pit' will not be significantly more than this value.

Although the life to first pit (LTP) decreases with increasing *D*-ratio for both sets of experiments (as expected from Figure 7.6), there seems to be no obvious dependence of the crack propagation time (CPT) upon that ratio. There *may* perhaps be an effect of using extreme-pressure additive blends compared with the base oils. This is indicated by the average values of 1.67×10^4 cycles for the CPT of the base oil experiments, and 3.88×10^4 cycles for the CPT of the additive blend experiments. Although there is almost 100% spread in the experimental points about these averages, this could indicate a possible ameliorative effect of the additive blends as regards crack propagation. We will return to this possibility later in this sub-section.

All research projects eventually become a problem of how one apportions one's time to the many possible approaches to the project. Research into pitting is no exception to the rule. With so many possible parameters and so many available physical techniques, the newcomer might well be discouraged from spreading his/her effort too thinly across the whole spectrum. However, the newcomer should resist the temptation common to some researchers in tribology, which is to concentrate on only *one* aspect of their tribo-system. He/she must be prepared to consider the mechanical, metallurgical, physical and chemical

Table 7.1. *Crack propagation times* (*CPT*) *for the base oils only* (*Phillips, 1979*)

D-ratio	CPT ($\times 10^4$)	LTP ($\times 10^5$)	(CPT/LTP)%
1.54	0.30	7.78	0.38
2.42	1.69	6.48	2.60
3.78	0.42	8.23	0.50
5.39	0.53	3.37	1.57
6.00	0.47	2.27	2.00
6.33	6.60	2.86	23.0

Table 7.2. *Crack propagation* (*CPT*) *for extreme-pressure additive blends* (*Phillips, 1979*)

D-ratio	CPT ($\times 10^4$)	LTP ($\times 10^5$)	(CTP/LTP)%
0.69	7.90	16.20	4.90
0.92	1.09	18.8	0.60
1.55	0.38	7.70	0.50
2.42	0.16	6.88	0.23
4.46	13.20	10.40	12.5
6.54	0.57	3.01	1.90

aspects that are nearly always present in *any* tribo-system failure, and particularly so in contact fatigue pitting. Phillips (1979) *did* carry out research into the contact fatigue pitting of line-contact tribo-elements which embodied much of the interdisciplinary approach implied in the previous sentence. In common with most tribologists, Phillips (1979) also found he had too many variables. Clearly, he would have liked to have thoroughly examined all *twenty* pairs of specimens used to deduce the regression line for the base oils in Figure 7.6 and all the *fourteen* pairs of specimens used for the regression line relating to the additive blend experiments. In fact, he chose the *twelve* experiments covered by Tables 7.1 and 7.2 (for his crack propagation time analysis), together with a further *six* experiments (for his Auger electron spectroscopy analyses). The choice of any *particular* set of specimens was determined by the D-ratio and whether the lubricant was a base oil or an extreme-pressure additive blend.

To illustrate how the *combined use* of the magnetic crack detector and the scanning electron microscope can be used in a typical contact fatigue problem, we will discuss the work of Phillips (1979), insofar as it relates to the topography of the pits formed on a steel disc (of D-ratio equal to 6.54) in the presence of an extreme-pressure lubricant. In fact, we have already seen the scanning electron micrographs of this surface some 2.11×10^4 cycles after the initial surface fatigue crack was detected (see Figure 2.30(*a*)). A very distinct arrow-head fatigue crack pattern can be seen in that figure, as also can the surface wear pattern (i.e. the

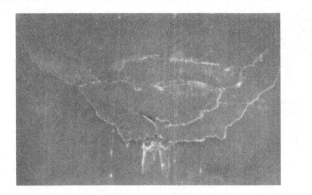

300 μm

Figure 7.15 *The appearance of the arrow-head shaped fatigue crack of Figure 2.30 (a) after a further 3.88 × 10³ cycles (D = 6.54)*

vertical lines) so noticeable on all of the scanning electron micrographs of disc surfaces produced at D-ratios greater than about 1.5 in the presence of the extreme-pressure additive blend.

The major point concerning this test was that the fatigue crack initiated from a surface flaw or furrow (at the point of the arrow-head) which had been detectable as a small but repetitive signal, from the beginning of the test. This was the only occasion that Phillips (1979) was able to pinpoint the location of an *impending* pitting failure prior to it being confirmed as a surface contact fatigue crack. This surface flaw was probably produced during the surface grinding procedure carried out during the fabrication of the specimen. This type of stress raiser has been noted by Littman and Widner (1966) and is classified by them as a 'Point Surface Origin' (PSO) mode of failure.

Figure 7.15 shows the condition of the crack of Figure 2.30(a) after a further 3.88 × 10³ cycles from the cycles to that figure. Here, the surface flaw can be seen quite clearly. Further interlaced surface cracking has occured and the leading edge itself seems to stand proud of the surface. *At the same time as Figure 7.15* was obtained, a second surface-breaking crack was observed a short distance behind the arrow-head surface cracking pattern (see Figure 7.16(a)). The crack measured approximately 1 mm across but, unlike the other fatigue cracks, it curved in the opposite direction, that is towards the rolling direction, instead of away from it. It seems likely that this second crack was where a sub-surface crack, extending possibly from the leading edge crack shown in Figure 2.30(a), had broken the surface and would subsequently form the *trailing* edge of the pitting failure. The detected signal (shown in Figure 7.16(b)) almost saturated the amplifier, thereby indicating the relatively large surface area contained by the crack. Some fine structure in the negative half-cycle of the signal was also evident.

After a further 3.08 × 10³ stress cycles from Figure 7.15, the leading edge surface broke up, as shown in Figure 7.17(a). The leading edge crack, however, was still unconnected to the surface crack of the trailing edge, which had changed little from its appearance in Figure 7.16(a). The surface cracking signal, however, had

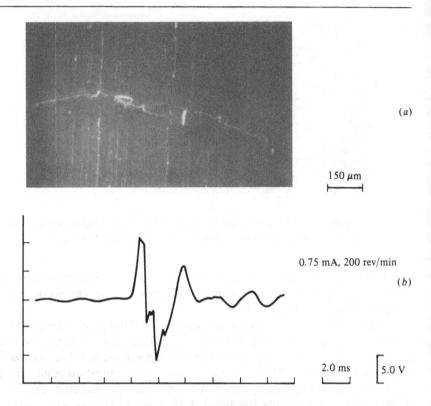

(a)

150 μm

0.75 mA, 200 rev/min

(b)

2.0 ms   5.0 V

Figure 7.16 (a) The second surface-breaking crack a short distance behind
Figure (7.15) (b) the detected signal in the magnetic crack detector

changed considerably (see Figure 7.17(b)). The amplitude of the signal had
reduced to one-third of its previous peak value and the leading edge was no
longer sharp. More importantly, the negative portion of the signal contained two
distinct peaks with an intermediate positive-going peak which was observed to
oscillate up and down.

During a further 4.6×10^3 cycles (making a total of 1.16×10^4 cycles from the
number of cycles to the detection of the first crack), more fragments became
removed (see Figure 2.30(b)). Extensive cracking connected the leading edge (at
the bottom of the micrograph) and the trailing edge (at the top of the micrograph)
originally shown in Figure 7.16(a). The trailing edge crack never appeared to
propagate, indicating that the orientation of the crack below the surface was in
the wrong direction. Large amplitude oscillations of the front half of the signal
coincided with the stage of development depicted in Figure 2.30(b). Figure 7.18(a)
and (b) show this effect with the signal in a 'high' and 'low' state respectively,
whilst Figure 7.18(c) shows three successive traces of the same signal. The gross
instability of the first 4 ms of the signal is in marked contrast to the repeatability
of the remainder of the signal. The exact physical significance of these oscillations
is not very clear. It is obvious, from Figure 2.30(b), that pitting is about to occur.

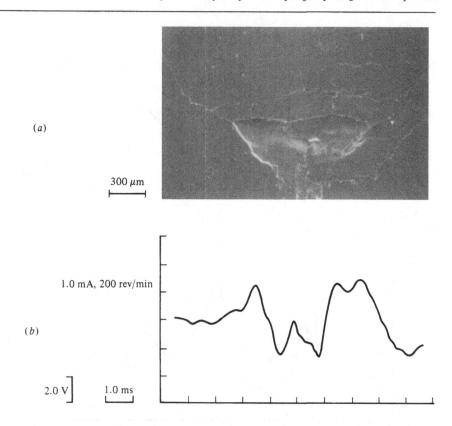

Figure 7.17 *(a) Break up of the leading edge of the crack system of Figure 7.15 some 3.08 × 10³ further cycles after that figure; (b) the change in the detected signal resulting in two negative-going peaks*

Hence, the oscillating nature of the signal can be taken as a signature of this fact. Perhaps the oscillations represent an oscillating pit fragment about to be removed from its parent disc? We will return to discuss how magnetic detection and scanning electron microscopy has provided us with new, and sometimes unexpected, facts about contact fatigue and pitting, later in this section. Let us now turn our attention to the application of another powerful surface analytical tool, namely Auger electron spectroscopy, to the study of pitting through contact fatigue.

(iii) Auger electron spectroscopy applied to pitting failures in the presence of extreme-pressure additives

Modern Auger electron spectrometers (AES) have a scanning facility, enabling the distribution of various chemical species across the surface to be analysed and compared with the actual topographical features. The examples to be given in this section involve the use of an AES without the scanning facility. The lack of examples with modern instruments need not worry us unduly, since we are

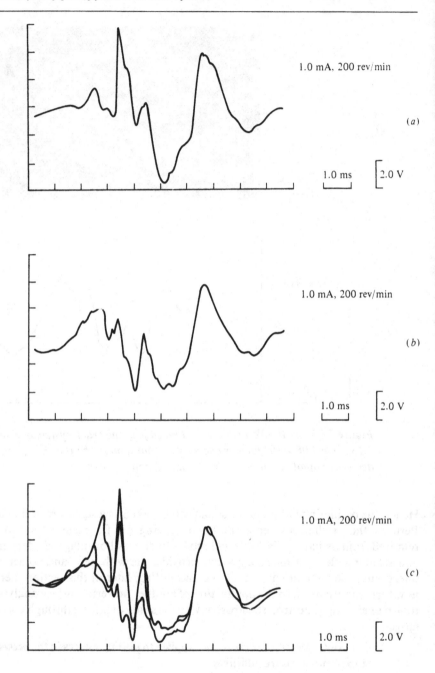

1.0 mA, 200 rev/min

(*a*)

1.0 ms   2.0 V

1.0 mA, 200 rev/min

(*b*)

1.0 ms   2.0 V

1.0 mA, 200 rev/min

(*c*)

1.0 ms   2.0 V

Figure 7.18 (*a*) The oscillating signal in the 'high' state; (*b*) the oscillating signal in the 'low' state; (*c*) successive traces of the same signal. [*All traces relate to the surface shown in Figure 2.30 (*b*)]*

mainly interested in *depth profiling* which, of course, destroys the initial topography. We will therefore describe the work of Phillips (1979) which concerned itself with the premature pitting failure of rolling/sliding discs in the presence of an extreme-pressure additive blend, compared with the pitting life under a typical base oil. The use of AES to examine the chemistry of the thin surface films formed under these conditions seemed rather obvious. The role of *D*-ratio in promoting or decreasing the tendency to form pits under contact fatigue conditions also needed investigating.

Phillips (1979) calibrated the energy scale of the Auger electron spectra from his pitted discs using a pure silver plate mounted on the specimen holder. The plate was atomically-cleaned by prolonged sputtering with a xenon ion beam and the spectra recorded using the same settings that were used for all the disc spectra. The silver Auger peaks were identified and assigned energies from the *Handbook of Auger Electron Spectroscopy* published by Physical Electronic Industries Incorporated. This method of calibration assured an accuracy of $\pm 2\,\text{eV}$ in measuring the energies of the Auger peaks from the discs.

Typical Auger spectra, taken from the running track surfaces of the discs, with and without the extreme-pressure additive, are shown in Figure 7.19. Both spectra were recorded as a function of $(\text{d}N(E)/\text{d}E)$, with the same spectrometer settings and before any xenon-ion sputtering. It should be emphasized that, prior to

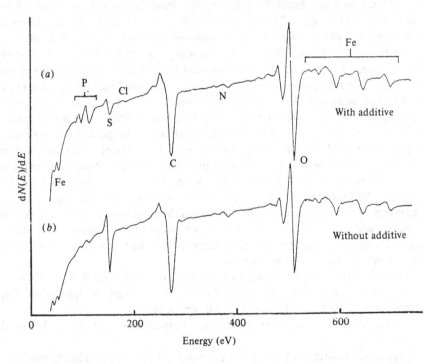

Figure 7.19 *Representative Auger spectra from the running tracks of (a) a disc tested in the extreme-pressure additive blend and (b) in the base oil only*

mounting in the spectrometer, the discs were ultrasonically cleaned and rinsed in a vapour bath to remove all traces of lubricating oil. The discs were also demagnetized, to remove any residual magnetism that would have degraded the resolution of the analyser. The disc holder was itself thoroughly cleaned. The disc specimens were never touched with bare hands. The spectra shown in Figure 7.19 are representative of the spectra obtained from the running tracks, the chamfers at each side of each disc (where no contact occurs between the rolling discs), and the actual failed area (the pits) obtained at all D-ratios.

Both traces show distinct carbon, oxygen and iron peaks. Traces of chlorine and nitrogen can also be seen. The low energy doublet at 43 and 52 eV, characteristic of a fully-oxidized iron surface is clearly evident (Wild, 1976; Suleman and Pattinson, 1973). Since the extreme-pressure additive contained sulphur and phosphorus, the region between 80 and 160 eV was of particular interest. Phosphorus, in GaP, produces a single peak at 120 eV whilst sulphur, in CdS, produces a peak at 152 eV (*Handbook of Auger Electron Spectroscopy*). Both spectra in Figure 7.19 show a sulphur peak at 151 eV (it is conventional to take the Auger peak energy at the position of the negative peak in the plots of $(dN(E)/dE)$ versus energy). Preceding the sulphur peak on both traces, we can see two peaks at 96 and 110 eV. These peaks are more prominent on the surface tested with the extreme-pressure additive. Phillips (1979) examined this part of the spectra in more detail, by reducing the modulation voltage from 8.5 volts (peak-to-peak) down to 3.8 Vpp, and found two more peaks, one at 103 eV and the other at 118 eV. Phillips (1979) assigned the peak at 118 eV to elemental phosphorus, as it was within ± 2 eV of the accepted value of 120 eV. In some work on surfaces obtained in the rotating-cantilever fatigue machine using these extreme-pressure additives, Phillips *et al.* (1976) showed that the two peaks at 96 and 110 eV were also attributable to phosphorus. In this instance, the peak at 103 eV was also thought to be due to phosphorus, although no real evidence exists for this conclusion. The occurrence of hitherto unreported Auger peaks is a feature which tribologists should be aware of. It is thought that the particular chemical environment of the phosphorus atoms on the disc surfaces might well cause additional Auger transitions not normally observed outside of the tribo-system. The four-peak multiplet structure was found in the running tracks, the chamfers and the pits of all the discs analysed that had been tested in the extreme-pressure additive blend. It was not always present in the surfaces run under the base oil only, suggesting that some contamination had occurred. We will return to this point later in this sub-section.

One of the most important features of AES is the facility for *depth profiling*, that is, the technique of alternate Auger electron analysis and ion sputtering. Depth profiling was used by Phillips (1979) in order to:

(a) examine the variation in elemental composition with depth below disc surfaces;

(b) reveal the existence of, and study the nature of, any surface films formed by the extreme-pressure additive; and

(c) determine if the surface composition altered as the D-ratio changed over two orders of magnitude.

The xenon ion potential of 400 eV was chosen as a compromise between an accelerating potential that would have produced a very slow surface material removal rate and one that would have removed material too fast to reveal the extent of any very thin surface films. In the following (selected) depth profiles, the approximate atomic percentage concentrations of the elements detected were calibrated using the *Relative Sensitivity Factors method* detailed in Section 4.5.4. For each element detected, a particular Auger peak was chosen and the atomic percentage concentration deduced from Equation (4.41), with I_i assumed to be proportional to the peak-to-peak height, $(1/a_i)$ assumed to be given by Figure 4.17, and the sum of the products such as $(a_j I_j)$, taken over all j constituents, assumed to be equal to 100. We have selected some typical profiles from the running tracks on the pitting discs at low, medium and high D-ratios, with and without the extreme-pressure additive.

Figure 7.20 shows the depth profiles obtained from the running tracks of steel discs exhibiting contact fatigue pitting after 4.02×10^6 cycles to first pit in the presence of a base oil only and 1.62×10^6 cycles in the presence of the same base oil plus 5 wt% extreme-pressure additive. The *D-ratios were both low*, being 0.53 for the base oil experiment and 0.69 for the experiment with the base oil and the extreme-pressure additive. It will be noted that the abscissae are in units of ion doses of (nA)(min). It has been calculated (Rosenberg and Wehner, 1962) that, for an iron surface bombarded with 400 eV xenon ions with a current density of $0.5 \, \mu A/cm^2$ and a sputtering yield of 0.8 atom/ion, then 0.1 nm (i.e. 1 Å) would be removed approximately for every 20(nA)(min). It turns out, however, that the track surfaces include oxides which, in general, sputter faster than the metal (Kelly and Lam, 1973). Taking all relevant factors into account, it can be shown (Phillips, 1979) that a final ion dose of 400(nA)(min) probably corresponds to a depth of somewhere in the region of 50 to 100 Å, that is an upper limit of 100 Å.

Returning to Figure 7.20, we can see that there is a thicker layer of surface carbon on the track of the base oil test ((Figure 7.20(*a*)) than on the track of the base oil + extreme-pressure additive test. In both surfaces, however, the rate of decrease in carbon content is very similar until each surface reaches its equilibrium content at around 100(nA)(min), that is at approximately 10 to 25 Å below the surface. Note that the oxide layer exists in both surfaces to depths at least as far as 400 (nA/min), that is, the oxide layer is at least 50 Å thick (probably much more!). It is interesting to note that the concentration profiles for intermediate D-ratios (i.e. $D = 2.42$) did not differ very much from Figure 7.20. For high D-ratios, however, there is a marked difference between the profiles for the base oil test and the test with the base oil + the extreme-pressure additive. Both are different from the profiles shown in Figure 7.20, as can be seen from Figure 7.21. In this new figure, we see that there was more surface carbon, and a thicker oxide film, on the running track of the test with the base oil than in the track of the test carried out under the same base oil + the extreme-pressure additive blend. The carbon profile decreased more rapidly in the track formed under the extreme-pressure additive and levelled off at a lower concentration. The oxygen and iron profiles from the surfaces formed at high D-ratios, followed each other closely and steadily increased as surface material was removed from

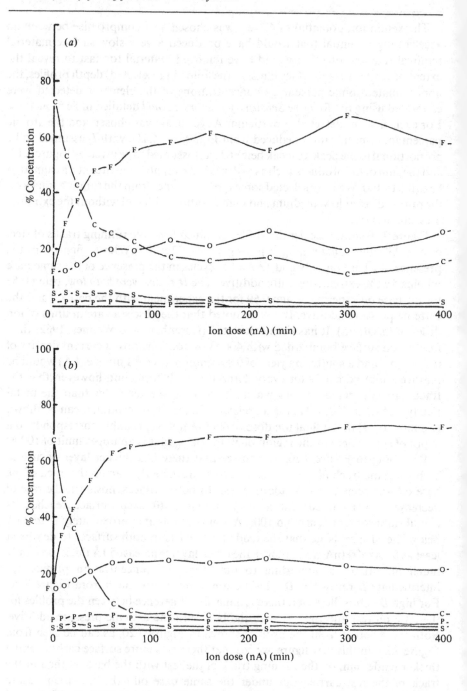

Figure 7.20 *Depth profiles obtained from running tracks of steel discs exhibiting contact fatigue pitting at D-ratios of (a) D=0.53 and (b) D=0.69 in the presence of a base oil and a base oil + an extreme pressure additive respectively*

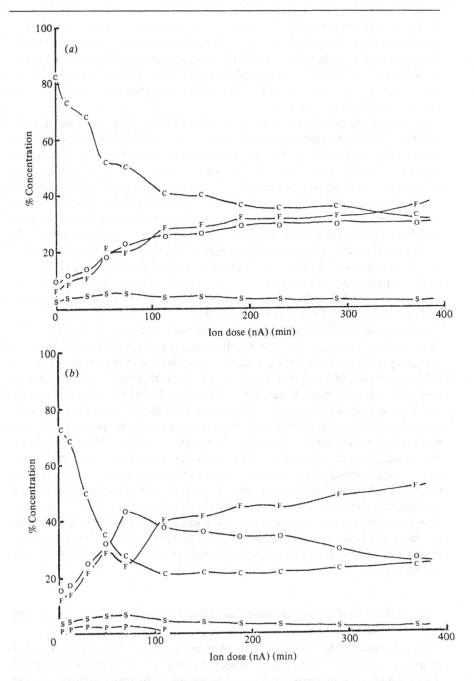

Figure 7.21 *Depth profiles obtained from running tracks on steel discs exhibiting contact fatigue pitting at D-ratios of (a) 11.47 and (b) 13.47 in the presence of a base oil and a base oil + extreme–pressure additive respectively*

the test with the base oil only. This should be compared with the oxygen and iron profiles on the surfaces formed under the extreme-pressure additive, where the oxygen and iron profiles left each other's company at approximately that depth ($\sim 60(\text{nA})(\text{min})$) which corresponded to a maximum in the sulphur profile. At this point, the oxygen reached a maximum and then decreased over the remainder of the depth sampled.

In view of the occasional deleterious effect of using some extreme-pressure additives (intended to increase the anti-wear activity of the oil) which sometimes seem to reduce the pitting life of the tribo-system, it is relevant to examine the depth profiles of Figures 7.20 and 7.21 in more detail as far as the sulphur and phosphorus contents are concerned. This is done in Figure 7.22, which also includes the intermediate D-ratio experiments at $D = 2.42$. It may be recalled that, as far as the iron, oxygen and carbon percentages were concerned, there were no significant differences between the low and intermediate D-ratio experiments. As far as the sulphur content is concerned, Figure 7.22 shows that, through the first few surfaces layers (20–50 Å), the level increased and then decreased to form a rounded peak that slowly decayed to an almost constant background level, indicating that a thin surface film containing sulphur existed on *all* running-track surfaces, regardless of the type of lubricant.

As far as the phosphorus profiles are concerned, we can see (from Figure 7.22) that, on the intermediate and low D-ratio surfaces formed in the presence of an extreme-pressure additive, the phosphorus profiles had a similar shape to the sulphur, but were of a higher concentration. In the high D-ratio surfaces, we see a much lower level of phosphorus, which disappeared after about 20 Å of surface had been sputtered away. These observations indicate that phosphorus (which can only come from the extreme-pressure additive) was always part of the surface film (as well as sulphur) formed on all the extreme-pressure lubricated surfaces, but on the intermediate and low D-ratio surfaces, it was present to a much greater depth. The phosphorus content of the surfaces formed under the base oil only (for medium and low D-ratio) is probably due to contamination during the examination process in the AES. Such artefacts often occur in surface analysis. The tribologist must be confident enough to know when a detected species is an artefact and when it is a species that has been generated in the tribo-system. It is hoped that this book will go some way toward providing the reader with such confidence.

(iv) The effect of extreme-pressure additives on the contact fatigue pitting of steels

Before discussing the application of physical analytical techniques to other pitting problems, it is fitting (at this stage) to summarize what has been revealed about the effect of extreme-pressure additives upon the contact fatigue pitting lives of steels in sliding and rolling contact, through the application of magnetic detection, scanning electron microscopy and Auger electron spectroscopy.

As far as 'lives to first pit' are concerned, we have seen that there is an inverse relationship (logarithmic) with respect to the D-ratio, that is the larger the D-ratio

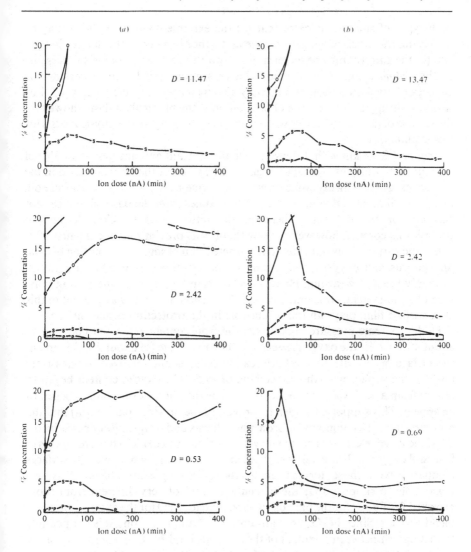

Figure 7.22 *Comparison of depth profiles for sulphur and phosphorus obtained from the running tracks of pitting steel discs at various D-ratios, lubrication being provided by (a) a base oil and (b) the same base oil plus an extreme-pressure additive*

the shorter the 'life to first pit'. Let us discuss the effects of extreme-pressure additives in terms of (1) high *D*-ratios and (2) intermediate to low *D*-ratios:

(1) *High D-ratios*

Under these conditions, the base oils give decreasingly shorter pitting lives than the extreme-pressure blends, as one increases the *D*-ratio. The physical analysis of both types of surface indicates that (a) the extreme-pressure additive may be increasing the crack propagation time and hence increasing the 'life to first pit' and (b) the additive may be increasing the severity of the wear of the surfaces due to the *sliding* component of their relative motion, and hence decreasing the chances of crack initiation and thereby also increasing the 'life to first pit'. For surfaces pitting in the presence of base oil only, the physical analysis shows that the reverse of (a) and (b) above will occur. Let us briefly give the physical basis for this explanation.

From AES we know that there is a small, constant amount (less than 5%) of sulphur in all the running tracks, regardless of whether the lubricant used in the contact fatigue pitting experiments was an extreme-pressure blend with a base oil, or a base oil alone. Obviously, the sulphur comes from the base oil and cannot account for the differences between the pitted surfaces. Considering the phosphorus content, however, we saw that it was present only for the first $20\,\text{Å}$ below the surface. This indicates a thin reacted film of some combination of iron, phosphorus and oxygen present only in the extreme-pressure blend surfaces, probably the anti-wear film already discussed in Section (5.3) for 'sliding only' type of tribo-systems. It is unlikely that the carbon, which is also present in this $20\,\text{Å}$ surface film, plays an important role in the protection of the surface.

Under extreme-pressure conditions, the iron content remained constant at about 40% whilst the oxygen reduced to the same value (of about 20%) found in the surface pitted with base oil lubrication only, as one removed progressively more of the surface with the ion etching process. The oxygen content began to reduce from a peak value of about 40% just as the phosphorus content began to disappear. This supports the idea (proposed in the previous paragraph) that the anti-wear film is a compound of iron, phosphorus and oxygen. For depths below $20\,\text{Å}$, the Auger electron spectrometry indicates an iron oxide. The percentages of iron and oxygen indicate different types of iron oxide in the base oil only surfaces compared with those formed under the extreme-pressure blend. It seems reasonable to suppose that the percentages are not very different from those expected for Fe_3O_4 for the base oil surfaces, and for FeO for the extreme-pressure blend surfaces. Since Fe_3O_4 contains both Fe^{2+} and Fe^{3+} ions, it is not possible to distinguish between mixtures of these oxides (and Fe_2O_3) from their energy spectra. Tentatively, we may assume that the extreme-pressure blend produces an oxide (FeO) which is not so beneficial as Fe_3O_4 (see Chapter 6). Hence, the amount of sliding wear will be more for the extreme-pressure blend surfaces than for the base oil surfaces. This, of course, means that the initiating cracks will be removed faster than with the base oil, thereby giving the extreme-pressure blend surfaces an increased pitting life.

(2) *Intermediate to Low D-ratios*

Under these conditions, the base oils give increasingly longer pitting lives than the extreme-pressure blends, as one decreases the *D*-ratio below about 2.0. Remembering, that this is the region where very little interaction occurs between the actual surfaces (due to the interposition of hydrodynamic or elasto-hydrodynamic films between the surfaces), then we are faced with the *corrosive* aspect of an extreme-pressure additive before it is modified by the interaction of opposing surface asperity contacts. Clearly, the extreme-pressure additive blend will initiate more cracks, which will then propagate in the classical hydraulic pressure process every time the crack passes through the contact zone. The base oil, without any of the corrosive extreme-pressure additive, will have less tendency to initiate cracks and hence will have a longer 'life to first pit' (as shown in Figure 7.6).

The physical basis for this is shown by the similarity between the depth profiles for the base oil and the extreme-pressure additive blends (see Figures 7.20(*a*) and (*b*) respectively). The phosphorus in the running track for the base oil is probably due to contamination arising from the method of viewing both types of surface together in the Auger spectrometer. The phosphorus in the surface of the extreme-pressure additive blend experiment is present (as also are iron and oxygen) at depths up to and greater than 100 Å. The percentages of iron and oxygen (60% and 20%) *could* be consistent with Fe_3O_4. However, it is much more likely that the oxygen is also chemically interacted with the phosphorus and the iron. Once again, the surfaces are covered with a thin (~ 20 Å) layer of carbon and sulphur.

This has been a classic example of the way in which a tribologist can be faced with so many variables, in this case:
(a) the type of lubricant,
(b) the thickness of the oil film,
(c) the viscosity of the oil film,
(d) the roughness of the surfaces and
(e) the crack propagation rate in the presence of the lubricant,
that he/she is loathe to add a few more dimensions to the problem by using several physical analytical techniques. The use of the *D*-ratio, however, has enabled us to combine (b) (c) and (d) into one parameter, turning the problem into one of only three variables. By combining related variables in this way, we can bring most tribological problems down to manageable proportions.

This is the last of our examples of the use of physical analytical techniques in the study of the contact fatigue pitting of tribosystems. The choice has been somewhat limited due to the fact that much previous work into pitting failures failed to use physical techniques. Let us now end the chapter by describing the present position as regards the contact fatigue pitting of tribo-systems.

7.6 The current position in contact fatigue pitting failures

The foregoing paragraphs have, it is hoped, shown that the present position as regards our basic knowledge of pitting in well-controlled line-contact systems, such as sliding and rolling discs, is very healthy. This position has not been attained without a significant contribution from the physical anaytical techniques of magnetic crack detection, scanning electron microscopy and Auger electron spectroscopy plus depth profiling. There is still much scope for more sophisticated analysis of the chemical states of the Fe^{2+} and Fe^{3+} in the oxides, as well as the full analysis of the thin surface (anti-wear?) film formed in the presence of extreme-pressure additive blends. We should be using the magnetic crack detector, in conjunction with scanning AES, to deduce the effects of chemistry on the crack propagation rates for oil-lubricated and water-emulsion-lubricated systems. The use of the D-ratio is obviously an ideal way of studying the effects of changing viscosity, film thickness and surface topography upon contact fatigue pitting.

In contrast to the position with well-controlled line-contact tribo-systems, our knowledge of point contact systems is confined mainly to studies of *overstressed* commercial ball-bearing packages, where the statistical B_{10} life is the main aim of the research. There is a real need for a more basic approach to point-contact systems and how they fail through contact fatigue pitting. The Unisteel tester could well be used in conjunction with a magnetic crack detector device, monitoring the pitting of the flat disc against which the ball bearings are sliding and rolling. Some attempts should also be made to use the D-ratio to investigate the pitting lives of various material/lubricant combinations. More extensive use of physical analysis of surfaces formed in point-contact contact fatigue testers might well repay the user.

Contact fatigue pitting has been likened to old age and the problems it brings to humans. For this reason, it has always been difficult to carry out realistic tests without either increasing the Hertzian stresses or reducing the hardnesses from those occurring in the service condition. We must try to modify the conditions so that a contact fatigue test should not last longer than (say) a week, if we are ever to understand the mechanisms whereby ball- and roller-bearings fail.

8 The analysis of oxidational wear in tribo-systems

8.1 Introduction

The publication of the definitive paper by Lim and Ashby (1987) was probably the most important step towards providing tribologists with an easily applied method for estimating the type of wear mechanism most likely to occur for a given set of conditions. It is clear that mild-oxidational wear and severe-oxidational wear are very widespread in the wear mechanism map for steel (see Figure 1.5). It seems that oxidational wear is one of those areas of tribological endeavour most likely to receive an increasing amount of attention, especially in view of the interest in developing tribo-systems which can function effectively (without conventional lubrication) in high temperature environments, for example the ceramic diesel engine.

We have already discussed the oxidational theory of wear in some detail (in Section 1.4.4). In Section 6.1.4, we described how heat flow analysis in conjunction with the oxidational wear theory can be used to obtain information about temperatures (T_c) occurring at the real areas of contact, the number N of asperity contacts within those real arcs of contact and the thickness ξ of the oxide film formed at the real areas of contact. In Section 6.2.3 it was shown how proportional analysis (by X-ray diffraction) of the wear debris produced in the oxidational wear of steels, led to independent estimates of the contact temperature. In Section 7.5.3, it was shown (by Auger electron spectroscopy) that oxygen (in the form of oxides) is an important element in the surfaces of rolling/sliding steel tribo-elements suffering from pitting and failure in the presence of a lubricant.

In this chapter, we will analyse the oxidational wear of several tribo-systems, the emphasis always being placed upon the use of physical methods of analysis. In the first sub-section, the oxidational wear of a low-alloy steel sliding against itself in a room-temperature air environment is analysed, via the transmission electron microscopy of replicas of the worn surfaces. We then examine the topography of worn surfaces (using scanning electron microscopy) formed during the sliding wear of high-chromium ferritic steel against austenitic stainless steel, the objective being to measure oxide plateaux heights and to use those measurements to provide information about the surface model used in the oxidational theory of mild wear. Our third tribo-system is 316 stainless steel sliding against itself. In this example, we will see how Auger electron spectroscopy (AES) and X-ray photoelectron spectroscopy (XPS) have been applied to study the chemistry involved in oxidational wear. It will also be shown that depth-profiling by AES can be used to show the oxidized nature of the plateaux of contact in the mild-oxidational wear of these stainless steels.

We will *not* be discussing *severe*-oxidational wear, that is, wear that occurs at loads beyond the transition load designated T_3 by Welsh (1965). This type of wear is probably best modelled by a combination of the mild-oxidational wear theory with one of the various severe wear models proposed by Suh and his co-workers (1977). At the time of writing, this combination has not been reported in the literature which is unfortunate, since loads greater than T_3 are not infrequently attained in practical tribo-systems. This is an area of tribological endeavour that is waiting to be exploited, especially by a researcher with an interest in, and some knowledge of, the application of physical methods of analysis to practical problems.

Let us now start our chapter on the analysis of the mild-oxidational wear of tribo-systems, by describing how electron microscopy has been applied to the study of wear in a typical low-alloy steel tribo-system.

8.2 The oxidational wear of a typical low-alloy steel tribo-system

8.2.1 Introduction

In this sub-section, we are going to discuss the application of transmission electron microscopy (TEM) to study the topographies of worn surfaces via the use of a shadowed replica technique, similar to that described in Section 2.2. This technique does have distinct advantages over the *more direct* method of scanning electron microscopy (SEM), advantages probably peculiar to tribology. We will discuss these advantages later.

The example we are about to describe arose from some pin-on-disc wear experiments designed to investigate the effect of chromium content upon the wear behaviour of low-alloy steels. AISI 4340 steel was used as the standard and a 3% Cr-$\frac{1}{2}$% Mo steel was used to see if there were any differences in the sliding behaviour of these two steels. The AISI 4340 steel has a similar composition to the steel normally used in the shafts of large *marine* turbines. The 3% Cr-$\frac{1}{2}$% Mo steel was similar to the steel tried by the turbine manufacturers as a replacement material which would have good corrosion resistance in the humid atmosphere of a ship's engine room. The resulting 'wire-wool' failures caused much economic hardship to the manufacturer and became the subject of much tribological endeavour in the 1960s (e.g. Dawson and Fidler, 1965). Although some tribologists thought these failures were due to foreign particles, for example metallic swarf, in the lubricant from the very start, I am convinced the failures were due to incompatibility between the chromium-steel shaft and the chromium-steel wear fragments embedded in the soft, white metal bearing pads in the journal. It may seem unlikely to the tribological newcomer, but it is quite common for soft materials to pluck out fragments from their hard counterfaces when starting up, and at other times when the hydrodynamic film cannot be maintained between the sliding surfaces.

Having gone into some detail regarding the choice of steels for investigating 'wire-wool failures', it is typical of the subject of tribology to report that there were no significant differences found between the wear behaviours of these two steels. Amongst the similarities found were the worn surface topographies as revealed by TEM of replicas. What was really important about the application of this physical technique to steel tribo-systems, was that a possible mechanism of mild-oxidational wear was clearly indicated by the electron micrographs (Quinn, 1968). This mechanism has been found again and again since those early days, and has become an accepted feature of the oxidational wear theory. Thus we have the anomaly of attempts to show *differences* bringing forth evidence of *similarities*, the latter evidence being much more important than it would have been if it had shown differences! The heuristic nature of much successful research, particularly experimental research, is something often overlooked by the more theoretically-inclined researcher. Tribology is very much a subject which is constantly providing the experimental researcher with *unexpected* results. Let us now describe the work reported in Quinn (1968) relating to the surface topographies of pins and discs of AISI 4340 steels worn under conditions of mild-oxidational wear.

8.2.2 Transmission electron microscopy of replicas of surfaces undergoing mild-oxidational wear

A two-stage replica method was used to obtain specimens from selected pin and disc surfaces. The technique was conventional, involving the taking of first-stage plastic replicas of the surfaces, 45° shadowing with Pt-Pd, 90° evaporation of carbon to form a second-stage carbon replica of the shadowed plastic replica, removal of the first-stage plastic replica by a suitable solvent, washing the carbon replica, and mounting it on an electron microscope grid. The replica was then examined in a conventional transmission electron microscope and several photographs taken of 'typical' and 'interesting' features. One should, of course, be careful in the selection of such features. It is essential to have a systematic approach to this problem. It is suggested that four or five replicas are obtained from macroscopically typical areas of the specimen to be examined. Each replica should be superficially scanned in order to deduce 'typical' and 'interesting' topographies. A sketch of the replica and its relation to microscope grid coordinates will help to 'fix' on any interesting features noticed during the initial scanning. In wear research, we are interested in wear debris and it is advisable to always try to use the *first* replica from a newly-worn surface, since this replica will remove most of the loose wear debris. In lubricated wear, it is unlikely that one finds any wear debris on the surfaces. However, it is advisable to thoroughly remove all traces of the actual lubricant physisorbed upon the worn surfaces before taking a replica, since this could lead to artefacts. Let us now discuss the interpretation of the transmission electron mirographs of the two-stage replicas obtained by the author (Quinn, 1968) from the surfaces of worn steel pins and discs.

It is usual to orient the electron micrographs of replicas so that the direction of shadowing lies from top to bottom. In some research studies, the first negative of the micrograph from a two-stage shadowed replica is used to form a reversed negative, which is then printed. This photographic technique will provide a dark final print, with the shadows also very dark. However, in the studies of fractured surfaces, where one is very much interested in cracks, it is usual to print directly

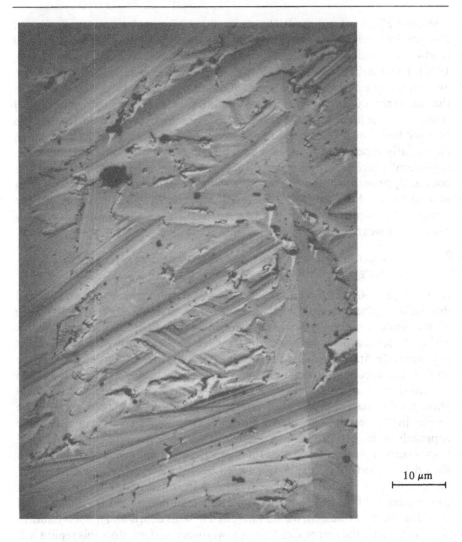

10 μm

Figure 8.1 *Electron micrograph of a replica from an unworn steel disc surface (Quinn, 1968)*

from the first negative. Under these conditions, the shadows due to shadowing will appear white and the cracks will be dark. Because of the more aesthetically pleasing aspect of the direct prints from the first negatives, we will follow the fractographers and present our electron micrographs so that the shadows appear white, and seem to originate from a source beyond the top edge of the print.

Figure 8.1 is an example of an electron micrograph from a two-stage, Pt–Pd shadowed, carbon replica of an unworn AISI 4340 steel disc surface. Note that the surface is criss-crossed with troughs. These features must be troughs, and not ridges, since they have their darkest edges towards the direction of shadowing (always from the top to bottom). This interpretation was discussed in Section 2.2. Because of the smoothness of the surface (about 1.27 μm centre line average, CLA), there are no gross white 'shadows'. Close inspection of Figure 8.1, however, will reveal that there are a few small dark features which cast white shadows below their position on the electron micrograph, showing that the direction of shadowing was indeed from top to bottom. Note that the magnification is not much more than one might expect from a very good optical microscope. However, note the extremely good contrast, large depth of focus and the very high resolution of Figure 8.1. The randomly oriented troughs were deliberately formed on the initial surfaces of the steel discs by a suitable lapping motion. This was done to exclude any possible directional effects (or effects of lay) on the wear behaviour of this tribo-system. The presence of the troughs is *not* unexpected, since the lapping operation must clearly involve the removal of surface material by the very much harder pieces of SiC powder in the lapping compound.

We will, initially, restrict our discussion to the study of the surfaces of the worn steel discs. Apart from the technical difficulties experienced in obtaining satisfactory replicas from the 6.2 mm diameter surface of the worn pins, one is more likely to find surfaces representing the various stages of a wear mechanism if one examines the worn discs, rather than the pins. This is obviously true for 'like sliding on like', such as is the case for the surfaces about to be described. In fact, the worn disc topographies could be classified into *three* main typical topographies. These classifications have since been obtained in many other oxidational wear sets of experiments. They are:

(a) *Rough* surfaces with wear particles, but no visible wear tracks;
(b) *Smooth* surfaces with wear tracks and cracks, but very few wear particles; and
(c) *Smooth islands* surrounded by rough, debris-packed areas like (a) above.

Figure 8.2 shows an electron micrograph of a replica (the initial one!) taken from a typical rough area of the surface. This surface topography was very common and covered about 80% of the nominal surface area. The black regions in the micrograph are, in fact, wear fragments which have been removed by the plastic replica from the original specimen surface. They will, of course, sink into the plastic as the first impression is being made. Hence, they will *not* stand proud of the surface when the shadowing material is evaporated on to the plastic replica and so no white shadows will be expected from them. In the final carbon replica, these debris particles, being opaque to electrons, produce regions of low density of photographic blackening on the negative. Hence they appear black in the final print. The topography associated with the debris-packed regions is rough, consisting mainly of small hillocks (~ 3 μm in size). The average size of the wear particles is about 1 μm. Note the absence of anything resembling a wear track.

It is worth mentioning at this stage that selected area electron diffraction examination of the wear debris is *not* generally feasible due to several reasons. Typically, an iron oxide particle must have a diameter less than 2000 Å, that is 0.2 μm, if the typical electron accelerating voltages (between 50 and 100 kV) are used to irradiate the transmission electron diffraction specimen. Electron diffraction maxima (from iron specimens) lose their contrast when the path

10 μm

Figure 8.2 *Replica of a typical rough region of a worn steel disc surface (mild wear)*

length in the specimen exceeds more than about 2 or 3 mean free paths (Halliday, 1960). Hence, with debris particles averaging about 10 times the mean free path, it is unlikely that any *usable* electron diffraction patterns, (even of the selected area variety described in Section 3.3.6), could be obtained from these 'extraction' replicas. Another problem is the presence of Pt and Pd from the shadowing operation. These will tend to 'overpower' any diffraction which *might* occur from the smallest debris particles. If one wishes to use the selected area electron diffraction method, one could omit the shadowing stage. Unfortunately, the contrast of the electron micrograph will then depend only on the very slight differences in path lengths of electrons passing through the carbon replica at various angles. This is normally insufficient for most instruments. However, the use of an image intensification device might possibly overcome this second (apparent) disadvantage.

   Figure 8.3 is an example of the typically smooth, cracked region of a steel surface worn under mild (oxidation) wear conditions at 21.6 N load and 6.50 m/s speed. Note how extremely smooth the surface must be. The wear tracks, from top left to bottom right, are very fine and shallow, being about 0.5 μm or less apart. This surface is very much smoother than the original surface (shown in Figure 8.1). Note the white shadow at the border between the rough debris-packed surface and the smooth, almost debris-free, tracked region. At the time of shadowing, the plastic replica of the rough region must have been *above* the smooth surface. Hence the *actual* rough surface must have been *below* the level of the smooth surface. This was the first indication of the formation of smooth, cracked plateaux during mild wear. Note the granular nature of the areas enclosed by the cracks – these areas could readily form the flakes known to be

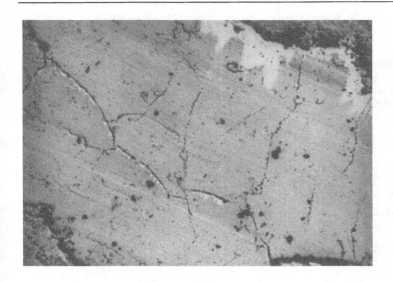

10 μm

Figure 8.3 *A typically smooth, cracked region of a steel surface formed under conditions of mild-oxidational wear (21.6 N, wear rate = $2 \times 10^{-13}\,m^3/m$, speed = 6.50 M/s)*

occasionally present in wear debris. Such flakes would have a short existence, being rapidly reduced to the particles so evident in Figure 8.2, by the action of the two opposing surfaces.

An example of the third typical topography, so characteristic of mild-oxidational wear, is given in Figure 8.4, which shows a smooth, cracked, tracked *island* in a region of rough, untracked surface, packed with wear debris particles. It is very important to establish whether the smooth regions of the wear surface were above or below the adjoining rough regions. Figure 8.4 shows white shadows in the top right corner. These are due to the fact that the replica of the rough region is above the surface of the replica of the smooth region. This is confirmed by the absence of white shadow at the lower edge of the smooth region, that is the edge farthest away from the shadowing source. It can be concluded that the smooth areas of the original specimen were above the general level of the adjoining rough area of the specimen. Since the angle of shadowing was about 45°, one can estimate that the smooth areas were about 4 μm *above* the rough areas on the actual surface concerned. The large black region to the left of Figure 8.4 is either a region of coagulated wear particles trapped by the motion of the disc surface towards the left top corner of the figure, or it could be a large (thick) flake of the original surface removed along with the replicating material. The latter possibility would be in accord with the prow mechanism of Cocks (1962), if the top left edge of the smooth area is the *leading* edge.

Figure 8.5 shows another smooth island region similar to Figure 8.4, except that now the smooth region appears to be in a slight trough (as indicated by its edge being darker nearest to the direction of shadowing). Alternatively, the island

405

Figure 8.4 *Electron micrograph of the replica of a smooth cracked island region of a steel surface worn under mild wear conditions*

could be sitting on the rough areas with its top right edge curled upwards away from the general surface. Note that edge to the right of the smooth surface shows *no* agglomeration of particles similar to the black mass in Figure 8.4, thereby showing that the edge noted in Figure 8.5 is, indeed, a trailing edge. Note also that the large area at the trailing edge of the smooth island seems ready for removal as a flake.

Very similar topographies were also found on the worn pin surfaces (Quinn, 1968). It is interesting to note that these topographies have been found in other tribo-systems exhibiting mild-oxidational wear, for example, see Quinn and Woolley (1970), Sullivan and Athwal (1983), Allen, Quinn and Sullivan (1985) and Quinn and Winer (1987), to name but a few. The mechanism of mild-oxidational wear (which was given in Section 1.4.3, without any theoretical or experimental backing) is based very much on the topographies revealed by TEM of replicas. As regards the oxidized nature of the plateaux of contact, however, it had been necessary to rely heavily on the *indirect* evidence of X-ray diffraction analysis of the wear debris (see Section 6.2), until recent *unambiguous direct* AES evidence (Sullivan and Saied, 1988) showed that the plateaux formed during the continuous sliding of low-alloy steel (EN8) against itself, are indeed oxidized *before* becoming detached and degraded into wear particles.

It is true to say that, for most purposes, SEM has tended to supersede the use of replicas in surface analysis. About the only good reasons for continuing to use the replica method are (a) if a scanning electron microscope is not available and (b) if one wishes to follow the sequential processes in the wear of a surface without removing either tribo-element. The first reason is obvious. The second reason is very important, since there are immense difficulties in carrying out *in situ* wear

10 μm

Figure 8.5 *Electron micrograph of the replica of the trailing edge of a smooth island region, showing an area about to be removed as a flake*

experiments in the *vacuum* of the scanning electron microscope. On the other hand, to stop the wear experiment whenever one wishes, taking a plastic replica of the surfaces of interest at each stop, is fairly easy. The plastic replicas can be stored and later processed in the usual way to give two-stage shadowed replicas of a surface as it wears. The considerable work (and skill) involved in taking such replicas has so far been prohibitive. As far as this author is aware, there is no *reliable* physical method for following a wear process at the required depth of focus and with the required resolution, *unless* that process occurs in the *vacuum* of the electron microscope (or other physical analytical instrument). Even inter-ferometry through a transparent tribo-element has not proved very informative. There is a real need for a physical method that will show us what is actually happening at the interface *during* sliding.

8.2.3 Summary

In this sub-section relating to the mild-oxidation wear of low-alloy steels, we have seen that the use of replicas, together with TEM, produced electron micrographs which led to the proposal of the mild-oxidation wear mechanism, involving dominant plateaux of contact being the sites for the myriad local contacts, which eventually (through oxidation at the plateaux of contact) reach a critical thickness of about 3 or 4 μm before breaking off as flakes (which eventually become wear particles of the same order of magnitude as the thickness of the oxide film). The topographies revealed by TEM of replicas of worn low-alloy steel surfaces have also been found with other steels, often by the use of scanning electron microscopy rather than by using replicas. Nevertheless, the replica technique is the best technique for following a wear mechanism *as it is happening in the normal atmosphere of the laboratory*. It can also be used to examine tribo-elements in equipment in *actual service*. Neither of these facilities is available to scanning electron microscopy. Let us now discuss a system in which SEM has been very successfully applied, namely the high-chromium ferritic steel sliding on austenitic stainless steel system.

8.3 The oxidational wear of high-chromium ferritic steel on austenitic stainless steel

8.3.1 Introduction

In Section 6.1.2, we saw how Sullivan and Athwal (1983) used an iterative technique, based on Equation (6.11) and the comparison of δ_{theory} with δ_{expt}, for deducing consistent values of N (the number of contacting asperities on the operative plateaux formed on the surfaces of AISI 52100 steel pins and discs undergoing mild-oxidational wear). They also deduced the radius a of each asperity contact, the temperature T_c of the contact and the thickness ξ of the oxidized plateaux. Sullivan and Athwal (1983) assumed that the plateaux thickness on the pin (ξ_p) equalled the thickness on the disc (ξ_d). Although they did not compare their values of ξ with any measured values of this quantity, their average value of $(4.3 \pm 1.2)\,\mu m$ (see Table 6.1) is very close to the value of about $4\,\mu m$ obtained with the low-alloy steels (see Section 8.2). If, however, Sullivan and Athwal had inserted their measured values of (ξ_d) and (ξ_p) into Equation (6.11), they could have deduced values of a entirely in terms of known constants and quantities obtained from the heat flow equations given in Section 6.1.2. Hence, they could have obtained values of T_c from either Equation (6.5) or (6.6) and, of course, values of N from Equation (6.12). In fact, this alternative method was used by Allen (1982) in his research into the wear of diesel engine exhaust valve materials. Let us discuss Allen's (1982) work in more detail.

8.3.2 Measurement of oxide plateaux heights by scanning electron microscopy

Allen (1982) was interested in the factors affecting the unlubricated wear of valve/valve seat tribo-elements at elevated temperatures. The valve is made of an austenitic stainless steel in a typical exhaust system in a diesel engine, whilst the valve seat is generally made of a high-chromium ferritic steel. These materials have to slide against each other at temperatures up to about 700 °C. The exhaust systems of such engines often suffer from *valve sinkage*, with consequent loss of performance and eventual breakdown. This is due to preferential wear of the valve seat (the high-chromium ferritic steel) compared with the actual valve (the austenitic stainless steel). Since the exhaust gases contain CO_2, NO_2, CO and water vapour, it was considered that oxidational wear was the most likely wear process to occur under these conditions. The added complexities of the real situation, in the form of trace elements from the fuel and the lubricating oil, had to be ignored by Allen (1982) in his laboratory simulation. His main objective was to understand the fundamental mechanisms by which these materials wear under continuous sliding conditions at various temperatures in *air* environments. This was intended to be the forerunner of a series of projects involving other operating variables, such as different gas environments and a change from continuous sliding to one of sliding plus impact. As is so usual with research funding at universities, these projects never materialized, due to economic changes which strongly influenced the diesel engine manufacturing industry in both Britain and the United States.

A selection of electron micrographs, taken in the SEM using the tilt correction facility, is shown in Figure 8.6. Perhaps we should say a few words about how the tilt facility can be used to measure oxide plateaux thicknesses. In a typical investigation, the angle of tilt is adjusted to a fairly high angle, say 45°, and the tilt correction set to this value. This unifies the magnification over the area of specimen under study. In regions where the oxide film has fractured leaving near-parallel edges, the vertical distance between the edges can be measured and converted into micrometres (using the 'bar' at the bottom left corner of the micrograph), since the vertical magnification is equal to the horizontal magnification under these conditions of tilt. A number of measurements should be taken over each specimen in order to obtain the average film thickness for that specimen. We will not say much about the topographies of the surfaces shown in Figure 8.6, except to mention that the surfaces of the high-chromium ferritic pins and the austenitic stainless steel discs were very similar. Typically, the wear topography of these steels was characterized by numerous oxide islands, which were all very smooth in appearance. It is obvious that the topographies and wear mechanisms originally noted in the sliding of low-alloy steels also occur in these very different steels. The proof that the island plateaux are indeed oxides is readily shown by glancing angle X-ray diffraction and/or AES. We have already discussed the glancing-angle, edge-irradiated X-ray diffraction film technique in Section 3.2.4. We saw how this technique can be used to help identify the surface chemistry of extreme-pressure lubricant films formed on steels in the presence of organo-sulphur compounds (see Section 5.1.4). Allen (1982) used this technique to show that, under conditions where mild-oxidation wear prevailed, the rhombohedral and spinel oxide phases were detected in the surface of the high-chromium ferritic pins. The ferrite lines were used as the standard for internal calibration purposes. The same oxides were also present in the surface of the worn austenitic stainless steel discs. For these X-ray diffraction patterns, the austenite lines were used as standard.

We have been speaking of mild-oxidational wear. It is interesting to note that, at a sliding speed of 2 m/s, the room temperature and 200 °C wear patterns (i.e. the plots of wear rate versus load) had transitions from mild-oxidational to severe-oxidational wear at loads of about 70 N and about 45 N respectively. The runs at 300, 400 and 500 °C (all at 2 m/s) were always in the state of mild-oxidational wear, as shown in Figure 8.7. We can see, from this figure, that the scanning electron micrographs shown in Figures 8.6(e), (f) and (g) relate to surfaces that have undergone mild (oxidational) wear, whilst Figures 8.6(h) and (i) are relevant to oxidational wear at the transition load. Figures 8.6(a) and (b) were also at the transition load for the room temperature runs at 1 m/s and Figures 8.6(c) and (d) were just into the severe oxidational wear regime.

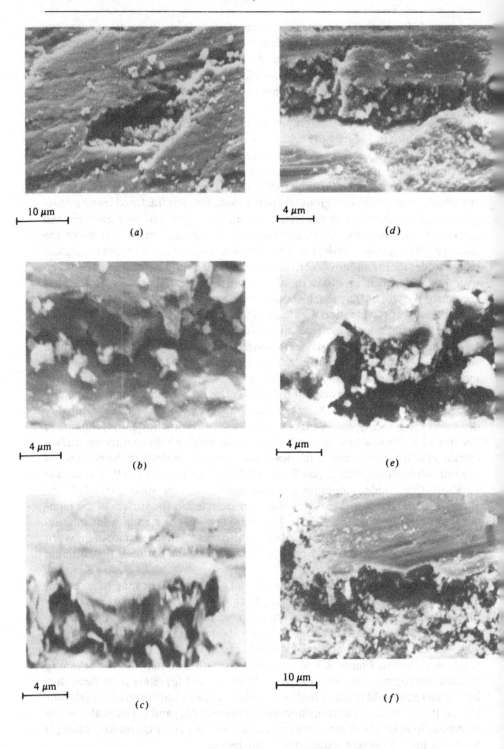

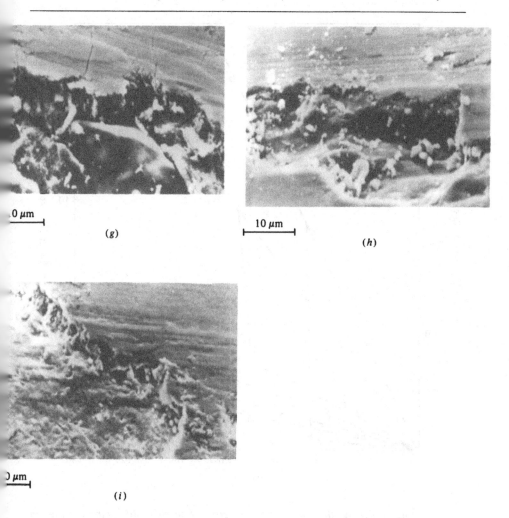

Figure 8.6 *Scanning electron micrographs of various pin and disc surfaces showing the thicknesses of the oxide plateaux. (a) Pin: 62.5 N, 1m/s room temperature; (b) disc: 62.5 N, 1 m/s room temperature; (c) pin: 56 N, 3.3 m/s room temperature; (d) disc: 56 N, 3.3 m/s room temperature; (e) pin: 100 N, 2 m/s, 400 °C; (f) disc: 62.5 N, 2 m/s, 300 °C; (g) pin: 69 N, 2 m/s, 400 °C; (h) pin: 75 N, 2 m/s room temperature; (i) disc: 75 N, 2 m/s room temperature*

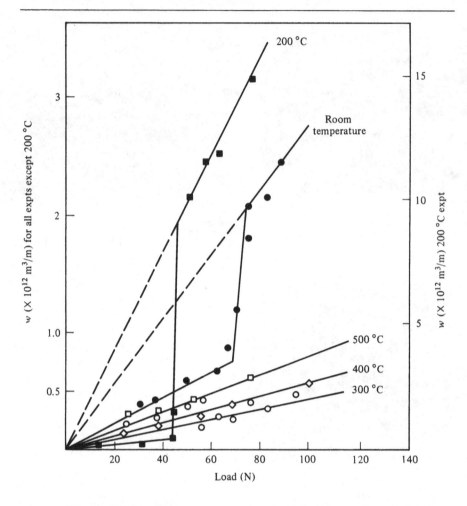

Figure 8.7 *Equilibrium wear patterns of ferritic steel pins sliding on austenitic stainless steel discs at various disc temperatures (2 m/s sliding speed)*

Returning to the main theme of this sub-section, which is the measurement of oxide plateaux heights, Table 8.1 gives a summary of all the measurements carried out in the SEM by Allen (1982) in his wear studies with diesel engine exhaust valve materials. This table shows that oxidational wear of some sort, whether it be mild-oxidational wear or severe-oxidational wear, was clearly occurring. This is indicated in the column headed 'Analysis of wear debris'. From this analysis, taken in conjunction with Figure 8.7, it seems reasonable to assume that the severe-oxidational wear of these materials is to be associated with metallic debris plus the rhombohedral oxide only. Mild-oxidational wear seems to provide only oxidized debris, namely the rhombohedral oxide plus the spinel oxide. Although the wear debris *could* have oxidized after removal, we will

Table 8.1. *Summary of the SEM measurements of plateau heights for high-chromium ferritic steel pins sliding on austenitic stainless steel discs* (*Allen, 1982*)

Experimental conditions	Plateaux heights (μm)		Analysis of wear debris*
	Pin (ξ_p)	Disc (ξ_d)	
(i) 0.23 m/s – all tests (No transition load)	1.5 ±0.5	1.5±0.5	F+A+R
(ii) 1 m/s – below transition	2.0 ±0.25	2.0±0.25	R+S
1 m/s – above transition (Transition load ≈ 60 N)	1.5 ±0.25	3.0±0.5	F+R+S+A
(iii) 2 m/s – below transition	3.25±0.25	3.5±0.25	R+S
2 m/s – at transition	2.5 ±0.25	3.5±0.25	A+F+R+S
2 m/s – above transition (Transition load ≈ 65 N)	2.0 ±0.25	4.0±0.5	A+F+S
(iv) 3.3 m/s – all tests (No transition load)	3.5 ±0.5	3.5±0.5	R+S
(v) 2 m/s – heated disc tests			
(a) 200 °C – below transition	3.5 ±0.25	3.5±0.5	R+S
(b) 200 °C – at transition	3.0 ±0.5	3.5±0.5	A+F+R+S
(c) 200 °C – above transition (Transition load ≈ 44 N)	<0.5	3.25±0.5	A+F+S
(vi) 2 m/s; 300, 400, 500 °C (No transition load)	5±1	5±1	R+S

*X-ray diffraction analysis of wear debris: F, Ferrite; A, Austenite; R, Rhombohedral oxide [$(Fe_2Cr)_2O_3$]; S, Spinel oxide [$FeFe_{(2-x)}Cr_xO_4$].

continue to assume that oxidation occurs at the actual real areas of contact, before removal as oxidized fragments of the surface. This assumption will be shown to be valid in a later sub-section, dealing with the application of depth profiling plus AES to surfaces undergoing oxidational wear (see Section 8.4). The transitions mentioned in Table 8.1 are, of course, the transition from mild-oxidational wear to severe-oxidational wear. If there was no transition, such as with the 0.23 m/s test, we must look at the debris (and the wear rate), to realize that this was severe-oxidational wear. The lack of transitions at 3.3 m/s and in the 300, 400 and 500 °C runs at 2 m/s was due to the conditions *not* being right for severe-oxidational wear. In other words, mild-oxidational wear was occurring.

8.3.3 Calculation of contact temperature (T_c), number of contacts (N) and radius of contact (a)

Inspection of Table 8.1 shows that the assumption made by Sullivan and Athwal (1983) regarding the equality of oxide plateau heights on both pins and discs was probably not far from the truth. Certainly, as regards the materials and conditions used by Allen (1982) we see that whenever mild-oxidational wear occurs (as indicated by the presence of *only* the rhombohedral and spinel oxides), this assumption is true. Let us now calculate the values of (T_c), (N) and (a) from Equation (6.11) for a typical experiment, namely the 62.5 N experiment carried out at 3.3 m/s, in which, from Table 8.1, we see that $(\xi_p) = (\xi_d) = 3.5 \,\mu\text{m}$. The experimental values of $(UF) = 60.74 \,\text{W}$, $H_1 = 5.52 \,\text{W}$, $\delta_{\text{expt}} = 9.1\%$, $(T_S)_p = 199 \,^\circ\text{C}$, $(T_S)_d = 70 \,^\circ\text{C}$, $(p_m W) = 5.65 \times 10^7/\text{m}^2$, $(K_S)_p = 26.8 \,\text{W/m K}$ and $(K_S)^d = 15.0 \,\text{W/m K}$. It should be noted that the bulk hardness (p_m) of the pin material was taken as the bulk temperature $(T_S)_p$ of the pin, using figures from the 'ASM Committee on Engine Valves' contribution to the *Metal Handbook* (volume 1, page 326) giving the hardness response with temperature of a material very similar to the high-chromium ferritic steel used by Allen (1982). $(K_S)_d$ was taken at $(T_S)_d$ and $(K_S)_d$ was assumed to be the mean value for the range 20 to 500 °C.

As regards the diffusivity (X_o) of the oxide, which is equal to $[K_o/(\rho_o c_o)]$, one can readily obtain sufficiently accurate values of ρ_o and c_o, from density and specific heat capacities of iron oxides and chromium oxides, as given in the 44th edition of the *Handbook of Chemistry and Physics* published by the Chemical Rubber Company, namely $(\rho_o c_o) = 4.5 \times 10^6 \,\text{J}/(\text{m}^3\text{-K})$. The conductivity of the oxide (K_o) is not easily assigned a value (a) because there is no available data on the thermal conductivities of $[(\text{Fe,Cr})_2\text{O}_3]$ and $(\text{Fe Fe}_{(2-x)}\text{Cr}_x\text{O}_4)$ and (b) because the values for Fe_2O_3 and Fe_3O_4 (as given by Molgaard and Smeltzer, 1971) can only be assigned if one knows the temperature of the oxides and the proportions of each oxide present. Hence, we will follow Allen (1982) and choose a 'starting value' for K_o equal to 4.0 W/(m–K) and then use an iterative technique to deduce (δ_{theory}) from Equations (6.13), (6.2), (6.3) and (6.11). If δ_{theory} is not close to δ_{expt}, then a new value of K_o is chosen until the theoretical and experimental values of the partition of heat are *sufficiently* close.

With K_o set equal to 4.0 W/(m–K), one obtains $X_o = K_o/(\rho_o c_o) = 8.826 \times 10^{-7} \,\text{m}^2/\text{s}$. Hence

$$A = \frac{0.1 U(UF - H_1)(p_m)}{2(X_o)(W)(K_S)_d} = 3.887 \times 10^{13}$$

$$B = \left(-\frac{p_m}{W}\right)\left[\frac{0.86(UF - H_1)}{(K_S)_d} - \frac{H_1}{(K_S)_p} - \frac{0.1 U(UF - H_1)(\xi_d)}{2(X_o)(K_o)}\right] = 2.95 \times 10^8$$

and

$$C = \left(\frac{-p_m}{W}\right)\left[\frac{0.86(UF - H_1)(\xi_d)}{K_o} - \frac{H_1\xi_p}{K_o}\right] + (T_S)_p - (T_S)_d = -1.846 \times 10^3$$

From these expressions, we get $4AC = -28.705 \times 10^{16}$, $B^2 = 8.703 \times 10^{16}$. This yields a value of the contact radius (a) equal to $4.07 \,\mu\text{m}$ and $\alpha_{P_e} = 0.86 - 0.10 U a 2X_o$ (from Equation (6.4)) $= 0.1$. Hence from Equation (6.2), (6.3) and (6.13), we get $\delta_{\text{theory}} = 10\%$.

This is fairly close to the experimental value of 9.1% for the division of heat at the pin/disc interface. If we now assume $K_o = 4.2 \,\text{W}/(\text{mK})$ this gives $A =$

Table 8.2. *The variation with load of surface parameters relating to the oxidational wear of high-chromium ferritic steel on austenitic stainless steel at 2.0 m/s (Allen, 1982)*

Load (N)	K_o (W/mK)	a	(δ_{expt}) (%)	(δ_{theory}) (%)	N	(T_c) (°C)
12.5	5.0	9.15	9.8	4.2	13	319
31.25	5.0	8.9	11.6	6.8	36	452
37.5	4.9	8.9	7.7	5.0	43	315
50.0	4.8	8.58	8.3	6.7	61	413
62.5	4.8	8.4	10.5	7.0	80	519
67.0	4.5	7.9	11.6	9.5	97	437
75.0	4.5	7.3	14.0	12.6	131	509
81.25	4.2	7.3	13.0	12.6	142	516
87.5	4.2	7.3	12.8	12.6	153	511

3.702×10^{13}, $B = 2.48 \times 10^8$ and $C = -1.752 \times 10^3$. This yields 4.3 μm for the contact radius which, in turn, gives $\delta_{theory} = 9.6\%$. This is considered good enough agreement with δ_{expt}, so the iteration process is halted. A contact radius value of 4.3 μm eventually yields $N = 305$ for the number of contacting asperities and $(T_c) = 509$ °C for the contact temperature, from Equations (6.12) and (6.5) or (6.6) respectively.

To give the reader some idea of the numbers involved in this investigation, Table 8.2 gives the variation with load of the various surface parameters for the room temperature experiments carried out by Allen (1982) at a sliding speed of 2.0 m/s. This is an interesting wear pattern since it has a transition at around 62.5 N. We have included a column giving the value of K_o that goes with the best fit between δ_{theory} and δ_{expt}. The trends indicated by Table 8.2 are shown in the graphs of the computed number of contacts N, the computed radius of a typical asperity contact (a) and the computed contact temperature (T_c) at this typical asperity versus load (see Figure 8.8). The general surface temperature of the pin $(T_S)_p$ is also plotted against load. This quantity has been calculated from heat flow measurements. It is interesting to note that the best line through the T_c points is parallel to the best line drawn through the $(T_S)_p$ points, indicating approximately constant difference of about 270 °C between these two temperatures at all loads. This is similar to the temperature difference of 200 °C found in the measurements made by an embedded thermocouple in the 6.25 m/s experiments with AISI 4340 steel discussed in Section 6.2.3 (which dealt with proportional analysis by X-ray diffraction) and which is embodied in Equation (6.41). Inspection of Table 6.4 in Section 6.1.3 shows that (for EN8 steels sliding at 2 m/s) heat flow analysis and the oxidational wear theory gave $(T_c - (T_S)_p)$ values of about 190 °C, for the range of loads (4–10 N); about 380 °C, for the range of loads between 15 and 30 N; and 430 °C, for the loads between 35 and 40 N. It seems reasonable that there should be a direct relation between the contact temperature

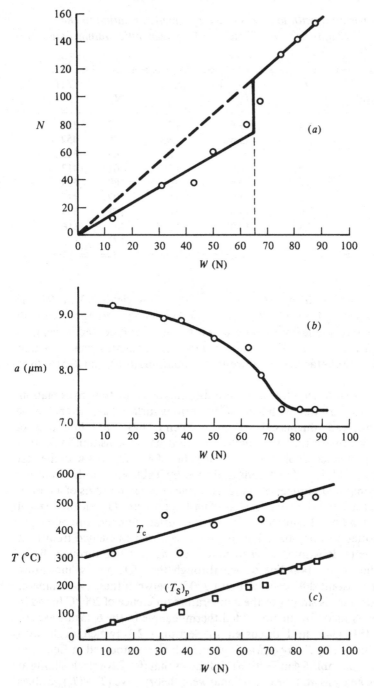

Figure 8.8 *Computed values of surface model parameters plotted against load (a) number of contacts (N), (b) radius of contact (a) and (c) contact temperature (T_c) (sliding speed: 2.0 m/s)*

(T_c) and the general surface temperature of the pin $(T_S)_p$. This constant difference between T_c and $(T_S)_p$ (and $(T_S)_d$) is an important result which could lead to approximate predictions of the contact temperature (T_c) from thermocouple measurements of the bulk temperature of the pin alone.

Figure 8.8 also shows a clear transition in the number of contacts N at a load around 65 N. If the reader will refer to Figure 8.7, he/she will see that this was the load at which the wear rate versus load curve also had a transition, namely a transition from mild-oxidational wear to severe-oxidational wear. Clearly N increases with load in a linear fashion for both types of wear, but at a much faster rate for severe-oxidational wear. It does seem that, for the severe wear, the contact radius remains constant.

8.3.4 Summary

This sub-section has shown that SEM can be used to confirm the oxidational wear mechanism, originally discovered through the use of TEM of replicas of worn steel surfaces. By far the most important aspect of SEM in oxidational wear is its facility for accurately measuring plateaux heights, so that one can then use those measurements to provide one with consistent values of the number of contacts (N), their radius (a) and, most important of all, their temperature (T_c). Let us now discuss the application of Auger electron spectroscopy, X-ray photoelectron spectroscopy and X-ray diffraction to the oxidational wear of 316 stainless steels.

8.4 The oxidational wear of 316 stainless steel tribo-systems

8.4.1 The practical background of the tribological problems involving 316 stainless steels

Stainless steels are often used in a system because of their excellent resistance to corrosion. As we have already seen (in Section 8.2.1), the manufacturers of marine diesel engines soon discovered that *increased corrosion resistance* is not always consistent with a tribologically satisfactory performance, when one has a tribo-system operating in a corrosive atmosphere! In the United Kingdom, the emergence of the advanced gas cooled reactor (AGR) in the 1970s coincided with several tribological problems, some of which were definitely attributable to the use of 316 stainless steels in systems not usually considered to be tribo-systems.

AGRs operate by using a highly-pressurized, continuously circulating, gas to convey the heat generated at the reactor core to the boilers, in which the water is thereby converted into steam. The use of a high-density and high-velocity fluid led to a number of unforeseen engineering and design problems in the commissioning stages of the AGRs. The basis of the problems has generally been identified as being due, either directly or indirectly, to vibration mechanisms caused by the fluid flow. An example of this is given in the paper by Chivers *et al.* (1978). These authors discuss the damage to a component within the gas-coolant circulators of an AGR as a result of mechanical vibration during a proving test. They showed, by tribological examination of the damaged surfaces, that the wear of the isolating dome flange and the sealing flange components of the circulators was caused by fretting (see Section 1.1.6).

Efficient heat transfer between the boiler tubes and the gas coolant requires that the gas flow must be *turbulent*. This is the cause of another unforeseen problem, due to the *vibration* of the boiler tubes against their supports. In some designs, the tubes are supported with clamps, and some vibrational energy is absorbed by frictional damping. If sufficient damping is not provided, then failure of the tubes can occur by means of fatigue or impact wear (Levy, 1975).

The flanges, and the tube support clamps mentioned above, were fabricated from stainless steel stock not very different from 316 stainless steel. In addition to the mechanical and fretting damage sustained by vibrating AGR components, under certain conditions, a so-called *duplex oxide* occurs on stainless steels in AGR environments, which can cause excessive oxidational damage to those parts also subjected to mechanical vibration. The morphology of duplex oxides has been found to be either uniform or nodular. Both forms of the oxide have been found to occur in the same surface. The uniform variety of the oxide, however, tends to nucleate in colonies, which eventually coalesce to form a contiguous film. The nodular variety of the oxide tends to nucleate singularly and thus produce partial coverage by isolated mounds of duplex oxide. These mounds tend to be, in general, hemispherical in shape, and, typically, between 10 and 40 μm in size. The mounds, or nodules, are very susceptible to mechanical removal, particularly via vibration. Thus we have *two* factors influencing the failure of stainless steel components in the AGR, firstly the fatigue and impact wear due to vibration and, secondly, the removal of 'duplex oxide' nodules (also due to vibration).

8.4.2 Research into the oxidation and wear of stainless steel under AGR conditions

Smith (1984), using Energy Dispersive Analysis by X-rays (EDX) in the SEM has shown that the duplex scales formed, at 650 °C and above, in a CO_2/CO mixture, consist of an outer layer of Fe_3O_4 and an inner layer of an Fe, Cr, Ni spinel oxide, which heals rapidly. By 'healing', one normally means that there is a progressively decreasing slope (k_p) of the 'weight gained squared' (i.e. the parabolic weight gain) versus 'time of oxidation'. After about 200 h, the decreasing slope attains an approximately constant value, namely $k_p = 9.1 \times 10^{-12} \, kg^2/m^4s$. The *initial* slope was $k_p = 8.3 \times 10^{-10} \, kg^2/m^4s$. The use of 650 °C and a CO_2/CO gas mixture was intended to simulate the *oxidational* environment of the AGR. Clearly, in order to understand the failure mechanisms operating in AGR components made of this stainless steel, one must simulate the mechanical and vibrational environment of the AGR, as well as the temperature and gas environment.

Although Smith (1985) has also done friction and wear experiments with 316 stainless steels, his simulation was confined to the type of motion (i.e. vibrational) and the temperature range likely to be encountered in the AGR (i.e. 20–500 °C). His gas environment was air (and not a mixture of CO_2 and CO), his load was 8 N and his reciprocating frequency was 2 Hz over a stroke length of 9 mm. The use of air instead of CO_2/CO gas mixtures is probably immaterial as far as oxidational wear is concerned (if it does, in fact, occur at this low load and low speed). The choice of load and speed was to ensure that surface heating induced by friction between his pin-on-flat tribo-elements was negligible compared with the temperature of the environment. The restricted conditions under which Smith (1985) carried out his experiments tend to reduce the relevance of his results (a) to the actual service situation and (b) to the basic literature on wear. Time and time again, the newcomer to tribology will find that the relevance of a wear paper suffers considerably in this respect. As pointed out by Welsh (1965), unless *wear patterns*, that is plots on log/log paper of wear rate (in units of m^3/m) versus load, are obtained for a wide range of loads, then the results in one load range will be very different from the results in another load range, especially if the ranges are on either side of a transition load. If one then wishes to see the effect of a third (independent) variable, then these plots can be placed on the same frame of reference. For example, if chromium content is of interest, then Figure 6.8 shows how to present the wear behaviour in terms of wear patterns at different chromium contents of the steels. If temperature of the environment is the third variable, then Figure 8.7 is an example of how to present one's results.

Whilst Smith (1984, 1985) was investigating the oxidational and wear aspects of the AGR duplex oxide problem through medium-term oxidation experiments in an AGR mixture of CO_2/CO gas in conjunction with reciprocating wear experiments in air, Quinn and Wallace (1985) were investigating these aspects by means of long-term oxidation experiments, backed up by continuous sliding experiments, both series of experiments being carried out in the typical AGR gas mixture. This was a classic example of cooperation between industry (in the form of the Central Electricity Generating Board) and university (Aston University), where the same industrial problem was approached by different ways by researchers at the industrial establishment (Smith was at the CEGB Research Laboratories at Berkeley in Gloucestershire, England) and at the university (Quinn and Wallace were members of the Tribology and Surface Analysis Research Group at Aston University, Birmingham, England). Both groups benefited from the cooperation and, hopefully, the problem associated with the AGR was brought closer to a successful resolution by having two points of view and a more complete coverage of the relevant parameters.

Quinn and Wallace (1985) measured weight gain as a function of time of oxidation up to 3000 h in their static oxidation tests on two typical stainless steels often used in AGR components, namely AISI 316 and AISI 310. It was decided to keep the temperatures constant at 650 °C (the normal working temperature of the components of interest in the AGR), due mainly to the time factor (Wallace was on a 3 y PhD degree course). They used the normal CO/CO_2 gas mixture used in the AGR and reported the following k_{p1} (initial oxidation rate) and k_{p2} (equilibrium oxidation rate between 1000 and 3000 h) values for these steels (see Table 8.3). It is interesting to note that k_{p1} for AISI 316, namely $(8.3 \pm 1.7) \times 10^{-12} kg^2/m^4 s$ is not very different from the so-called equilibrium k_{p2} value, reported by Smith (1984), of $9.1 \times 10^{-12} kg^2 m^4 s$, indicating that Smith's estimate of 200 h as the critical time of oxidation for healing to occur may be a considerable *under-estimate*.

Table 8.3. *Parabolic oxidation rate constants for AISI 316 and 310 stainless steels at 650 °C in an AGR CO/CO_2 gas mixture*

	Parabolic oxidation rate constants	
	AISI 316 $kg^2/(m^4\text{–}s)$	AISI 310 $kg^2/(m^4\text{–}s)$
Initial rate (k_{p1})	$(8.3\pm1.7)\times10^{-12}$	$(3.1\pm0.9)\times10^{-12}$
Equilibrium rate (k_{p2})	$(3.7\pm1.4)\times10^{-12}$	$(1.2\pm0.7)\times10^{-12}$

Quinn and Wallace (1985) produced only a limited number of wear patterns for both AISI 316 and 310. We will only consider the data for AISI 316. The wear patterns (i.e. plots of wear rate versus load) for room temperature, 300 and 500 °C environments of CO_2 obtained at 0.5 m/s indicated that 316 stainless steel tribo-systems will exhibit mild-oxidational wear at loads up to 15 N, after which severe-oxidational wear will ensue. In experiments *not* described in this book, it was found that, above this load, all the wear rate data points lie on the same straight line when plotted against the load (on log–graph paper) regardless of the temperature of the environment. Clearly, the large amount of chromium (17.2%) in 316 stainless steel suppresses the oxidational component of severe-oxidational wear at increased ambient temperatures.

It is unfortunate that, apart from using X-ray diffraction to analyse the debris and SEM together with EDX to analyse the statically-oxidized surfaces and duplex oxide nodules, neither Smith (1984, 1985) nor Quinn and Wallace (1985) used the (more relevant) techniques of AES and XPS. These techniques have become a pre-requisite for any tribology research programme. In the next sub-section, we will discuss some of the research carried out in the Tribology and Rheology Research Laboratories of Georgia Institute of Technology in Atlanta, USA, relevant to the continuous sliding wear of AISI 316 stainless steel. In this research, AES and XPS were used as a *routine* surface inspection method.

8.4.3 The application of physical methods of analysis to the wear of stainless steels

(i) Introduction

The research we are about to discuss formed part of a PhD programme designed to study the effects of load and speed on the sliding wear behaviour of 316 stainless steel, without lubrication, and in a normal atmosphere of air, (Hong, 1986). X-ray diffraction, XPS and AES formed a large part of Hong's (1986) programme. Before we discuss the application of the above techniques to a stainless steel tribo-system, we should briefly outline the mechanical behaviour of his system. In essence, Hong (1986) confirmed the transition first noted by Quinn and Wallace (1985) at 0.5 m/s, namely a change from mild- to severe-oxidational wear as the load was increased, the transition load being 15 N. Actually, Hong

(1986) obtained these transitions at loads of 19, 16, 14 and 13 N in the wear patterns (i.e. wear rate versus load curves) for 1.75, 2.0, 3.0 and 3.5 m/s respectively. It can also be shown, by careful analysis of Hong's (1986) wear rates, that the specific wear rate (i.e. the wear rate per unit load) for the 2.0 and 1.76 m/s experiment was $(0.51 \pm 0.09) \times 10^{-13} \, \text{m}^3/\text{mN}$, which is almost a factor of two *more* than the specific wear rate for the 3.0 and 3.5 m/s, namely $(0.27 \pm 0.09) \times 10^{-13} \, \text{m}^3/\text{mN}$. Let us start with the X-ray diffraction analysis of the wear debris, since it has some bearing on the above described differences in mild-oxidational wear rates.

(ii) X-ray diffraction analysis

Wear debris, collected from tribo-systems worn at loads *below* the transition loads mentioned above, was examined by X-ray diffraction in a conventional X-ray diffractometer, using a CuK$_\alpha$ X-ray source (wavelength = 1.54 Å). There seemed to be a difference in the compositions of the wear debris from the two lower speed (higher specific wear rate) experiments compared with the two higher speed (lower specific wear rate) experiments. For the higher wear rates, the debris produced (after an initial running-in period, during which martensite and austenitic particles were found in the debris) was mainly the rhombohedral oxide $(\text{Fe, Cr}_2\text{O}_3)$. For the lower wear rates, the debris was mainly the spinel oxide $(\text{Fe}, \text{Fe}_{(2-x)} \text{Cr}_x\text{O}_4)$. Clearly, mild-oxidational wear occurred at all loads below the transition load, the magnitude of that wear depending upon the type of oxide produced at the real areas of contact.

It is interesting to note that the evidence for a change of oxide from rhombohedral to spinel (as one increases the speed), consisted of diffraction maxima at $2\theta = 78.2°$ for the rhombohedral oxide and $2\theta = 56.8°$ and $75.0°$, for the spinel oxide. These were the only maxima which could *not* be attributed to either martensite or austenite. The credibility of the identification of the oxides clearly needed confirmation by an alternative technique. This is where XPS was used to help, as we shall see in the next sub-section.

(iii) X-ray photoelectron spectroscopy

Hong (1986) used the SSL Model SSX 100-06, small spot, X-ray photoelectron spectrometer, which features a monochromatized Al K$_\alpha$ X-ray source, a 180° hemispherical analyser, and flexible sample handling capability. During the analysis, the background pressure was around 7×10^{-10} torr. For the sputtering, the pressure increased to 3.5×10^{-7} torr due to primary argon. The electron binding energies were calibrated against the adventitious carbon 1s peak (typically present in most XPS spectra), assuming a binding energy of 284.6 eV. The expected instrumental error in determining the binding energy of the elements was less than ± 0.1 eV.

The general XPS spectra of the as-received 316 stainless steel and the worn surfaces before and after sputtering are shown in Figures 8.9, 8.10 and 8.11 respectively. Before sputtering, these spectra show peaks of C, O, Fe, Cr and Ni. After sputtering, the carbon 1s peak disappears indicating that this peak really is due to an extremely thin carbonaceous layer, due to contamination in the

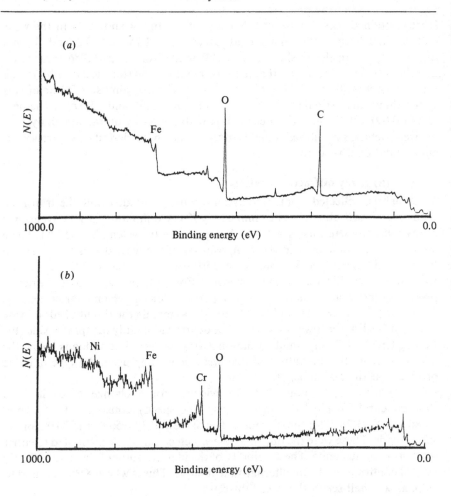

Figure 8.9 *The general XPS spectra of as-received 316 stainless steel (a) before and (b) after Ar ion sputtering to a depth of 9 Å*

instrument. The oxygen peak, however, still remains, indicating the continued existence of an oxide layer in all of the surfaces examined.

High-resolution XPS spectra of the $Fe(2p_{3/2})$ and $Cr(2p_{3/2})$ peaks obtained from (a) an as-received specimen, (b) a wear-tested sample at 1.75 m/s and a load of 7.8 N and (c) a wear-tested sample which had been running at 3.0 m/s (also at a load of 7.8 N) are shown in Figure 8.12(a), for $Fe(2p_{3/2})$ and Figure 8.12(b) for the $Cr(2p_{3/2})$ peak. The binding energies of the iron and chromium peaks are listed in Table 8.4, together with the accepted values for the (2p) electron peak energies of the most probable constituents. For instance, in the as-received conditions, Figure 8.12 reveals that the major components of the $Fe(2p_{3/2})$ peaks occurred at 707.1 and 709.4 eV. According to Table 4.6, these peaks correspond to metallic Fe (706.5 eV) and Fe in the divalent oxide state (FeO), which has a binding energy of

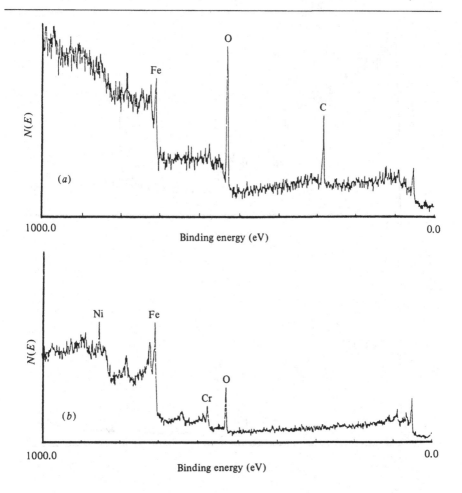

Figure 8.10 *The general XPS spectra of wear-tested 316 stainless steel pin at a 1.75 m/s sliding speed (a) before and (b) after Ar ion sputtering to a depth of 51 Å*

709.6 eV. The chromium values have been obtained from the *Handbook of X-Ray Photoelectron Spectroscopy* published by Physical Electronics, namely 574.1 eV (for metallic chromium) and 576.6 eV (for Cr_2O_3), both energies being for the $Cr(2p_{3/2})$ electrons. Further confirmation comes from the presence of a peak at 585.9 eV, which corresponds to the $Cr(2p_{1/2})$ electron energy of 586.3 eV.

The presence of Fe_2O_3 and Cr_2O_3 in the XPS spectra from the surfaces worn at 1.75 m/s, under a load of 7.8 N, is consistent with the presence of the rhombohedral oxide $[(Fe, Cr)_2O_3]$ detected by X-ray diffraction in the wear debris from the low-speed experiments. The presence of Fe_3O_4 in the XPS spectra from the surfaces worn at 3.0 m/s (under the same load) is consistent with the presence of the spinel oxide $(Fe, Fe_{(2-x)}Cr_xO_4)$ detected by X-ray diffraction in the wear debris from the high-speed experiments.

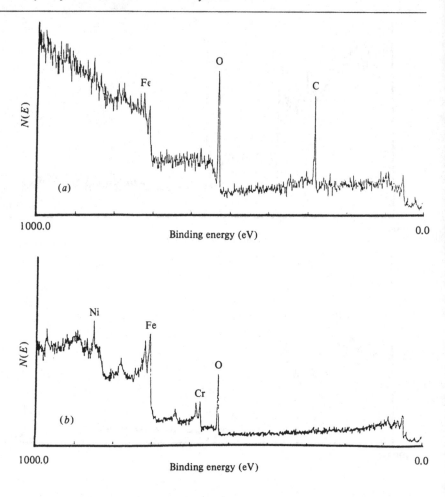

Figure 8.11 *The general XPS spectra of wear-tested 316 stainless steel pin at a 3.0 m/s sliding speed (a) before and (b) after Ar ion sputtering to a depth of 51 Å*

From the above, we see that XPS can give definite information about the *chemistry* of the surface, and sub-surface, regions of worn tribo-elements. It is not so good as AES as regards being able to *locate* where the elements are, relative to any topograhical features, such as the oxide plateaux formed in mild-oxidation wear. Let us now see how Hong (1986) applied AES to the 316 stainless steel tribo-systems mentioned above.

(iv) Auger electron spectroscopy

A Perkin–Elmer Model PHI 600 Scanning Auger Microprobe was used to determine the depth-concentration profiles of Fe, Cr and O for the as-received stainless steel surfaces and the wear-tested surfaces (Hong, 1986). The background pressure was 6×10^{-10} torr. This pressure increased to 2.4 ×

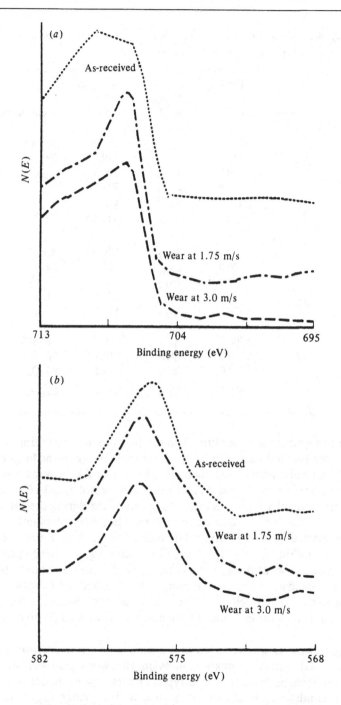

Figure 8.12 *The high-resolution XPS spectra from the as-received and wear-tested specimens showing the variations of (a) Fe(2p$_{3/2}$) and (b) Cr(2p$_{3/2}$) electrons*

Table 8.4. *Binding energies of Fe and Cr electron levels in 316 stainless steel before and after being subjected to mild-oxidational wear*

Specimen	Element	Binding energy (eV)	Energy level	Accepted value (eV)	Identification
Unworn AISI 316 stainless steel pin (9 Å sputter)	Fe	707.0	$2p_{3/2}$	706.5	Metallic
		719.8	$2p_{1/2}$	719.7	Fe
		709.4	$2p_{3/2}$	709.6	FeO
		722.8	$2p_{1/2}$	722.9	
	Cr	576.3	$2p_{3/2}$	576.6	Cr_3O_3
		585.9	$2p_{1/2}$	586.3	
AISI 316 Stainless steel worn at 1.75 m/s and 7.8 N (51 Å sputter)	Fe	707.0	$2p_{3/2}$	706.5	Metallic
		720.0	$2p_{1/2}$	719.7	Fe
		710.6	$2p_{3/2}$	710.7	Fe_2O_3
	Cr	574.6	$2p_{3/2}$	574.1	Metallic Cr
		576.7	$2p_{3/2}$	576.6	Cr_2O_3
AISI 316 Stainless steel, worn at 3.0 m/s and 7.8 N (51 Å sputter)	Fe	706.9	$2p_{3/2}$	706.5	Metallic
		720.0	$2p_{1/2}$	719.7	Fe
		723.4	$2p_{1/2}$	723.5	Fe_3O_4
	Cr	576.1	$2p_{3/2}$	576.6	Cr_2O_3

10^{-7} torr during the sputtering procedure. The electrons used for exciting the atoms of the surfaces (so that they give off the increased energy in the form of Auger electrons) had been accelerated through 15 kV. The AES results were obtained by a cylindrical mirror analyser having on energy-resolution ($\Delta E/E$) of 0.6%. The sputtering rate of Ta_2O_5 was taken as the standard during the depth profiling. It was assumed that the sputtering rate was effectively constant.

Auger electron spectra from as-received 316 stainless steel, before and after sputtering, are shown in Figure 8.13. Note that, after sputtering, the carbon peak was significantly reduced, as also was the oxygen peak. Figure 8.14 shows the depth-concentration profiles of iron, chromium, carbon, nickel and oxygen for the as-received specimen. The thickness of the oxide was determined as being about 35 Å, which indicates the existence of the normal–air-formed passivating oxide.

Using the Auger electron spectrometer as a scanning electron microscope, Hong (1986) identified typical plateau regions in the worn stainless steel surfaces. The Auger electron spectra from a typical plateau, worn under mild-oxidational wear conditions, is shown in Figure 8.15(*a*). After argon ion sputtering to a depth of 1200 Å, one can see (from Figure 8.15(*b*)) that the oxygen is still present, whereas the carbon contaminant has completely disappeared. This fact alone tells us that the plateau consists of *at least* 0.12 μm of oxide. By making

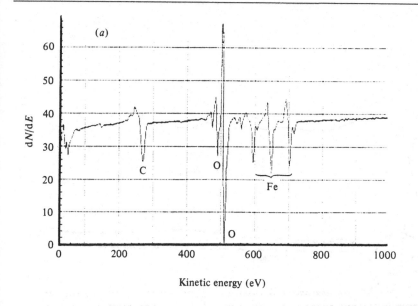

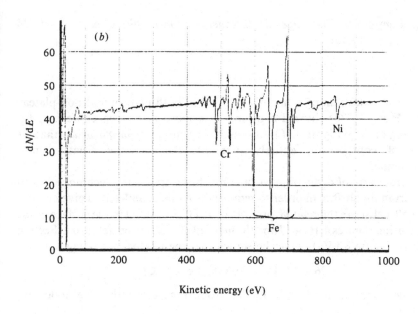

Figure 8.13 *Auger electron spectra from as-received 316 stainless steel(a) before and (b) after sputtering with argon ions*

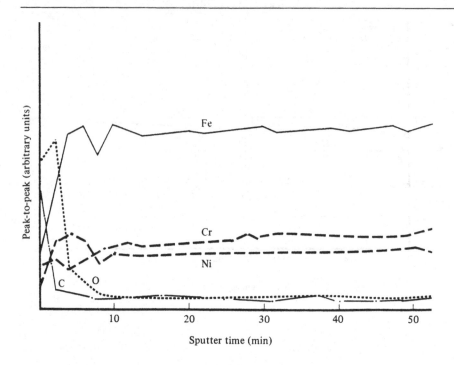

Figure 8.14 *The Auger depth concentration profiles of as-received 316 stainless steel*

the instrument scan along a line which takes in both plateau and non-plateau regions, Hong (1986) showed a dramatic increase in oxygen content in the plateau region, (see Figure 8.16). Although we cannot say *unequivocally* that the plateau of contact are entirely oxidized, the evidence seems almost overwhelming.

A possible way of deducing just how much the sputtering technique has to remove from the surface in order to remove the oxide completely is given below. First of all, calculate the relative proportions of C, Fe and O in the surface of the plateau, using Figures 4.16, 4.17 and the method of calculation detailed in Section 4.5. From Figure 8.15(*a*), it can be shown that the relative percentages are:

$$26\% \text{ C}; 41\% \text{ O}; 33\% \text{ Fe}; 0\% \text{ Cr}$$

Using the same method, we obtain, from Figure 8.15(*b*), the following percentages:

$$0\% \text{ C}; 16\% \text{ O}; 59\% \text{ Fe}; 25\% \text{ Cr}$$

It seems that it takes 1200 Å of sputtering to reduce the oxygen percentage in the plateau surface from 41 to 16%. In other words, a 25% reduction in oxygen percentage is equivalent to 1200 Å of plateaux height. In order to remove the

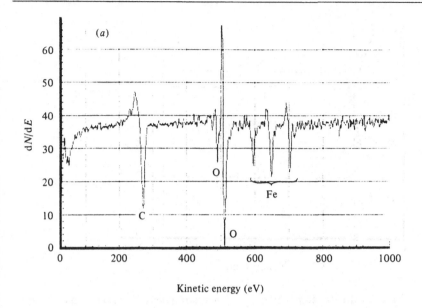

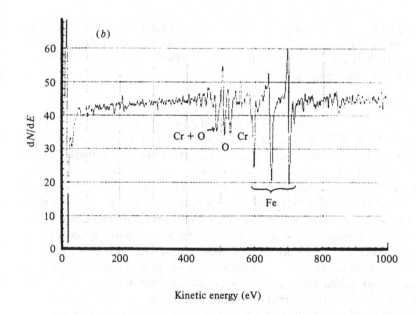

Figure 8.15 *Auger electron spectra from a typical plateau region of a worn stainless steel surface (a) before and (b) after argon ion sputtering to a depth of 1200 Å*

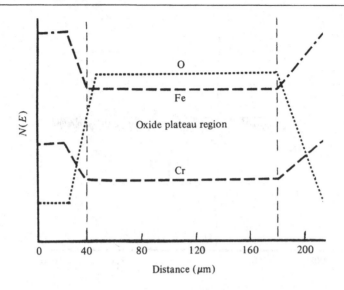

Figure 8.16 *An Auger line scan showing the increase of oxygen content in the region of a contact plateau formed during the wear of stainless steel*

remaining 16%, we would require (16/25) of 1200 Å, that is 768 Å. Hence, the total thickness of the plateau is 1968 Å, that is about 2000 Å or 0.2 µm. This is a little low compared with the values obtained by Allen (1982) (see Table 8.1) but is, nonetheless, an encouraging result.

8.4.4 Summary

In this sub-section on the oxidational wear of AISI 316 stainless steels, we have seen that the use of these steels in the nuclear industry has sometimes led to premature failure of systems not normally thought of as tribo-systems. This was due to the vibrational motion always present in a nuclear reactor such as the AGR. Research into the oxidation and the wear of these steels, both by industrial and academic research laboratories, has not revealed much information about possible solutions to the tribological problems involved in using stainless steels. Nevertheless, the mere existence of the problem has spawned a considerable amount of basic research into the dry wear of stainless steels. Most of this research has involved the use of physical methods of analysis to confirm the existence of mild-oxidational wear below a transition load of around 15 N. These physical methods have shown that the plateaux of contact in this form of wear are indeed mainly oxide, possibly with a critical thickness (ξ) somewhat less than the 3 or 4 µm found in low-alloy steel systems. XPS has been especially useful in that it has *confirmed* the chemistry of the worn surfaces *suggested* by the X-ray diffraction analysis. Once again, we have seen *lower wear* associated with the *spinel oxide*, compared with the wear associated with the rhombohedral oxide.

8.5 Concluding remarks

Oxidational wear, whether it be of the mild or severe kind, is a widespread feature of the wear mechanism maps for steels (Lim and Ashby, 1987). More and more practical situations are occurring where no conventional lubrication is possible, due to the high temperature environment, for example the ceramic diesel. This is where mild-oxidational wear would be an acceptable condition. In this chapter, we have seen how physical methods of analysis have been used to confirm that the mild-oxidational wear mechanism is relevant to the wear of low-alloy steels and stainless steels. We have seen that electron microscopy (either SEM or the TEM of replicas), XPS and AES have shown:

(a) that the mating asperities on opposing surfaces undergoing mild-oxidational wear are concentrated on a few plateaux, each plateau being of the order of the size expected from the Bowden and Tabor (1954) equation;

(b) that the plateaux seems to consist entirely of oxide;

(c) that the spinel form of oxide is more conducive to low wear than the rhombohedral form; and

(d) that SEM measurements of oxide plateau thicknesses can be used, in conjunction with the heat flow equations given in Section 6.1.2, to give estimates of the contact temperature (T_c), the number (N) of contacts occurring at the *dominant* plateau of contact, and the size (a) of the individual asperity contact between opposing plateaux.

Although these important findings have mainly been confined to steel tribo-systems, there is no reason why physical methods of analysis should not bring about another breakthrough in the wear of other tribo-systems, undergoing other forms of wear than mild-oxidational wear. This chapter has shown how powerful the use of SEM, X-ray diffraction, XPS and AES can be, when applied to a form of wear which involves the chemical, physical and mechanical properties of the tribo-elements.

The next (and final) chapter of this book on the use of physical analytical techniques for studying tribo-systems is concerned with a very important kind of tribo-system, namely one involving ceramic interfaces.

9   The application of physical techniques to selected ceramic tribo-systems

9.1   Introduction

The hardness and chemical stability of ceramics *should* make them ideal materials for tribo-elements that have to function under severe wear and high temperature conditions. Unfortunately, large variations in both friction and wear resistance can occur when ceramics are slid against themselves, or against a metal counterface, in the unlubricated condition. These variations have tended to prevent the widespread use of ceramics in tribo-systems. However, the recent interest in developing the ceramic diesel has tended to bring about an increase in fundamental research into the tribological properties of ceramic/ceramic and ceramic/metal tribo-systems, some of which looks very promising.

Some of the earliest work on the tribology of ceramics was carried out by Seal (1958) on diamond. Other materials include ionic crystals (Steijn, 1963), sapphire (Steijn, 1961), silicon carbide (Miyoshi and Buckley, 1979a), boron nitride (Buckley, 1978), manganese zinc-ferrite (Miyoshi and Buckley, 1981b), titanium (Nutt and Ruff, 1983), silicon nitride (Dalal, Chiu and Rabinowicz, 1975) and many other materials too numerous to mention. Since we are mainly interested in the application of physical analytical techniques to ceramic tribology, we will be *very selective* and discuss silicon nitride, silicon carbide and sapphire tribo-systems. Other systems may be more relevant to the reader's interest, but at least we know that various physical analytical techniques have been applied to tribo-systems with these materials as tribo-elements.

9.2   Silicon nitride tribo-systems

9.2.1   Introduction

From the many papers published on the tribological properties of Si_3N_4, the papers by Fischer and Tomizawa (1985), Tomizawa and Fischer (1986) and Jahanmir and Fischer (1986) have been chosen for this short review. These authors used a pin-on-disc tribometer in which a hot-pressed Si_3N_4 pin was loaded (at 9.8 N) on a horizontal, rotating Si_3N_4 disc, the speed of sliding at the pin being very slow (1 mm/s). The choice of load and speeds was to make sure that the frictional heating would be insignificant (about 1 to 5 °C above room temperature).

Fischer and Tomizawa (1985) found that, when Si_3N_4 was slid against itself in the presence of dry gases, the surfaces wore by *micro-fracture*. When slid in a very humid environment, a tribo-chemical reaction occurred between the water molecules and the wearing silicon nitride, to produce an amorphous layer of *silicon dioxide* on the surface. This layer seemed to be responsible for reducing the wear by two orders of magnitude, that is a factor of 100. No wear particles were detected in this tribo-system, presumably because the wear occurred by dissolution of the material. Under very humid conditions, the coefficient of friction was stable at 0.75 (in a gaseous environment) or 0.7 (in water). When sliding silicon nitride at temperatures between 150 to 800 °C, Tomizawa and Fischer (1986) found that interaction with water vapour caused a decrease in the friction coefficient to values as low as 0.2. This is quite remarkable, since one of the disadvantages of using ceramics in tribo-systems lies in the generally high (~ 0.7) friction coefficients.

Tomizawa and Fischer (1986) suggest that the large variations in friction and wear of silicon nitride were not due to poor reproducibility of the material, as previously suspected, but to a complicated mode of wear involving mechanical and chemical phenomena. In an attempt to bring together all the factors affecting the friction and wear of silicon nitride, Jahanmir and Fischer (1986) carried out experiments with hot-pressed Si_3N_4 sliding on itself at 1 mm/s under a load of 9.8 N in humid air, water, pure hexadecane and in hexadecane with 0.5% stearic acid. Their experiments are of particular relevance to this book since they applied several physical methods of analysis to support their measurements of friction and wear under the various environments. For instance, the worn regions were examined by scanning electron microscopy (SEM) and Auger electron spectroscopy (AES), to assess the wear mechanisms and to deduce the chemical composition of the surface layers in the wear tracks. The used lubricating solutions were analyzed by emission spectroscopy (see Section 4.2.4) to see if any changes had occurred in their chemical composition. Surface layers were stripped from the wear tracks of selected discs, and their morphology and crystallography determined by transmission electron microscopy (TEM) and selected area electron diffraction (see Section 3.3.6).

9.2.2 Friction and wear experiments with silicon nitride

Jahanmir and Fischer (1986) report equilibrium friction coefficients of 0.7 (for humid air), 0.5 (for water), 0.2 (for pure hexadecane) and 0.1 (for hexadecane with stearic acid). The result for pure hexadecane is most surprising, when one considers that this is a chemically inert fluid which does *not* provide boundary lubrication for *metal* sliding on metal (Fischer and Luton, 1983), where a friction coefficient of 0.65 is quite common!

Figure 9.1 summarizes the wear results of Jahanmir and Fischer (1986), from which it can be seen that (over the short distance slid of 200 m) the wear of Si_3N_4 under humid air or water is linear with respect to distance slid, giving a wear rate of 1.7×10^{-13} m^3/m. This is about the same rate as in the high speed (2 m/s) unlubricated, room temperature, wear experiments shown in Figure 8.7 for ferritic steel sliding on austenitic stainless steel, thereby indicating no real gain would be obtained if one were to use Si_3N_4 in place of these exhaust valve/valve seat materials under conditions where no lubrication is possible. It is when one considers the pure hexadecane and hexadecane plus stearic acid that one sees a

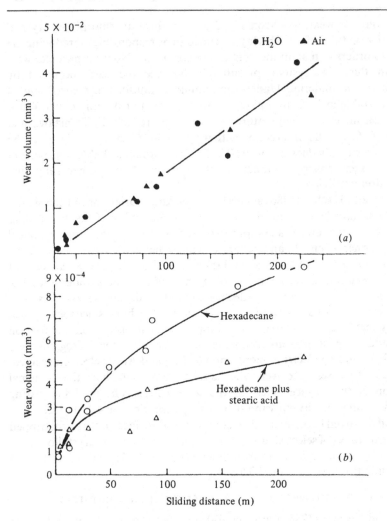

Figure 9.1 *Wear of* Si_3N_4 *sliding against itself at a speed of 1 mm/s and a load of 9.8 N (a) in air (35% relative humidity) and water and (b) in pure hexadecane and hexadecane with 0.5% stearic acid*

dramatic decrease (over 100 m sliding distance). The shape of the curves in Figure 9.1(*b*) indicates a running-in (similar to that occurring in mild-oxidational wear) followed by an equilibrium wear rate of 1.0×10^{-15} m³/m and 2.5×10^{-15} m³/m for the hexadecane plus stearic acid and the pure hexadecane, respectively. It is indeed unfortunate that typically we require ceramic tribo-elements to function in hostile environments, particularly those of high ambient temperatures, otherwise the extremely low wear rates (together with low friction coefficients) obtained with hexadecane plus stearic acid would make Si_3N_4 a very good candidate for a ceramic tribo-material. Clearly, these lubricants could not co-exist with an elevated temperature environment. Nonetheless, this particular tribo-system deserves to be investigated, in the hope that some of the basic principles involved in the friction and wear of Si_3N_4 under hexadecane plus stearic acid could be applied to the search for a high temperature lubricant for Si_3N_4.

9.2.3   Scanning electron microscopy of worn silicon nitride

Jahanmir and Fischer (1986) made an extensive study (using SEM) of the surfaces of the wear tracks formed under the four environments mentioned above. Most of their electron micrographs reminded me of the oxidized plateaux formed on steel surfaces that have undergone oxidational wear. Although one cannot precisely use the 1 µm bars placed on the micrographs by Jahanmir and Fischer to measure the thicknesses of the plateaux, in a direction perpendicular to the bar, one can assume that the tilting of the specimen surface would be less than 45°, so that there would be a foreshortening. Hence any measurements of plateaux thicknesses would be a minimum value. In fact, the measurements were:
(a) Humid air environment: $(1.0 \pm 0.2)\,\mu m$.
(b) Water environment: $(1.3 \pm 0.2)\,\mu m$.
(c) Pure hexadecane: $(0.4 \pm 0.2)\,\mu m$.
(d) Hexadecane + stearic acid: $(2.2 \pm 0.2)\,\mu m$.
In fairness to Jahanmir and Fischer, it should be pointed out they they had other interpretations. It is interesting to note, however, that for the hexadecane + stearic acid experiments, these authors say that the topography of the worn material suggests the presence of a film. They also say that their electron micrographs of this fourth group of surfaces clearly shows that material, worn away from the higher crystallites, fills the lower regions of the surface. They go on to say that these micrographs show a region where the flat tops of the polished silicon nitride crystallites are surrounded by wear material that filled the valleys. This description is very similar to that given in Sections 8.2 and 8.3, where we were discussing the characteristic topography of mild-oxidational wear, with its oxidized plateaux of contact about 1 to 4 µm in height.

In an attempt to identify the surface material, Jahanmir and Fischer (1986) stripped off material from the Si_3N_4 discs in the region of the wear track. From their TEM, they deduced that most of the removed material was amorphous. This was confirmed by the lack of any diffraction maxima when the instrument was used in the selected area electron diffraction mode of operation. There were a few darker pieces in the thin grayish amorphous material. These pieces gave rise to a single crystal spot pattern when viewed by selected area electron diffraction. This pattern could be indexed as that to be expected from a single crystallite of Si_3N_4. The identity of the *amorphous* material was tentatively shown to be silicon and oxygen by the Electron Energy Loss Spectrum (EELS). We have not mentioned this technique in our survey of physical methods in view of the fact that it has hardly ever been applied to tribo-systems. The reason for this lack of interest is probably due to the problems always associated with preparing thin transmission specimens from solid surfaces. If one can examine a surface film in situ, one has less chance of introducing artefacts and, of course, the examination is virtually non-destructive. The identity of the dark (crystalline) pieces was shown (by

EELS) to be silicon and nitrogen, thereby confirming the selected area electron diffraction evidence.

9.2.4 Auger electron spectroscopy of worn silicon nitride

Although *scanning electron microscopy* was used to examine the worn surface topography of *all* the specimens, Jahanmir and Fischer (1986) were *very selective* as regards the application of *other physical methods of analysis*. For instance, TDM, EELS and HEED were *not* applied to the humid air and water experiments mentioned above. AES, on the other hand, was applied only to the pure hexadecane experiments. This must be because of the time factors and logistics involved. The SEM was applied to every specimen, because it is reasonably easy to do so, and because the interpretation is fairly straightforward. To apply *five* physical analytical techniques to *four* groups of specimens entails *at least* twenty analyses. To be sure that what one finds is typical (and not just peculiar to one particular specimen), one should take at least three analyses for each specimen, making a total of sixty for the experiments carried out by Jahanmir and Fischer (1986). Obviously, this is not possible. Hence, one must be *very selective* both as regards *method of analysis* and as regards *choice of specimen*. The application of physical analysis to tribology calls for a mature and knowledgeable approach, to ensure one does not waste time or resources!

To return to the AES analysis by Jahanmir and Fischer (1986), these authors performed their measurements on a specimen that had been covered with gold in order to carry out the SEM mentioned above. This device prevents the electron beam charging up the surface by providing a conduction path from the surface to the specimen holder. Although this may be necessary for non-conducting specimens, in general the tribologist should *not* treat his/her surfaces in any way that could change the results of the tribo-chemical or tribo-physical reactions. Since this specimen was ion-milled (with Ar ions) for as long as 800 s, the effects of the gold film were rapidly obliterated by the sputtering action of the ions. Figure 9.2 shows the concentration of the various elements found in the surface of the Si_3N_4 disc as a function of sputtering time. The rate of material removal was about 2 Å/ps (on the average).

In Figure 9.2.(a), one can see the elemental concentrations as a function of the cumulative sputtering time. After removal of the gold overlayer in the first 50 s or so, the Auger spectra from the surface of the Si_3N_4 disc *outside of the wear track*, contained peaks corresponding to silicon, carbon, nitrogen and oxygen. Jahanmir and Fischer determined the elemental concentrations in the way already described in Section 5.4.3, namely by measuring the amplitude of the peaks in the (dN/dE) spectrum, multiplying these amplitudes by the relative sensitivity factors published in the *Physical Electronics Auger Handbook*, and by a scale factor which ensures that the sum of the concentrations adds up to 100%. The slow decrease of the carbon and oxygen concentrations, both of which elements are always found as a contaminant in all surfaces, is the result of the sputtering mechanism. Note that the nitrogen-to-silicon ratio, as measured after bombardment with 5 ke V Ar ions, is larger than the ratio corresponding to Si_3N_4. This is due to ion and electron bombardment excitation which cause preferential

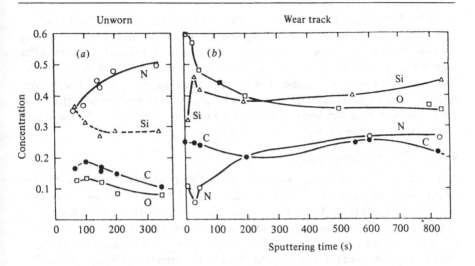

Figure 9.2 *Composition of* Si_3N_4 *disc (from Auger electron spectra) after bombardment with 5 keV Ar ions for the times shown in abscissa; (a) outside the wear track and (b) inside wear track, after sliding 200 m in hexadecane at 1 mm/s under a load of 9.8 N (Jahanmir and Fischer, 1986)*

sputtering of one element and electron migration of another according to Fransen *et al.* (1985). Because of these phenomena one really cannot use AES for a *truly quantitative* analysis method for Si_3N_4.

In Figure 9.2(*b*), which relates to the wear track itself, one can see that, even at the start of the sputtering, there was no trace of gold from the initial overlay. This indicates that an amount of material had been removed from the wear track during the examination of the 'off-track' surface. This illustrates a point about surface analysis that I have been emphasizing for many years, namely, any analyses of surfaces outside of the wear tracks should always be carried out with great caution. This is because of the tribo-chemical and tribo-physical effects, which can often cause reactions in those parts of the surface away from the nominal contact region. It is false economy to have several wear scars on the same specimen, since every time one runs an experiment, the *whole* of the specimen surface becomes subjected to the effects of tribo-physical and -chemical reactions at the (very small) areas of real contact. Jahanmir and Fischer (1986) would have been better placed (as regards understanding their experiments) if they had viewed the wear track first and then compared it with a new, virgin piece of Si_3N_4 specimen.

From Figure 9.2(*b*), it can also be seen that the wear track *initially* contains a very high concentration of oxygen, about half as much silicon, an appreciable amount of carbon, and very little nitrogen. As sputtering proceeds further, the oxygen content decreases quite rapidly, the nitrogen content increases, and the silicon content remains approximately the same as initially. Note the high concentration of carbon ($\sim 20\%$), which is much higher than in the unworn area.

A study of the *shape* of the carbon Auger peak revealed that the carbon is *not* present in the *carbidic* form, indicating that no silicon carbide is formed during the tribo-chemical reaction. It seems reasonable to assume that the carbon in the wear track corresponds to hexadecane from the lubricant that is mixed in with the material. Jahanmir and Fischer (1986) reported that the sizes of the silicon, nitrogen, and oxygen Auger peak-to-peak heights were consistent with the TEM which showed that the wear track was a mixture of silicon oxide and small crystallites of silicon nitride.

9.2.5 Concluding remarks

Although the Si_3N_4 tribo-system is still far from being fully understood, the use of several physical methods of analysis has led to the conclusion that frictional sliding of silicon nitride in the presence of reactive fluids (humid air, water and hexadecane plus stearic acid) tends to be accompanied by tribo-chemical reactions that lead to amorphous silicon dioxide. It also happens when Si_3N_4 is slid in the presence of (unreactive) hexadecane, showing that it is most likely that the tribo-oxidation is due to the common factor in all the experiments of Jahanmir and Fischer (1986) namely the oxygen in the air that is always around, albeit in an entrapped form, in the lubricant.

The fact that the silicon dioxide is *amorphous* would normally mean that one would not be able to detect its presence using any of the diffraction techniques. The evidence for its amorphous nature is somewhat tenuous. Although the proportions of silicon and oxygen were said to be consistent with the presence of silicon dioxide, the authors must have found it difficult to distinguish between the silicon from silicon dioxide, and silicon from the silicon nitride of the specimen. Possibly X-ray photoelectron spectroscopy could have indicated more precisely the presence of silicon from silicon dioxide (with its 2p peak at 99.4 eV binding energy) and silicon from Si_3N_4 (with its 2p peak at 101.9 eV binding energy). These values have been taken from Briggs and Seah (1983).

Clearly, the work of Jahanmir and Fischer (1986) has presented us with some very interesting problems. What is the mechanism whereby the hexadecane, although normally considered to be inert insofar as *metal* surfaces are concerned, seems to provide very effective boundary lubrication when silicon nitride is slid against itself? From the plateaux seen by SEM and the silicon dioxide tentatively identified in the worn surfaces, I suspect that this could be another manifestation of mild-oxidational wear! Only the *critical* use of the appropriate physical techniques might provide the solution to the problem of determining the mechanisms of friction and wear of silicon nitride.

By concentrating our attention on the work of Fischer and his co-workers we have, of course, overlooked the very good work of other investigators into the friction and wear of silicon nitride, for instance, the work of Ishigaki, Kawaguchi, Isawa and Toiba (1985), Page and Adewoye (1978) and Kumura, Okada and Enomoto (1989). The results of Kimura *et al.* (1989) are particularly interesting since they show that Si_3N_4 sliding against itself at very low speeds (2.5 mm/s) gives rise to topograhical features very similar to those found by the author (Quinn, 1968) in his electron microscopy of replicas taken from *steel* surfaces undergoing oxidational wear. Kimura *et al.* (1989) found cracks running *perpendicular* to the direction of sliding. These cracks were visible in both optical and scanning electron micrographs. By scanning along a line *parallel* to the direction of sliding with two-channel secondary electron detectors they obtained a surface profile (using the method reported by Suganuma, 1985), which indicated a 'rising' of the surface *in front* of the foremost crack line. The height of this rising was about 0.3 µm. Is this a plateau of contact, or it is truly a buckling of the surface layers prior to breaking up into fragments? Once again, we find that the oxidational wear mechanism (or something that looks very much like it) could possibly be operating in the wear of ceramics. Let us now discuss the application of physical methods of analysis to another important ceramic, namely silicon carbide.

9.3   Silicon carbide tribo-systems

9.3.1   Introduction

The high strength, creep resistance and oxidation resistance of silicon carbide makes it an important material for high temperature mechanical applications in severe environments. Hence its use for the manufacture of gas turbine blades, a situation in which high-speed rotor blades often slide, momentarily, against the stator blades, at temperatures in excess of 600 °C and in a very hostile environment. It is also used as an abrasive for grinding where, as a result of the extremely small chip size and the high wheel speed involved, the instantaneous temperatures at the tip of the silicon carbide particle are extremely high. Sometimes, these temperatures may reach the melting point of the metal workpiece. Clearly, it is a ceramic with some very interesting properties, not the least of which is the high reactivity of freshly-formed silicon carbide surfaces with metals. There is a real need for its surface properties to be studied from a fundamental point of view.

In fact, a considerable amount of work along these lines has already been carried out by Dr Buckley and his co-workers. They have also looked at the tribological properties of silicon carbide/metal tribo-systems sliding at very light loads, very slow speeds, over very short distances and under very high vacuum (Miyoshi and Buckley, 1979a,b, 1980, 1981a,b, and 1982a,b). Some of the above running conditions could possibly be simulating the conditions that a tribo-system must face in outer space. Sometimes, these authors use a range of temperatures from room temperature up to 1500 °C, presumably in an attempt to simulate some of the conditions likely to be encountered when a space probe is subjected to the sunlgiht. The actual application of the knowledge gained from the very limited range of running conditions is never discussed by Miyoshi and Buckley. In their paper on the tribological properties and surface chemistry of silicon carbide, Miyoshi and Buckley (1982a) state that their research is aimed at understanding both the fundamentals of the surface chemistry involved *and* the tribological properties of silicon carbide tribo-systems.

It is unfortunate that these authors have been so much more involved in the surface chemistry per se than with relating that surface chemistry to the tribological properties of their systems. In the experiments carried out with polycrystalline pins of pure iron sliding against single crystal α-silicon carbide (Miyoshi and Buckley, 1982a), a large proportion of the physical analysis was carried out *before* the specimens were subjected to relative movement against each other in the vacuum of the apparatus containing the Auger electron spectrometer. The object of the analysis was to characterize the surface chmistry *before* friction and wear took place. This was done by AES and XPS examination of silicon carbide surfaces at room temperature and at 250, 400, 600, 800, 900, 1000 and 1500 °C. All their wear experiments were carried out in a vacuum of 10 nPa, using silicon carbide single crystal specimens that had been pre-heated up to 1200 °C. The lower temperature was to avoid melting the iron. The only analysis of their specimens *after wear*, was by using SEM to examine the topography of the wear tracks, with special reference to metal transfer from the pin and fracture pits in the tracks.

The reason for the intensive use of physical analysis *before* wear is probably because the experiments carried out by Miyoshi and Buckley (1982a) were intended to investigate the *initial stages of wear*. These stages will, of course, depend upon the state of the silicon carbide surface before sliding. The processes of wear are so complex that there is very little chance of fundamental investigations producing very much useful information, if they are applied to single crystals. Within a very short time after starting to slide a polycrystalline metal (e.g. iron) against a single crystal (e.g. silicon carbide), especially with very clean surfaces in a vacuum, one will have a deterioration in the crystallinity of the single crystal. This may not be very noticeable over the very short distances covered by Miyoshi and Buckley (1982a) but, in continuous repeated sliding contacting conditions, it is most likely that either transferred metal particles are wearing against the metal of the pin, or a surface film of mixed metal and silicon carbide grains is wearing against a similar contact film on the surface of the pin. One suspects that Miyoshi and Buckley tend to concentrate their analysis on the surfaces before wear partly because there is nothing much to detect in the surfaces after wearing over such a short distance, at such a low speed, and such a low load! What is the meaning of a wear rate over such a small sliding distance? How can a wear pattern (i.e. a wear rate versus load graph) be established for such a tribo-system?

Although the work of Dr Buckley and his co-workers may not be relevant to the *macroscopic* world of tribo-elements in continuously sliding contact conditions at high speeds, heavy loads and almost infinitely long running times, the fine work coming out from his laboratory deserves a place in this book. Consequently, we will use the remainder of this sub-section to summarize the results of Miyoshi and Buckley (1982a) in their experiments with the silicon carbide/iron tribo-system.

9.3.2 Wear experiments with silicon carbide/iron tribo-elements

In the experiments carried out by Miyoshi and Buckley (1982a), the pin was fabricated from 99.99% pure iron. The radius of the pin was 0.79 mm, which gave

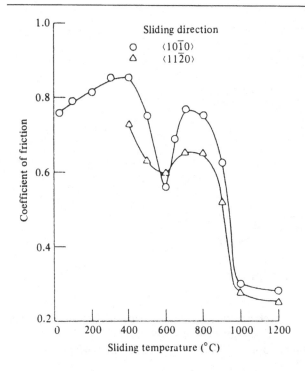

Figure 9.3 *The effect of temperature on the coefficient of friction for an iron pin sliding against the {0001} surface of an 'as-received' single crystal of silicon carbide (load: 0.2 N, vacuum: 10 nPa)*

a pressure of 0.8 GPa at the constant load 0.2 N used in these experiments. The time of sliding was also held constant for all experiments at an arbitrarily chosen value of 30 s. The single-crystal α-silicon carbide used in these experiments was a 99.9% pure compound of silicon and carbon. Silicon carbide is a hexagonal, closed-packed crystal structure, with unit cell dimensions of $a_{uc} = 3.082$ Å and $c_{uc} = 15.118$ Å (see Section 3.1.7 regarding the definition of a close-packed hexagonal structure). The {0001} plane was nearly parallel to the sliding surface of the sample, which consisted of silicon carbide platelets.

The friction-force traces were generally characterized by a 'stick–slip' behaviour. Hence, the friction values shown in the next two figures relate to the *static* friction. Figure 9.3 shows the variation in the static coefficient of friction for an iron pin sliding against the {0001} surface of an 'as-received' single crystal of silicon nitride. The silicon nitride had actually been baked out in the vacuum system and the iron specimen cleaned by sputtering with argon ions. It was then heated to the sliding temperature before the friction experiment was started. Very different graphs were obtained in the friction experiments with silicon carbide crystals that had been pre-heated to 800 and 1500 °C, as shown in Figure 9.4. We will discuss these friction results later, after summarizing the physical analysis of these specimens, *before* the wear actually took place.

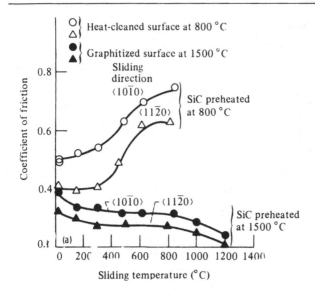

Figure 9.4 *The effect of temperature on the coefficient of friction for an iron pin sliding against the {0001} surface of single crystals of silicon carbide pre-heated to 800 and 1500 °C (load: 0.2 N, vacuum: 10 nPa)*

9.3.3 Auger electron spectroscopy of silicon carbide

A typical Auger electron spectrum of the silicon carbide {0001} plane *before* heating is shown in Figure 9.5. This figure has been taken from the paper by Miyoshi and Buckley (1982a). A contaminant carbon peak is evident in Figure 9.5(*a*), as well as an oxygen peak. The spectrum for surfaces heated at 800 °C (see Figure 9.5(*b*)) indicates all the contamination has gone. In addition to the silicon peak, there is evidence of carbide-type carbon on the surface of silicon carbide (see the peaks labelled A_0, A_1 and A_2 in Figure 9.5(*b*)). When heated to 1500 °C (see Figure 9.5(*c*)), two features are revealed by the Auger spectrum. These are (a) a graphite-type carbon peak at 271 eV and (b) a decrease of the silicon peak with increasing temperature. The latter is due to preferential evaporation of silicon from the silicon carbide. The graphite is formed by the collapse of two successive carbon layers into one layer of carbon hexagons, after evaporation of the silicon atoms from the intermediate layers present in the room-temperature structure of silicon carbide.

Before discussing these results any further, let us consider what Miyoshi and Buckley (1982a) discovered in their X-ray photoelectron analysis of the same specimens.

9.3.4 X-ray photoelectron spectroscopy of silicon carbide

Typical XPS spectra from pre-heated silicon carbide single crystals are shown in Figure 9.6. These were taken from the paper by Miyoshi and Buckley (1982a).

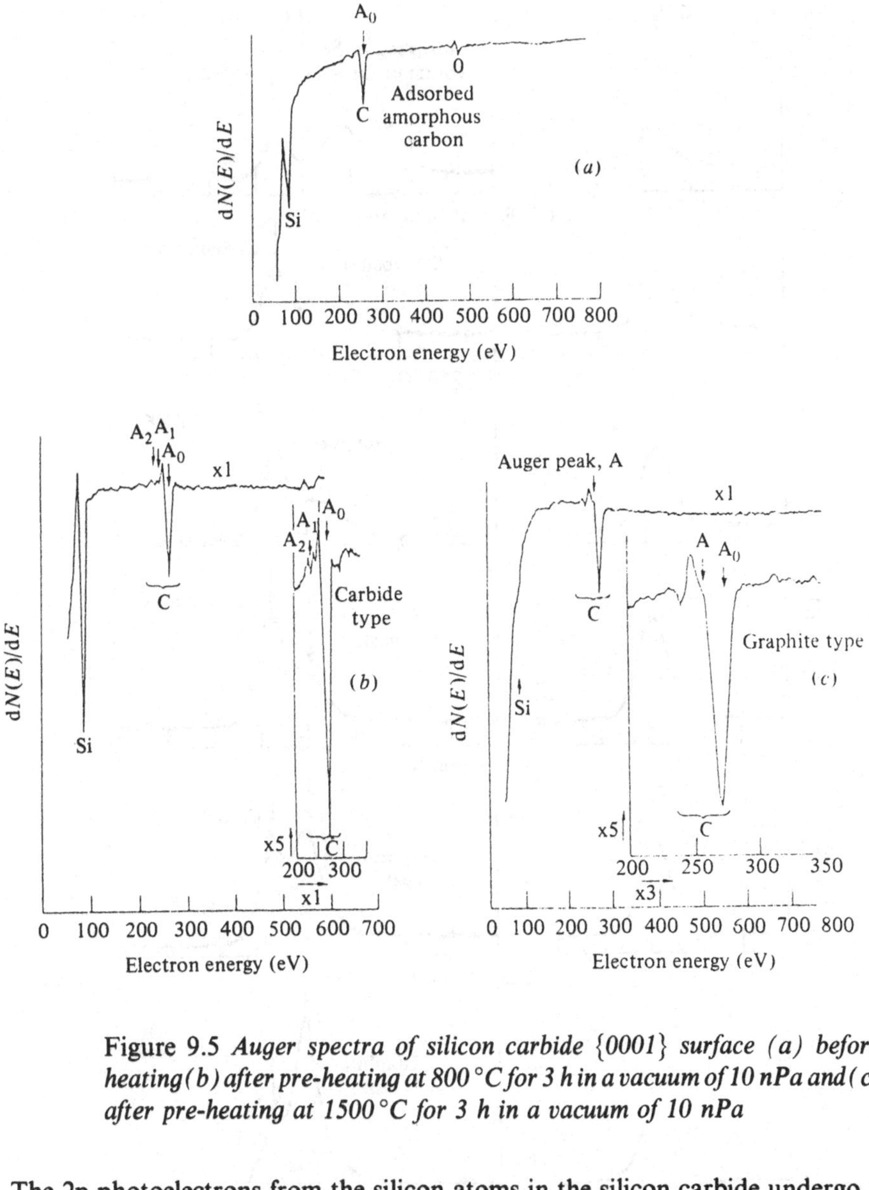

Figure 9.5 *Auger spectra of silicon carbide {0001} surface (a) before heating (b) after pre-heating at 800 °C for 3 h in a vacuum of 10 nPa and (c) after pre-heating at 1500 °C for 3 h in a vacuum of 10 nPa*

The 2p photoelectrons from the silicon atoms in the silicon carbide undergo a gradual change in energy from 100.4 eV (at the bake-out temperature of 250 °C) to 100.0 eV (at the 1500 °C pre-heating temperature). The photoelectron emission lines for the 1s levels of the silicon carbide split asymmetrically into doublet peaks. These peaks are due to two distinguishable types of carbon, namely (a) a carbon contamination peak and a carbide peak at the room temperature and (b) the graphite and carbide peaks at temperatures from 400 to 1500 °C. The carbide peak reaches a maximum at 800 °C, above which it reduces in size, whilst the graphite peak continually increases to a dominant position at 1500 °C. Clearly,

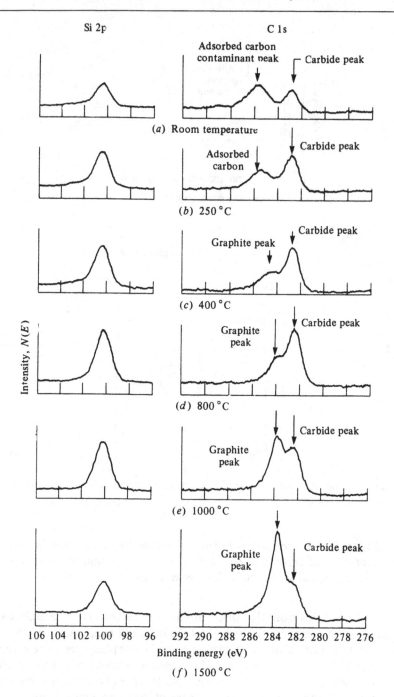

Figure 9.6 *Representative silicon (2p) and carbon (1s) peaks obtained by XPS analysis of the {0001} surface of silicon carbide, pre-heated to various temperatures up to 1500 °C*

the changes in peak sizes are important results which require an explanation. The line shifts of the 2p silicon peaks are, however, of the order of instrumental error and therefore cannot be taken too seriously. Any hypothesis based on these line shifts would have to be examined very critically indeed.

9.3.5 Concluding remarks

This sub-section on the silicon carbide tribo-systems, illustrates some of the problems that are often associated with the application of physical analytical techniques to tribology, particularly the surface analytical techniques. So much of the time and effort that is spent on *characterizing* the surface *before* any wear takes place is wasted as soon as the first contact is made. More effort and time should be spent on the *forensic* use of physical analytical techniques, whether it be on surfaces that have been formed under service conditions or under laboratory conditions. In the work reported by Miyoshi and Buckley (1982a), the only *post-wear* analysis was of the topography, using SEM. Although some interesting spherical and hexagonal particles were found in *some* fracture pits in the silicon carbide (see Figure 9.7), it is not clear if these features are typical of the silicon carbide/iron tribo-system, or merely an interesting 'one-off' event.

The main conclusion to be drawn from the work of Miyoshi and Buckley (1982a) is that pre-heating the single-crystal silicon carbide up to 1500 °C causes it to graphitize, so that any subsequent tribological experiments carried out on

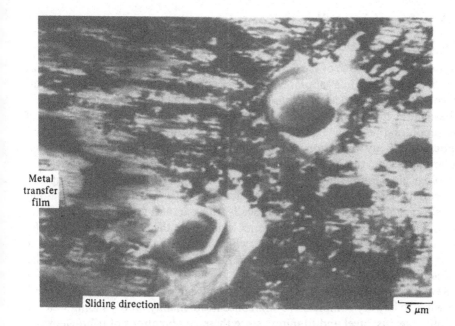

Figure 9.7 *Scanning electron micrograph of wear track with fracture pits on silicon carbide {0001} surface as a result of a single pass of iron at 800 °C in a vacuum of 10 nPa*

the pre-heated silicon carbides will always provide increasingly lower coefficients of friction as they are subsequently heated to higher temperatures of sliding. Preheating to only 800 °C or less, will tend to give increasingly higher friction coefficients up to about 800 °C, after which it would appear that there is a tendency for the friction to eventually reduce to the lower values (~ 0.2) obtained after pre-heating to 1500 °C.

The above-mentioned conclusion drawn from the work of Miyoshi and Buckley (1982a) is probably useful insofar as the *initial stages* of wear between iron and silicon carbide are concerned. It is difficult, however, to see where the next steps should be taken along the route towards a general understanding of the iron/silicon carbide tribo-system. For this work to be complete, we should know more about the wear processes involved in *more extended sliding* under various atmospheres (including air and vacuum). More efforts should be made towards using the powerful techniques and AES and XPS *after* the wear has taken place. There is a temptation for surface analysts to put too much effort into the surface analysis of a very limited number of specimens obtained in a few uncritical experiments, unrelated to the *general wear* behaviour of the tribo-system. It is sincerely believed (by this author) that this book could form the basis of (a) educating surface analysts about tribology and (b) educating tribologists about surface analysis. Perhaps the last system to be discussed in this chapter (namely the sapphire/titanium tribo-system) will succeed in illustrating how a happy compromise *can* be made between good tribology and good surface analysis.

9.4   Sapphire tribo-systems

9.4.1   Introduction

Duwell (1966) was one of the first researchers to study sapphire single crystal as one of the elements in a sapphire/steel tribo-system. His work is of interest to us, since he showed an effect of crystal orientation on the wear rate of the sapphire. The identification of the main crystallographic axis with respect to the surface of the sapphire sphere that Duwell (1966) used as a wearing member, was found by optical and X-ray diffraction methods not explicitly covered in this book. In fact, polarimetry and the X-ray diffraction method of Laue are so rarely used in tribological situations that it was not worth mentioning in the earlier chapters. Essentially, light is *polarized* in a characteristic way when it is passed along certain crystallographic axes of an optically transparent crystal, such as sapphire. Also, X-rays are *diffracted* into patterns which have the same symmetry as the symmetry properties of the crystal, when the X-ray beam passes along a prominent crystal axis. In fact, continuous 'white' X-radiation is used in the diffraction method of Laue.

Several investigators have investigated the wear of α-Al_2O_3 (sapphire) and shown it to be very complex. We will concentrate mainly on the sliding wear of sapphire against steel and titanium, since these combinations of tribo-elements seem to provide results which are more readily understood in terms of existing theories of wear. They are also of practical interest in view of the recent trend towards using ceramic interfaces in practical tribo-systems, such as the ceramic diesel. Both tribo-systems have also been investigated using physical methods of analysis.

9.4.2 Sapphire–steel tribo-systems

In this sub-section, we will be concerned with the wear of sapphire spheres against AISI 1045 steel discs which had been hardened and tempered to a hardness of Rockwell C60 (i.e. 700 VPN). By using this geometry, Duwell (1966) was able to concentrate most of the wear of the tribo-system on to the sapphire single crystal. The reader will probably be aware that this is the usual way for measuring the wear of one of a pair of different materials. In fact, some work has been done whereby the wear is concentrated on the steel member by making it the pin of a pin-on-disc machine with sapphire as the disc (Quinn and Winer, 1987). These authors were mainly interested in the oxidational wear of their tool steel and the 'hot-spots' as seen through the sapphire (see Section 6.3.3).

Duwell (1966) showed there was a marked effect of orientation upon the wear rate of the sapphire sphere. To understand this effect, reference must be made to Figure 9.8. The direction of sliding is indicated by the angle (ϕ_s), and the deviation of the C-axis from the normal to the slide interface is indicated by θ. The point on the sphere where the C-axis emerges normal to a plane that is tangential to the surface is called B. Hence, B is the pole of the C-axis on the sphere. If B is projected to the plane of sliding, and O is the initial contact point on the sphere, the direction OB in the indicated coordinate system defines the slide direction (angle ϕ_s). The results of the wear experiments at 1.3 m/s and 20.2 m/s are shown in Table 9.1.

Clearly, 90° is a high wear orientation compared with 270°, for the 1.3 m/s experiments. There is a suggestion that the friction coefficient is also significantly greater in the 90° orientation. These effects were not apparent at 20.2 m/s. Duwell also presented evidence that the orientation of the C-axis (i.e. angle θ) has an important effect on the wear, as shown in Figure 9.9(a). This figure shows how important θ can be, especially at the lower speeds where the combined effect of $\theta = 4.5°$ and (ϕ_s) = 90° causes almost a two-orders of magnitude increase in specific wear rate compared with the experiments with $\theta = 22°$ and (ϕ_s) = 270°. Note, however, that the friction coefficient is not affected very significantly at the very high speeds covered by the range of Figure 9.9.

It is interesting to note that α-Al$_2$O$_3$ has a similar orientation for low wear as does graphite. We saw, in Section 5.2.4, how graphite twins naturally at 22° when subjected to sliding. Graphite is hexagonal with $a = 2.464$ Å, $c = 6.736$ Å and $(c/a) = 2.734$. Sapphire (i.e. α-Al$_2$O$_3$) has $a = 4.768$ Å, $= 12.99$ Å and $(c/a) = 2.730$. Is it possible that, with *exactly* the same (c/a) ratio as graphite, sapphire will twin to give low wearing orientations in much the same way as graphite? Inspection of the pole figure given in Figure 5.22 reveals a preponderance of basal plane normals (i.e. C-axes) at about 20° lying in a direction towards the direction of sliding. This is where B will be located for (ϕ_s) = 270° in Figure 9.8(*b*). Once more, we see how important crystal direction and orientation can be with materials that take up preferred orientations during sliding.

The high speeds used by Duwell (1966) are very much more than one usually

Table 9.1. *Specific wear rates of sapphire as a function of speed and orientations* $(\theta = 20°)$

(ϕ_s) (deg)	Speed (m/s)	Specific wear rate $(m^3/m\,kg)\ (\times 10^{-15})$	Friction coefficient
90	1.3	11.3	0.54
270	1.3	0.4	0.49
90	1.3	13.7	0.66
270	1.3	0.6	0.51
90	20.2	7.5	0.23
270	20.2	6.0	0.20

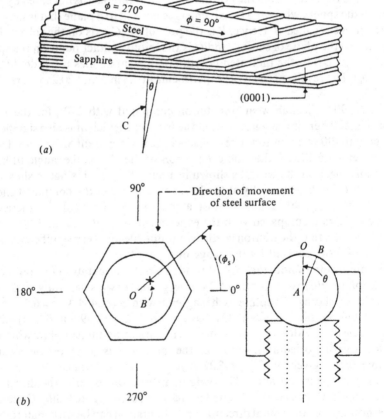

Figure 9.8 *(a) Schematic diagram of Al_2O_3/Fe sliding interface; (b) orientation of sapphire slider (after Duwell, 1966)*

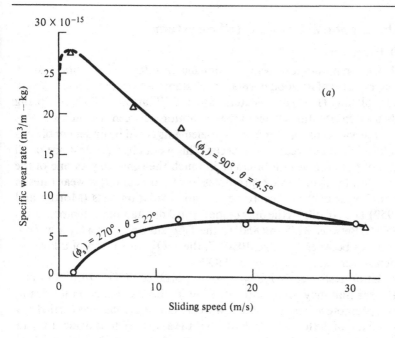

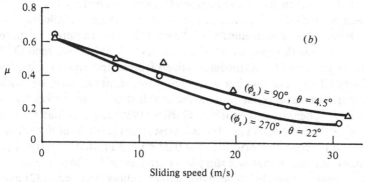

Figure 9.9 (a) *Specific wear rate as a function of sliding speed on a high* $(\theta = 4.5° \ (\phi_s) = 90°)$ *and low* $(\theta = 22°, \ (\phi_s) = 270°)$ *wear orientation (load 6.9 N); (b) coefficient of friction for experiments in Figure 9. (a) (Duwell, 1966)*

finds in rotating or reciprocating machinery. Nevertheless, the high temperatures generated in these experiments presumably accelerate what happens at the slower (more normal) speeds. Duwell (1966) suggested that, at these high temperatures, the interface between the Al_2O_3 and the oxidized steel produces spinels, which act as lubricants. Although not explicitly mentioned by Duwell, it is obvious that he had in mind something very much like *oxidational* wear occurring in his experiments. Let us now consider the sapphire–titanium tribo-system.

9.4.3   The sapphire–titanium tribo-system

(i) Introduction

Titanium has the reputation of being a poor material to use in a tribo-system. Nevertheless, because of its good corrosion resistance and high specific strength [Miller and Holladay (1985) and Waterhouse and Wharton (1974)], it is being increasingly used in situations where relative motion between titanium surfaces must occur. One way of using titanium tribo-elements could be under conditions whereby mild-oxidational wear would occur. This would be feasible if one or all of the oxides of titanium were lubricious, in much the same way as one of the oxides of iron (i.e. the spinel oxide) promotes low friction and low wear rates in the oxidational wear of steel. According to static oxidation tests (Morton and Baldwin, 1952) TiO_2 (rutile) is the only oxide formed on titanium at temperatures less than 800 °C. Between 825° and 850 °C, the TiO_2 sits on top of a layer of TiO. For temperatures between 875 and 1050 °C, the TiO_2 sits on top of a layer of Ti_2O_3, which in turn sits on a layer of TiO.

Static oxidation tests clearly indicate the importance of TiO (rutile) in any oxidational wear that may occur with titanium. Is this oxide a good lubricant, that is, does it promote low friction and low wear? Gardos (1989) has carried out an extensive review of the literature related to rutile as a lubricious oxide (LO). In fact, rutile is often found in its non-stoichiometric form, namely TiO_{2-x}. This means that there are vacant anion sites in the crystal structure of rutile, so that each Ti^{4+} ion is *not always* surrounded by the six O^{2-} ions required for pure stoichiometry, even though there are always three Ti^{4+} ions around each O^{2-} ion. There are at least eight intermediate sub-stoichiometric titanium oxides between Ti_2O_3 and TiO_2. The number of oxygen anion vacancies increases with increasing temperatures and reductions in the ambient pressure, especially in those layers of TiO_{2-x} nearest to the surface. Gardos (1989) suggests that, for x in the range $0.02 < x < 0.07$, rutile will exhibit its lowest stear rate on its surface. This means the oxide will indeed be lubricious. For $0.07 < x < 0.11$, that is for $TiO_{1.91}$ to $TiO_{1.89}$, it has been shown that *new* slip planes are created by the formation of oxygen anion vacancies, namely planes with Miller indices between (132) and (121). The natural cleavage planes of rutile are (110) (101) and (100). As x increases from 0.22 up to 0.34, that is for $TiO_{1.91}$ to $TiO_{1.66}$, the {121} planes dominate. Since these new slip planes are at lower angles than the natural cleavage planes, this means it is easier to glide across these (more abundantly available) planes. Thus one sees that TiO_{2-x} (rutile) could well be a lubricious oxide, provided one can arrange for x to be within certain well-defined ranges for given environmental conditions.

Gardos (1989) has provided one of the few publications in the tribology literature that gives an insight into the systematic changes which are known to occur in the stoichiometry, purity and crystal structure of an oxide as a function of the environment. The fact that he has concentrated upon rutile (TiO_{2-x}) titanium oxide is mainly because he had been associated with the research carried out at Georgia Institute of Technology on the wear of titanium upon sapphire (Hong and Winer, 1988). Let us summarize some of this work, and see how the review by Gardos (1989) fits in with their results involving analysis of the tribologically-formed oxides present in the wear debris and in the worn surfaces.

(ii) The sapphire/titanium wear experiments and analyses

Hong and Winer (1988) used pins of commercially-pure titanium, machined down from rods. The composition was 99.6% titanium with iron, oxygen and carbon accounting for most of the impurity content. The sapphire discs were also of commercial quality. The pin-and-disc assembly could be enclosed within a high-temperature chamber fitted with a sapphire window, to enable a scanning infra-red detector to look at the interface between the titanium pin and the sapphire disc, and thereby provide data from which the hot-spot temperatures could be deduced. Other physical techniques used in this investigation included SEM, XPS, scanning AES, and EDX.

Figure 9.10 is based on Hong and Winer's (1988) experiments with a titanium pin sliding against a sapphire disc (without lubrication and at a speed of 4 m(s) in which no external heating was applied to the system. The points were taken from Hong and Winer's (1988) Table 3 and, by plotting on *linear* scales (instead of the logarithmic scales used by these authors for their Figure 2), it can be seen that there is probably a transition in the wear rate at loads somewhere in the region of 14 N, when the magnitude of the wear rate abruptly changes from about 10×10^{-14} m^3/m to about 80×10^{-14} m^3/m. This implies a change in the *k*-factor

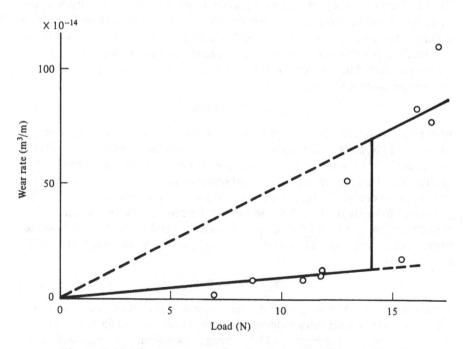

Figure 9.10 *The wear rate of a titanium pin on a sapphire disc as a function of load (no external heating and a sliding speed of 4 m/s) (from Hong and Winer, 1988)*

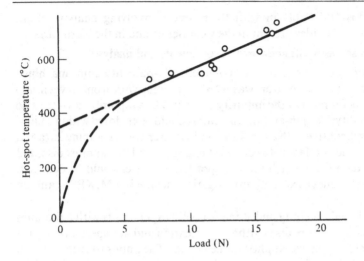

Figure 9.11 *Hot-spot temperatures (as calculated from infra-red detector outputs) between titanium pins and sapphire discs as a function of load (no external heating and a sliding speed of 4 m/s) (from Hong and Winer, 1988)*

(i.e. the factor of proportionality between wear rate and load) from about 1.0 m³/(m–N) *below* the transition load of about 14 N to about 5.0 m³/(m–N) for loads greater than 14 N. The authors also deduce hot-spot temperatures between the titanium pin and the sapphire disc, from the infra-red radiation and their results are plotted in Figure 9.11. Here we see a linear relationship between hot-spot temperature and load, namely:

$$T_c = 350 + 21.0\,W \qquad (9.1)$$

where T_c is in degrees Celsius and W is in newtons. This equation only relates to $7\,N < W < 17\,N$, since it is obvious that, for W approaching zero, then T_c will tend to approach the general surface temperature (T_s) which was equal to about 23 °C in this case. There is no significant transition in the T_c versus W graph. It is interesting to compare Figure 9.11 with Figure 8.8(*c*), where we see that T_c (as calculated from thermal analysis and oxide plateaux height measurements for high-chromium ferritic steel sliding on austenitic stainless steel at 2.0 m/s) also varies linearly with load. The equation for the T_c versus W curve of Figure 8.8(*c*) is:

$$T_c = 300 + 2.8\,W \qquad (9.2)$$

Note that, for both Figures 9.11 and 8.8(*c*), there is *no* transition in T_c around the loads at which the wear rates suddenly increase, namely 14 and 65 N respectively. It should be noted that Figure 9.11 has been drawn from data provided in the paper by Hong and Winer (1988), in much the same way as Figure 9.10. Let us see how these new figures help with the interpretation of their results.

The electron micrographs of the titanium pin surface were obtained on the

SEM. Several large plateaux were observed, reminiscent of the oxide plateaux so characteristic of oxidational wear. Auger depth-concentration profiles from the plateaux regions showed that they were oxidized down to at least 2000Å below the surface of the plateaux. An example of a typical depth concentration profile is shown in Figure 9.12 for a titanium pin which had been worn agaist sapphire *without* any external heating being supplied to the system. Figure 9.12 relates to the concentrations of O, Ti and C in the region of the plateaux seen in the scanning electron micrographs. Although the main thrust of Hong and Winer's (1988) paper relates to the wear of their sapphire/titanium system *without* external heating, it also contains some data related to the *elevated temperature wear* of this system. It is relevant to our discussion to show two more depth–concentration profiles obtained from their titanium pin surfaces after being worn at 400 °C against sapphire, namely Figures 9.13(a) and (b).

Now typically, air-oxidized titanium surface films formed at room temperatures are about 30 Å thick (Hauffe, 1965). Note that, even after being subjected to 400 °C for some considerable time, (i.e. time enough to establish the equilibrium wear rate), the off-plateau regions give rise to profiles indicating relatively thin (< 600 Å) oxide (see Figure 9.13(b)), compared with the thickness at which oxygen is detected in the plateau regions, namely at least 2000 Å (for the room temperature wear experiments) and at least 3000 Å (for the 400 °C, elevated temperatures, wear experiments). Clearly, oxide thicknesses formed in the sapphire/titanium

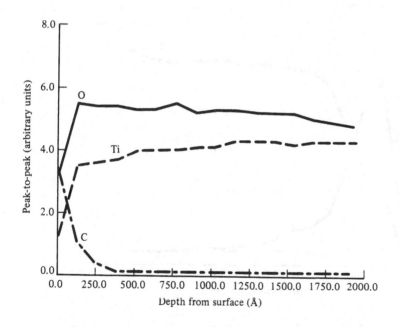

Figure 9.12 *Auger depth-concentration profiles from a plateau in a titanium pin surface which had previously been worn against sapphire, without external heating (from Hong and Winer, 1988)*

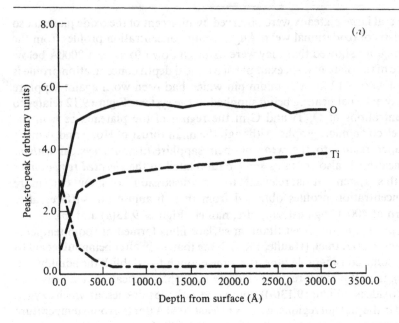

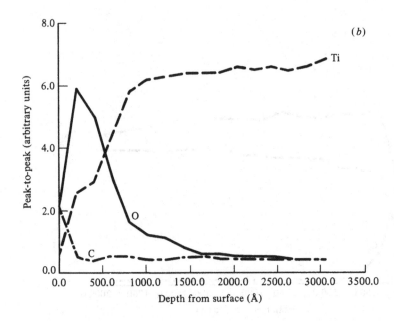

Figure 9.13 *Auger depth-concentration profiles from a titanium pin surface which had previously been worn against a sapphire disc, at 400 °C. (a) Profiles from the plateau region; (b) profiles from the off-plateau region (from Hong and Winer, 1988)*

Table 9.2. *X-ray diffraction patterns obtained by irradiating wear debris collected from experiments with a titanium/sapphire tribo-system at 23 and 400°C*

Interplanar spacing	Intensity		Identification
(d) (Å)	23 °C	400 °C	
3.46	–	Weak	Ti_3O_5
2.69	–	Weak	Ti_3O_5
2.55	Medium	–	Ti
2.34	Medium	–	Ti
2.24	Strong	–	Ti
2.07	Medium	Medium	TiO
1.84	–	Weak	Ti_3O_5
1.72	Medium	–	Ti
1.53	–	Weak	Ti_3O_5
1.46	Weak	Weak	$Ti + Ti_3O_5$

tribo-system are of similar magnitude to those formed in the oxidational wear of steels.

AES cannot readily tell us what type of titanium oxide had been present on the pin surface. On the other hand, XPS should be able to distinguish between TiO, TiO_2 or Ti_3O_5 from the differences in line positions of the ($2p_{3/2}$) peaks of titanium. XPS cannot tell us, however, very much about small changes in the numbers of anion vacancies that might be present in these oxides. In fact, Hong and Winer (1988) showed that, for the room temperature experiments, there were peaks at 453.8 and 458.4 eV, corresponding to Ti and TiO_2 only. The oxide could well have been TiO_{2-x}, with x very small. For the 400 °C experiments, there was only one peak (around 458.5 eV) due to TiO_2.

In an attempt to sort out the important oxide as regards tribological behaviour, Hong and Winer (1988) use X-ray diffraction to examine the wear debris collected during two typical wear experiments, one at room temperature, the other at 400 °C (see Table 9.2). It is clear, from this table, that the stable oxide formed in the room temperature experiments was TiO. This agrees with the findings of Nutt and Ruff (1983). At 400 °C it would seem that Ti_3O_5 is also present. Let us now discuss these results in terms of how far they support the concept of the oxidational wear of titanium upon sapphire.

(iii) The oxidational wear of titanium on sapphire

Although it is evident that oxidation plays an important role in the wear of titanium at high ambient temperatures, we have insufficient data available (at the present time) to examine that role in detail. Instead, we will concentrate our attention on the oxidational wear of titanium in a room-temperature ambient atmosphere of air. All the evidence suggests that the titanium wears against the sapphire by an oxidative wear mechanism, involving the production of TiO, with just a thin layer of the lubricious oxide TiO_2 (or TiO_{2-x}, with x very small) between the TiO and the sapphire. The conclusion about the thin nature of the TiO_2 layer is based on the fact that it was not possible to detect its presence in the X-ray diffraction pattern from the wear debris. Clearly, if it was present, it must have been a very minor constituent. Although TiO_2 (or TiO_{2-x}, with x very small) may well be a lubricious oxide, it does not seem to be a major product in tribo-oxidation. This conclusion is based on very few experiments and a small amount of analysis. It is possible that the wear of titanium under temperatures less than 800 °C and/or in vacuum could produce this lubricious oxide in place of the low temperature oxide (TiO). The subject is still wide open.

Hong and Winer (1988) use the values of the hot-spot temperatures indicated by Figure 9.11, to explain the oxidational wear of titanium in terms of a modification of the oxidational wear mechanism which assumes a *linear* dependence of oxide film thickness upon the time of oxidation, instead of the more conventional *parabolic* dependence. Without further work into the static-oxidation and tribo-oxidation of titanium, along the same lines as those already carried out for the mild-oxidational wear of steels, there is no point in discussing the validity of this assumption any further at this stage.

9.5   Concluding remarks on the application of physical techniques to ceramic tribo-systems

As implied in the title of this chapter, the choice of ceramic tribo-systems discussed has been very selective. As always, I have been aware that the good tribologist is one who is not afraid of complex tribo-systems. The time has passed when simple theories, and/or simple explanations in qualitative terms, are applied to a whole host of complex situations (such as one finds in tribology) regardless of their relevance to those situations. Now that we have access to computers with huge memory storage capacities, we must be prepared to model the real situation more closely than we could possibly have done before the advent of the digital computer. Ceramic tribo-systems depend upon very complex crystalline structures, such as, for instance, SiO_{2-x}. This chapter has taken a few ceramic tribo-systems, shown where physical analysis can help, but has made no attempt to bring the results together in a unified discussion. With these systems, the tribologist must be aware of their crystalline properties as well as their mechanical properties.

It is hoped that this chapter has provided a fairly painless introduction into ceramic tribology. From even the small amount of data given in this chapter it should be obvious that there are some similarities between the wear of some ceramics and the mild-oxidational wear of steel. For the first time, we have seen that mild-oxidational wear definitely occurs in ceramic tribo-systems. It is probably too early to try to apply the oxidational mild wear theory to ceramics. We must wait for the basic oxidational parameters, for (say) titanium, to be produced from static and tribo-oxidational experiments and carefully analysed, before we can apply the appropriate oxidational wear theoretical expression to our results, especially at high temperatures and/or low vacuum. Physical analytical instruments will take a leading role in the search for the ideal ceramic/ceramic or (more probably) the ideal metal/ceramic combination to give *both low friction and low wear.*

References

CHAPTER 1

Allen, C. B., Quinn, T. F. J. and Sullivan, J. L. (1985), The Oxidational Wear of High-Chromium Ferritic Steel on Austenitic Stainless Steel, *JOT Trans. ASME*, **108**, p. 215.

Archard, J. F. (1953), Contact and Rubbing of Flat Surfaces, *J. Appl. Phys.*, **24**, p. 981.

Archard, J. F. (1957), Elastic Deformation and the Laws of Friction, *Proc. Roy. Soc.*, **A 243**, p. 190.

Archard, J. F. (1959a), The Wear of Metals, *New Scientist*, **5**, p. 1299.

Archard, J. F. (1959b), The Temperature of Rubbing Surfaces, *Wear*, **2**, p. 438.

Archard, J. F. (1961), Single Contacts and Multiple Encounters, *J. Appl. Phys.*, **32**, p. 1420.

Archard, J. F. and Hirst, W. (1956), The Wear of Metals under Unlubricated Conditions, *Proc. Roy. Soc.*, **A 236**, p. 397.

Barber, J. R. (1967), The Influence of Thermal Expansion on the Friction and Wear Process, *Wear*, **10**, p. 155.

Bickerstaff, A. E. (1969), The Real Area of Contact between a Carbon Brush and a Copper Collector, *Proceedings of the 15th Holm Seminar on Electric Contact Phenomena*, Illinois Institute of Technology, Chicago, p. 241.

Blok, H. (1937), Theoretical Study of Temperature Rise at Surfaces of Actual Contact under Oiliness Lubricating Conditions, Proceedings of a General Discussion on Lubrication and Lubricants, *Proc. Inst. Mech. Engrs.*, **2**, p. 222.

Bowden, F. P. and Tabor, D. (1954), Friction and Lubrication of Solids; Part I, Clarendon Press, Oxford.

Burwell, J. T. and Strang, C. D. (1952a), on the Empirical Law of Adhesive Wear, *J. Appl. Phys.*, **23**, p. 18.

Burwell, J. T. and Strang, C. D. (1952b), Metallic Wear, *Proc. Roy. Soc.*, Series A, **212**, p. 470.

Cabrera, N. and Mott, N. F. (1949), Theory of the Oxidation of Metals, *Rep. Prog. in Phys.*, **12**, p. 93.

Clark, W. T., Pritchard, C. and Midgley, J. W. (1967), Mild Wear of Unlubricated Hard Steels in Air and Carbon Dioxide, *Proc. Inst. Mech. Engrs.*, **182**, p. 97.

Crook, A. W. (1962), Some Physical Aspects of Lubrication and Wear, *Contemp. Phys.*, **3**, p. 257.

Czichos, H. and Salomon, G. (1974), The Application of Systems Thinking and Systems Analysis to Tribology, *BAM Bericht*, No. 30, Berlin.

Davies, D. E., Evans, U. R. and Agar, N. J. (1954), The Oxidation of Iron at 175 °C to 350 °C, *Proc. Roy. Soc.*, **A 225**, p. 443.

Greenwood, J. A. and Williamson, J. B. P. (1966), The Contact of Nominally Flat

Surfaces, *Proc. Roy. Soc.*, **A 295**, p. 300.

Halliday, J. S. (1957), The Application of Reflection Electron Microscopy to the Study of Wear, *Proceedings of the First International Conference on Lubrication and Wear*, Inst Mech. Engrs, London (Paper 40).

Holm, R. (1946), *Electrical Contacts*, Gebers, Stockholm, Sweden.

Hong, H., Hochman, R. F. and Quinn, T. F. J. (1988), A New Approach to the Oxidational Theory of Mild Wear, *STLE* (formerly *ASLE*) *Trans.*, **31**, p. 71.

Kubaschewski, O. and Hopkins, B. E. (1962), *Oxidation of Metals and Alloys*, Butterworths, London.

Lancaster, J. K. (1957), The Influence of Temperature upon Metallic Wear, *Proc. Phys. Soc. (London)*, **B70**, p. 122.

Lim, S. C. and Ashby, M. F. (1987), Wear Mechanism Maps, *Acta Metall.*, **35**, p. 1.

Molgaard, J. and Srivastava, V. K. (1975), Apparatus for the Study of Oxidative Wear of Unlubricated Surfaces, *Wear*, **33**, p. 179.

Molgaard, J. and Srivastava, V. K. (1977), The Activation Energy of Oxidational Wear, *Wear*, **41**, p. 263.

Naylor, H. (1973), Viscosity and Rheology, *Tribology Handbook*, Section F-4, Butterworths, London.

Neale, M. J. (1973), Editor, *Tribology Handbook*, Butterworths, London.

Quinn, T. F. J. (1962), The Role of Oxidation in the Mild Wear of Steels, *Brit. J. Appl. Phys.*, **13**, p. 33.

Quinn, T. F. J. (1967), The Effect of 'Hot-Spot' Temperatures on the Unlubricated Wear of Steel, *ASLE Trans.*, **10**, p. 158.

Quinn, T. F. J. (1968), The Dry Wear of Steel as Revealed by Electron Microscopy and X-Ray Diffraction, *Proc. Inst. Mech. Engrs.*, **182**, Pt 3N.

Quinn, T. F. J. (1969), An Experimental Study of the Thermal Aspects of Sliding and their relation to the Unlubricated Wear of Steel, *Proc. Inst. Mech. Engrs.*, **183**, Pt 3P, p. 129.

Quinn, T. F. J. (1978), The Division of Heat and Surface Temperatures at Sliding Steel Interfaces and their relation to Oxidational Wear, *ASLE Trans.*, **21**, p. 78.

Quinn, T. F. J. (1980), The Classifications, Laws, Mechanisms and Theories of Wear, *Fundamentals of Tribology*, Editors N. P. Suh and N. Saka, The MIT Press, Cambridge, Mass., p. 477.

Quinn, T. F. J. (1981), The Effects of 'Out-of-Contact' Oxidation on the Oxidational Wear of Steels, *Proceedings of the Third International Tribology Congress*, Warsaw, Poland, September.

Quinn, T. F. J. (1983), Review of Oxidational Wear, Part I: The Origins of Oxidational Wear, *Tribology International*, **16**, p. 257.

Quinn, T. F. J. and Winer, W. O. (1987), An Experimental Study of the 'Hot-Spots' occurring during the Oxidational Wear of Tool Steel on Sapphire, *JOT, ASME Trans.*, **109**, p. 315.

Rabinowicz, E. and Tabor, D. (1951), Metallic Transfer between Sliding Metals, *Proc. Roy. Soc.*, **A208**, p. 455.

Razavizadeh, K. and Eyre, T. S. (1983), Oxidative Wear of Aluminium Alloys, *Wear*, **87**, p. 261.

Sexton, M. D. (1984), A Study of Wear in Cu–Fe Systems, *Wear*, **94**, p. 275.

Stott, F. H., Glascott, J. and Wood, G. C. (1984), Factors Affecting the Progressive Development of Wear-Protective Oxides in Iron-based Alloys during Sliding at Elevated Temperatures, *Wear*, **97**, p. 93.

Stott, F. H., Glascott, J. and Wood, G. C. (1985), Models for the Generation of Oxides During Sliding Wear, *Proc. Roy. Soc.*, **A402**, p. 167.

Stott, F. H., Lin, D. S. and Wood, G. C. (1973), The Structure and Mechanism of Formation of the Glaze Oxide Produced on Nickel-based Alloys During Wear at High Temperatures, *Corrosion Science*, **13**, p. 449.

Stott, F. H. and Wood, G. C. (1978), The Influence of Oxides on the Friction and Wear of Alloys, *Tribology International*, **11**, p. 211.

Suh, N. P. (1977), The Delamination Theory of Wear, *Wear*, **44**, p. 1.

Sullivan, J. L. and Granville, N. W. (1984), Reciprocating Sliding Wear of Steel in Carbon Dioxide at Elevated Temperatures, *Tribology International*, **17**, p. 63.

Sullivan, J. L., Quinn, T. F. J. and Rowson, D. M. (1980), Developments in the Oxidational Theory of Mild Wear, *Tribology International*, **12**, p. 153.

Tao, F. F. (1969), A Study of Oxidational Phenomena in Corrosive Wear, *ASLE Trans.*, **12**, p. 97.

Timoshenko, S. (1934), *Theory of Elasticity*, McGraw-Hill, New York.

Welsh, N. C. (1965), The Dry Wear of Steels, *Proc. Roy. Soc., Phil. Trans.*, **A257**, p. 31.

Wilson, J. E., Stott, F. H. and Wood, G. C. (1984), The Development of Wear-Protective Oxides and their Influence on Sliding Friction, *Proc. Roy. Soc.*, **A369**, p. 557.

Yoshimoto, G. and Tsukizoe, T. (1957), On the Mechanism of Wear between Metal Surfaces, *Wear*, **1**, p. 472.

CHAPTER 2

Barton, J. R. and Kusenberger, F. N. (1971), Fatigue Damage Detection, *STP* 495, *ASTM*, p. 193.

Betz, C. E. (1967), *Principles of Magnetic Particle Testing*, Magnaflux Corp.

Bowden, F. P. and Tabor, D. (1954), *The Friction and Lubricating of Solids*, Part I, Clarendon Press, Oxford.

Breton, B. C., Thong, J. T. L. and Nixon, W. C. (1986), A Dynamic Real-Time 3-D Measurement Technique for IC Inspection, *Microelectronic Engineering*, **5**, p. 541.

Carter, T. L., Butler, R. H., Bear, H. R. and Anderson, W. J. (1958), Investigation of Factors Governing Fatigue Life with the Rolling-Contact-Fatigue Spin Rig, *ASLE Trans.*, **1**, p. 23.

Crook, A. W. (1958), The Lubrication of Rollers, *Phil. Trans. Roy. Soc.*, Series A, **250**, p. 232.

Crook, A. W. (1961), Elastohydrodynamic Lubrication of Rollers, *Nature*, London, **190**, p. 1182.

Dawson, P. H. (1961), The Pitting of Lubricated Gear Teeth and Rollers, *Power Transmission*, **30**, p. 208.

Dawson, P. H. (1962), The Effect of Metallic Contact on the Pitting of Lubricated Rolling Surfaces, *J. Mech. Engrg Sci.*, **4**, p. 16.

Dawson, P. H. (1963), The Effect of Metallic Contact on the Shape of the S–N curve for Pitting Fatigue, Symposium on Fatigue in Rolling Contact, Inst. Mech. Engrs, London, p. 41.

Furey, M. J. (1961), Metallic Contact and the Friction between Sliding Surfaces, *ASLE Trans.*, **4**, p. 1.

Hack, K. and Feller, H. G. (1970), Friction and Wear Characteristics of Gold Single Crystals at Small Pressures, (in German), *Z. Metallkunde*, **61**, p. 394.

Halliday, J. S. (1960), The Contrast, Breadths and Relative Intensities of Electron Diffraction Rings, *Proc. Roy. Soc.*, Series A, **254**, p. 30.

Halliday, J. S. and Quinn, T. F. J. (1960), Contrast of Electron Micrographs, *Brit. J. Appl. Phys.*, **II**, p. 486.

Kannel, J. W., Bell, J. C. and Allen, C. M. (1965), Methods for determining pressure distributions in Lubricated Rolling Contact, *ASLE Trans.*, **8**, p. 250.

Littman, W. E. and Widner, R. L. (1966), Propagation of Contact Fatigue from Surface and Sub-surface Origins, *J. Basic Engrg, Trans. ASME* (Series D), **88**, p. 624.

Parker, R. J. (1975), Correlation of Magnetic Perturbation Inspection Data with Rolling-Element Beaming Fatigue, *JOLT, Trans. ASME*, **97**, p. 151.

Phillips, M. R. and Chapman, C. J. S. (1978), A Magnetic Method for Detecting the Onset of Surface Fatigue, *Wear*, **49**, p. 265.

Phillips, M. R. and Quinn, T. F. J. (1978), The Effect of Surface Roughness and Lubricant Film Thickness on the Contact Fatigue Life of Steel Surfaces lubricated with a Sulphur–Phosphorus type of Extreme-pressure Additive, *Wear*, **51**, p. 11.

Quinn, T. F. J. (1964), The Preferred Orientations within a Contact Film formed by the Repeated Sliding of an Electrographite Brush on a Copper Ring, *Brit. J. Appl. Phys.*, **15**, p. 513.

Rabinowicz, E. (1977), The Formation of Spherical Wear Particles, *Wear*, **39**, p. 312.

Scott, D. and Mills, G. H. (1970), A Scanning Electron Microscope Study of Fracture Phenomena Associated with Rolling Contact Surface Fatigue Failure, *Wear*, **16**, p. 234.

Scott, D. and Mills, G. H. (1973), Spherical Debris – Its Occurrence, Formation and Significance in Rolling-Contact Fatigue, *Wear*, **24**, p. 235.

Seifert, W. W. and Westcott, V. C. (1972), A Method for the Study of Wear Particles in Lubricating Oil, *Wear*, **21**, p. 27.

Smith, N. J. (1936), Preparation of Thin Oxide Films, *J. Am. Chem. Soc.*, **58**, p. 513.

Spreadborough, J. (1962), The Frictional Behaviour of Graphite, *Wear*, **5**, p. 18.

Swain, M. V. and Jackson, R. E. (1976), Wear-like Features on Natural Fault Faces, *Wear*, **37**, p. 63.

Tolansky, S. (1970), *Multiple Beam Interference Microscopy of Metals*, Cambridge University Press, London.
Way, S. (1935), Pitting due to Rolling Contact, *J. Appl. Mech.*, **2**, p. A49.
Way, S. (1936), Gear Tooth Pitting, *Electr. J.*, (London), **33**, p. 175.
Westcott, V. C. and Seifert, W. W. (1973), Investigation of Iron Content of Lubricating Oil Using a 'Ferrograph' and an Emission Spectrometer, *Wear*, **23**, p. 239.

CHAPTER 3

Cullity, B. D. (1962), *Elements of X-Ray Diffraction*, Addison-Wesley, London.
Dillon, J. A., Schlier, R. E. and Farnsworth, H. E. (1959), Some Surface Properties of Si–C Crystals, *J. Appl. Phys.*, **30**, p. 675.
Ehrenberg, W. (1934), Visual Observation of Slow Electrons by Crystals, *Phil. Mag.*, **18**, p. 878.
Germer, L. H. and Hartman, C. D. (1960), Improved Low-Energy Electron Diffraction Apparatus, *Rev. Sci. Instrum.*, **31**, p. 784.
Henry, N. F. M., Lipson, H. and Wooster, W. A. (1951), *The Interpretation of X-Ray Diffraction Photographs*, Macmillan, London.
Isherwood, B. J. and Quinn, T. F. J. (1967), The Application of a Glancing-Angle, X-ray Diffraction Film Technique to the Study of Low-Temperature Oxidation of Iron–Chromium Alloys, *Brit. J. Appl. Phys.*, **18**, p. 717.
MacRae, A. U. (1963), Low-Energy Electron Diffraction: Improved Experimental Methods Provide New Information on the Structure of Surfaces of Solids, *Science*, **139**, p. 379.
Peisser, M. S., Rooksby, H. P. and Wilson, A. J. C. (Editors) (1955), *X-Ray Diffraction by Polycrystalline Materials*, Institute of Physics, London.
Quinn, T. F. J. (1970), The Edge-Irradiated, Glancing-Angle X-Ray Diffraction Technique, *J. Phys. D.: Appl. Phys.*, **3**, p. 210.
Quinn, T. F. J. (1971), *The Application of Modern Physical Techniques to Tribology*, Van Nostrand and Reinhold, New York.
Rymer, T. B. (1970), *Electron Diffraction*, Methuen, London.
Whitaker, A. (1967), A Small Bulk Specimen Holder for X-Ray Powder Camera, *J. Sci. Instr.*, **44**, 32.

CHAPTER 4

Auger, P. (1925), On the Secondary β-radiation Produced in a Gas by X-rays, *Comptes Rendus*, **180**, p. 65.
Baldwin, J. M. (1968), Bibliography of Laser Publications of Interest to Emission Spectroscopists, *USAEC Reports* IN-1219 and IN-1262.
Bertin, E. P. (1971), The Electron Probe Microanalyser, In *Principles and Practice of X-ray Spectrometric Analysis*, Plenum Press, New York, p. 605.

Bouwman, R. (1973), Auger Electron Spectroscopy for the Analysis of Solid Surfaces, *Ned. Tijdstshrift voor Vacuumtechnik*, **II**, p. 37.

Briggs, S. and Seah, M. P. (1983), Appendix 4, in *Practical Surface Analysis by Auger and X-ray Photoelectron Spectroscopy*, John Wiley, Chichester.

Burwell, J. T. and Strong, C. D. (1952a), On the Empirical Law of Adhesive Wear, *J. Appl. Phys.*, volume 23, p. 18.

Burwell, J. T. and Strong, C. D. (1952b), Metallic Wear, *Proc. Roy. Soc.*, Series A, volume 212, p. 470.

Chang, C. C. (1975), General Formalism for Quantitative Auger Analysis, *Surf. Sci.*, **48**, p. 9.

Cline, J. E. and Schwarz, S. (1967), Determination of the Thickness of Aluminium on Silicon by X-ray Fluorescence, *J. Electrochem. Soc.*, **114**, p. 605.

Duncumb, P., Shields-Mason, P. K. and da Casa, C. (1969), Accuracy of Atomic Number and Absorption Corrections in Electron Probe Microanalysis, In *Fifth International Conference on X-ray Optics and Microanalysis*, Editors, Möllenstedt, G. and Gaukler, K. H. Springer, Berlin, p. 146.

Dunne, J. A. (1963), Continuous Determination of Zn Coating Weights on Steel by X-ray Fluorescence, in *Advances in X-ray Analysis*, **6**, p. 345, Plenum Press, New York.

Francis, J. M. and Jutson, J. A. (1968), The Application of Thin-layer X-ray Fluorescence Analysis to Oxide Composition Studies on Stainless Steel, *J. Sci. Instrum.*, **E1**, p. 772.

Heinrich, K. F. J. (1967), The Absorption Correction Method for Microprobe Analysis, *Proceedings for the Second National Conference on Electron Probe Analysis*, paper 7.

Hutchins, G. A. (1974), Electron Probe Microanalysis, in *Characterization of Solid Surfaces*, Editors, Kane, P. F. and Larrabee, G. B., Plenum Press, New York.

Kane, P. K. and Larrabee, G. B. (Editors) (1974), *Characterization of Solid Surfaces*, Plenum Press, New York.

Koh, P. K. and Caugherty, B. (1952), Metallurgical Applications of X-ray Fluorescent Analysis, *J. Appl. Phys.*, **23**, p. 427.

Lancaster, J. K. (1964), A Review of Radioactive Tracer Applications in Friction, Lubrication and Wear, *Tech. Note CPM64*, Royal Aircraft Establishment.

Lim, S. C. and Ashby, M. F. (1987), Wear Mechanism Maps, *Acta. Metall.*, **35**, p. 1.

Meitner, L. (1922), On the Relation Between β- and γ-rays, *Z. Phys.*, **9**, p. 145.

Meitner, L. (1923), On the β-radiation Spectrum of UX, and Its Interpretation, *Z. Phys.*, **17**, p. 54.

Meyer, F. and Vrakking, J. J. (1972), Quantitative Aspects of Auger Electron Spectroscopy, *Surf. Sci.*, **33**, p. 271.

Moratibo, J. M. (1975), A First-Order Approximation to Quantitative Auger Analysis in the Range 100 to 100 eV Using the CMA analyser, *Surf. Sci.*, **49**, p. 318.

Palmberg, P. W. (1976), Quantitative Auger Electron Spectroscopy Using Elemental Sensitivity Factors, *J. Vac. Sci. Technol.* **13**, p. 214.

Palmberg, P. W. *et al.* (1972), *Handbook of Auger Electron Spectroscopy*, Physical Electronic Industries.

Philibert, J. (1962), A Method for Calculating the Absorption Correction in Electron Probe Microanalysis, in *X-ray Optics and X-ray Microanalysis*, Editors, Pattee, H. H., Coslett, V. E. and Engström, A., Academic Press, New York, p. 379.

Reed, S. J. B. (1965), Characteristic Fluorescence Correction in Electron Probe Microanalysis, *Brit. J. Appl. Phys.*, **16**, p. 913.

Rhodin, T. N. (1955), Chemical Analysis of Thin Films by X-ray Emission Spectroscopy, *Anal. Chem.*, **28**, p. 1857.

Schreiber, T. P., Ottolini, A. C. and Johnson, J. L. (1963), X-Ray Emission Analysis of Thin Films Produced by Lubricating Oil Additives, *Appl. Spectrom.* **17**, p. 17.

Sewell, P. B., Mitchle, D. F. and Cohen, M. (1969), High Energy Electron Diffraction and X-ray Emission Analysis of Surfaces and their Reaction Products, *Develop. Appl. Spectrom.*, **7A**, p. 61.

Sietmann, R. (1988), False Attribution – a Female Physicist's Fate, *Physics Bulletin*, **39**, p. 316.

Singer, I. L. (1985), Surface Analysis, Ion Implantation and Tribological Processes Affecting Steels, *Appl. Surf. Sci.*, **18**, p. 28.

Smuts, J., Plug, C. and Van Niekirk, J. (1967), Coating Thickness Determination of Tin Plate by X-ray Methods, *J. S. African Inst. Mining Met.*, April 1967, p. 462.

Strasheim, A. and Blum, F. (1971), The Effect of Spark Parameters in the Use of Spectrochemical Analysis of Surfaces, *Spectrochim. Acta*, **26B**, p. 685.

Taylor, N. J. (1969), Resolution and Sensitivity Considerations of an Auger Electron Spectrometer based on LEED Optics, *Rev. Sci. Instum.*, **40**, p. 792.

Vrakking, J. J. and Meyer, F. (1974), Electron Impact Ionization Cross-sections of Inner Shells Measured by Auger Electron Spectroscopy, *Phys. Rev.*, **A9**, p. 1932.

Vrakking, J. J. and Meyer, F. (1975), Measurement of Ionization Cross-sectioning and Backscattering Factors for Use in Quantitative AES, *Surf. Sci.*, **47**, p. 50.

Wagner, C. D., Riggs, W. M., Davis, L. E., Moulder, J. F. and Muilenberg, G. E. (1979), *Handbook of X-ray Photoelectron Spectroscopy*, Perkin-Elmer Corporation, Eden Prairie.

CHAPTER 5

Aird, R. T. and Forgham, S. L. (1971), The Lubricating Quality of Aviation Fuels, *Wear*, **18**, p. 361.

Allum, K. G. and Forbes, E. S. (1968), The Load-Carrying Properties of Organic Sulphur Compounds: Application of Electron Probe Microanalysis, *ASLE Trans.*, **11**, p. 162.

Archard, J. F., and Cowking, E. W. (1965), Elastohydrodynamic Lubrication at Point Contacts, *Proc. Inst. Mech. Engrs.* **180**, Pt3B, p. 47.

Archard, G. D., Gair, F. C. and Hirst, W. (1961), The Elasto-hydrodynamic Lubrication of Rollers, *Proc. Roy. Soc.*, **A262**, p. 51.

Archard, J. F. and Kirk, M. T. (1961), Lubrication at Point Contacts, *Proc. Roy. Soc.*, **A261**, p. 532.

Azouz, A. (1982), The Application of Physical Methods of Analysis to the Study of Surfaces Formed During Wear by Organo-Sulphur Compounds, PhD Thesis, Aston University, England.

Baldwin, B. A. (1975), Chemical Characterization of Wear Surfaces using X-ray Photoelectron Spectroscopy, *Lubrication Engineering*, **23**, p. 125.

Bernal, J. D. (1924), The Structure of Graphite, *Proc. Roy. Soc.*, **A106**, p. 749.

Bird, R. J. and Galvin, G. D. (1976), The Application of Photoelectron Spectroscopy to the study of EP films on Lubricated Surfaces, *Wear*, **37**, p. 143.

Bjerk, R. O. (1973), Oxygen – an Extreme-Pressure Agent, *ASLE Trans.*, **16**, p. 97.

Bowden, F. P. and Young, (1951), The Frictional Properties of Carbon, Graphite and Diamond, *Proc. Roy. Soc.*, **A208**, p. 444.

Bragg, W. L. (1928), *An Introduction to Crystal Analysis*, Bell, London, p. 64

Braithwaite, E. R. (1966), Friction and Wear of Graphite and Molybdenum Disulphide, *Scientific Lubrication*, **18**, p. 3.

Braithwaite, E. R. and Rowe, G. W. (1963), Principles and Applications of Lubrication with Solids, *Scientific Lubrication*, **15**, p. 92.

Buckley, D. H. (1974), Oxygen with a Clean Iron Surface and the Effect of Rubbing Contact on these interactions, *ASLE Trans.*, **17**, p. 206.

Cameron, A. and Gohar, R. (1966), Theoretical and Experimental Studies of the Oil Film in Lubricated Point Contacts, *Proc. Roy. Soc.*, **A291**, p. 520.

Cheng, H. S. (1970), A Numerical Solution of the Elastohydrodynamic Film Thickness in an Elliptical Contact, *JOLT Trans. ASME.*, **F93**, p. 349.

Coy, R. C. and Jones, R. B. (1981), The Thermal Degradation and E.P. Performance of Zinc Dialkyldithiophosphate Additives in White Oil, *ASLE Trans.*, **24**, p. 77.

Coy, R. and Quinn, T. F. J. (1975), The use of Physical Methods of Analysis to Identify Surface Layers Formed by Organo-sulphur Compounds in Wear Tests, *ASLE Trans.*, **18**, 163.

Crook, A. W. (1961a), Elastohydrodynamic Lubrication of Rollers, *Nature*, **190**, p. 1182.

Crook, A. W. (1961b), The Lubrication of Rollers, I: Film Thickness with relation to viscosity and speed, II: A Theoretical Discussion of Friction and the Temperatures in an Oil Film, *Phil. Trans. Roy. Soc.*, **A254**, p. 223.

Datta, K. K. (1984), The Physical Methods of Analysis Applied to the Friction and Wear of Graphites and Carbons, MSc Thesis, University of Aston, Birmingham.

Datta, K. K., Sykes, A. and Quinn, T. F. J. (1990), The Effect of Crystallinity upon the friction and wear of Graphites, paper in preparation for *Wear*.

Davey, W. and Edwards, E. D. (1957), The Extreme-Pressure Lubricating Properties of some Sulphides and Disulphides, in Mineral Oil, as Assessed by the Four-Ball Machine, *Wear*, **1**, p. 291.

Deacon, R. F. and Goodman, J. F. (1958), Lubrication by Lamellar Solids, *Proc. Roy. Soc.*, **A243**, p. 464.

Debies, T. P. and Johnston, W. G. (1980), Surface Chemistry of some Anti-Wear Additives as determined by Electron Spectroscopy, *ASLE Trans.*, **23**, p. 289.

Dowson, D. and Higginson, G. R. (1966), *Elastohydrodynamic Lubrication*, Pergamon Press, London.

Fisher, J. (1973), The Electrical and Mechanical Properties of Electrographite/Metal Interfaces, PhD Thesis, University of Aston, Birmingham.

Frieze, E. J. and Kelly, A. (1963), Deformation of Graphite Crystals and the Formation of the Rhombohedral Form, *Phil. Mag.*, **8**, p. 1519.

Foord, C. A., Wedeven, L. D., Westlake, F. J. and Cameron, A. (1969), Optical Elastohydrodynamics, *Proc. Inst. Mech. Engrs*, **184**, Part I, p. 1.

Forbes, E. S., Allum, K. G. and Silver, H. B. (1968), The Load-Carrying Properties of Metal Dialkyl Dithiophosphates: Application of Electron Probe Microanalysis, *Proc. Inst. Mech. Engrs*, **183**, p. 37.

Furey, M. F. (1959), Film Formation by an Anti-Wear Additive in an Automotive Engine, *ASLE Trans.*, **2**, p. 91.

Furey, M. (1966), Metallic Contact and Friction between Sliding Surfaces, *ASLE Trans.*, **4**, p. 1.

Georges, J. M., Martin, J. M., Mathia, T., Kapsa, P. L., Meille, G. and Montes, H. (1979), Mechanism of Boundary Lubrication with Zinc dithiophosphate, *Wear*, **53**, p. 9.

Godfrey, D. (1962), Chemical Changes in Steel Surfaces During Extreme-Pressure Lubrication, *ASLE Trans.*, **5**, p. 57.

Greenhill, E. B. (1948), The Lubrication of Metals by Compounds Containing Sulphur, *J. Inst. Petroleum*, **34**, p. 659.

Grubin, A. N. and Vinogradava, I. E. (1949), *Investigation of Scientific and Industrial Research*, Central Research Institute for Technology and Mechanical Engineering, Book No. 30, p. 115.

Halliday, J. S. and Quinn, T. F. J. (1960), The Contrast of Electron Micrographs, *Brit. J. Appl. Phys.*, **11**, p. 486.

Isherwood, B. J. and Quinn, T. F. J. (1967), The Application of a Glancing-Angle, X-ray Diffraction Technique to the study of low-temperature Oxidation of Iron–Chromium Alloys, *Brit. J. Appl. Phys.*, **18**, p. 717.

Jahanmir, S. (1987), Wear Reduction and Surface Layer Formation by a ZDDP Additive, *JOT, Trans. ASME*, **109**, p. 577.

Jenkins, R. O. (1934), Electron Diffraction Experiments with Graphite and Carbon Surfaces, *Phil. Mag.*, **17**, p. 45.

Kannel, J. W., Bell, J. C. and Allen, C. W. (1965), Methods for Measuring Pressure Distributions in Lubricated Rolling Contact, *ASLE Trans.*, **8**, p. 250.

Larson, R. (1958), The Performance of Zinc Dithiophosphates as Lubricating Oil Additives, *Scientific Lubrication*, **10**, p. 12.

Laves, F. and Baskin, Y. (1956), On the formation of the rhombohedral modification of graphite, *Z. Kristallogr.*, **107**, p. 337.

Lenz, F. (1954), Zur Streuung mittelschneller Elektron in kleinstre Winkel, *Z. Naturforschung*, **9a**, p. 185.

Lim, S. C. and Ashby, M. F. (1987), Wear Mechanism Maps, *Acta Metall.*, **35**, p. 1.

Lipson, H. and Stokes, A. R. (1943), The Structure of Graphite, *Proc. Roy. Soc.*, **A181**, p. 101.

Loeser, E. H., Wiquist, R. C. and Twiss, S. B. (1959), Cam and Tappet Lubrication IV: Radio-active Study of Sulphur in the EP Film, *ASLE Trans.*, **2**, p. 199.

Midgley, J. W. and Teer, D. G. (1961), Orientation of Graphite and Carbon During Sliding, *Nature*, **189**, p. 735.

Murakami, T., Sakai, T., Yamamoto, Y. and Hirano, F. (1983), Lubricating Performance of Organic Sulphides under Repeated Rubbing Conditions, *ASLE Trans.*, **28**, p. 363.

Nichols, F. (1988), Editor, *Proceedings of Workshop on Mechanistic Modelling of Wear processes at Argonne National Laboratory*, organized by US Department of Energy, Energy Conversion and Utilization Technologies Division.

Poole, W. and Sullivan, J. L. (1979), The Wear of Aluminium–Bronze on Steel in the presence of Aviation Fuel, *ASLE Trans.*, **22**, p. 154.

Poole, W. and Sullivan, J. L. (1980), The Role of Aluminium Segregation in the Wear of Aluminium–Bronze/Steel interfaces under Conditions of Boundary Lubrication, *ASLE Trans.*, **23**, p. 401.

Porgess, P. V. and Wilman, H. (1960), Surface Re-orientation, Friction and Wear in Sliding Graphite, *Proc. Phys. Soc.*, **76**, p. 513.

Quinn, T. F. J. (1963), A Topographical and Crystallographic Study of the Surfaces of Rubbed Electrograhite, *Brit. J. Appl. Phys.*, **14**, p. 107.

Quinn, T. F. J. (1964), The Preferred Orientations within a Contact Film formed by the Repeated Sliding of an Electrographite Brush on a Copper Ring, *Brit. J. Appl. Phys.*, **15**, p. 513.

Quinn, T. F. J. (1971), *The Application of Modern Physical Techniques to Tribology*, Van Nostrand & Reinhold, New York, p. 156.

Quinn, T. F. J. (1982), The Role of Oxide Films in the Friction and Wear Behaviour of Metals, in *Microscopic Aspects of Adhesion and Lubrication*, Editor, J. M. Georges, Elsevier, Amsterdam, p. 579.

Quinn, T. F. J. (1984), The Use of Electron Diffraction Analysis in the Study of Contact Films Formed by Sliding Graphite on Copper, *Proceedings of the First Indian Carbon Conference, New Delhi, December 1983*, Indian Carbon Society.

Rowe, G. W. (1960a), Frictional Behaviour of Boron Nitride and Graphite, *Wear*, **3**, p. 274.

Rowe, G. W. (1960b), The Friction and Strength of Graphite at high temperatures, *Wear*, **3**, p. 454.

Rounds, F. (1976), Some Factors Affecting the Decomposition of Three Commercial Zinc Organic Phosphates, *ASLE Trans.*, **18**, p. 78.

Rounds, F. (1985), Contribution of Phosphorus to the Anti-Wear Performance of Zinc Dialkyl-dithiophosphates, *ASLE Trans.*, **28**, p. 475.

Sakurai, T. and Sato, K. (1966), Study of Corrosivity and Correlation between Chemical Reactivity and Load-Carrying Capacity of Oils containing Extreme-Pressure Agents, *ASLE Trans.*, **9**, p. 77.

Savage, R. H. (1948a), Graphite Lubrication, *J. Appl. Phys.*, **19**, p. 1.
Savage, R. H. (1948b), Physically and Chemically Adsorbed films in Lubrication of Graphite, *Trans. ASME*, **70**, p. 497.
Sibley, L. B. and Orcutt, F. K. (1961), Elastohydrodynamic Lubrication of Rolling contact Surfaces, *ASLE Trans.*, **4**, p. 234.
Smith, N. J. (1936), Preparation of Thin Oxide Films, *J. Am. Chem. Soc.*, **58**, p. 173.
Snidle, R. W. and Archard, J. F. (1972), Experimental Investigation of elasto-hydrodynamic lubrication at Point Contacts, *EHL Symposium*, Inst. Mech. Engrs, London.
Sullivan, J. L. (1986), The Role of Oxidation in the Protection of Sliding Metal Surfaces, PhD Thesis, University of Aston, Birmingham.
Sullivan, J. L. and Wong, L. F. (1986), The Influence of Surface Films on the Protection of Metal Surfaces under Boundary Lubriction Conditions, *Surface and Interface Analysis*, **9**, p. 493.
Thomson, G. P. and Cochrane, T. (1939), *Theory and Practice of Electron Diffraction*, Macmillan, London.
Tomaru, M., Hironaka, S. and Sakurai, T. (1977), Effects of Oxygen on the Load-Carrying Action of some Additives, *Wear*, **41**, p. 117.
Toyoguchi, M. and Takai, Y. (1962), Effect of Oxidation on Reaction between Iron and Sulphur Compounds, *ASLE Trans.*, **8**, p. 1.
Van Brunt, C. and Savage, R. H. (1944), Carbon Brush Contact Films: Part I, *General Electrical Review*, **47**.
Wedeven, L. D., Evans, D. and Cameron, A. (1971), Optical Analysis of Ball-Bearing Starvation, *JOLT, Trans. ASME*, **93**, p. 349.
Wheeler, D. R. (1978), X-ray Photoelectron Spectroscopic Study of Surface Chemistry of Dibenzyl Disulphide on Steel under Mild and Severe Wear Conditions, *Wear*, **47**, p. 243.

CHAPTER 6

Allen, C. B., Quinn, T. F. J. and Sullivan, J. L. (1986), The Oxidational Wear of High-Chromium Ferritic Steel on Austenitic Stainless Steel, *JOT, Trans. ASME*, **107**, p. 172.
Archard, J. F. (1959), The Temperature of Rubbing Surfaces, *Wear*, **2**, p. 438.
Archard, J. F. (1961), Single Contacts and Multiple Encounters, *J. Appl. Phys.*, **32**, p. 1420.
Archard, J. F. and Hirst, W. (1956), The Wear of Metals under Unlubricated Conditions, *Proc. Roy. Soc.*, **A236**, p. 397.
Athwal, S. S. (1982), PhD Thesis, University of Aston, Birmingham.
Averbach, B. L. and Cohen, M. (1948), X-ray Determination of Retained Austenite by Integrated Intensities, *Metals Technology*, T.P.2342.
Baig, A. R. (1968), A Study of the Microscopic Crystallographic and Oxidational Processes Involved in the Sliding Over of Unlubricated Steel Surfaces, PhD Thesis, Brunel University, Uxbridge, Middlesex.

Bair, S., Griffioen, J. A. and Winer, W. O. (1986), The Tribological Behaviour of an Automotive Cam and Flat Lifter System, *JOT, Trans. ASME*, **108**, p. 478.

Blok, H. (1937), Surface Temperatures under Extreme-pressure Lubricating Conditions, *Second World Petroleum Conference*, Paris, June 1937, **3**, section 4.

Bowden, F. P. and Ridler, K. E. W. (1935), The Surface Temperature of Sliding Metals, *Proc. Cambridge Phil. Soc.*, **31**, p. 431.

Bowden, F. P. and Ridler, K. E. W. (1936), Physical Properties of Surfaces, Part III: The Surface Temperature of Sliding Metals, the Temperature of Lubricated Surfaces, *Proc. Roy. Soc.*, **A164**, p. 640.

Bowden, F. P. and Tabor, D. (1954), *The Friction and Lubrication of Solids*, Part I, Clarendon Press, Oxford.

Caplan, D. and Cohen, M. (1966), The Effect of Cold Work on the Oxidation of Iron from 100 °C to 650 °C, *Corrosion Science*, **6**, p. 321.

Dayson, C. (1967), Surface Temperatures at Unlubricated Sliding Contacts, *ASLE Trans.*, **10**, p. 169.

Furey, M. J. (1964), Surface Temperatures in Sliding Contact, *ASLE Trans.*, **1**, p. 133.

Griffioen, J. A. (1985), Temperature Measurements in Tribo-Contacts by means of Infra-red Radiometry, PhD Thesis, School of Mechanical Engineering, Georgia Institute of Technology, Atlanta.

Grosberg, P., McNamara, A. B. and Molgaard, J. (1965), The Performance of Ring Travellers, *J. Text. Inst.*, **56**, p. T24.

Grosberg, P. and Molgaard, J. (1966–67), Aspects of the Wear of Spinning Travellers: The Division of Heat at Rubbing Surfaces, *Proc. Inst. Mech. Engrs.*, **181**, p. 16.

Herbert, E. G. (1926), The Measurement of Cutting Temperatures, *Proc. Inst. Mech. Engrs*, **1**, p. 289.

Jaeger, J. C. (1942), Moving Sources of Heat and the Temperature of Sliding Contacts, *Proc. Roy. Soc. NSW*, **56**, p. 203.

Kubaschewski, O. and Hopkins, B. E. (1962), *Oxidation of Metals and Alloys*, 2nd Edition, Butterworths, London.

Lim, S. C. and Ashby, M. F. (1987), Wear Mechanism Maps, *Acta Metall.*, **35**, p. 1.

Ling, F. F. and Pu, S. L. (1964), Probable Interface Temperatures in Solids in Sliding Contact, *Wear*, **7**, p. 23.

Meinders, M. A., Wilcock, D. F. and Winer, W. O. (1984), Infra-red Measure ments of a reciprocating Seal Test, *Proceedings 9th Leeds-Lyon Conference.* Editors, D. Dowson and C. M. Taylor, IPC Science and Technology, Guildford, p. 321.

Nagaraj, H. S. (1976), Investigation of Some Temperature-Related Phenomena in Elasto-hydrodynamic Contacts, including Surface Roughness Effects, PhD Thesis, School of Mechanical Engineering, Georgia Institute of Technology, Atlanta.

Nagaraj, H. S., Sanborn, D. M. and Winer, W. O. (1977), Effects of Load, Speed and Surface Roughness on Sliding EHD Contact Temperatures, *JOLT, Trans. ASME*, **99**, p. 321.

Poole, W. and Sullivan, J. L. (1979), The Wear of Aluminium–Bronze on Steel in the Presence of Aviation Fuel, *ASLE Trans.*, **22**, p. 154.

Poole, W. and Sullivan, J. L. (1980), The Role of Aluminium Segregation in the Wear of Aluminium–Bronze/Steel Interfaces Under Conditions of Boundary Lubrication, *ASLE Trans.*, **23**, p. 401.

Quinn, T. F. J. (1962), The Role of Oxidation in the Mild Wear of Steel, *Brit. J. Appl. Phys.*, **13**, p. 33.

Quinn, T. F. J. (1967), The Application of X-Ray Diffraction Techniques to the Study of Wear, *Advances in X-Ray Analysis*, **10**, p. 311, Plenum Press, New York.

Quinn, T. F. J. (1968), An Experimental Study of the Thermal Aspects of Sliding Contacts and their relation to the Unlubricated Wear of Steel, *Proc. Inst. Mech. Engrs.*, **183**, (Part 3p), p. 129.

Quinn, T. F. J. (1978), The Division of Heat and Surface Temperatures at Sliding Steel Interfaces and their relation to Oxidational Wear, *ASLE Trans.*, **21**, p. 78.

Quinn, T. F. J. (1983(a)), Review of Oxidational Wear, Part I: The Origins of Oxidational Wear, *Tribology International*, **16**, p. 257.

Quinn, T. F. J. (1983(b)), Review of Oxidational Wear, Part II: Recent Developments and Future Trends in Oxidational Wear Research, *Tribology International*, **16**, p. 305.

Quinn, T. F. J., Baig, A. R., Hogarth, C. A. and Müller, H. (1973), Transitions in the Friction Coefficients, Wear Rates and the Compositions of the Wear Debris Produced in the Unlubricated Sliding of Chromium Steels, *ASLE Trans.*, **16**, p. 239.

Quinn, T. F. J., Rowson, D. M. and Sullivan, J. L. (1980), Application of the Oxidational Theory of Mild Wear to the Sliding Wear of Low-Alloy Steel, *Wear*, **65**, p. 1.

Quinn, T. F. J. and Winer, W. O. (1985), The Thermal Aspects of Oxidational Wear, *Wear*, **102**, p. 67.

Quinn, T. F. J. and Winer, W. O. (1987), An Experimental Study of the 'hot-spots' occurring during the Oxidational Wear of Tool Steel on Sapphire, *JOT, Trans. ASME*, **109**, p. 315.

Rowson, D. M. (1982), Private communication, King Edward School, Edgbaston, Birmingham.

Rowson, D. M. and Quinn, T. F. J. (1980), Frictional Heating and the Oxidational Wear Theory, *J. Phys. D*, **13**, p. 208.

Santini, J. J. and Kennedy, F. E. Jr. (1975), An Experimental Investigation of Surface Temperatures and Wear in Disk Brakes, *Trans. ASLE*, **18**, p. 210.

Shore, H. (1925), Thermoelectric Measurement of Cutting Tool Temperature, *J. Wash. Acad. Sci.*, **15**, p. 85.

Sullivan, J. L. and Athwal, S. S. (1983), Mild Wear of a Low-Alloy Steel at temperatures up to 500 °C. *Tribology International*, **16**, p. 123.

Sullivan, J. L. and Granville, N. W. (1984), Reciprocating Sliding Wear of 9% Cr steel in carbon dioxide at Elevated Temperatures, *Tribology International*, **17**, p. 63.

Sullivan, J. L., Quinn, T. F. J. and Rowson, D. M. (1980), Developments in the Oxidational Theory of Mild Wear, *Tribology International*, **12**, p. 153.
Tabor, D. (1959), *Proc. Roy. Soc.*, **A251**, p. 378.
Turchina, V., Sanborn, D. M. and Winer, W. O. (1974), Temperature Measurements in Sliding Elastohydrodynamic Point Contacts, *JOLT, Trans. ASME*, **96**, p. 464.
Welsh, N. C. (1965), The Dry Wear of Steels, *Phil. Trans. Proc. Roy. Soc.*, **A257**, p. 31.
Wymer, D. G. and MacPherson, P. B. (1975), An Infra-red Technique for the Mesurement of Gear Tooth Surface Temperature, *Trans. ASLE*, **18**, p. 32.

CHAPTER 7

Archard, J. F. and Cowking, E. W. (1965–66), Elastohydrodynamic Lubrication at Point Contacts, *Proc. Inst. Mech. Engrs*, **180**, pt. 3B, p. 47.
Archard, J. F. and Hirst, W. (1956), The Wear of Metals under Unlubricated Conditions, *Proc. Roy. Soc.*, **A936**, p. 397.
Armstrong, E. L., Leonardi, S. J., Murphy, W. R. and Wooding, P. S. (1978), Evaluation of Water-Accelerated Bearing Fatigue in Oil-Lubricated Ball Bearings, *Lubrication Engineering*, **34**, p. 15.
Bell, J. C. and Kannel, J. W. (1970), Simulation of Ball-Bearing Lubrication with a Rolling Disc Apparatus, *JOLT Trans. ASME*, **92**, p. 1.
Blok, H. (1937), Theoretical Study of Temperature Rise at Surfaces of Actual Contact under Oiliness Lubricating Conditions, *Proceedings of General Discussion on Lubrication and Lubricants*, London, 1937, Institution of Mechanical Engineers, p. 222.
Blok, H. (1970), The Postulate About the Constancy of Scoring Temperature, Editor P. M. Ku, Interdisciplinary Approach to Lubrication of Concentrated Contacts, NASA Rep. SP-237 (National Aeronautics and Space Administration).
Cantley, R. E. (1977), The Effect of Water in Lubricant Oil on Bearing Fatigue Life, *ASLE Trans.*, **20**, p. 244.
Czichos, H. (1980), Fundamentals of the Systems Approach to Tribology and Tribo-testing, *Fundamentals of Tribology*, Editors N. P. Suh and Saka, N. Massachusetts Institute of Technology Press, Cambridge, MA, p. 1171.
Dalal, H. H., Chiu, Y. P. and Rabinowicz, E. (1975), Evaluation of Hot-Pressed Silicon Nitride as a Rolling Bearing Material, *ASLE Trans.*, **18**, p. 244.
Dawson, P. H. (1962), The Effect of Metallic Contact on the Pitting of Unlubricated Surfaces, *J. Mech. Engrg Sci.*, **4**, p. 13.
Dawson, P. H. (1968), Rolling Contact Fatigue Crack Initiation in 0.3% Carbon Steel, *Proc. Inst. Mech. Engrs.*, **183**, p. 16.
Dawson, P. H. and Fidler, F. (1965–66), Wire-Wool Failures, *Proc. Inst. Mech. Engrs*, **180**, p. 528.
Diaconescu, E. N., Kerrison, G. D. and MacPherson, P. B. (1975), A New

Machine for Studying the Effects of Sliding and Traction on the Fatigue Life of Point Contacts, with Initial Test Results, *ASLE Trans.*, **18**, p. 210.

Dowson, D. and Higginson, G. R. (1966), *Elastohydrodynamic Lubrication*, Pergamon Press, London.

Kelly, R. and Lam, N. Q. (1973), The Sputtering of Oxides, *Radiation Effects*, **19**, p. 39.

Kenny, P. (1977), The Development of IP Test Method 305/74 Tentative – The Assessment of Lubricants by measurement of Their Effect on the Rolling Fatigue Resistance of Bearing Steels using the Unisteel Machine, *Proceedings of Symposium on Rolling Contact Fatigue*, p. 47, The Institute of Petroleum, London.

Kenny, P. and Yardley, E. D. (1972), The Use of Unisteel Rolling Fatigue Machines to compare the Lubrication Properties of Fire Resistant Fluids, *Wear*, **20**, p. 105.

Knight, G. C. (1977), The Effect of Water in Lubricant Oil on Bearing Fatigue Life, *ASLE Trans.*, **20**, p. 244.

Lester, W. G. C. (1973), Predicting the Reliability of Mechanical Assemblies Containing a Multiplicity of Rolling Contact Bearings, *Proceedings of 1st European Tribology Convention*, Inst. Mech. Engrs, London, 173.

Littman, W. E. and Widner, R. L. (1966), Propagation of Contact Fatigue from Surface and Subsurface Origins, *Trans. ASME, J. Basic Engrg*, **188**, p. 624.

Michan, B., Berthe, D. and Godet, M. (1974), Observations of Oil Pressure Effects in Surface Crack Development, *Tribology International*, **1**, p. 119.

Muller, R. (1966), The Effect of Lubrication on Cam and Tappet Performances, *MTZ*, **27**, p. 58.

Naylor, H. (1967), Cams and Friction Drives, *Second International Lubrication & Wear Conference*, Inst. Mech. Engrs, London, 1967, Paper 15.

Neale, M. J. (Editor) (1973), *Tribology Handbook*, Butterworths, London.

Onions, R. A. (1973), Pitting Failures of Gears, PhD Thesis, University of Leicester.

Onions, R. A. and Archard, J. F. (1975), Pitting of Gearings and Discs, *Proc. Inst. Mech. Engrg*, **188**, p. 673.

Phillips, M. R. (1979), Physical Analysis of Pitting Failure with an Extreme-Pressure Additive in Rolling and Sliding Discs, PhD Thesis, The University of Aston, Birmingham.

Phillips, M. R. and Chapman, C. J. S. (1978), A Magnetic Method for Detecting the Onset of Surface Contact Fatigue, *Wear*, **49**, p. 265.

Phillips, M. R., Dewey, M., Hall, D. D., Quinn, T. F. J. and Southworth, H. N. (1976), The Application of Auger Electron Spectroscopy to Tribology, *Vacuum*, **26**, p. 451.

Phillips, M. R. and Quinn, T. F. J. (1978), The Effect of Surface Roughness and Lubricant Film Thickness upon the Contact Fatigue Life of Steel Surfaces lubricated with a Sulphur-Phosphorus type of Extreme-Pressure Additive, *Wear*, **51**, p. 11.

Quinn, T. F. J., Baig, A. R., Hogarth, C. A. and Müller, H. (1973), Transitions in the Friction Coefficients, Wear Rates and Compositions of the Wear Debris

Produced in the Unlubricated Sliding of Chromium Steels, *ASLE Trans.*, **16**, p. 239.

Rabinowicz, E. (1972), Exoelectrons, *Scientific American*, p. 74.

Rosenberg, D. and Wehner, G. K. (1962), Sputtering Yields for low energy He^+, Kr^+ and Xe^+ ion bombardment, *J. Appl. Phys.*, **33**, p. 1842.

Saunders-Davies, D. L., Richards, M. N. and Galvin, G. D. (1980), Factors Affecting the Performance of Tanker-Propulsion Gearing – An Operator's Viewpoint, *Trans. Inst. Mar. Engrs*, **92**, p. 2.

Scott, D. (1963), The Effect of Material Properties, Lubricant and Environment on Rolling Contact Fatigue, *Proceedings of Symposium on Fatigue in Rolling Contact*, Inst. Mech. Engrs, p. 103.

Scott, D. (1968), Ball-Bearing Steels: Factors Influencing Rolling Contact Fatigue Resistance, *NEL Report No.* 360, Ministry of Technology, August 1968.

Scott, D. and Blackwell, J. (1971), NEL Rolling Contact Tests – Accelerated Service Simulation Tests for Lubricants and Materials for Rolling Elements, *Wear*, **17**, p. 321.

Scott, D. and Scott, H. M. (1960), The Application of Electron Microscopy to the Pitting Failure of Ball Bearings, *Proceedings of European Conference of Electron Microscopy*, **1**, p. 539.

Suleman, M. and Pattinson, E. B. (1973), Changes in Auger Spectra of Mg and Fe due to Oxidation, *Surf. Sci.*, **35**, p. 75.

Sullivan, J. L. and Middleton, M. R. (1985), The Pitting and Cracking of SAE 52100 Steel in Rolling/Sliding Contact in the Presence of an Aqueous Lubricant, *ASLE Trans.*, **28**, p. 431.

Swahn, H., Becker, P. C. and Vingsbo, O. (1976), Electron Microscope Studies of Carbide Decay During Contact Fatigue on Ball Bearings, *Metal Science*, p. 35.

Syniuta, W. D. and Carrow, C. J. (1970), A Scanning Electron Microscope Fractographic Study of Rolling Contact Fatigue, *Wear*, **15**, p. 187.

Tallian, T. E., Brady, E. F., McCool, J. I. and Sibley, L. N. (1965), Lubricant Film Thickness and Wear in Rolling Point Contact, *ASLE Trans.*, **8**, p. 411.

Tallian, T. E., Chiu, Y. P., Huttenlocher, D. F., Kamenshine, J. A., Sibley, L. B. and Sindlinger, N. E. (1964), Lubricant Films in Rolling Contact of Rough Surfaces, *ASLE Trans.*, **7**, p. 109.

Timoshenko, S. P. and Goodier, J. N. (1951), *Theory of Elasticity*, McGraw-Hill, New York.

Way, S. (1935), Pitting Due to Rolling Contact, *Trans. ASME*, **57**, p. A49.

Wild, R. K. (1976), Electron Spectroscopy Applied to the Study of Metal Oxidation, *Vacuum*, **26**, p. 441.

Wu, Y. L. (1980), Physical Analysis of Pitting in Rolling and Sliding Discs, PhD Thesis, Aston University, Birmingham.

CHAPTER 8

Allen, C. B. (1982), The Oxidational Wear of Diesel Engine Materials, PhD Thesis, Aston University, England.

Allen, C. B., Quinn, T. F. J. and Sullivan, J. L. (1985), The Oxidational Wear of High-Chromium Ferritic Steel on Austenitic Stainless Steel, *JOT Trans. ASME*, **107**, p. 172.

Bowden, F. P. and Tabor, D. (1954), *The Friction and Lubrication of Solids*, Part I, Oxford University Press.

Caplan, D. and Cohen, M. (1966), The Effect of Cold Work on the Oxidation of Iron from 100° to 650°, *Corrosion Science*, **6**, p. 321.

Chivers, T. C., Gordelier, S. C., Roy, J. and Wharton, M. (1978), *Proceedings of British Nuclear Energy Society Conference on Vibrations in Nuclear Energy Plant.*

Cocks, M. (1962), Interaction of Sliding Metal Surfaces, *J. Appl. Phys.*, **33**, p. 2152.

Dawson, P. H. and Fidler, F. (1965), Wire-Wool Type Bearing Failures: The Formation of the Wire Wool, *Proc. Inst. Mech. Engrs*, **180**, Part I, p. 513.

Halliday, J. S. (1960), The Contrast, Line Breadths and Relative Intensities of Electron Diffraction Ring Patterns from Vacuum-Evaporated Films of Iron, *Proc. Roy. Soc.* **A254**, p. 30.

Hong, H. (1986), A Metallurgical Study of the Oxidational Theory of Mild Wear in Stainless Steel and Surface-Modified Stainless Steel, PhD Thesis, Georgia Institute of Technology, Atlanta, Georgia, USA.

Levy, G. (1975), Patterns of Tribology, *Proceedings of the 4th International Tribology Conference*, Paisley, Scotland.

Lim, S. C. and Ashby, M. F. (1987), Wear Mechanism Maps, *Acta Metall.*, **35**, p. 1.

Molgaard, J. and Smeltzer, W. W. (1971), Thermal Conductivity of Magnetite and Haematite, *J. Appl. Phys.*, **42**, p. 3644.

Quinn, T. F. J. (1968), The Dry Wear of Steel as revealed by Electron Microscopy and X-Ray Diffraction, Paper 24; Tribology Convention, Pitlochry, Scotland, 1968. Published in *Proc. Inst. Mech. Engrs*, **182**, Pt. 3N, p. 201.

Quinn, T. F. J. and Wallace, L. R. (1985), The Oxidational and Tribological Behaviour of Austenitic Stainless Steel under a CO_2-based Environment, *Proceedings of ASME Conference on Wear of Materials*, Vancouver, April 1985.

Quinn, T. F. J. and Winer, W. O. (1987), An Experimental Study of the 'hot-spots' occurring during the Oxidational Wear of Tool Steel on Sapphire, *JOT Trans. ASME*, **109**, p. 315.

Quinn, T. F. J. and Woolley, J. N. (1970), The Unlubricated Wear of $3\% \, Cr{-}\frac{1}{2}\% \, Mo$ Steel, *Lubrication Engineering*, **27**, p. 312.

Smith, A. F. (1984), The Duplex Oxidation of Vacuum-Annealed 316 Stainless Steel in CO_2/CO gas Mixtures between 500 and 700 °C, *Corrosion Science*, **24**, p. 629.

Smith, A. F. (1985), The Sliding Wear of 316 Stainless Steel in air in the temperature range 20–500 °C, *Tribology International*, **18**, p. 35.

Sullivan, J. L. and Athwal, S. S. (1983), Mild Wear of a Low-Alloy Steel at Temperatures up to 500 °C, *Tribology International*, **16**, p. 213.

Sullivan, J. L. and Saied, S. O. (1988), A Study of the Development of Tribological Oxide Protective Films Mainly by Auger Electron Spectroscopy, *Surface and Interface Analysis*, **12**, p. 541.

Suh, N. P. (1977), The Delamination Theory of Wear, *Wear*, **44**, p. 1.

Welsh, N. C. (1965), The Dry Wear of Steels, *Proc. Roy. Soc., Phil. Trans.* A257, p. 31.L.

CHAPTER 9

Adewoye, O. O. and Page, T. F. (1981), Frictional Deformation and Fracture in Polycrystalline SiC and Si_3N_4, *Wear*, **70**, p. 37.

Briggs, D. and Seah, M. P. (1983), Editors, *Practical Surface Analysis by Auger and X-Ray Photoelectron Spectroscopy*, John Wiley, London.

Buckley, D. H. (1978), Friction and Transfer Behvaviour of Pyrolytic Boron Nitride in Contact with various metals, *ASLE Trans.*, **21**, p. 118.

Dalal, H. M., Chiu, Y. P. and Rabinowicz, E. (1975), Evaluation of hot-pressed Silicon-Nitride as a Rolling Bearing Material, *ASLE Trans.*, **18**, p. 211.

Duwell, E. J. (1966), The Effect of Sliding Speed on the Rate of Wear of Sapphire on Steel, *Wear*, **9**, p. 363.

Fischer, T. E. and Luton, M. J. (1983), Effect of Simple Lubricants on Deformation and Wear in Concentrated Sliding Contact, *ASLE Trans.*, **26**, p. 31.

Fischer, T. E. and Tomizawa, H. (1985), Interaction of Tribochemistry and Microfracture in the friction and Wear of Silicon Nitride, *Wear*, **105**, p. 29.

Fransen, F., Vandenberghe, R., Vlaeminck, R., Hinoul, M., Remmerie, J. and Maes, H. E. (1985), Electron and Ion Beam Degradation Effects in AES Analysis of Silicon Nitride Films, *Surface Interface Analysis*, **7**, p. 79.

Gardos, M. N. (1989), The Effect of Anion Vacancies in the Tribological Properties of Rutile (TiO_{2-x}), *STLE* (formerly *ASLE*) *Trans.*, **32**, p. 25.

Hauffe, K. (1965), *Oxidation of Metals*, Plenum Press, New York.

Hong, H. and Winer, W. O. (1988), A Fundamental Tribological Study of Ti/Al_2O_3 Contact in Sliding Wear, paper presented at ASME/STLE Tribology Conference, October 1988, in Baltimore, USA (ASME Paper No. 88-Trib-48).

Ishigaki, H., Kawaguchi, I., Isawa, M. and Toiba, Y. (1985), Friction and Wear of Hot-Pressed Silicon Nitride and Other Ceramics, *Wear of Materials*, Editor, K. C. Ludema, ASME, New York, p. 13.

Jahanmir, S. and Fischer, T. E. (1986), Friction and Wear of Silicon Nitride Lubricated by Humid Air, Hexadecane and Hexadecane plus 0.5 percent Stearic Acid, ASLE Preprint No. 86-TC-2D-1. Paper presented at ASLE/ASME Tribology Conference, Pittsburgh, October 1986.

Kimura, Y., Okada, K. and Enomoto, Y. (1989), Sliding Damage of Silicon Nitride, in *Plane Contact, Wear of Materials*, Editor, K. C. Ludema, ASME, New York, p. 25.

Miller, P. D. and Holladay, J. W. (1985), Friction and Wear Properties of Titanium, *Wear*, **82**, p. 113.

Miyoshi, K. and Buckley, D. H. (1979a), Friction and Fracture of Single-Crystal Silicon Carbide in Contact with itself and Titanium, *ASLE Trans.*, **22**, p. 146.

Miyoshi, K. and Buckley, D. H. (1979b), Friction and Wear Behaviour of Single-Crystal Silicon Carbide in Sliding Contact with Various Metals, *ASLE Trans.*, **22**, p. 245.

Miyoshi, K. and Buckley, D. H. (1980), The Friction and Wear of Metals and Binary Alloys in Contact with an Abrasive Grit of Single-Crystal Silicon Carbide, *ASLE Trans.*, **23**, p. 460.

Miyoshi, K. and Buckley, D. H. (1981a), The Adhesion, Friction and Wear of Binary Alloys in contact with Single-Crystal Silicon Carbide, *Trans. ASME*, **103**, p. 180.

Miyoshi, K. and Buckley, D. H. (1981b), Friction and Wear of Single-Crystal Manganese-Zinc Ferrite, *Wear*, **66**, p. 157.

Miyoshi, K. and Buckley, D. H. (1982a), Tribological Properties and Surface Chemistry of Silicon Carbide at Temperatures to 1500 °C, *ASLE Trans.*, **26**, p. 53.

Miyoshi, K. and Buckley, D. H. (1982b), XPS, AES and Friction Studies of Single-Crystal Silicon Carbide, *Appl. Surf. Sci.*, **10**, p. 357.

Morton, P. H. and Baldwin, W. M. (1952), Scaling of Titanium in Air, *Trans. ASME* **44**, p. 1004.

Nutt, S. R. and Ruff, A. W. (1983), A Study of the Friction and Wear Behaviour of Titanium under Dry Sliding Conditions, *Wear of Materials*, ASME, New York, p. 426.

Page, T. F. and Adewoye, O. O. (1978), Hardness and Wear Behaviour of Silicon Carbide and Silicon Nitride Ceramics, *Proc. Brit. Ceramics Society*, **26**, p. 193.

Quinn, T. F. J. (1968), The Dry Wear of Steel as revealed by Electron Microscopy and X-Ray Diffraction, paper 24; Tribology Convention, Pitlochry, Scotland, 1968. Published in *Proc. Inst. Mech. Engrs*, **182**, Pt 3N, p. 201.

Quinn, T. F. J. and Winer, W. O. (1987), An experimental study of the 'hot-spots' occurring during the Oxidational Wear of Tool Steel on Sapphire, *JOT, Trans. ASME*, **109**, p. 315.

Seal, M. (1958), The Abrasion of Diamond, *Proc. Roy. Soc.*, **A248**, p. 379.

Steijn, R. P. (1961), On the Wear of Sapphire, *J. Appl. Phys.*, **32**, p. 1951.

Steijn, R. P. (1963), Sliding and Wear in Ionic Crystals, *J. Appl. Phys.*, **34**, p. 419.

Suganuma, T. (1985), Measurement of Surface Topography using SEM with Two Secondary Electron Detectors, *J. Electron Microscopy*, **34**, p. 328.

Tomizawa, H. and Fischer, T. E. (1986), Friction and Wear of Silicon Nitride at 150 °C to 800 °C, *ASLE Trans.*, **29**, p. 35.

Waterhouse, R. B. and Wharton, M. M. (1974), Titanium and Tribology, *Ind. Lub. Tech.*, **26**, p. 20.

Index

aberration
  astigmatic, 38
  chromatic, 37
  spherical, 37
aberrations, in optical lenses, 37
abrasion, 4
absorption coefficient
  for light, 263
  linear, 161
  mass, 161
absorption correction, X-ray, 173
absorption edges, for X-rays, 162
absorption factor for X-rays, 121
absorption X-ray spectra, 160
activation energy, for parabolic oxidation,
  29, 298, 299
additive
  disulphide, 202
  extreme-pressure, 74
additives
  anti-oxidant, 276
  lubricity, 276
adhesion, 4
adhesive wear, 5
adiabatic diesel, 229
advanced gas-cooled reactor (AGR), 417
adventitious carbon peak (XPS), 421
AES analysis of anti-wear films of ZDDP,
  280
AES analysis of worn silicon carbide, 442
AES analysis of worn silicon nitride, 436
AES analysis of worn stainless steel, 424
AES applied to pitting failure under e.p.
  additives, 387
AGR (Advanced Gas-cooled Reactor), 4
Aldermaston School of Surface Physics, 9,
  24
α-Al_2O_3 (sapphire), 446
α-Fe R-values, 331
aluminium/electrographite tribo-system, 321
aluminium R-values, 324
aluminium bronze/steel sliding system, 284
amorphous contrast of electron micrographs,
  140, 267
amplitude density function, 57, 59
amplitude factor for X-rays, 119

analyser, electron energy, 177
  electrostatic, 179
  magnetic, 178
analysis
  of amorphous wear tracks, silicon nitride,
    436
  of lubricant films, 202
  of oxidational wear in tribo-systems, 399
  of solid lubricant films, 224
angular momentum, 149
anti-oxidant additives, 276
anti-wear additives, 137, 202, 275
  definitions, 203, 277
anti-wear region of lubrication, 203
anti-wear surface films, XPS analysis of, 277
aperture, numerical, 40
aqueous lubricant
  Stribeck curve, 374
  and pitting, 368
Archard and Cowking equation, 371
Archard Wear Law, 29
Archard's hypothesis of friction, 14
area density of contact spots, 34
area of contact
  apparent, 10, 34
  real, 9
Arrhenius constant
  for static oxidation, 298
  for tribological oxidation, 29, 299
associated Auger electron emission, 176
astigmatism, 38
Aston University, Tribology Group, 383, 419
atomic number correction, 171
atomic scattering factor for X-rays, 118, 323
atomic sensitivity factor, XPS, 181
Auger electron emission, associated with
  XPS, 176
Auger electron spectroscopy, (AES), 190
Auger lines, in XPS spectra, 183
Auger Microprobe, Scanning, 424
Auger parameter, modified, (XPS), 188
Auger process, 190
austenite, proportion of retained, 318
autoradiographic technique, 24
average roughness height, 57
aviation fuel, lubricity of, 283